Lecture Notes in Computer Science 5450

Commenced Publication in 1973
Founding and Former Series Editors:
Gerhard Goos, Juris Hartmanis, and Jan van Leeuwen

Hyoung-Joong Kim   Stefan Katzenbeisser
Anthony T.S. Ho (Eds.)

Digital Watermarking

7th International Workshop, IWDW 2008
Busan, Korea, November 10-12, 2008
Selected Papers

 Springer

Volume Editors

Hyoung-Joong Kim
Korea University
Graduate School of Information Management & Security, CIST
Seoul 136-701, Korea
E-mail: khj@korea.ac.kr

Stefan Katzenbeisser
Technische Universität Darmstadt
Security Engineering Group
Hochschulstr. 10, 64289 Darmstadt, Germany
E-mail: skatzenbeisser@acm.org

Anthony T.S. Ho
University of Surrey
School of Electronics and Physical Sciences
Department of Computing
Guildford, Surrey, GU2 7XH, UK
E-mail: a.ho@surrey.ac.uk

Library of Congress Control Number: 2009935463

CR Subject Classification (1998): E.3, K.6.5, D.4.6, I.5, I.3, K.4.1, K.5.1

LNCS Sublibrary: SL 4 – Security and Cryptology

ISSN   0302-9743
ISBN-10   3-642-04437-9 Springer Berlin Heidelberg New York
ISBN-13   978-3-642-04437-3 Springer Berlin Heidelberg New York

springer.com

© Springer-Verlag Berlin Heidelberg 2009
Printed in Germany

Typesetting: Camera-ready by author, data conversion by Scientific Publishing Services, Chennai, India
Printed on acid-free paper   SPIN: 12755608   06/3180   5 4 3 2 1 0

Preface

It is our great pleasure to present in this volume the proceedings of the 7th International Workshop on Digital Watermarking (IWDW) which was held in Busan, Korea, during November 10-12, 2008. The workshop was hosted by the by Korea Institute of Information Security and Cryptology (KIISC) and sponsored by MarkAny, BK21 CIST (Korea University), ETRI.

Since its birth in the early 1990s, digital watermarking has become a mature enabling technology for solving security problems associated with multimedia distribution schemes. Digital watermarks are now used in applications like broadcast monitoring, movie fingerprinting, digital rights management, and document authentication, to name but a few. Still, many research challenges remain open, among them security and robustness issues, reversibility, and authentication. Continuing the tradition of previous workshops, IWDW 2008 also featured—besides papers dealing with digital watermarking—contributions from other related fields, such as steganography, steganalysis, and digital forensics.

The selection of the program was a challenging task. From more than 62 submissions (received from authors in 15 different countries) the Program Committee selected 36 as regular papers. At this point we would like to thank all the authors who submitted their latest research results to IWDW 2008 and all members of the Program Committee who put significant effort into the review process, assuring a balanced program. In addition to the contributed papers, the workshop featured three invited lectures delivered by Y. Q. Shi, C.-C. Jay Kuo, and Miroslav Goljan; summaries of their lectures can be found in this proceedings volume.

We hope that you will enjoy reading this volume and that it will be a catalyst for further research in this exciting area.

November 2008

H.-J. Kim
S. Katzenbeisser
Anthony TS Ho

Organization

General Chair

Hong Sub Lee   KIISC, Korea

Organizing Committee Chair

Jeho Nam   ETRI, Korea

Technical Program Chair

H. J. Kim   Korea University, Korea
S. Katzenbeisser   Technische Universität Darmstadt, Germany
Anthony TS Ho   University of Surrey, UK

Technical Program Committee

M. Barni	University of Siena, Italy
J. Bloom	Thomson, USA
C. C. Chang	Feng-Chia U., Taiwan
J. Dittmann	University of Magdeburg, Germany
J.-Luc Dugelay	Eurecom, France
M. Goljan	SUNY Binghamton, USA
B.Jeon	SKKU, Korea
T. Kalker	HP, USA
M. Kankanhalli	NUS, Singapore
A. Ker	Oxford University, UK
Alex Kot	NTU, Singapore
C. C. Jay Kuo	USC, USA
I. Lagendijk	Delft University of Technology, The Netherland
H.-Kyu Lee	KAIST, Korea
C. T. Li	University of Warwick, UK
Z. Lu	Sun Yat-sen University, China
B. Macq	UCL, Belgium
K. Martin	Royal Holloway University London, UK
N. Memon	Polytechnic University, USA
K. Mihcak	Bogazici University, Turkey
M. Miller	NEC, USA
Z. Ni	WorldGate Communications, USA
J. Ni	Sun Yat-sen University, China
H. Noda	Kyushu Inst. of Tech., Japan
J.-S. Pan	NKUAS, Taiwan
F. Perez-Gonzalez	University of Vigo, Spain

I. Pitas	University of Thessaloniki, Greece
A. Piva	University of Florence, Italy
Y.-M. Ro	ICU, Korea
A.-R. Sadeghi	University of Bochum, Germany
H. G. Schaathun	University of Surrey, UK
K. Sakurai	Kyushu University, Japan
Q. Sun	Inst. Infocomm Research, Singapore
H. Treharne	University of Surrey, UK
S. Voloshynovskiy	University of Geneva, Switzerland
S. Wang	University of Shanghai, China
M. Wu	University of Maryland, USA
S. Xiang	Sun Yat-sen University, China
G. Xuan	Tongji University, China
H. Zhang	Beijing University of Technology, China
Dekun Zou	Thomson, USA

Secretariat

Amiruzzaman Md.

Reviewers

M. Amiruzzaman	K. Mihcak
M. Barni	M. Miller
J. Bloom	Z. Ni
J. Briffa	J. Ni
C. C. Chang	H. Noda
J. Dittmann	J.-S. Pan
J.-Luc Dugelay	F. Perez-Gonzalez
M. Goljan	I. Pitas
A. TS Ho	A. Piva
B. Jeon	Y.-M. Ro
T. Kalker	A.-R. Sadeghi
M. Kankanhalli	V. Sachnev
S. Katzenbeisser	H. G. Schaathun
A. Ker	K. Sakurai
H. J. Kim	Y. Q. Shi
A. Kot	Q. Sun
C. C. Jay Kuo	P. Sutthiwan
I. Lagendijk	H. Treharne
H.-Kyu Lee	S. Voloshynovskiy
C. T. Li	S. Wang
S. Lian	M. Wu
Z. Lu	S. Xiang
B. Macq	G. Xuan
K. Martin	H. Zhang
N. Memon	D. Zou

Table of Contents

A Robust Watermarking Scheme for H.264*

Jian Li[1], Hongmei Liu[1], Jiwu Huang[1], and Yongping Zhang[2]

[1] School of Information Science and Technology
Sun Yat-sen University, Guangzhou, China, 510006
[2] Research Department, Hisilicon Technologies CO., LTD Beijing, China, 100094
isshjw@mail.sysu.edu.cn

Abstract. As H.264 receives many applications, it becomes more and more important to develop copyright protection methods for the new standard. Watermarking technology combined with video codec is considered a possible solution. In this paper we propose a new robust watermarking scheme for H.264, which makes the following contributions: 1) embed the watermark in a H.264 new syntax element named reference index; 2) propose a modifying the current block algorithm to improve the robustness of the scheme; 3) modify the current block by means of a geometry method, which makes the modifying algorithm degrade the video quality as slightly as possible. Experimental results have demonstrated the good performance of our scheme.

1 Introduction

H.264 is gradually accepted to be a dominant video coding standard. It is time to develop copyright protection methods appropriate to it. Watermarking is widely considered a possible solution. Since digital video is generally stored and distributed in compressed format, it is time-consuming to decode and re-encode video for embedding watermark. The approach of hiding watermark in video bitstream is more reasonable to practical applications. However such scheme is closely related with specific video coding standard. In most cases, applying existing algorithms to a new standard is difficult.

A few literatures have discussed the issues mentioned above. Ho et al. [1] proposed a hybrid watermarking scheme. Authors embedded the robust watermark into special DCT coefficients of I-frames, and extracted it from video bitstream directly. Literature [2] extended the method to P-frames. But Ho's scheme is sensitive to some attacks followed by recompression. The reason is that syntax elements for embedding watermark tend to be changed or lost in the second coding process. Noorkami and Mersereau in literatures [3] and [4] gave a robust algorithm scheme with controllable detection performance. To reduce the complexity of the algorithm, authors embedded watermark in the residual of I-frames and thus avoided decompressing the video; to gain robustness against prediction mode changing, like [5], they extracted watermark from the decoded video sequence. They also extended the method to P-frames in [6]. Noorkami's scheme was more robust, but its detection process was very complex.

* This work was supported by NSFC (90604008, 60633030), 973 Program (2006CB303104) and NSF of Guangdong (06023191).

H.J. Kim, S. Katzenbeisser, and A.T.S. Ho (Eds.): IWDW 2008, LNCS 5450, pp. 1–15, 2009.

The most challenging issue of designing video watermarking in compressed domain is how to control the tradeoff between robustness and imperceptibility. Redundancy in compressed video is much less than that in uncompressed one, which means embedding watermark there is difficult. Furthermore, the remainder after compression is critical to video quality, slight change probably leading to great degradation. In this paper, we propose a robust video watermarking scheme for H.264. The watermark is embedded in a H.264 syntax element named reference index during video encoding process, and extracted from video bitstream directly. When changing the reference index for embedding watermark, we meanwhile modify the current block to improve the scheme robustness. The changed reference indexes during watermarking process would not be found when re-encoding video, so the watermark isn't resistant to strong attacks yet. Existing countermeasures mainly focus on improving embedding strength [7]. However the embedding strength is limited usually in compressed domain. We propose a novel method which slightly modifying the current block, so that the changed reference index becomes *better* than the original. Our goal is if the video undergoes attacks accompanied with recompression, the encoder still tends to choose the watermarked reference index. It is a new attempt to enhance robustness of the watermarking that is combined with codec. Experimental results show the effectiveness of this method.

The rest of this paper is organized as follows. In Section 2, we first introduce H.264 simply, and then describe the process of watermark embedding. Section 3 demonstrate how to extract watermark from video bitstream. The simulation results and discussions are presented in Section 4, followed by conclusions in Section 5.

2   The Proposed Watermarking Scheme for H.264

2.1   Related Issues of H.264

H.264 introduces multi-picture motion-compensated prediction (henceforth MCP) technique to enable efficient coding. It allows an encoder to select, for motion compensation purpose, among a number of pictures that have been decoded and stored in buffers [9]. Figure 1 illustrate what MCP is by comparing with MPEG-2. The current block is highlighted in dark in the middle of the figure. There is only one picture stored for prediction in MPEG-2 standard. So it is no need to assign a number to stored picture, motion vector (MV) itself is enough. On the contrary, in H.264 more than one picture is stored. So the stored picture number is needed additionally to specify the prediction signals. In Fig. 1, the numbers are denoted by $R_0, R_1, \cdots$. H.264 records the picture number by means of a syntax element named index of reference picture. It is also termed reference index for short. In Fig. 1, the solid arrow line above the current block represents reference index and the associated MV. In the example, reference index is equal to R_0.

Current macroblock is divided into two blocks in Fig. 1. There are actually more than one partition type for inter-prediction in H.264. Different partition types are corresponding to different inter modes, and thus different coding results. Encoder searches all possible modes to get the best one by means of R-D (rate-distortion) optimization approach. The process is termed mode-decision.

For convenience of presenting our algorithm later, we give the method of getting reference index, which is used in most present encoders [8]. Assume REF is a variable

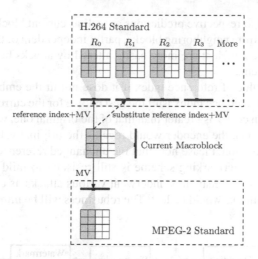

Fig. 1. Multipicture motion-compensated prediction

representing an arbitrary reference index, $REF \in [0, L)$, where L is integer greater than 0. Encoder selects the optimal REF which minimizes:

$$J(REF|\lambda_{Motion}) = D(s, c(REF, \mathbf{m}(REF))) +$$
$$\lambda_{Motion} \cdot (R(\mathbf{m}(REF) - \mathbf{p}(REF)) + R(REF)) \quad (1)$$

where J denotes Lagrangian cost, consisting of distortion measurement D and bit rate measurement R. D is the distance between current block s and its prediction signal c. As mentioned above, prediction signal is associated with reference index REF and MV $\mathbf{m}$. R, scaled by λ_{Motion}, calculates the bits needed to code a syntax element. Expression $\mathbf{m} - \mathbf{p}$ represents MVD, the difference between MV $\mathbf{m}$ and the predicted vector $\mathbf{p}$. Another variable λ_{Motion} termed Lagrangian multiplier. It has the following relations with quantization parameter (QP) for I/P-frames:

$$\lambda_{Motion,I,P} = \sqrt{\lambda_{Mode}} = \sqrt{0.85 \times 2^{(QP-12)/3}}. \quad (2)$$

2.2 Embedding Watermark into Reference Index

Reference index is suitable to embed watermark. Even though its payload is a bit smaller than that of DCT coefficient or MV, the watermarking capacity is still very high owing to large quantity of video data. Moreover, reference index has two advantages over other syntax elements. 1) Extracting watermark from reference index has lower complexity than from the DCT coefficient, which is very important for real-time applications. The reference index is coded by Exp-Golomb codes, which is much simpler than context-adaptive variable length coding (CAVLC) used by DCT coefficient. So decoding reference index is far more convenient. 2) Reference index is written into bitstream directly without the need of prediction. On the contrary, MV subtracts the predicted vector and the difference (MVD) is encoded; DCT coefficient is transformed from prediction residual, the difference between current block and prediction signal. Both syntax elements are related to respective predictions as well as current block. Prediction value is generally derived from neighboring blocks, partly independent of the syntax element. So MVD and DCT coefficient are more easily altered by attacks because neighboring blocks also affect them.

We change the value of reference index if it dose not fit the embedding watermark bit. But the changed reference index isn't the best choice for the current block. Its corresponding Lagrangian cost $\hat{J}$ is greater than that of the original one, which is denoted by J. According to Eq. (1), the encoder would regains the original value when bitstream undergoing recompression. Please note that the unchanged reference index is robust to recompression. The watermarking scheme is still resistant to mild recompression attack. But if watermarked bitstreams underwent various attacks as described in Fig. 2, the embedded watermarks would be lost. The robustness will be improved greatly if we

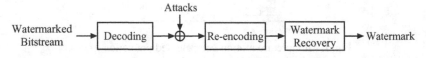

Fig. 2. Attack process

have $\hat{J} < J$. According to Eq. (1) J has two terms, $D(s, c)$ and $\lambda \cdot R(REF)$. The second term can't be altered when the variable REF is certain. Moreover, compared with the first term its value is very small. The first term can be changed by modifying the current block. So we also modify the current block if its reference index is changed when embedding watermark. In the following subsections, we will discuss the watermarking framework in detail, and then give the algorithm of modifying blocks.

2.3 Proposed Watermarking Framework

The watermark embedding process is presented in Fig. 3. *Embedding Control* determines how to watermark reference index and modify the current block. We use the last bit of reference index to embed watermark bit. In order to describe our embedding algorithm more clearly, we have the following notations.

i) *best reference index*: The reference index before watermarking, chosen by encoder according to Eq. (1). It is denoted by $best\_ref\_idx$.

ii) *watermarked reference index*: The reference index after embedding watermark. It is denoted by wm_ref_idx.

iii) *best prediction signal*: The prediction signal associated with best reference index.

iv) *second best prediction signal*: The prediction signal associated with substitute reference index which will be defined in 6.

v) *REF*, **m** and *J*: Their definitions are same as those in Eq. (1).

vi) *substitute reference index*: Denoted by sub_ref_idx. A reference index prepared for substituting the *best reference index*. In our scheme, it is get by

$$sub\_ref\_idx = \arg\min_{r \in S}\{J(r \mid \lambda)\} \tag{3}$$

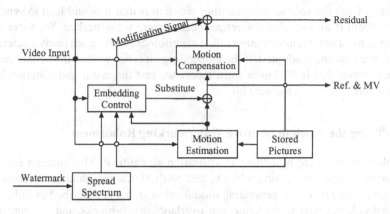

Fig. 3. Watermarking Framework

where

$$S = \{REF \mid REF \bmod 2 \neq best\_ref\_idx \bmod 2\} \tag{4}$$

We prepare a substitute for each reference index. The last bit of *substitute reference index* is different with that of *best reference index* according to Eq. (4). If *best reference index* does not fit the watermark bit, we can replace it with *substitute reference index*. We give an example in Fig. 1 to illustrate *substitute reference index* more clearly. Assume the value domain of REF is $[0, 6)$, so $R_0, R_1, \cdots, R_5 = 0, 1, \cdots, 5$. If *best reference index* is equal to R_0, there are R_1, R_3 and R_5 have different parity with it. *Substitute reference index* is the one with least Lagrangian cost among them. Assume R_1 is chosen, the dashed arrow line in Fig. 1 represents the substitute reference index and MV associated.

We use a bi-polar vector as the watermark sequence. Denote an arbitrary bit as $a_j \in \{-1, 1\}, j \in N$. The embedding watermark w is get by spread-spectrum method

$$w_i = b_i \cdot p_i \tag{5}$$

where

$$b_i = a_j, j \cdot F \leq i < (j+1) \cdot F \tag{6}$$
$$p_i \in \{-1, 1\}, i \in [1, F] \tag{7}$$

F is the spread-spectrum factor. For each hiding bit w_i,

$$wm\_ref\_idx = \begin{cases} sub\_ref\_idx, & \text{for } (best\_ref\_idx \bmod 2) \neq w_i, \\ best\_ref\_idx, & \text{for } (best\_ref\_idx \bmod 2) = w_i. \end{cases} \tag{8}$$

Not all P-frames in a GOP (group of pictures) are suitable for embedding watermark. In fact, the first P-frame can only get prediction from its previous I-frame. Hence, in this frame reference index only has one value. We don't watermark this frame, instead of embedding from the second one. Another problem is that it would lead to synchronization problem if all available reference indexes were watermarked. We solve it by embedding at most one bit information into a macroblock. If there are many a reference index, only the first one is adopted. If the embedding bit is not written in stream because macroblock is encoded in P_skip or Intra mode, we just discard it and watermark the following macroblock with the next bit.

2.4   Modifying the Block to Improve Watermarking Robustness

Current block is modified by adding modification signal to it. Modification signal is determined by 3 parameters, current block, best prediction signals and second best prediction signal. The process of generating modification signal is described as follows.

We first divide the current block into non-overlapping sub-blocks, and generate modification signals for each sub-block separately. As we know that DCT and Hadamard transformation both are applied to 4×4 block in H.264, so we set sub-block size to be 4×4 accordingly. For each 4×4 block, assume that itself, its corresponding best prediction and second best prediction signals are denoted by O, B and S, respectively. Each of them has 16 independent elements, so they can be taken as three vectors in $\mathbb{R}^{16}$ space. In order to make the modifying algorithm degrade the video quality as slightly as possible, we transform the modifying process into a geometry problem. O, B and S are regarded as three points in coordinate system and expressed by:

$$O = (P_{O_1}, P_{O_2}, \cdots, P_{O_{16}}) \tag{9}$$

$$B = (P_{B_1}, P_{B_2}, \cdots, P_{B_{16}}) \tag{10}$$

$$S = (P_{S_1}, P_{S_2}, \cdots, P_{S_{16}}) \tag{11}$$

where P_{O_i}, P_{B_i} and P_{S_i} are equal to the ith pixel values of current 4×4 block, best prediction and second best prediction signals, respectively.

The geometrical relationships between O, B and S are specified in Figure 4, where M is the midpoint of segment BS, and P is the projection of O on Hyperplane PM which is the perpendicular bisector of BS. For each point on PM, the distance from

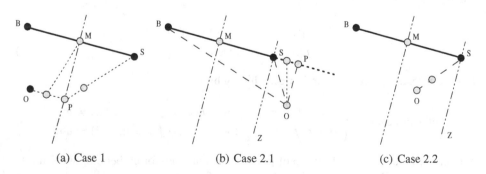

(a) Case 1   (b) Case 2.1   (c) Case 2.2

Fig. 4. Geometry depiction of modifying the 4x4 block

it to point B is equal to point S. In our algorithm, we always use L_2 norm to model distance, which is defined by

$$D(O, B) = (\sum_{i=1}^{16} (P_{O_i} - P_{B_i})^2)^{\frac{1}{2}}.$$ (12)

PM divides the space into two separate parts: S_1 and S_2. In our algorithm, points in PM are included in S_1. Any point X in space satisfies:

$$\begin{cases} D(X, B) \leq D(X, S), & \text{for } X \in S_1 \\ D(X, B) > D(X, S), & \text{for } X \in S_2 \end{cases}$$ (13)

Two cases in Eq. (13) are discussed separately. The first case involves that O, B and S satisfy

$$D(O, B) \leq D(O, S), \quad O \in S_1.$$ (14)

D is the same as the first term of Eq. (1), and here it gives the distance between two 4×4 blocks. We aim to modify the 4×4 block such that it get closer to second best prediction signal. Hence, the equivalent operation is moving point O from S_1 to S_2. Another requirement is the 4×4 block should be modified as slightly as possible, and equally the moving distance of O should be as short as possible. Summing up two above points, the goal is developing a shortest path, along which we drive O from S_1 to S_2. Hyperplane PM is the boundary between S_1 and S_2, and thus the critical state is moving O to PM. Obviously, the shortest distance between point O and hyperplane PM is OP. Hence P is the critical point we try to find. Based on the discussion above, we can get the following two equations:

$$\overrightarrow{OP} = k\overrightarrow{BS}$$ (15)

$$\overrightarrow{BS} \cdot \overrightarrow{PM} = 0$$ (16)

where Eq. (15) states that OP is parallel with BS, and if O belongs to S_1, $k > 0$. Equation (16) states that BS is perpendicular to vector PM. O, B and S have been given, and M is the midpoint of BS, which can be calculated by

$$M = \frac{1}{2}(B + S)$$ (17)

So P can be get by searching. We rewrite Eq. (15) and Eq. (16)

$$P = k(S - B) + O$$ (18)

$$sum = (P - M) \cdot (B - S)$$ (19)

Algorithm 2.1. Calculation process of P

i) Initializing. $k = 0$; k is increased by a step of k_{step}.
ii) Searching. Starting with O, P is calculated repeatedly by Eq. (18). Every time we increase k by k_{step} until Eq. (16) is hold. Because it's hard equal to 0 exactly, a threshold

β is predefined. We stop search process when $sum < \beta$.

iii) Output the final value of P.

The difference between P and O is the modification signal for current 4×4 block. In fact, there is a same operation for Case 2. We generate a new value for the current 4×4 block, and the modification signal is given by subtracting the original value from the new one. After all available 4×4 blocks have been modified, we splice all modification signals.

In the second case three points satisfy

$$D(O, B) > D(O, S), \quad O \in S_2. \tag{20}$$

We further divide case 2 into two subcases, correspondingly, separating S_2 into two subsets, denoted by $S_{2.1}$ and $S_{2.2}$. $S_{2.1}$ contains all the points on the *right* side of SZ as shown in Fig. 4(b), where SZ is a hyperplane perpendicular with BS. We should note that there is no *right* or *left* terms in $\mathbb{R}^{16}$ space. In fact, any point X in S_2, if satisfying the following inequality:

$$\overrightarrow{XS} \cdot \overrightarrow{BS} < 0 \tag{21}$$

is an element of $S_{2.1}$. And the rest points of S_2 belong to $S_{2.2}$.

Suppose O belongs to $S_{2.1}$, the projection of it onto segment BS dose not intersect BS into its interior, but the elongation. And the intersection is denoted by P. According to triangle inequality we can get:

$$D(B, S) \geq |D(O, B) - D(O, S)| \tag{22}$$

where the equality holds when O is on the elongation of segment BS. Hence, now the equivalent operation of modifying 4×4 block is moving O to BS. Constrained by the least modification requirement, the optimal moving approach is the projection of O onto elongation of BS, and intersection P is the move destination. So vector OP and BS is orthogonal to each other, and thus their dot product is given by:

$$\overrightarrow{OP} \cdot \overrightarrow{BS} = 0 \tag{23}$$

And because P is on the elongation of BS, we have

$$P = k(S - B) + S \tag{24}$$

According to Eq. 23, we can get

$$sum = (P - O) \cdot (S - B) \tag{25}$$

Equation (24) and (25) are almost same as Eq. (18) and (19) except input variables. So P can be calculated by *Algorithm 2.1* with a little changes. First the start point of P is not O but S. Secondly, we move P along the direction of $\overrightarrow{BS}$, and stop once Eq. (23) holds. The other operations of both are same.

It is comparatively simple when O belongs to $S_{2.2}$. Combining Eq. (21) and (13), a point in $S_{2.2}$ should satisfy:

$$\overrightarrow{OS} \cdot \overrightarrow{BS} \geq 0 \tag{26}$$

$$\overrightarrow{OM} \cdot \overrightarrow{BS} < 0 \tag{27}$$

which means that there is a point X within segment MS satisfying:

$$\overrightarrow{OX} \cdot \overrightarrow{BS} = 0 \tag{28}$$

Hence the projection of O onto BS lies in its interior. According to Eq. (22), we also need move O to a point in segment BS. But now the destination isn't X but S instead.

3 Watermark Recovery

Watermark can be retrieved from bitstream directly. We get the embedding bit from watermarked reference index, and derive the final watermark bit from a correlation sum. The recovery algorithm is described as follows.

i) Beginning with the second P-frame of a GOP, the watermarked reference index is extracted from the first partition block of each macroblock.

ii) To calculate the spread-spectrum watermark by

$$\hat{w}_i = \begin{cases} 1, & \text{for reference\_index(i) } mod\ 2 = 1 \\ -1, & \text{for reference\_index(i) } mod\ 2 = 0 \end{cases} \tag{29}$$

If no motion information available, such as P_skip or Intra macroblock, $\hat{w}_i$ is set equal to 0.

iii) To compute the correlation sum between $\hat{w}_i$ and predefined w_i

$$c_j = \sum w_i \cdot \hat{w}_i, \ j \cdot F \leq i < (j+1) \cdot F \tag{30}$$

iv) To determine the retrieved watermark, we need predefine a threshold ξ such that

$$\hat{b}_j = \begin{cases} 1, & \text{for } c_j \geq \xi \\ -1, & \text{for } c_j < \xi \end{cases} \tag{31}$$

4 Experimental Results

We have the watermarking scheme incorporated in H.264 reference software JM12.4 [12]. Encoder is set to be Baseline profile. Frame rate is set to be 30 frame/s. In order to get extensive and reliable conclusions, more than 10 QCIF (172x144) video sequences are used for test. The parameters in *Algorithm 2.1* are set: $k_{step} = 0.1, \beta = 0.05$.

Table 1. Imperceptibility Test

Video Sequence	PSNR Decrease (dB)	Bit rate Increase (%)
Coastguard	1.3606	3.68%
Foreman	1.0036	6.18%
Carphone	1.1425	4.24%
Highway	0.7673	3.61%
Salesman	0.5399	2.14%
Silent	0.9555	3.31%
Mother	0.5393	4.35%
Akiyo	0.2925	1.21%
Grandma	0.2652	2.66%
Claire	0.5398	2.15%
Hall	0.9016	1.97%
Average	**0.755**	**3.22%**

4.1 Imperceptibility Test

We use the changes of PSNR and bit rate show the imperceptibility of watermarking. If the video sequence is encoded with watermark embedded, the PSNR decreases and bit rate increases, as shown in Table 1, where the average values under different QP (22:2:32) for all sequences are given. The GOP structure is IPPPPPI. In a GOP, except I-frame and the first P-frame we have four other P-frames watermarked. From Table 1 we can find if a video is plenty of motion information, such as Coastguard, it has bigger changes in PSNR and bit rate. Because more watermark bits are embedded into them. Figure 5 illustrates the perceptual quality of watermarked video. The example frame is get from *Coastguard* which PSNR is decreased most.

| (a) Original frame | (b) Compressed frame | (c) Watermarked frame |

Fig. 5. Comparison of perceptual quality

Comparing with the existing schemes, the decrease of PSNR is smaller than that in [2], which average value is 1.1dB. But authors don't provide us the increase value of bit rate. If only P-frames are watermarked [6], Noorkami's method increases the video bit rate by 1.54% . But we should notice that authors adopt a GOP structure of IBPBPBI, so there are only two frames watermarked.

4.2 Robustness Test

The robustness is illustrated by bit error rate (BER). In the experiments, we split the video sequence into 20-frame clips, and embed 8-bit watermark into each of them. For a video sequence, the BER is defined by the average of 11 clips. When embedding watermark, we set $QP = 22$. In Eq. 31 we set $\xi = 0$. The test items and detailed descriptions are demonstrated in Table 2, some of which are learned from literature [13].

Table 2. Description of Robustness Test Items

Test Item	Description and parameter
Gaussian Blurring	Standard Deviation: $\sigma_f = 0.4 : 0.4 : 2$
AWGN	Standard Deviation: $\sigma_a = 2 : 2 : 16$
Brightness Increase	Increasing the video luminance with $v_i = 10 : 10 : 50$
Brightness Decrease	Decreasing the video luminance with $v_d = -10 : -10 : -50$
Sharpening Spatial Filtering	Filter: $fspecial('unsharp', \alpha)$, $\alpha = 0.2 : 0.2 : 1$
Contrast Decrease	Decrease factor: $f = 0.9 : -0.1 : 0.4$
Rotation	Degree: $\beta = 1 : 1 : 5$
Requantization	Quantization Parameter: QP $= 24 : 2 : 32$

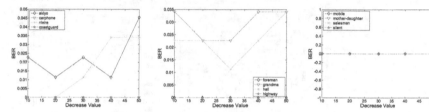

Fig. 6. Robustness to Brightness Decrease

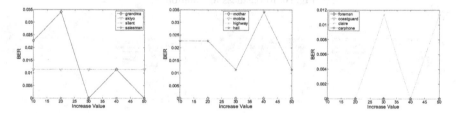

Fig. 7. Robustness to Brightness Increase

The attack process has been described in Fig. 2. Assume spread-spectrum signal is independent with reference index, the BER is expected to be 0.5 for non-watermarked video. So it is acceptable if BER below 0.25. Generally speaking, the videos with little motion information, such as Hall, Claire and so on, embed fewer watermark bits per frame. Consequently, they are more sensitive to attacks. On the contrary, the videos with more motion information achieve robustness to very strong attacks.

Our scheme is very robust to video brightness and contrast changes (Fig. 6 - Fig. 9). Such attack operations change all frames in a same manner. Reference index is mainly

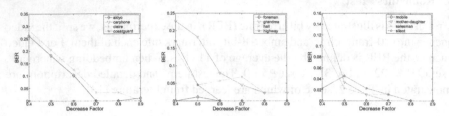

Fig. 8. Robustness to Contrast Decrease

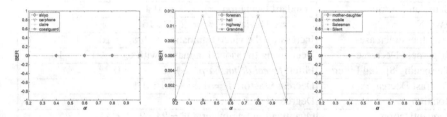

Fig. 9. Robustness to Contrast Increase

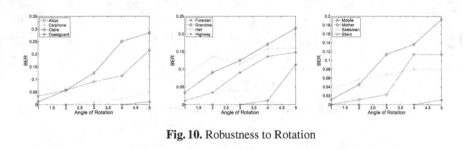

Fig. 10. Robustness to Rotation

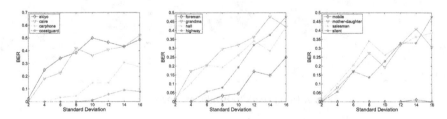

Fig. 11. Robustness to AWGN

decided by the differences betweens frames. So most watermarks embedded can be maintained in bitstream. From Fig. 10 we find watermarking is acceptably resistant to rotation. When the video is rotated 5 degrees, BER is below 0.25 in most case. After rotation the pixels deviate from their positions because of rotation, but most of them are still in the original macroblocks, and thus the watermark can not be removed. Additive white gaussian noise (AWGN for short), Gaussian blurring and recompression change

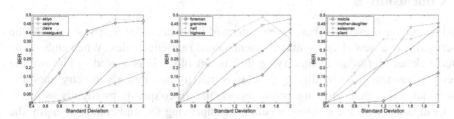

Fig. 12. Robustness to Gaussian Blurring

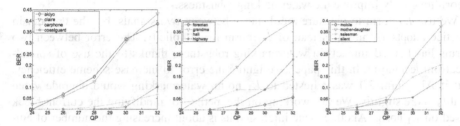

Fig. 13. Robustness to Recompression

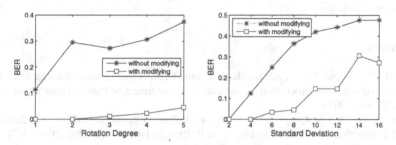

Fig. 14. Comparison of the BER of our scheme with the scheme without modifying the current blocks

the difference between frames greatly. Our watermarking can still survive AWGN at 32dB at least, with *Standard Deviation* = 6 in Fig. 11. And it is resistant to 5 × 5 Gaussian blurring, with *Standard Deviation* = 0.8 in Fig. 12. If video bit rate decreases to more than 50%, corresponding to QP = 28 in Fig. 13, all sequences are still have BER below 0.25. Especially for sequences with abundant motion, like Mobile, watermarking achieves robustness to very strong attacks.

To show the effectiveness of modifying algorithm, we also compare the robustness of the watermarking scheme with and without modifying the current block. Comparison results of carphone sequence are illustrated in Fig. 14. We can find that thanks to modifying algorithm BER decrease by 0.26 on average for rotation attack and by 0.2 on average for AWGN attack.

5 Conclusion

In this paper, we present a robust video watermarking scheme for H.264. We embed the watermark into a new H.264 syntax element named reference index. When embedding watermark, an algorithm of modifying the current block is proposed to improve the scheme robustness. The modifying process is transformed into a geometry question, which makes the modifying algorithm degrade the video quality as slightly as possible. Comparing with the existing methods of improving the embedding strength, the modifying current block algorithm directly increases the probability of the embedded watermarks being found again when re-encoding. The experimental results show the algorithm greatly improve the watermarking robustness.

We use L_2 norm to measure the distance between two signals, but the encoder in practice adopts L_1 norm instead of L_2 norm for simplicity. The error between two norms has limited impact on watermarking robustness thanks to the use of spread-spectrum technique. In this paper, we ignore the error to increase scheme efficiency. But if Algorithm 2.1 was adjusted to L_1 norm, watermarking would degrade video quality more slightly. We are working on the method of modifying the current block based on L_1 norm. Another issue interested us is about increasing the number of embedded watermark bits. In this paper, watermark is only embedded into P-frames. It would improve the watermarking payload greatly if we watermarked B-frames, also with reference indexes.

References

1. Qiu, G., Marziliano, P., Ho, A.T.S., He, D.J., Sun, Q.B.: A hybrid watermarking scheme for H.264/AVC video. In: Proceedings - International Conference on Pattern Recognition, vol. 4, pp. 865–868 (2004)
2. Zhang, J., Ho, A.T.S., Qiu, G., Marziliano, P.: Robust video watermarking of H.264/AVC. IEEE Transactions on Circuits and Systems II: Express Briefs 54(2), 205–209 (2007); 2(1), 14-23 (March 2007)
3. Noorkami, M., Mersereau, R.M.: Towards robust compressed-domain video watermarking for H.264. In: Proc. of SPIE-IS&T Electronic Imaging, vol. 6072, 60721A. SPIE, San Jose (2006)
4. Noorkami, M., Mersereau, R.M.: A framework for robust watermarking of H.264-encoded video with controllable detection performance. IEEE Transactions on Information Forensics and Security
5. Hartung, F., Girod, B.: Watermarking of uncompressed and compressed video. Signal Processing 66(3) (May 1998)
6. Noorkami, M., Mersereau, R.M.: Digital video watermarking in P-frames. In: Proceedings of SPIE-IS&T, vol. 6505, 65051E (2007)
7. Jordan, F., Kutter, M., Ebrahimi, T.: Proposal of a watermarking technique for hiding/retrieving data in compressed and decompressed video. ISO/IEC Doc. JTC1/SC29/WG11 MPEG97/M2281 (July 1997)
8. Lim, K.P., Sullivan, G.J., Wiegand, T.: Text Description of Joint Model Reference Encoding Methods and Decoding Concealment Methods. Joint Video Team (JVT) of ISO/IEC MPEG and ITU-T VCEG Busan, Korea (April 2005)
9. Thomas, W., Sullivan, G.J.: Overview of the H.264/AVC video coding standard. IEEE Transactions on Circuits and Systems for Video Technology 13(7) (July 2003)

10. Sullivan, G.J., Wiegand, T.: Rate-distortion optimization for video compression. IEEE signal Process. Mag. 15, 74–90 (1998)
11. ITU-T H.264(05/03) Series H: Audiovisual And Multimedia Systems: Advanced video coding for generic audiovisual services
12. H.264/AVC Software Coordination (2007), http://iphome.hhi.de/suehring/tml/
13. Coskun, B., Sankur, B., Memon, N.: Spatio-Temporal Transform Based Video Hashing. IEEE Transaction on Multimedia 8(6) (December 2006)

Detection of Double MPEG Compression Based on First Digit Statistics

Wen Chen and Yun Q. Shi

New Jersey Institute of Technology, Newark, NJ, USA 07102
{wc47,shi}@njit.edu

Abstract. It is a challenge to prove whether or not a digital video has been tampered with. In this paper, we propose a novel approach to detection of double MEPG compression which often occurs in digital video tampering. The doubly MPEG compressed video will demonstrate different intrinsic characteristics from the MPEG video which is compressed only once. Specifically, the probability distribution of the first digits of the non-zero MPEG quantized AC coefficients will be disturbed. The statistical disturbance is a good indication of the occurrence of double video compression, and may be used as tampering evidence. Since the MPEG video consists of I, P and B frames and double compression may occur in any or all of these different types of frames, the first digit probabilities in frames of these three types are chosen as the first part of distinguishing features to capture the changes caused by double compression. In addition, the fitness of the first digit distribution with a parametric logarithmic law is tested. The statistics of fitting goodness are selected as the second part of the distinguishing features. We propose a decision rule using group of pictures (GOP) as detection unit. The proposed detection scheme can effectively detect doubly MPEG compressed videos for both variable bit rate (VBR) mode and constant bit rate (CBR) mode.

Keywords: first digit distribution, double MPEG compression.

1 Introduction

To prove the originality and integrity of multimedia data, a variety of active authentication techniques have been proposed, such as watermarking techniques. The active detection requires authentication data be embedded into multimedia data for the late use of authentication. However, it is not common in practice to embed authentication data when the medium is generated at the first place. This gives rise to the need of passive detection which only relies on the intrinsic characteristics of a medium itself instead of authentication data.

In many applications, digital videos are available in the compressed format such as MPEG. However, tampering operations are often performed in the uncompressed domain, as shown in Fig. 1. Due to its large size, the tampered video always needs to be re-encoded and re-saved in the compression format. As a result, the double compression occurs. If the quantization parameters in the

H.J. Kim, S. Katzenbeisser, and A.T.S. Ho (Eds.): IWDW 2008, LNCS 5450, pp. 16–30, 2009.
© Springer-Verlag Berlin Heidelberg 2009

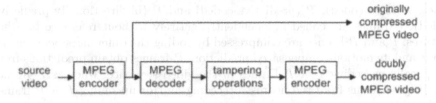

Fig. 1. MPEG video tampering operations

second encoder are different from those in the first encoder, the double compression may reveal the occurrence of tampering. Therefore,the detection of MPEG double compression can be used to authenticate MPEG video.

In the past several years, little work has been done to detect video double compression [1] while most of the studies were focused on JPEG image double compression [2, 3, 4]. In [1], MPEG double compression is detected by examining the periodic artifacts introduced into the histograms of I frames during the double MPEG compression. The approach works for MPEG video with variable bit rate (VBR) control mode. However, as pointed out by the authors, this approach will fail for MPEG video with constant bit rate (CBR) control mode.

It has been shown in [4] that the probability distribution of the first digits of non-zero JPEG quantized AC coefficients will be disturbed if the JPEG image is doubly compressed with two different quantization factors. Except for motion compensation, as a lossy compression algorithm for video, MPEG uses somehow JPEG-like compression scheme for a video frame in terms of methodology. The similarity between MPEG and JPEG inspires us to investigate the first digit distribution of non-zero MPEG quantized AC coefficients. By exploiting the statistical disturbance in the first digit distribution, we propose a novel detection approach to detection of double MPEG compression in both VBR and CBR videos. To the best of our knowledge, this is the first effective approach to detection of double MPEG compressed CBR videos.

The rest of this paper is organized as follows. In Section 2, an overview of MPEG compression is given. The video dataset is described in Section 3. In Section 4, the first digit distribution of MPEG videos are investigated. In Section 5, we propose the detection scheme. The experimental results are reported in Section 6. In Section 7, the conclusions are drawn.

2   Overview of MPEG Compression

2.1   Layered Structure of MPEG Video

MPEG standard has specified the layered structure of video sequence. An MPEG video is partitioned into a hierarchy having six layers: sequence, group of pictures (GOP), picture, slice, macroblock, and block.

The top layer is the video sequence. Each video sequence is divided into one or more GOPs, and each GOP is composed of pictures of three different coding types: I (intra-coded), P (predictive-coded) and B (bi-directionally predictive coded). I frames are coded independently, entirely without reference to other pictures. P and B frames are compressed by coding the differences between the frame and its motion-compensated prediction. P-frames obtain predictions from temporally preceding I or P frames in the video sequence, whereas B frames obtain predictions from the nearest preceding and/or upcoming I or P frames in the video sequence.

Each frame is composed of one or several slices. The slice consists of a contiguous sequence of macroblocks. Each macroblock consists of a 16x16 sample array of luminance samples (Y) and two 8x8 blocks of chrominance samples (Cr and Cb). These 8x8 blocks of data from the frame itself (for I frame) or the frame prediction residual (for P and B frames) are the basic coding unit. The 8x8 blocks of data are first transformed into DCT coefficients which are then quantized and entropy encoded to generate MPEG video bitstream.

2.2 Quantization

Quantization is basically a process for reducing the precision of the DCT coefficients in order to save data. The quantization process involves division of the DCT coefficient values by integer quantizing values. The integer quantizing values (QV) are the product of quantization table values (Q) and quantization scales Q_Scale, i.e. $QV = Q \times Q\_Scale$.

The quantization scale takes on values in the range of [1, 31] in MPEG-1 [5]. During the encoding, the quantization scale is fixed if the VBR control is enabled. If CBR control is enabled, the quantization scale varies so that the target bit rate remains constant.

The target bit rate is very sensitive to the quantization scale. Variable quantization is the most important and most effective technique for controlling bit rate. Although quantization scale varies during encoding, there is a reciprocal relationship between the bit rate and average quantizing value. Roughly speaking, the higher the bit rate, the smaller the average quantizing value is, and vice versa. If the target bit rate for the doubly MPEG compressed video is different from the originally MPEG compressed video, as illustrated in Section 4, the statistical disturbance will appear in the doubly MPEG compressed video due to the change in the quantizing values.

3 Video Dataset

3.1 YUV Source Video Sequences

Several widely known raw YUV sequences in CIF (*Common Intermediate Format*) format are selected as source video sequences. CIF defines a video sequence with a resolution in pixels of 352 x 288 (width x height). These YUV sequences include *Akiyo*, *Football*, *Bus*, *Coastguard*, *Flower*, *Foreman*, *Hall*, *Stefan*, *Tempete*, and *Waterfall*. The contents and motion complexity of the YUV sequences vary. The total number of frames in each of the sequences is 300, 90, 150, 300,

Fig. 2. YUV source video sequences

250, 300, 300, 90, 260, and 260, respectively. The first frame of each sequence is illustrated in Fig. 2.

3.2 MPEG Video Clips

Two publicly available MPEG encoders are used to generate MPEG-1 video clips. They are Berkeley MPEG-1 Video Encoder developed at Berkeley Multimedia Research Center [6] and MPEG-2 Video Codec developed by the MPEG Software Simulation Group ($MSSG$) [7]. The Berkeley MPEG-1 Video Encoder provides VBR and CBR control mode. The $MSSG$ MPEG-2 Video Codec only provides CBR option.

Berkeley MPEG-1 Video Encoder was used to generate VBR video clips. It is known that the integer quantizing values are the product of quantization table values and quantization scales. Denote quantization scale in I, P and B frame as IQ, PQ, and BQ, respectively. The quantization table values, PQ and BQ are fixed during the encoding. PQ and BQ is 10 and 25, respectively. IQ takes on one of the nine following integral values: 2, 3, 4, 5, 6, 7, 8, 9, and 10 in the generation of originally MPEG compressed video clips. To simulate the double compression, each of the originally compressed MPEG video clips with IQ equal to 6 is first decompressed. The decompressed YUV sequence is then compressed with IQ taking on one of the integral values: 2, 3, 4, 5, 7, 8, 9, and 10 to generate doubly compressed video clips.

Both Berkeley MPEG-1 Video Encoder and $MSSG$ MPEG-2 Video Codec were used to generate CBR video clips. To generate originally MPEG compressed video clips, each of the raw YUV sequences were compressed with the following seven different bit rates (bits/s): 806400, 921600, 1036800, 1152000, 1267200, 1382400, and 1497600, respectively. To simulate the double compression, each of the originally MPEG video clips with bit rate equal to 1152000 is first decompressed. The decompressed YUV sequence is then compressed with each of the six bit rates 806400, 921600, 1036800, 1267200, 1382400, and 1497600 to generate doubly compressed video clips.

Totally there are three groups of MPEG-1 video clips: CBR video clips encoded by *MSSG* MPEG-2 Video Codec; CBR video clips encoded by Berkeley MPEG-1 Encoder; and VBR video clips encoded by Berkeley MPEG-1 Encoder.

4 First Digit Distribution in MPEG Compression Video

4.1 Background

If a given JPEG image is compressed for only once, it has been shown in [4] that the probability distribution of the first digits of non-zero JPEG quantized AC coefficients will follow the parametric logarithmic law (also called generalized Bendford's Law):

$$p(x) = Nlog_{10}(1 + \frac{1}{s + x^q}), x = 1, 2, \cdots, 9 \qquad (1)$$

where x represents the first digits of non-zero JPEG AC coefficients, N, s and q are parameters to precisely describe the distribution. Moreover, it is shown that the first digit distribution will violate the law when the double compression occurs.

As said in Introduction, MPEG uses somehow JPEG-like compression scheme for a video frame in terms of methodology. Specifically, both MPEG and JPEG use block-based DCT, quantization, and entropy coding as shown in Fig. 3.

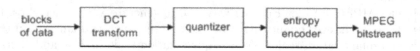

Fig. 3. Compression of 8x8 blocks data

The similarity between MPEG and JPEG encourages us to investigate how well the first digits of non-zero MPEG quantized AC coefficients follow the parametric law. Because the video sequence is in color, each picture must have three channels including a luminance channel and two chrominance channels. In the study here, only the luminance channel is considered.

4.2 Visual-Comparison Test

How well does the parametric logarithmic law model the first digit distribution of MPEG video? The natural first test to carry out is a visual comparison of the actual first digit distribution and the fitted first digit distribution. The Matlab Curve Fitting toolbox was used to find the fitted distribution. In the experiment, the values of parameter N, q, s in Equation (1) are limited to [0.1, 3], [0.1, 3] and [-1, 1], respectively. We found that the good fitting results can be achieved with such range of parameter values. The visual comparisons are demonstrated via Akiyo and Football video sequence. The reason for choosing Akiyo and Football

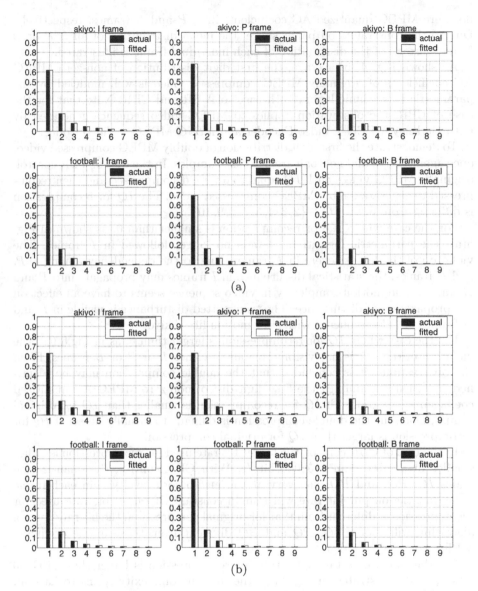

Fig. 4. The mean first distribution of originally MPEG compressed video. (a) VBR ($IQ = 6$); (b) CBR(Berkeley, target bit rate = 1152000 bits/s).

for illustration is that they contain rather different motion complexity, i.e., the former has less motion while the latter has fast motion.

The first digit distributions of the originally MPEG compressed video sequence using VBR and CBR modes are illustrated in Fig. 4 (a) and (b), respectively. There the horizontal axes stand for the first digits 1, ... , 9, and the vertical axes represent the mean actual and fitted first digit distribution of non-zero MPEG quantized AC coefficients in I, P and B frames, respectively. Due to the constraints on paper space, the mean distribution of all I, P or B frames instead of the distribution of each individual frame is illustrated.

As shown Fig. 4, in average, the first digit distribution of non-zero MPEG AC coefficients of originally MPEG compressed video is well modeled by the parametric logarithmic law in both cases of VBR and CBR. Note that for the case of CBR, the observation applies to MPEG video sequence generated by both Berkeley MPEG-1 Video Encoder and *MSSG* MPEG-2 Video Codec.

To demonstrate the first digit distribution of doubly MPEG compressed video, once again, Akiyo and Football are used as examples. In the case of VBR control, originally MPEG videos with IQ equal to 6 were decompressed and then recompressed to generate doubly MPEG compressed video. The IQ for recompression is different from 6, i.e. $IQ \in \{2, 3, 4, 5, 7, 8, 9, 10\}$.

For the case where IQ for recompression is smaller than 6, as shown in Fig. 5(a), the parametric logarithmic law is no longer followed in I frames. The violation of the law is severe in I frames. Due to the dependency among I, P, and B frames, the statistical disturbance in I frames may propagate into P and B frames. The motion complexity in video sequence seems to have an effect on the propagation of disturbance. The transmitted disturbance is visible in P and B frames of video sequence with slow motion like Akiyo.

For the case where IQ for recompression is larger than 6, as in Fig.5 (b), the violation of the law is not visibly obvious in either I, or P or B frames. From the observations, it is concluded that the double compression may be identified by visually examining the first digit dist of non-zero MPEG AC quantized coefficients in I frames if IQ for recompression is smaller than IQ for original compression, but the double compression is difficult to visually detect if IQ for recompression is larger than IQ for original compression.

In the case of CBR control, the target bit rate used for recompression is either smaller or greater than 1152000 bits/s. With CBR control, the quantization scales IQ, PQ, and BQ may vary at the macroblock level within each frame such that the output bit rate will remain constant. If target bit rate used for recompression is different from that for original compression, the bit rate control algorithm will choose IQ, PQ and BQ different from those in the originally MPEG compressed video.

For the case where target bit rate for recompression is larger than 1152000 bits/s, as demonstrated in Fig. 6(a), the motion complexity seems to have an influence on the first digit distribution. For video sequence with fast motion in neighboring frames like Football, it is obvious that the first digit distributions in I, P, and B frames are not modeled by the logarithmic law. For video sequence with slow motion in neighboring frames like Akiyo, the violation of the logarithmic law is not so visually obvious.

For the case where target bit rate for recompression is smaller than 1152000 bits/s, the first digit distributions of doubly MPEG compressed videos are shown in Fig. 6(b). Although it seems that the first digits have the tendency to follow the logarithmic law, a closer look at the figure can find the abnormality at certain

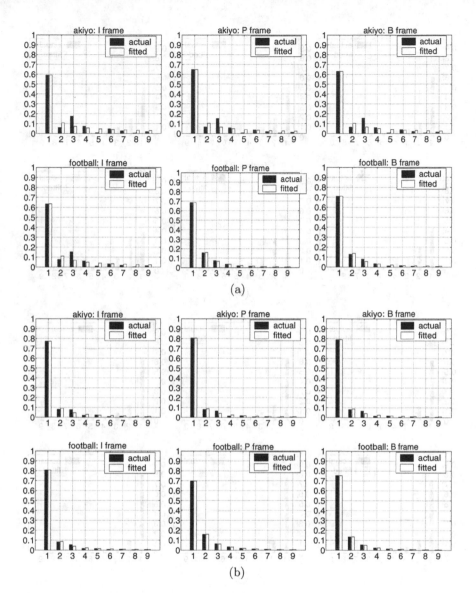

Fig. 5. The mean first distribution of doubly MPEG compressed video - VBR. (a) IQ = 4; (b) $IQ = 9$.

digits, for example, the probabilities of digits 8 and 9 in Football are very close to zeros, and the probabilities of digits 7 and 8 are almost equal in Akiyo.

4.3 Chi-square Test

The chi-square test [8] is a more formal way of carrying out the above comparison. In the chi-square test, we consider the null hypothesis that the first digits of

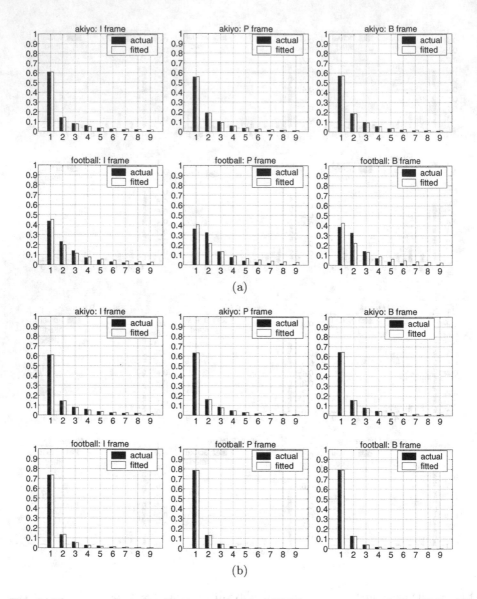

Fig. 6. The mean first distribution of doubly MPEG compressed video - CBR. (a) target bit rate = 1497600 bits/s; (b)target bit rate = 806400 bits/s.

non-zero MPEG quantized AC coefficients in each frame follows the parametric logarithmic law. The significant level is set to be 1% and the degree of freedom is set to to be 8. The entries in the Table 1 represent how many percentages of frames which do not reject the null hypothesis. For each IQ or target bit rate,

Table 1. The percentage of frames following the logarithmic law

	I		P		B	
	originally	**doubly**	originally	**doubly**	originally	**doubly**
VBR	0.6989	**0.0153**	0.6857	**0.1347**	0.7111	**0.1209**
CBR[1]	0.6494	**0.1682**	0.5552	**0.2296**	0.5853	**0.2056**
CBR[2]	0.7336	**0.2418**	0.5987	**0.2219**	0.5676	**0.1909**

there are total 164 I frames, 613 P frames and 1523 B frames in the originally or doubly MPEG compressed video clips, respectively. CBR[1] and CBR[2] stand for Berkeley and *MSSP* CBR videos, respectively.

In the case of VBR, about 70% of I frames, 69%of P frames and 71% of B frames in originally MPEG compressed videos do not reject the null hypothesis. However, only about 1.5% of I frames, 13% of P frames and 12% of B frames in doubly MPEG compressed videos do not reject the null hypothesis.

In the case of CBR[1], about 65% of I frames, 56% of P frames and 59% of B frames in originally MPEG compressed videos do not reject the null hypothesis. However, only about 17% of I frames, 23% of P frames and 21% of B frames in doubly MPEG compressed video do not reject the null hypothesis.

In the case of CBR[2], about 73% of I frames, 60% of P frames and 57% of B frames in originally MPEG compressed videos do not reject the null hypothesis. However, only 24% of I frames, 22% of P frames and 19% of B frames in doubly MPEG compressed videos do not reject the null hypothesis.

The chi-square test results indicate that most of I, P, and B frames in the originally MPEG videos do not reject the null hypothesis, but only a few of I, P, and B frames in doubly MPEG videos do not reject the null hypothesis, which is consistent with the visual comparison.

5 Detection of Double MPEG Compression Video

From Section 4, it is found that the first digit distribution in doubly compressed MPEG video does not follow the parametric logarithmic law as well as originally compressed MPEG video does. The disturbance in the first digit distribution is a good indication of occurrence of double compression.

To detect double MPEG compression, it is intuitive to examine the first digit distribution by visual comparison or chi-squared test. However, such detection is not reliable. For example, in the VBR case, when IQ for recompression is larger than that for original compression, the violation of the law is not visibly obvious, thus hard to detect. Also, it was found from Section 4.3 that rather high percentage of originally MPEG compressed video rejects the null hypothesis. To make the detection more reliable, we consider using machine learning framework.

Instead of using frame as detection unit, the GOP is proposed as detection unit to fully catch the statistical disturbance which may appear in I, P, or B frames. We further propose the following decision rule:

*The MPEG video is classified as being doubly compressed if the ratio D/N exceeds
a specific threshold T, where N is the total number of GOPs in a given MPEG
video, D is the number of GOPs detected as being doubly compressed and T is a
value $\in [0, 1]$.*

Based on the decision rule, the detection of each GOP is the key to the
accurate classification of MPEG video as being doubly compressed or not. The
framework of detection of each GOP is shown in Fig. 7.

Fig. 7. The detection of each GOP

A typical GOP is composed of one I frame, several P frames and certain
number of B frames. In the feature analysis, the features for each GOP will be
extracted from I, P, and B frames. Specifically, the first digit probabilities of
the non-zero MPEG AC quantized coefficients and the goodness-of-fit statistics
form the feature vector of each GOP. To calculate the goodness-of-fit statistics,
the fit of the first digit distribution is tested with the parametric logarithmic law
in Equation (1), using the Matlab Curve Fitting toolbox. The toolbox support
the following goodness-of-fit statistics: sum of squares due to error (SSE), root
mean squared error ($RMSE$) and R-*square*. Their definitions are as follows:

$$SSE = \sum_{i=1}^{n} \frac{(p_i - \hat{p}_i)^2}{p_i} \tag{2}$$

$$RMSE = \sqrt{\frac{\sum_{i=1}^{n} (p_i - \hat{p}_i)^2}{n}} \tag{3}$$

$$R - square = 1 - \frac{\sum_{i=1}^{n} \frac{(p_i - \hat{p}_i)^2}{p_i}}{\sum_{i=1}^{n} \frac{(p_i - \bar{p}_i)^2}{p_i}} \tag{4}$$

where p_i and $\hat{p}_i$ are the actual probability and the fitted value for the digit $i =$
1, 9, respectively, and $\bar{p}$ is the mean of probabilities. A value closer to 0 for SSE
and $RMSE$, and closer to 1 for R-*square* indicate a better fit.

The procedure of feature extraction is illustrated in Fig. 8. The feature of I
frame includes nine first digit probabilities of non-zero MPEG quantized AC coefficients and three goodness-of-fit statistics. For P and B frames, the individual
frame's first digit probability and goodness-of-fit statistics are first computed as
in Fig.8 (b), and then their mean is finally included in the feature vector for
each GOP as in Fig.8 (a). The dimension of feature is 36 (12x3).

If the MPEG video is compressed only once, the first digits of non-zero MPEG
quantized AC coefficients in I, P and B frames generally follow the parametric
logarithmic law. For P and B frames, their mean first digit distribution in each
GOP will also follow the parametric law. However, if the MPEG video is doubly

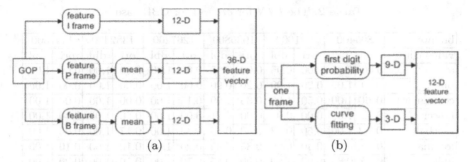

Fig. 8. Flowchart of feature extraction (a)feature extraction of GOP; (b)feature extraction of frame

compressed with the different quantization scale, the first digit distribution of individual frame will be disturbed. As a result, if the fit of the first digit distribution with the parametric function is tested, the fitting results would most likely become worse. That is, SSE and $RMSE$ would be away from 0, and $R\text{-}square$ would not be closer to 1.

6   Experimental Results

To classify each GOP of a given MPEG video, Support Vector Machines (SVM) classifier with poly kernel is implemented using the libraries from LIBSVM [9]. The program to read the MPEG quantized AC coefficients was developed based on the Matlab routine "mpgread.m" which was written by David Foti for MPEG-1 video [10]. In the experiment, we excluded the VBR video clips with IQ equal to 6, and CBR video clips with target bit rate equal to 1152000 bits/s. There are three distinct groups of video clips: CBR video clips encoded by $MSSG$ MPEG-2 Video Codec denoted as $group$-1; CBR video clips encoded by Berkeley MPEG-1 Encoder denoted as $group$-2; and VBR video clips encoded by Berkeley MPEG-1 Encoder denoted as $group$-3. For each group, there are ten categories: akiyo, bus, coastguard, flower, football, foreman, hall, stefan, tempete and waterfall. Here one category refers to a collection of video clips with the same video content but different quantization parameters in the same group.

For each group, the GOP feature vectors from nine categories are used to train the classifier, the remaining one category not involved in the training is used for testing. For example, if akiyo is used for testing, then it will not be involved in the training. After training the classifier, each GOP of the testing video clip is tested by the SVM classifier. Each GOP will be labeled by the classifier as being originally MPEG compressed or being doubly MPEG compressed. Once all GOPs in the testing video clips have been labeled, we count the number of the GOPs classified as being doubly MPEG compressed and calculate the D/N. Since each of the ten video categories has to test in turn, we have to perform ten training-testing cycles for each group.

Table 2. The D/N for *group*-1 for CBR case

bitrate	806400		921600		1036800		1267200		1382400		1497600	
video clip	ori	dbl	ori	dbl	ori	dbl	ori	dbl	ori	dbl	ori	dbl
akiyo	0.00	0.95	0.00	0.95	0.00	0.95	0.00	1.00	0.00	1.00	0.00	1.00
bus	0.40	1.00	0.20	1.00	0.20	0.90	0.00	1.00	0.00	1.00	0.00	1.00
coastguard	**0.60**	1.00	0.50	1.00	0.35	1.00	0.15	1.00	0.00	1.00	0.00	1.00
flower	0.12	1.00	0.00	0.94	0.00	0.71	0.00	1.00	0.00	1.00	0.00	1.00
football	**0.83**	1.00	0.50	0.83	0.50	0.83	0.33	1.00	0.17	1.00	0.00	1.00
foreman	0.50	0.95	0.40	0.90	0.35	0.95	0.20	1.00	0.10	1.00	0.10	1.00
hall	0.10	0.85	0.00	1.00	0.00	0.85	0.10	1.00	0.10	1.00	0.20	1.00
stefan	0.17	1.00	0.17	0.83	0.17	**0.50**	0.00	1.00	0.00	1.00	0.00	1.00
tempete	0.00	0.67	0.00	0.78	0.00	**0.50**	0.00	1.00	0.00	1.00	0.00	1.00
waterfall	0.00	**0.50**	0.00	0.61	0.00	0.67	0.00	1.00	0.00	1.00	0.00	1.00

Table 3. The D/N for *group*-2 for CBR case

bitrate	806400		921600		1036800		1267200		1382400		1497600	
video clip	ori	dbl	ori	dbl	ori	dbl	ori	dbl	ori	dbl	ori	dbl
akiyo	0.30	1.00	**0.95**	1.00	0.45	1.00	0.20	1.00	0.35	1.00	0.40	1.00
bus	0.30	1.00	0.20	1.00	0.00	0.80	0.00	1.00	0.00	1.00	0.00	1.00
coastguard	**0.60**	1.00	0.30	1.00	0.20	0.95	0.05	1.00	0.00	1.00	0.00	1.00
flower	0.06	0.94	0.00	0.88	0.00	**0.41**	0.00	1.00	0.00	1.00	0.00	1.00
football	0.50	0.83	0.33	1.00	0.17	1.00	0.33	1.00	0.17	1.00	0.17	1.00
foreman	0.15	0.70	0.20	0.60	0.10	0.90	0.05	1.00	0.00	1.00	0.00	1.00
hall	0.00	**0.35**	0.00	0.75	0.00	1.00	0.00	0.95	0.00	0.95	0.05	0.95
stefan	0.00	0.67	0.00	0.67	0.00	0.67	0.00	1.00	0.00	1.00	0.00	1.00
tempete	0.06	0.94	0.00	0.78	0.06	0.78	0.00	1.00	0.00	1.00	0.00	1.00
waterfall	0.00	0.94	0.00	0.56	0.00	**0.28**	0.00	1.00	0.00	1.00	0.00	1.00

The D/N values for *group*-1, *group*-2 and *group*-3 are listed in Table 2, 3, and 4, respectively, where ori stands for originally compressed video, and dbl stands for doubly compressed video. Except for the first two rows, each row represents the D/N values of one video category. The results for $IQ = 10$ is not included in Table 4 due to the constraint of paper space.

It is found that most of D/N values of doubly MPEG compressed video clips are closer to 1 and most of D/N values of originally MPEG compressed video clips are closer to 0. Based on the proposed decision rule, video clips will be classified as being doubly MPEG compressed if D/N is greater than a threshold T. If we use simple majority voting rule and set T to be 0.5, it is shown that, in *group*-1, only two originally and three doubly compressed video clips are misclassified; in *group*-2, two originally and three doubly compressed video clips are misclassified; in *group*-3, only one originally compressed video clip is misclassified. The D/N values of misclassified video clips are highlighted in black in Table 2, 3 and 4.

Table 4. The D/N for *group*-3 for VBR case

IQ	2		3		4		5		7		8		9	
video clip	ori	dbl	ori	dbl	ori	dbl	ori	dbl	ori	dbl	ori	dbl	ori	dbl
akiyo	0.05	1.00	0.05	1.00	0.05	1.00	0.05	1.00	0.05	0.95	0.00	1.00	0.00	1.00
bus	0.00	1.00	0.00	1.00	0.00	1.00	0.00	1.00	0.00	0.80	0.00	0.90	0.00	1.00
coastguard	0.00	1.00	0.00	1.00	0.00	1.00	0.00	1.00	0.00	1.00	0.05	1.00	0.10	1.00
flower	0.00	1.00	0.00	1.00	0.00	1.00	0.00	1.00	0.00	1.00	0.00	0.53	0.00	1.00
football	0.00	1.00	0.00	1.00	0.00	1.00	0.00	1.00	0.00	1.00	**0.67**	1.00	0.5	1.00
foreman	0.00	1.00	0.00	1.00	0.00	1.00	0.00	1.00	0.00	0.95	0.10	1.00	0.00	1.00
hall	0.00	1.00	0.00	1.00	0.00	0.85	0.00	0.75	0.00	0.90	0.00	1.00	0.00	1.00
stefan	0.00	1.00	0.00	1.00	0.00	1.00	0.00	1.00	0.00	1.00	0.00	1.00	0.00	1.00
tempete	0.00	1.00	0.00	1.00	0.00	0.94	0.00	1.00	0.00	0.89	0.00	1.00	0.00	1.00
waterfall	0.00	1.00	0.00	1.00	0.00	1.00	0.00	1.00	0.06	0.94	0.06	1.00	0.06	1.00

Table 5. The detection performance

group-1 for CBR case			*group*-2 for CBR case			*group*-3 for VBR case		
TNR	TPR	accuracy	TNR	TPR	accuracy	TNR	TPR	accuracy
0.967	0.950	0.958	0.967	0.950	0.958	0.988	1.000	0.993

The average classification performance is shown in Table 5 where TNR (true negative rate) is detection accuracy for originally MPEG compressed video, TPR (true positive rate) is doubly MPEG compressed video, and accuracy is equal to (TN+TP)/2. The results demonstrate that the proposed detection scheme is very effective in detecting the double compression for VBR and CBR video.

7   Conclusion

We propose a novel approach to detection of double MPEG compression. The first digit distribution of non-zero MPEG quantized AC coefficients in I, P, and B frames follows the parametric logarithmic law well if the video is MPEG compressed only once. The distribution is disturbed if the video is doubly MPEG compressed. To catch the disturbance, the 36-D features of the first digit probabilities and goodness-of-fit statistics are formed. We proposed a decision rule on the basis of GOP as detection unit. The experiments have demonstrated the effectiveness of the approach for both variable and constant bit rate control.

The unique existing prior art, which just extends double JPEG detection scheme to I frames of MPEG video for double MPEG compression detection [1], can only work for VBR video. The propose approach is more general since it can detect double MPEG compression not only for VBR video but also for CBR video. Instead of relying on specific observation, it uses machine learning framework with effective distinguishing features. Hence, it is more reliable and efficient.

References

1. Wang, W.H., Farid, H.: Exposing digital forgeries in video by detecting double MPEG compression. In: ACM Multimedia and Security Workshop, Geneva, Switzerland (2006)
2. Lukas, J., Fridrich, J.: Estimation of primary quantization matrix in double compressed JPEG images. In: Digital Forensic Research Workshop, Cleveland, Ohio, USA (2003)
3. Popescu, A.C.: Statistical tools for digital image forensics. Ph.D. Dissertation, Department of Computer Science, Dartmouth College (2005)
4. Fu, D., Shi, Y.Q., Su, W.: A generalized Benford's law for JPEG coefficients and its application in image forensics. In: IS&T/SPIE 19TH Annual Symposium, San Jose, California, USA (2007)
5. Mitchell, J.L., Pennebaker, W.B., Fogg, C.E., LeGall, D.J.: MPEG video: compression standard. Chapman & Hall, New York (1997)
6. http://bmrc.berkeley.edu/frame/research/mpeg/mpeg_encode.html
7. http://www.mpeg.org/MPEG/video/mssg-free-mpeg-software.html
8. Leon-Garcia, A.: Probability and Random Processes for Electrical Engineering, 2nd edn. Addison Wesley, Reading (1994)
9. Hsu, C.-W., Chang, C.-C., Lin, C.J.: LIBSVM: a library for support vector machines (2001)
10. Matlab Central File Exchange, http://www.mathworks.com

A Time Series Intra-Video Collusion Attack on Frame-by-Frame Video Watermarking

Sam Behseta[1], Charles Lam[2], Joseph E. Sutton[3], and Robert L. Webb[4]

[1] Department of Mathematics, California State University, Fullerton, CA 92834, USA
sbehseta@fullerton.edu
[2] Department of Mathematics, California State University, Bakersfield,
CA 93311, USA
clam@csub.edu
[3] Department of Computer Science, California State University, Bakersfield,
CA 93311, USA
joseph.e.sutton.csub@gmail.com
[4] Department of Computer Science, California Polytechnic State University,
San Luis Obispo, CA 93407, USA
webb@calpoly.edu

Abstract. Current methods of digital watermarking for video rely on results from digital watermarking for images. However, watermarking each frame using image watermarking techniques is vulnerable to an intra-video collusion attack because it provides grounds to make statistical inference about information within consecutive frames. An algorithm using bootstrapped time series is proposed to exploit this vulnerability. Experimental results demonstrate that this algorithm produces a video with a significantly lower similarity to the given watermarked video using the standard watermark detector.

1 Introduction

A watermark is a secondary signal embedded into a host signal that is used for identification purposes. The requirements of such a watermark are: (1) it does not perceptually alter the host signal in its usable form, (2) it survives any transformation that the host signal would commonly encounter, and (3) it would be difficult to remove without significantly degrading the fidelity of the original signal. In the literature these requirements are known as imperceptibility, robustness, and security.

We employ the watermarking method originally described by Cox, et al [6]. This method requires no knowledge of the host signal. The discrete cosine transform (DCT) is used to determine the parts of the signal that are altered by the watermark. The watermark itself is simply a signal of samples from a Gaussian distribution. This noise is then added to the some subset of DCT coefficients. The specific watermark signal is kept in a database by the content owner for comparison using a similarity metric.

H.J. Kim, S. Katzenbeisser, and A.T.S. Ho (Eds.): IWDW 2008, LNCS 5450, pp. 31–44, 2009.

In [6] the watermarking method is demonstrated on an image. The method is shown to be imperceptible and robust. The authors also showed that this method is resistant to many common security attacks, including collusion by averaging.

The goal of a collusion attack is to construct this new signal in such a way as to guarantee that it cannot be traced back to any of the colluding signals. In general, the most common collusion attack is to collect n independently watermarked signals and average together their attributes. This has been shown to be naive in that it does not obfuscate the watermarks unless n is prohibitively large [6,12]. The inter-video collusion attack can be formalized as follows: given n independently watermarked signals $\mathbf{F}^1, ..., \mathbf{F}^n$, the attack is defined by a function C such that

$$\hat{\mathbf{F}} = C(\mathbf{F}^1, ..., \mathbf{F}^n) \tag{1}$$

where $\hat{\mathbf{F}}$ is a new signal using information from all contributing signals.

A video signal is a sequence of images. There have been two proposed methods for watermarking such a signal. One method is to view the signal as a three dimensional object and add the watermark to the coefficients of its three dimensional discrete cosine transform (DCT3) [19,31]. This method has the drawback that multiple frames are necessary for the recovery of the watermark. In other words, the watermark cannot be recovered with a single frame of the video, and the watermark can be defeated by interspersing frames from several independently watermarked copies. For this reason, a frame-by-frame method is used.

A frame-by-frame method specifies that each frame is represented as its two dimensional discrete cosine transform coefficients (DCT2) and a watermark is added to these coefficients. Several different methods based on this idea have been published. See [11,25] for examples.

In this paper, we propose a new intra-video collusion attack utilizing correlation between frames. This results in a considerable reduction of the similarity metric.

There are several special considerations that apply to video watermarking. For a discussion of the specific issues that arise in applying these methods to video, see [12]. First of all, is the watermark placed in the video independent of time? For example, when the watermark alters a given frame, does the alteration of the subsequent frame depend on the alteration of the previous frame? Several papers have been written about different types of video watermarks. The highlights of each method are discussed in [11]. For a sample of papers that explore dependant watermarks see [31,7].

Several other uses of watermarks have been proposed, and while they are not the focus of this paper, we include references to the literature. An application for watermarking is steganography where a detailed signal is hidden in a host signal. Vila-Forcén, et al [26] have used collusion attacks to extract information about the host signal and original signal in this setting. The term "digital watermark" has been used for many other purposes (see Furon [16], Delaigle, et al. [8], Cannons and Moulin [2], Soriano, et al. [23] and references therein). For a thorough overview of digital watermarking see [20].

This paper deals specifically with collusion attacks. Others to consider the same problem for images include [1,5,6,15]. An effective attack using averaging is presented in Ergun, et al [15]. However, there are perceptible negative effects on the host signal when this attack is used. Applying the bootstrap method in a collusion attack on images to generate confidence intervals for the attributes of the original signal was explored in [1].

The problem of intra-video collusion attacks, i.e. where information from a single watermarked video is used to produce an unwatermarked video or a video with an obscured watermark is discussed in [11]. Other papers that address intra-video collusion attacks are [24,25,27,28]. Specifically, [11] suggests 2 attacks: (1) averaging across the temporal range of the watermarked film, but this results in ghosting artifacts when the film includes motion and (2) "Watermark Estimation Remodulation", a process in which each watermarked frame is compared to its low frequency components to estimate the watermark, this estimated watermark then averaged across the temporal space to improve this estimate, which is then removed. Method (2) was shown to be ineffective when the watermarks are frame-by-frame independent. Vinod and Bora [27,28] further extended the idea of temporal averaging in [11] to reduce the ghosting effects produced by films that include a lot of motion.

The work presented in [24] describes the concept of "statistical invisibility" which means that the variation between watermarked frames is approximately the same as the variation between the original frames. Further, [25] proposes that the watermark only be placed in areas of the video where the variation between frames is high. This type of video watermarking is not considered in our paper.

This paper presents a variation on the temporal averaging attack where each frame is watermarked by adding independent normal noise to low-frequency DCT2 coefficients. However, the attack is relevant to any situation where the components of each frame that are watermarked are related to previous and subsequent frames. Instead of averaging temporally-close frames, an autoregressive of order one [AR(1)] time series is used to estimate the value of each frame. Subsequently, the bootstrap is used to estimate the parameters of the AR(1) model. This results in a considerable reduction of the similarity metric.

The rest of the paper is organized as follows. In Section 2, we describe the statistical theory, and the general mathematical setup of our methods. In Section 3, the main algorithm and some ad hoc alterations are presented. Statistical results from these algorithms are summarized in 4. Future work is discussed in Section 5.

2   Background and Setup

2.1   The AR(1) Model

To take into account the inter-dependency of coefficients associated with each frame, we consider fitting an AR(1) or an Auto Regressive Model of order 1. Let I_t be the information at frame t, where $t = 0, ..., n - 1$. Consider the model

$I_t = \beta_0 + \beta_1 I_{t-1} + e_t$, where e_t is the white noise series with mean zero and variance σ_e^2. In a sense, this is a simple regression model with a lag-1 relation between the predictor and the response variable. It is worth noting that $E(I_t|I_{t-1}) = \beta_0 + \beta_1 I_{t-1}$, and $Var(I_t|I_{t-1}) = Var(e_t) = \sigma_e^2$. In other words, given the information at frame $t-1$, the current information is centered around $\beta_0 + \beta_1 I_{t-1}$, with the uncertainty of σ_e^2. This is the so-called Markov property.

Suppose that $E(I_t) = E(I_{t-1}) = \mu$. Then, $E(I_t) = \frac{\beta_0}{1-\beta_1}$. Also, it is easy to show that $Var(I_t) = \frac{\sigma_e^2}{1-\beta_1^2}$. The expected value relationship suggests that the mean of I_t exists for every $\beta_1 \neq 1$, and the mean of $I_t = 0$ if and only if $\beta_0 = 0$. The second relationship implies that as long as $0 < \beta_1 < 1$, $Var(I_t)$ remains bounded and non-negative. Most importantly, these two relationships reveal that every AR(1) model is in essence weakly stationary. That is, the first and the second moment of I_t are time invariant. The weak stationarity property of the AR(1) model provides a highly desirable infrastructure to construct further statistical inferences, chief among which, the estimation of the model parameter β_1.

Generally speaking, the conditional least squares method can be used to estimate the parameters of any AR(p) model. In the particular case of AR(1), the least squares solution for $\hat{\beta}_1$ is $\hat{\beta}_1 = \frac{\sum_{t=2}^n I_t I_{t-1}}{\sum_{t=2}^n I_{t-1}^2}$. Consequently, it is easy to show that $\sqrt{n}(\hat{\beta}_1 - \beta_1) \rightarrow N(0, (1-\beta_1^2))$.

2.2 The Bootstrap

Let $X_1^*, ... X_n^*$ be an independently drawn and identically distributed sample from $\hat{F}$, the empirical cumulative distribution function of a dataset $X_1, ..., X_n$. $X_1^*, ... X_n^*$, or the Bootstrap sample, can be acquired by sampling with replacement from $X_1, ..., X_n$. Also, suppose that T_n is an estimator of some unknown parameter θ (for example, $T_n = \bar{X}$). The Bootstrap sample is nonparametric in a sense that since it is obtained from the dataset, it makes no assumptions regarding the underlying statistical model and its parameters. By generating repeated Bootstrap samples, one can obtain a probability distribution for T_n^*, hence being able to assess its uncertainty.

Let $U_F(x) = Pr(T_n \leq x)$ denote the distribution of T_n. By replacing F with $\hat{F}$, we have $U_{\text{boot}}(x) = U_{\hat{F}}(x) = Pr[T_n^* \leq x|(X_1, ..., X_n)]$, the conditional distribution of T_n^* given data. Let ρ_∞ be a metric generated by a sup-norm[1] on the space of all distributions in R^p, for an integer p, representing the dimensionality of the distribution space. Then the following results hold (See Shao and Tu [21] for details):

(1) If T is continuously ρ_∞-Hadamard differentiable, then the Bootstrap estimator U_{boot} is strongly consistent. In other words, $\rho_\infty(U_{\text{boot}}, U_n) \rightarrow 0$ (almost surely).
(2) If T is continuously ρ_r-Frechet differentiable, then the Bootstrap estimator U_{boot} is strongly consistent.

---

[1] $\| h \|_\infty$ is the sup-norm of a function h on R^p, if $\| h \|_\infty = \sup_x |h(x)|$.

These results guarantee the fact that the Bootstrap distribution U_{boot} is consistent for many estimators such as the mean and the median.

2.3 Residual Bootstrap of AR(1)

Suppose that $e_t = I_t - \beta_1 I_{t-1}$, $t = 2, ..., n$. Let $\tilde{e}_t$ represent the standardized e_t, such that $\frac{1}{n-1} \sum \tilde{e}_t = 0$, and $\frac{1}{n-1} \sum \tilde{e}_t^2 = 1$. Then, by sampling $t = 1, ..., n$ many e_t^* with replacement from $\tilde{e}_t$, and letting $I_1^* = e_1^*$, and $I_t^* = \hat{\beta}_1 I_{t-1}^* + e_t^*$, a bootstrap time series based on the AR(1) model is formed. Thus, $\hat{\beta}_1^* = \frac{\sum_{t=2}^{n} I_t^* I_{t-1}^*}{\sum_{t=2}^{n} (I_{t-1}^*)^2}$ is the least squares estimator of the bootstrap model. It can be shown that the bootstrap least squares estimator $\hat{\beta}_1^*$ converges in probability to the original least squares estimator $\hat{\beta}_1$. The proof of the last statement is rather rigorous (see Shao and Tu [21]).

2.4 Notations

Consider a film $\mathbf{F}$ with frames $\{\mathbf{F}_0, \mathbf{F}_1, \mathbf{F}_2, \ldots, \mathbf{F}_{n-1}\}$. We can assume each frame $\mathbf{F}_i$ is a vector of real values. We consider the case in which watermark is inserted frame-by-frame so that the watermarked film $\hat{\mathbf{F}} = [\hat{\mathbf{F}}_0, \hat{\mathbf{F}}_1, \hat{\mathbf{F}}_2, \ldots, \hat{\mathbf{F}}_{n-1}]$ is specified by

$$\hat{\mathbf{F}}_t = \mathbf{F}_t + \mathbf{W}_t(K), \quad \mathbf{W}_t(K) \sim \mathcal{N}(0, \alpha)$$

where α is a scaling factor and $\mathbf{W}_t(K)$ is a watermark based on a secret key K. To abbreviate, we will write

$$\mathbf{W} = [\mathbf{W}_0(K), \mathbf{W}_1(K), \ldots, \mathbf{W}_{n-1}(K)] = [\mathbf{W}_0, \mathbf{W}_1, \ldots, \mathbf{W}_{n-1}].$$

Let $\hat{\mathbf{F}}^{(i)}$ be the i-th watermarked film using watermark $\mathbf{W}^{(i)}$. Hence, $\mathbf{W}_t^{(i)} = \hat{\mathbf{F}}_t^{(i)} - \mathbf{F}_t$. Let $\mathbf{W}^{(i)} = [\mathbf{W}_0^{(i)}, \mathbf{W}_1^{(i)}, \mathbf{W}_2^{(i)}, \ldots, \mathbf{W}_{n-1}^{(i)}]$ and $\mathbf{W}^{(j)} = [\mathbf{W}_0^{(j)}, \mathbf{W}_1^{(j)}, \mathbf{W}_2^{(j)}, \ldots, \mathbf{W}_{n-1}^{(j)}]$ be two distinct watermarks used for embedding. The similarity of $\mathbf{W}^{(j)}$ to $\mathbf{W}^{(i)}$ is

$$\text{sim}(\mathbf{W}^{(i)}, \mathbf{W}^{(j)}) = \frac{1}{n} \sum_{k=0}^{n-1} \frac{\mathbf{W}_k^{(i)} \cdot \mathbf{W}_k^{(j)}}{\mathbf{W}_k^{(i)} \cdot \mathbf{W}_k^{(i)}}.$$

The measurement is 1 if a film is compared to itself, and close to 0 if two uncorrelated watermarks were compared. Provided that watermark entries are normally distributed, it is expected that $||\mathbf{W}_k^{(i)}|| \approx ||\mathbf{W}_k^{(j)}||$. Therefore, $\text{sim}(\mathbf{W}^{(i)}, \mathbf{W}^{(j)}) \approx \text{sim}(\mathbf{W}^{(j)}, \mathbf{W}^{(i)})$.

3 Methods

Consider a watermarked film $\hat{\mathbf{F}}$ with n frames. Without loss of generality, we will analyze watermarked DCT2 coefficients only. Next, we implement the method described in Section 2.3 m times on each sequence of watermarked coefficients at the same location in all consecutive frames.

For each location s in each frame of $\hat{\mathbf{F}}$, we extract the coefficients $\{\hat{F}_{0,s}, \hat{F}_{1,s}, \ldots, \hat{F}_{n-1,s}\}$. Let $\mathbf{Z}$ be an $n \times m$ matrix of coefficients. Let $\mathbf{Z}_{t,i}$ denote the value at the i-th iteration at the t-th frame. We run the following algorithm on each watermarked location s, resulting in a film that we call $\hat{\mathbf{F}}'$.

Algorithm 1. The Main Algorithm

1: Initialize $\beta_{1,0}$, where

$$\beta_{1,0} = \frac{\sum_{t=1}^{n-1} \hat{F}_{t,s} \hat{F}_{t-1,s}}{\sum_{t=0}^{n-1} \hat{F}_{t,s} \hat{F}_{t,s}} \tag{2}$$

2: Initialize the first iteration of $\mathbf{Z}$. Let $Z_{t,0} = \beta_{1,0} \hat{F}_{t-1,s}$, $\forall t, 1 \leq t \leq n-1$.
3: Create normalized vector $\mathbf{Z}^{(std)} = [Z_0^{(std)}, Z_1^{(std)}, \ldots, Z_{n-1}^{(std)}]$ for random sampling. Let

$$Z_t^{(std)} = \frac{Z_t' - \mu'}{\sigma'}$$

   where μ' is the mean of $\mathbf{Z}'$, σ' is the standard deviation of $\mathbf{Z}'$, and $Z_t' = \hat{F}_{t,s} - \beta_{1,0} \hat{F}_{t-1,s}$.
4: For i from 1 to m do:

    4.1: For each i-th iteration, select a uniform random value $r \in_R [1, n-1]$. Obtain the next candidate,

$$Z_{t,i} = \beta_{1,i-1} Z_{t,0} + Z_r^{(std)}$$

    where $0 \leq t \leq n-1$.
    4.2: Evaluate $\beta_{1,i}$ using

$$\beta_{1,i} = \frac{\sum_{t=1}^{n-1} Z_{t,i} Z_{t-1,i}}{\sum_{t=0}^{n-1} Z_{t,i} Z_{t,i}} \tag{3}$$

5: Sort the iteration values $[Z_{t,0}, Z_{t,1}, \ldots, Z_{t,m-1}]$ for each frame t.
6: Select the median value Z_t' of each sorted frame in $[Z_{t,0}, Z_{t,1}, \ldots, Z_{t,m-1}]$. For example, if 100 iterations are executed, then $Z_t' = Z_{t,49}$.
7: Set $\hat{F}_{t,s}' = Z_t'$, for $0 \leq t \leq n-1$.

Implementing this algorithm may produce a ghosting effect due to the existence of extreme coefficient values. To address the ghosting effect, we employ a series of ad hoc alterations as explained below.

We compare the coefficients in $\hat{\mathbf{F}}'$ with $\hat{\mathbf{F}}$. In cases in which the discrepancy is large, we adjust the coefficient in $\hat{\mathbf{F}}'$ using one of the following ways. Let τ be a threshold value. Without loss of generality, τ is taken to be one standard deviation of the Gaussian distribution of the watermark.

Type 1: If $(\hat{F}_{t,s}' - \hat{F}_{t,s}) > 2\tau$, then $\hat{F}_{t,s}' = \hat{F}_{t,s} + 2\tau$.
    Otherwise, if $(\hat{F}_{t,s}' - \hat{F}_{t,s}) < -2\tau$, then $\hat{F}_{t,s}' = \hat{F}_{t,s} - 2\tau$.

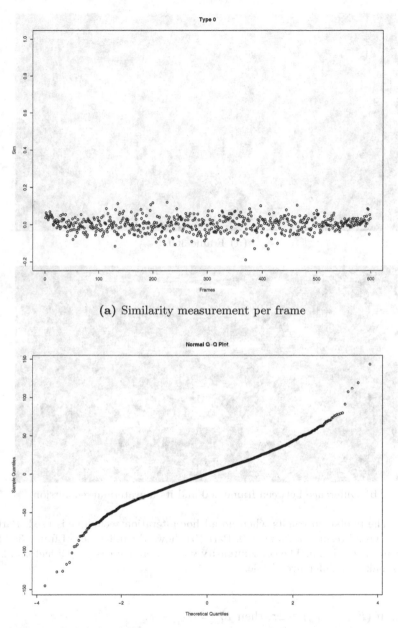

(a) Similarity measurement per frame

(b) QQ-plot of the differences between the modelled coefficient and its unwatermarked counterpart

Fig. 1. The bootstrap results when no ad hoc alteration technique is used. Part (a) is a plot of similarity measurement. An approximately 90% reduction of watermark evidence is achieved across all frames. Part (b) shows the QQ-plot of the differences between the modelled coefficient and its unwatermarked counterpart.

(a) Frame 360.

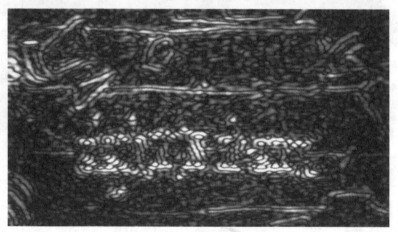

(b) Difference between frame 360 and its unwatermarked version.

Fig. 2. The bootstrap results when no ad hoc alteration technique is used. Part (a) shows the visual results of frame 360. Part (b) shows the difference of frame 360 from the unwatermarked film. The color intensity values were increased by a factor of 20, in order to make the difference visible.

Type 2: If $(\hat{F}'_{t,s} - \hat{F}_{t,s}) > 2\tau$, then $\hat{F}'_{t,s} = \hat{F}_{t,s} + \tau$.
  Otherwise, if $(\hat{F}'_{t,s} - \hat{F}_{t,s}) < -2\tau$, then $\hat{F}'_{t,s} = \hat{F}_{t,s} - \tau$.

Type 3: For each s-th coefficient, select a uniform random value $r \in_R [\tau, 2\tau]$.
  If $(\hat{F}'_{t,s} - \hat{F}_{t,s}) > 2\tau$ or $(\hat{F}'_{t,s} - \hat{F}_{t,s}) < -2\tau$, then $\hat{F}'_{t,s} = \hat{F}_{t,s} + r$.

Type 4: For each s-th coefficient, select a Gaussian random value $r \in_R [\tau, 2\tau]$.
  If $(\hat{F}'_{t,s} - \hat{F}_{t,s}) > 2\tau$ or $(\hat{F}'_{t,s} - \hat{F}_{t,s}) < -2\tau$, then $\hat{F}'_{t,s} = \hat{F}_{t,s} + r$.

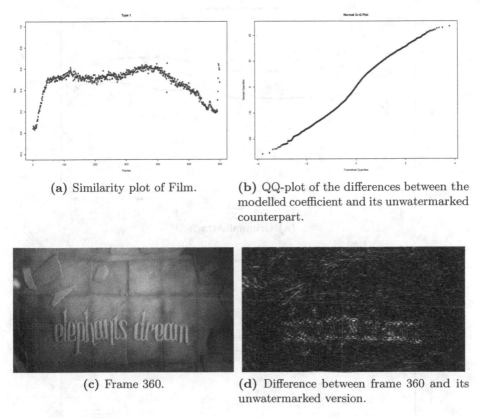

(a) Similarity plot of Film.

(b) QQ-plot of the differences between the modelled coefficient and its unwatermarked counterpart.

(c) Frame 360.

(d) Difference between frame 360 and its unwatermarked version.

Fig. 3. The bootstrap results when ad hoc alteration Type I is used

4   Results

We demonstrate the effect of our proposed models in Figures 1, 2 and 3. In Figures 1 and 2, where the results of the main algorithm is shown, we include the similarity values per frame, the QQ-plot of the differences between the modelled coefficient and its unwatermarked counterpart, a sample frame from the modelled film, and its difference from the unwatermarked film. As shown in Figure 1, the solid majority of frames have a similarity measurement between -0.1 and 0.1. This is an extremely promising finding. Note, however, that the presence of some extreme values create a deviation from normality, as shown in the QQ-plot, which created some distortion of the original images. This is precisely the reason that we implemented our ad hoc adjustments.

In Figure 3, we demonstrate the effectiveness of our ad hoc algorithm resulted from Type 1 alteration. The QQ-plot confirms the normality of the residual noise. The trade off to normality adjustment is an increase in the similarity values.

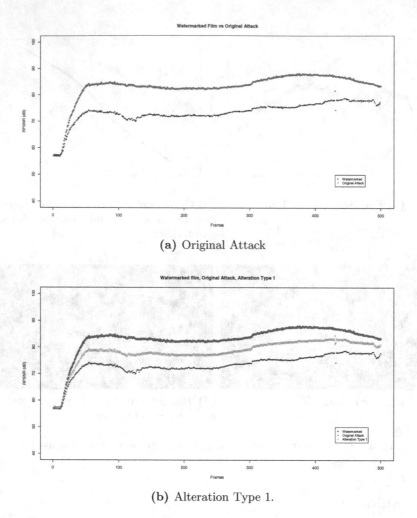

(a) Original Attack

(b) Alteration Type 1.

Fig. 4. RPSNR Values of Watermarked Film vs. Various Attacks. In (a), the RPSNR values of watermarked film vs. original attack is shown. The RPSNR values of the original attack is shown to be slightly lower than that of the original watermarked film. In figure (b) the RPSNR values of the attack with Alteration Type 1 is shown, as compared with the RPSNR values of the original watermarked film and the results of the original attack.

However, there is still an approximately 40% reduction in the evidence of the original watermark. The perceptual picture quality is preserved.

To illustrate the picture quality after the attack, we show results of RPSNR values [3,18,22] in Figures 4 and 5. The frame-by-frame RPSNR values of the original watermarked video is compared with the results from the original algorithm, and Type 1, 3, and 4 alterations, respectively.

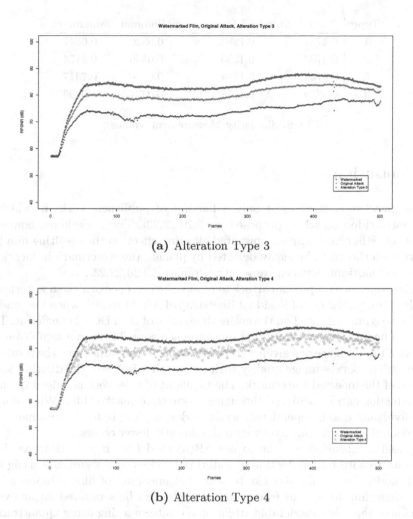

(a) Alteration Type 3

(b) Alteration Type 4

Fig. 5. RPSNR Values of Watermarked Film vs. Various Attacks. In figures (a) and (b), the RPSNR values of the attacks with Alteration Types 3 and 4 are shown, respectively, as compared with the RPSNR values of the original watermarked film and the results of the original attack. It can be easily seen that the RPSNR values for different Alteration Types are higher than then values from the original attack and lower than that of the watermarked Film.

In Figure 6, the table summarizes the statistical measurement of the similarity metric of frames in each of the alteration types. As shown in this table, alteration methods still result in a significant reduction in the similarity values. Additionally, the low standard deviation values suggest that these methods are highly consistent across all frames.

The film segments resulting from all algorithms are available at http://www.cs.csubak.edu/~mlg/research/IWDW-08/.

Type	Mean	Standard Deviation	Minimum	Maximum
1	0.4754	0.1280	0.0320	0.6547
2	0.6733	0.1350	0.0536	0.8174
3	0.5742	0.1304	0.0446	0.7317
4	0.5759	0.1658	0.0337	1.0205

Fig. 6. Similarity Measurement Values

5   Conclusion

To our knowledge, there exist only a handful of published works in which a single intra-video attack is proposed [8,19,20,22,23]. These methods, however, suffer from either imposing an unwanted ghosting effect on the resulting film [8], or have been shown to be easily defeated by placing the watermark in locations with lots of motions between consecutive frames [19,20,22,23].

In this paper, we propose an attack to reduce the presence of a given watermark in a film through the use of Residual Bootstrap of AR(1) model, when the modeling scheme is implemented on the entire time series of each DCT2 coefficient. The empirical findings demonstrate that our approach contributes to a significant reduction of the evidence of a given watermark, while at the same time, the resulting output still preserves image quality with a noise that imitates the distributional features of the inserted watermark. The highlight of this work is indeed the fact that reduction can be achieved through a single watermarked film. When multiple individually and independently watermarked copies are used, we achieve the equivalent of an averaging attack with significantly fewer copies.

A possible extension may be to use AR(n) model for $n > 1$. However, it is important to note that reduction of watermark evidence is significant using the AR(1) model. We would also like to note that encoding of film is independent of our algorithm, as long as frames from the film can be separated for analysis.

Recently, there is considerable attention on watermarking using Quantization Index Modulation methods (QIM) [4]. To the authors' knowledge, there has not been a detailed analysis on the noise distribution in the frequency domain after QIM has been applied. However, it is worth investigating the effect of this attack on QIM watermarked documents.

We have demonstrated that our method works well with a watermark that follows a Gaussian distribution. It is important to note that the proposed method will also work when the noise follows other distributions, such as the uniform distribution, or orthonormal noise, where estimators may be re-calculated depending on the nature of the noise. Our next step is to apply Bayesian time series modeling to acquire the posterior distribution associated with the consecutive signals. We then summarize the signal information with some central statistic for each posterior distribution. Currently, we are examining a host of non-informative priors on the DCT2 coefficients to provide an automatic paradigm for the frame-by-frame attacks.

Acknowledgements

We would like to thank CSUB students Christopher Gutierrez, Kevin Velado, and Max Velado[2] for their help in the preparation of this paper.

References

1. Behseta, S., Lam, C., Webb, R.L.: A Bootstrap Attack on Digital Watermarks in the Frequency Domain. In: Chen, L., Ryan, M.D., Wang, G. (eds.) ICICS 2008. LNCS, vol. 5308, pp. 361–375. Springer, Heidelberg (2008)
2. Cannons, J., Moulin, P.: Design and Statistical Analysis of a Hash-Aided Image Watermarking System. IEEE Trans. Image Processing 13(10), 1393–1408 (2004)
3. Chemak, C., Bouhlel, M.-S., Lapayre, J.-C.: New Watermarking Scheme for Security and Transmission of Medical Images for PocketNeuro Project. In: Spolecnost Pro Radioelektronicke Inzenyrstvi, pp. 58–63 (2007)
4. Chen, B., Wornell, G.W.: Quantization Index Modulation: A Class of Provable Good Methods for DigitalWatermarking and Information Embedding. IEEE Trans. Info. Theory 47(4), 1423–1443 (2001)
5. Comesaña, P., Pérez-Freire, L., Pérez-González, F.: The Return of the Sensitivity Attack. In: Barni, M., Cox, I., Kalker, T., Kim, H.-J. (eds.) IWDW 2005. LNCS, vol. 3710, pp. 260–274. Springer, Heidelberg (2005)
6. Cox, I.J., Kilian, J., Leighton, F.T., Shamoon, T.: Secure spread spectrum watermarking for multimedia. IEEE Trans. Image Processing 6(12), 1673–1687 (1997)
7. Deguillaume, F., Csurka, G., O'Ruanaidh, J., Pun, T.: Robust 3D DFT Digital Video Watermarking. In: Proceedings SPIE, vol. 3657, pp. 113–124 (1999)
8. Delaigle, J.F., De Vleeschouwer, C., Macq, B.: Watermarking algorithm based on a human visual model. Signal Processing 66, 319–335 (1998)
9. Dong, P., Brankov, J.G., Galatsanos, N.P., Yang, Y., Davoine, F.: Digital Watermarks Robust to Geometric Distortions. IEEE Trans. Image Processing 14(12), 2140–2150 (2005)
10. Doërr, G., Dugelay, J.: Countermeasures for Collusion Attacks Exploiting Host Signal Redundancy. In: Barni, M., Cox, I., Kalker, T., Kim, H.-J. (eds.) IWDW 2005. LNCS, vol. 3710, pp. 216–230. Springer, Heidelberg (2005)
11. Doërr, G., Dugelay, J.: Security Pitfalls of Frame-by-Frame Approaches to Video Watermarking. IEEE Trans. Signal Processing 52(10), 2955–2964 (2004)
12. Doërr, G., Dugelay, J.: A Guide Tour of Video Watermarking. Signal Processing: Image Communication 18, 263–282 (2003)
13. Efron, B.: The Jackknife, the Bootstrap and other Resampling Plans. Society for Industrial and Applied Mathematics, Philadelphia (1982)
14. Efron, B., Tibshirani, R.J.: An Introduction to the Bootstrap. Chapman & Hall, New York (1993)
15. Ergun, F., Kilian, J., Kumar, R.: A Note on the Limits of Collusion-Resistant Watermarks. In: Stern, J. (ed.) EUROCRYPT 1999. LNCS, vol. 1592, pp. 140–149. Springer, Heidelberg (1999)

[2] Student support is provided by CSUB Marc U*Star Program through National Institute of Health grant number 5 T34 GM069349-05, and the Louis Stokes Alliance for Minority Participation through National Science Foundation grant number HRD-0331537 and the CSU Chancellor's office.

16. Furon, T.: A Survey of Watermarking Security. In: Barni, M., Cox, I., Kalker, T., Kim, H.-J. (eds.) IWDW 2005. LNCS, vol. 3710, pp. 201–215. Springer, Heidelberg (2005)
17. Khayam, S.A.: The Discrete Cosine Transform (DCT): Theory and Application. Michigan State University (2003)
18. Kuttera, M., Petitcolas, F.A.P.: A fair benchmark for image watermarking systems. In: Proc. of SPIE Security and Watermarking of Multimedia Contents, pp. 226–239 (1999)
19. Li, Y., Gao, X., Ji, H.: A 3D Wavelet based Spatial-temporal Approach for Video Watermarking. In: ICCIMA 2003. Fifth International Conference on Computational Intelligence and Multimedia Applications, pp. 260–265 (2003)
20. Podilchuk, C.I., Delp, E.J.: Digital Watermarking: Algorithms and Applications. Signal Processing Magazine 18(4), 33–46 (2001)
21. Shao, J., Tu, D.: The Jackknife and Bootstrap. Springer, New York (2005)
22. Simpson, W.: Video Over IP. Focal Press (2005)
23. Soriano, M., Fernandez, M., Cotrina, J.: Fingerprinting Schemes. Identifying the Guilty Sources Using Side Information. In: Barni, M., Cox, I., Kalker, T., Kim, H.-J. (eds.) IWDW 2005. LNCS, vol. 3710, pp. 231–243. Springer, Heidelberg (2005)
24. Su, K., Kundur, D., Hatzinakos, D.: A Novel Approach to Collusion-resistant Video Watermarking. In: Proceedings SPIE, vol. 4675, pp. 491–502 (2002)
25. Su, K., Kundur, D., Hatzinakos, D.: A Content-Dependent Spatially Localized Video Watermark for Resistance to Collusion and Interpolation Attacks. In: Proceedings. 2001 International Conference on Image Processing, vol. 1, pp. 818–821 (2001)
26. Vila-Forcén, J.E., Voloshynovskiy, S., Koval, O., Pérez-González, F., Pun, T.: Practical Data-Hiding: Additive Attacks Performance Analysis. In: Barni, M., Cox, I., Kalker, T., Kim, H.-J. (eds.) IWDW 2005. LNCS, vol. 3710, pp. 244–259. Springer, Heidelberg (2005)
27. Vinod, P., Bora, P.K.: A New Inter-Frame Collusion Attack and a Countermeasure. In: Barni, M., Cox, I., Kalker, T., Kim, H.-J. (eds.) IWDW 2005. LNCS, vol. 3710, pp. 147–157. Springer, Heidelberg (2005)
28. Vinod, P., Bora, P.K.: Motion-compensated Inter-Frame collusion attack on video watermarking and a countermeasure. IEEE Proceedings on Information Security 153(2), 61–73
29. Wang, J., Gao, X., Shong, J.: A Video Watermarking based on 3-D Complex Wavelet. In: ICIP 2007, IEEE International Conference on Image Processing, vol. 5, pp. 493–496 (2007)
30. Wolfgang, R.B., Podilchuk, C.I., Delp, E.J.: Perceptual Watermarks for Digital Images and Video. Proceedings of IEEE 87(7), 1108–1126 (1999)
31. Zhu, W., Xiong, Z., Zhang, Y.: Multiresolution Watermarking for Images and Video. IEEE Transactions on Circuits and Systems for Video Technology 9(4), 545–550 (1999)

A Novel Real-Time MPEG-2 Video Watermarking Scheme in Copyright Protection*

Xinghao Jiang[1,2], Tanfeng Sun[1,2], Jianhua Li[1,2], and Ye Yun[1]

[1] School of Information Security Engineering, Shanghai Jiaotong University,
[2] Shanghai Information Security Management and Technology Research Key Lab
Shanghai, P. R. China, 200240
{xhjiang@sjtu.edu.cn,tfsun,lijh888}@sjtu.edu.cn

Abstract: In this paper, an efficient video watermarking scheme is presented through modifying the third decoded luminance differential DC component in each selected macro block. The modification is implemented by binary dither modulation with adaptive quantization step. The proposed scheme is based on the observation that luminance differential DC components inside one macro block are generally space correlated, so the quantization step can be adjusted according to adjacent differential components, to utilize properties of human visual system (HVS). Experimental results show that it can be implemented in real time with better visual quality.

Keywords: Video watermarking, real-time, luminance differential DC coefficient, human visual system, adaptive quantization step.

1 Introduction

Video watermarking is gaining popularity in information hiding community, along with rapid development of video industry. A good few video watermarking schemes are proposed for different applications such as error resilience [1], bit-rate control [2], interception detection [3]. Although some of them utilize video characteristics like motion vector [3] and temporal property [4], many others follow the way of image watermarking, e.g., LSB(least significant bit) [5], DCT(discrete cosine transform) coefficient [1], and DWT(discrete wavelet transform) coefficient [2] modification in selected frames.

However, some image-level schemes are not preferable when it comes to real-time embedding situations such as user labeling in VOD (video on demand) service, which requires instant operation on compression coded data with tolerable delay. The scheme proposed in paper [3] makes use of motion vector's LSB parity and therefore is only suitable for fragile watermark. In paper [6], AC coefficients' VLC(variable length code) code words are replaced or discarded to achieve the desired labeling objective. The scheme proposed in [7] adopts the DC component adjusting idea, but has to resume the original (quantized) DC coefficient values and the embedding intensity depends on

---

* Supported by the National Natural Science Foundation of China (No.60702042, 60802057), the 863 Hi-tech Research and Development program of China(No. 2009AA01Z407).

H.J. Kim, S. Katzenbeisser, and A.T.S. Ho (Eds.): IWDW 2008, LNCS 5450, pp. 45–51, 2009.

all the other DC coefficients inside the same region for fidelity assurance. The modification is implemented by performing binary dither modulation on one differential DC coefficient, and the quantization step is made self-adaptive according to the neighboring unchanged two inside the same macro block.

2 Differential DC Coefficient Modification Model

Certain video compression codec standards, such as MPEG-2, MPEG-4, H.264, apply DCT to original sequential pictures on block level, and encode AC coefficients with RLE (run-level coding), while DC coefficients are handled with DPCM (differential pulse code modulation) before being encoded into VLC code words. Take MPEG-2 for example, Fig.1 shows the predicting relation of four luminance (Y component) DC coefficients (after quantization), denoted as *DC1, DC2, DC3, DC4*, inside one 16×16 macro block, and *diff1, diff2, diff3, diff4* represent their predicted difference values.

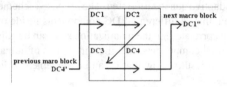

Fig. 1. Predicting relation of four luminance DC coefficients inside one macro block, with corresponding VLC code words representing diff1=DC1-DC4', diff2=DC2-DC1, diff3=DC3-DC2, diff4=DC4-DC3

   The differential DC coefficient modification model is constructed based on the observation that, in an I frame, luminance differential DC components inside one macro block are generally space correlated. Namely, luminance values within a small area usually change regularly in consistent directions, so luminance difference of two locations can be approximately estimated by the difference in surrounding areas. Since luminance DC coefficients indicate average luminance values of corresponding sub-blocks, their difference values reflect luminance changing state in different directions, according to the position of each sub-block. For an explicit expression, the relationship of three selected luminance differential DC components, *DC1-DC2*, *DC3-DC2*, and *DC4-DC2*, in one macro block is formulated with the following equation. Experimental verification of this estimation is provided in section 4.

$$DC3 - DC2 \approx \alpha * ((DC4 - DC2) + (DC1 - DC2)) \tag{1}$$

which is equivalent to

$$diff\,3 \approx \alpha * (diff\,4 + diff\,3 - diff\,2) \tag{2}$$

where α is a proportional factor. Fig.2 gives a more vivid depiction of this estimation.

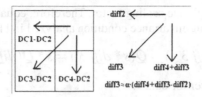

Fig. 2. Estimation of luminance differential DC components inside one macro block

This space correlation motivates the originality of modifying differential DC components to perform real-time watermarking on video data. Binary dither modulation is chosen as the modification method for its provable robustness. Provided with above estimating relation, it is feasible to make the quantization step self-adaptive and reduce the embedding-induced distortion due to uniform quantization. In this scheme, two luminance differential DC components, $diff3$ and $diff4$, in each selected macro block in an I frame, are modified as follows:

$$q = |q0 * (diff4 + diff3 - diff2)| \qquad (3)$$

where $q0$ is a pre-determined scaling factor and $|x|$ denotes the absolute value of x. q is the quantization step larger than zero, i.e., if the value of q is too small, e.g., smaller than 1.0, this macro block will be skipped. In case q is too large, upper bound can also be set to minimize the distortion, as demonstrated in the simulation section. If one bit of the binary message to be embedded is '0', then

$$diff3' = < q* < diff3/q >> \qquad (4)$$

else if '1' is to be embedded,

$$diff3' = < q * (< diff3/q + 1/2 > -1/2) > \qquad (5)$$

$<\cdot>$ is the round-off operation. Thus $diff3$ is moved to the nearest quantized position denoting '0' or '1'. After modifying $diff3$, $diff4$ is modified to keep the original value of $DC4$ and not to influence other DC coefficients in following macro blocks:

$$diff4' = diff4 + diff3 - diff3' \qquad (6)$$

In this way, the quantization step, q, is constrained by $(diff4+diff3)$ and $-diff2$, as illustrated in Fig.2. Both fidelity assurance and blind extraction are achieved, since only $DC3$ is actually modified and the quantization step can be accurately retrieved by the difference value of other unchanged three, i.e., $DC1$, $DC2$ and $DC4$. Another advantage to apply dither modulation to differential DC components is that it is very robust to gain attacks concerned in paper [7], because amplitude scaling will have the same effect on $diff2$, $diff3'$, $diff4'$ and hence q:

$$diff4'_s + diff3'_s - diff2_s = s * (diff4' + diff3' - diff2) \qquad (7)$$

$$q_s = |q0 * s * (diff4' + diff3' - diff2)| = s * q \qquad (8)$$

where s denotes the amplitude scaling factor. Therefore, equations (4) and (5) are well preserved, and the scaling-invariance condition analyzed in [17] is satisfied,

$$Q(diff\,3'_s) = Q(s * diff\,3') = s * Q(diff\,3) \tag{9}$$

Here $Q(x)$ is the quantizer for data x.

3 Watermark Embedding and Extracting

To enhance the robustness and security of the proposed scheme, templates containing random '0' and '1' signals, are generated for host I frames. From the first I frame, one bit information is embedded in each frame through repetitious embedding. Each template is of the same size with the number of macro blocks in an I frame. Only macro blocks with corresponding signal '1' in the template are modified. Detailed embedding procedure is described as follows.

1) Obtain L-bit binary original message (shorter than the total amount of I frames), and encrypt it by exclusive-or operation with the template generating seed, S, to get the watermark data to be embedded, W.

2) Use S to generate L templates containing random '0' and '1' signals, each template contains N signals, the same size with the macro block number of each I frame.

3) For the i-th I frame to be embedded, select macro blocks with signal '1' in the same position in the i-th template.

4) For each selected macro block, decode the three luminance differential DC coefficients' VLC code words to get $diff2, diff3, diff4$.

5) Calculate the quantization step q according to equation (3).

6) Modify $diff3$ and $diff4$ according to the i-th bit value of W, and equations (4)-(6).

7) Encode $diff3'$, $diff4'$ back to VLC code words and produce the embedded video data.

The reason to embed one bit in an I frame is that, in the following presented experiment, the embedded data is character and requires accurate restoration, rather than the error-tolerated restoration when the embedded data is image and embedding one bit in one macro block for high capacity is reasonable. For security concerns, since the embedding positions are private, the secret message can only be correctly extracted and decrypted using right scaling factor, $q0$, and the template generating seed, S. Further more, the secret message can be encrypted with more complex algorithm and modulated by pseudorandom binary sequence [10], instead of simple repetition, according to different application. As to the robustness of this dither modulation scheme, the watermark can never be attacked without introducing additional distortion .

The watermark extracting procedure is similar. Since $diff3'+diff4'$ is equal to $diff3+diff4$ by equation (6), and $diff2$ is unchanged, the quantization step, q, is retrieved reliable by equation (3). The embedded bit value for each selected macro block is determined by the minimum distance criterion . And then the i-th bit value of W is identical to the majority of these bit values extracted from the i-th I fame.

4 Simulation

This section presents two-step experimental results. The first step is to verify the estimation equation formulated in section 2. A large sum of data is collected and analyzed using MATLAB 6.5. The second step is to test the embedding effect of the watermarked video with the proposed scheme and make some comparison. The tested video file is in MPEG-2 format. This scheme can be easily applied to other video format with similar compression coding mechanism.

4.1 Verification of Luminance Differential DC Component Estimation

Four YUV sequence files: *Flower garden* (704×480), *Mobile* (704×480), *Susan* (352×240) and *Table tennis* (352×240), each with 90 Y, U, V pictures, are tested to verify equation (1), through calculating the value of the proportional factor, α, for every 16×16 macro block inside Y (luminance component) pictures. The AV (average value) and MAE (mean absolute error) value of α are computed to test α's stability. Table 1 lists the calculation result of α's AV/MAE value for some Y pictures from the tested YUV files.

The data in Table 1 show that, inside each Y picture, α varies mildly for most macro blocks, with a MAE less than 1.5, indicating a concentrated distribution of α.

Table 1. AV/MAE value of α for Y pictures, average AV/MAE is listed in the last row

No.	Flower garden	Mobile	Susan	Table tennis
1	0.649 / 0.955	0.480 / 1.324	0.688 / 0.584	0.640 / 0.398
10	0.300 / 1.171	0.629 / 0.988	0.841 / 0.860	1.055 / 0.828
Avg.	0.566 / 0.981	0.705 / 1.115	0.749 / 0.585	0.733 / 0.519

Take the first Y picture of *Flower garden* for example. The whole picture is divided into 1320 (44×30) 16×16 macro blocks, and after DCT is performed on the four 8×8 sub-blocks inside each macro block, the DC coefficients are used to calculate the differential components, *diff2, diff3, diff4*, and the ratio of *diff3* to *diff4+diff3-diff2*, α, as illustrated in Fig.2. The value of α for most macro blocks stays closely to the average value, 0.649, as indicated by the small MAE value, 0.955. For all the Y pictures, the value of α takes on similar properties. Hence the ratio of *diff3* to *diff4+diff3-diff2* can be represented by a constant for approximate calculation. For each tested file, poles around the average value of α (about 0.7) take up a large proportion of the whole, proving that the ratio of *diff3* to *diff4+diff3-diff2* is rather stable. So it is reasonable that if the quantization step for *diff3* is adjusted by *diff4+diff3-diff2* in each macro block, better visual quality can be attained than conventional dither modulation scheme using uniform quantization step. The following comparing experiment validates the superiority of the proposed scheme.

4.2 Implementation of The Proposed Scheme and Comparison

This section illustrates implementation of the proposed scheme. Four MPEG-2 video sequences, *Flower garden*(704×480, 6M bps), *Mobile*(704×480, 6M bps), *Susan*(352×240, 1.5M bps) and *Table tennis*(352×240, 1.5M bps), compression result of previous tested YUV files, are selected as experimental objects. Table 2 shows the performance of the scheme with quantization step varies from 0.1 to 0.4.

Table 2. MSE and PSNR(dB) calculation with re-converted embedded and non-embedded Y pictures or the proposed scheme and the uniform-quantizing scheme

Y pictures		Flower garden		Mobile		Susan		Table tennis	
		q0=0.1	q0=0.4	q0=0.1	q0=0.4	q0=0.1	q0=0.4	q0=0.1	q0=0.4
Adaptive	MSE	17.344	25.321	23.403	34.663	2.432	3.103	6.422	7.005
	PSNR	35.739	34.096	34.438	32.732	44.271	43.212	40.054	39.677
Uniform	MSE	25.213	38.870	35.601	59.101	3.144	4.749	6.369	10.562
	PSNR	34.115	32.235	32.616	30.415	43.156	41.365	40.090	37.894

The experimental data in Table 2 show that the visual quality of the embedded pictures with the proposed scheme is satisfying, with smaller MSE and higher PSNR (about 2dB) than conventional one using uniform quantization step.

4.3 Qualification

Above experiments does not set upper bound for the quantization step, and the visual quality of the embedded video might be severely degraded due to several exceptional large numbers. Below experimental results display the visual quality after qualification is set to the quantization step.

Table 4. MSE and PSNR(dB) calculation for embedded Susan when q0=2.0 and q is within different bounds, embedded MB(macro block) amount is provided in the last column

Bounds	MSE/PSNR	amount
$1.0 \leq q$	5.876/40.4399	1714
$1.0 \leq q \leq 100.0$	3.766/42.372	1437
$1.0 \leq q \leq 30.0$	2.277/44.556	816
$1.0 \leq q \leq 10.0$	1.517/46.322	416

5 Conclusion

In this paper, an efficient MPEG-2 video watermarking scheme is presented with luminance differential DC coefficient modification. The modification is implemented by binary dither modulation without resorting to original DC coefficients, and the quantization step is made self-adaptive, according to the modified component's two neighbors. Thus visual quality of the embedded video is well maintained, while the provable robustness of dither modulation technology is reserved. Experimental results verify the estimation relation of adjacent luminance differential DC components and display the superiority of the proposed scheme to conventional one using uniform quantization step. Watermark embedding and extracting procedures are based on independent macro blocks, and can easily be integrated with the video encoder and decoder. Therefore, it is suitable for real-time applications such as user labeling in VOD service, multimedia retrieval, and so on.

References

1. Lie, W.-N., Lin, T.C.-I., Lin, C.-W.: Enhancing video error resilience by using data-embedding techniques. IEEE Transactions on Circuits and Systems for Video Technology 16(2), 300–308 (2006)
2. Giannoula, A., Hatzinakos, D.: Integrating compression with watermarking on video sequences. In: International Conference on Information Technology: Coding and Computing, 2004. Proceedings. ITCC 2004, vol. 2, pp. 159–161 (2004)
3. Wang, C.-C., Lin, Y.-C., Ti, S.-C.: Satellite interference detection using real-time watermarking technique for SMS. In: Third International Conference on Information Technology and Applications, 2005. ICITA 2005, July 4-7, vol. 2, pp. 238–241 (2005)
4. Swanson, M.D., Zhu, B., Tewfik, A.H.: Multiresolution Scene-Based Video Watermarking using perceptual Models. IEEE Journal on Selected Areas in Communications 16(4) (May 1998)
5. van Schyndel, R.G., Tirkel, A.Z., Osborne, C.F.: A digital watermark. In: IEEE International Conference on Image Processing, 1994. Proceedings. ICIP 1994, November 13-16, vol. 2, pp. 86–90 (1994)
6. Langelaar Gerrit, C., Lagendijk Reginald, L., Jan, B.: Real-time labeling of MPEG-2 compressed video. Journal of Visual Communication and Image Representation 9(4), 256–270 (1998)
7. Zou, F., Lu, Z., Ling, H., Yu, Y.: Real-time video watermarking based on extended m-sequences. In: 2006 IEEE International Conference on Multimedia and Expo, July 2006, pp. 1561–1564 (2006)

Reversible Data Hiding Based on H.264/AVC Intra Prediction

Mehdi Fallahpour and David Megías

Estudis d'Informàtica, Multimèdia i Telecomunicació
Universitat Oberta de Catalunya- Rambla del Poblenou, 156, Barcelona, Spain
Tel: (+34) 933 263 600, Fax: (+34) 933 568 822
{mfallahpour,dmegias}@uoc.edu

Abstract. This paper proposes a novel high capacity reversible image data hiding scheme using a prediction technique which is effective for error resilience in H.264/AVC. In the proposed method, which is based on H.264/AVC intra prediction, firstly the prediction error blocks are computed and then the error values are slightly modified through shifting the prediction errors. The modified errors are used for embedding the secret data. The experimental results show that the proposed method, called shifted intra prediction error (SIPE), is able of hiding more secret data while the PSNR of the marked image is about 48 dB.

Keywords: Lossless data hiding, H.264/MPEG-4 AVC, intra prediction.

1 Introduction

New findings of data hiding in digital imaging open wide prospects of new techniques in modern imaging science, secure communication and content management. Data hiding has been proposed as a promising technique used for security, authentication, fingerprint, video indexing, error resilient coding, etc.

H.264/AVC [1] is the newest international video coding standard providing many techniques to improve the coding efficiency of intra and inter frames. Among many new techniques, the intra prediction technique is considered as one of the most important features in the success of H.264/AVC. This technique, which is used in the proposed method, increases the dependence of the neighbouring blocks. An error resilient method that embeds information into image or video itself is another technique used in H.264/AVC. Once an error is detected, the error resilient technique extracts the hidden information and recovers the error block. Using reversible information embedding in the error resilient causes the original digital content to be completely restored in the decoder and also results in a lossless extraction of the embedded data.

Reversible data hiding [2] is a novel category of data hiding schemes. The reversibility is essential to some sensitive applications such as medical diagnosis, remote sensing and law enforcement. The methods reported in [4-10] are considered among the best schemes in lossless data hiding. In [3], a high capacity lossless data hiding method was proposed based on the relocation of zeros and peaks of the histogram of the image blocks to embed the data. Recently, Lin and Hsueh [4] presented a reversible

H.J. Kim, S. Katzenbeisser, and A.T.S. Ho (Eds.): IWDW 2008, LNCS 5450, pp. 52–60, 2009.

data hiding method based on increasing the differences between two adjacent pixels to obtain a stego-image with high payload capacity and low image distortion. Among recent lossless methods performed on the transform domain, the schemes based on the integer wavelet transform domain are more notable. Tian [5] used the integer Haar wavelet transform and embedded the secret message into high-frequency coefficients by difference expansion. Kamstra and Heijmans [6] improved Tian's method by using the information in the low-frequency coefficients to find suitable expandable differences in the high-frequency coefficients. Xuan et al. [8] reported the lossless embedding algorithm carried out in the integer wavelet transform domain. Xuan et al. [7] proposed a lossless data hiding scheme based on optimum histogram pairs in the wavelet domain. Recently, a few prediction based data hiding methods have been proposed [9-10]. Thodi et al. [9] expanded the difference between a pixel and its predicted value in the context of the pixel for embedding data. In Kuribayashi et al.'s algorithm [10], a watermark signal is inserted in the LSB of the difference values between pixels.

The method proposed in this paper, called SIPE, is based on increasing the differences between pixels of the cover image and their intra prediction values. The prediction error at which the number of prediction errors is at a maximum is selected to embed the message. The prediction errors larger than the selected error are increased by "1". Furthermore, the selected prediction error is left unchanged and increased by "1" if the embedded bit is "0" and "1", respectively.

The SIPE method is able to embed a huge amount of data (15-120 kbits for a 512×512×8 greyscale image) while the PSNR of the marked image versus the original image is about 48 dB. In addition, simplicity and applicability to almost all types of images and H.264 video coding make this method superior to most of existing reversible data hiding techniques. Although the proposed lossless data hiding technique is applied to still images, it is very useful for H.264/AVC because the insertion of additional information only needs the shifting and embedding steps in coding steps. Furthermore, this lossless technique will not degrade the video quality.

2   H.264/AVC Intra Prediction

In H.264/AVC intra prediction method [11], a prediction block is formed based on previously reconstructed blocks. There are nine prediction modes for each 4×4 block. The sixteen elements in the 4×4 block (labelled from a to p in Fig. 1.(a)) are predicted by using the boundary pixels of the upper and left blocks which are previously obtained (labelled from A to M). These boundary elements are therefore available in the encoder and decoder to form a prediction reference.

For each 4×4 block, one of nine prediction modes can be selected by the encoder. In addition to DC prediction type, numbered as mode 2, where all elements are predicted by $(A + B + C + D + I + J + K + L)/8$, eight directional prediction modes are specified as shown in Fig. 1.(b). For mode 0 (vertical prediction), the elements above the 4×4 block are copied into the prediction block as indicated by arrows, Fig. 1 (c). Other modes copy adjacent pixels into the prediction block based on their prediction directions.

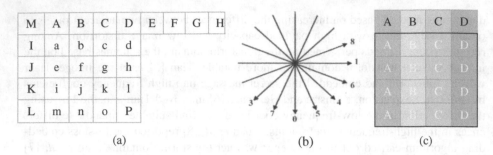

Fig. 1. 4×4 intra prediction mode (a) labeling of prediction samples. (b) 4×4 intra prediction mode direction. (c) Vertical (mode 0) prediction.

The rate distortion optimisation (RDO) technique [12] is used to take full advantage of the mode selection regarding maximising coding quality and minimising data bits. The RDO is applied to all of the 4×4 block intra-prediction modes to find the best one. This approach can achieve the optimal prediction mode decision. The only drawback for using RDO is the computational complexity. Recently, there is more focus on developing the 4×4 intra-prediction mode decision techniques with lower complexity.

3 Suggested Scheme

The SIPE method consists of an embedding and a extracting procedure. The embedding process includes both computing the prediction errors and embedding the information bits in the shifted prediction errors. Moreover, the data extraction is the reverse of data embedding. The proposed method is explained in the following two subsections.

3.1 Embedding

The embedding algorithm is as follows.

1. The prediction type is selected. It can be selected by RDO or other 4×4 intra-prediction mode decision techniques.
2. The prediction blocks are computed from the cover image by using an intra prediction algorithm (as described above). For the blocks without upper or left blocks, the coder uses their upper or left pixels for prediction.
3. The prediction error (PE) blocks are calculated by subtracting the predicted blocks from the cover image block, $e = I - P$.
4. The number of prediction errors in PE blocks equal to d is denoted by $D(d)$. The value M is found such that $D(M)$ is at a maximum. The following steps (5-6) are carried out for each 4×4 block completely and then iterated for the next block.
5. In the shifting stage, the modified PE block is derived from the PE block by this approach: for each PE block element $e_{i,j}$ (expect top-most row and the

left-most column of the cover image; $i \neq 1$ and $j \neq 1$), if $e_{i,j}$ is larger than M, then the modified PE $e'_{i,j}$ equals $e_{i,j} + 1$, otherwise $e'_{i,j} = e_{i,j}$.

6. In the embedding stage, each $e'_{i,j}$ ($i \neq 1$ and $j \neq 1$) with a value of M is increased by one if the corresponding bit of the data (to be embedded) is one, otherwise it will not be modified. After concealing data in $e'_{i,j}$, the embedded PE $e''_{i,j}$ is obtained.

7. Finally, the marked image I' is achieved by $I' = P + e''$.

In fact, the pixels in the top-most row and the left-most column of a cover image are preserved without carrying any hidden data. These pixels are used for recovering the original image and extracting the embedded data from the marked image. These row and column are the same in both the cover and the marked images. It is worth mentioning that, in the coder and the decoder, the raster scan order is used. The gray value of M, the prediction mode and the block size will be treated as side information that needs to be transmitted to the receiving side for data retrieval.

3.2 Detection

The following process is used for extracting the secret message from a marked image and also losslessly recovering the cover image. Let the marked image I' be the received image at the decoder. The following steps (1-4) are carried out for each block completely and then iterated for the next block.

1. The prediction block P of I can be obtained by using the intra prediction algorithm using its upper and left blocks which have been already restored.

2. If the embedded PE block element, $e''_{i,j} = I'_{i,j} - P_{i,j}$, is equal to $M + 1$, it is concluded that the embedded bit is "1". In this case, $e''_{i,j}$ should be decreased by one to obtain the modified PE block element, $e'_{i,j} = e''_{i,j} - 1$. If $e''_{i,j}$ is equal to M the embedded bit is "0" and $e'_{i,j} = e''_{i,j}$, otherwise there is no embedded data bit and again $e'_{i,j} = e''_{i,j}$.

3. If $e'_{i,j} > M$, then the prediction error $e_{i,j}$ is calculated by decreasing $e'_{i,j}$ by one, $e_{i,j} = e'_{i,j} - 1$, otherwise $e_{i,j} = e'_{i,j}$.

4. Finally, the $e_{i,j}$ should be added to prediction value $P_{i,j}$ to recover the original cover image pixel, $I_{i,j} = P_{i,j} + e_{i,j}$.

Fig. 2 shows an example of a 4×4 block greyscale image. The encoder scans the cover image block, Fig. 2(c1) pixel by pixel and subtracts the vertical prediction pixels, Fig. 2(c2), from the cover image pixels. In the PE (prediction error) block, Fig. 2(c3), following the computation of all prediction blocks the obtained M is equal to 0 and $D(M)$ in the current block equals 6. Suppose that the bit stream to be embedded is 101001. The encoder scans the PE block and all elements larger than 0 are increased by one, Fig. 2(c4), then modified prediction errors equal to 0 are chosen for embedding data.

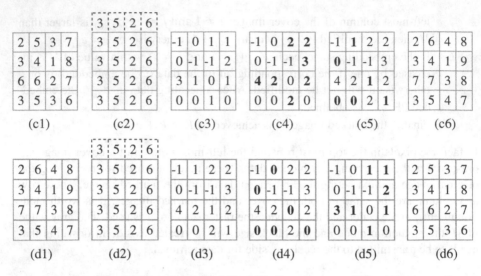

Fig. 2. Embedding steps (c1) - (c6) , Detection steps (d1) – (d6)

If the corresponding bit of the secret data is one, the modified prediction value is added by one, otherwise it will not be modified, as shown in Fig. 2(c5). The marked image, Fig. 2(c6), is obtained by adding the embedded prediction errors, Fig. 2(c5), to the prediction pixels, Fig. 2(c2). As described above, the value M (in this example equals to zero) and the prediction mode are treated as side information. The decoder extracts the secret data and obtains the original image block from the marked image block, Fig. 2(d1). The prediction block is computed based on the restored cover image blocks, Fig. 2(d2). The decoder scans the embedded PE block, Fig. 2(d3), which is obtained by subtracting the prediction block from the marked block. If the embedded PE element is equal to 1 and 0, the embedded data bit is "1" and "0", respectively. In case the embedded PE is equal to "1" or "0", the modified PE Fig. 2(d4), equals "0", otherwise it equals the embedded PE. In order to get PE, Fig. 2(d5), if the modified PE is larger than 0, it must be decreased by one, otherwise the PE equals the modified PE. Finally, restored cover image pixel, Fig. 2(d6), is computed by adding PE, Fig. 2(d5), to the prediction pixel, Fig. 2(d2).

This example clarifies the simplicity and reversibility of this method. In fact we have three steps in embedding and detecting: calculating prediction error, shifting and embedding. In the detector the prediction block is calculated based on restored cover image blocks and, thus, the prediction blocks in the embedding and detecting procedures are the same. In the encoder, the marked image is achieved by adding the embedded PE to the prediction image block and then, in the decoder, the embedded PE is achieved by subtracting the prediction block from the marked image. When we have the embedded PE block, the modified PE and the PE can be obtained very easily by decreasing the value of the block pixels. Finally, adding PE to the prediction block results in the restored cover image. Hence, all steps are reversible.

4 Experimental Results

The SIPE algorithm was implemented and tested on various standard test images from the UWaterloo database [13]. Also, the results of Lin and Hsueh's [4], Kamstra and Heijmans's [6], and Xuan et al's. [7] methods were compared with the results obtained with this method. This comparison shows that the SIPE method is able of hiding more secret data than almost all methods mentioned in the literature for the same (above 45dB) PSNR. The experimental results of the SIPE method show that the embedded data remains invisible, since no visual distortion can be revealed. It is worth pointing out that the embedded data were generated by the random number generator in MATLAB on a personal computer.

Table 1 summarizes the experimental results obtained by this method for the Mandrill, Lena, Peppers, Zelda, Goldhill, Barbara, and Boat images using vertical prediction. For the sake of brevity, only the simple case of one maximum of $D(d)$ is described, because the general cases of multiple maxima of $D(d)$ can be decomposed as a few maximum cases.

The payload capacity is increased by using multiple maximum of $D(d)$. In the Right Shifted type, M is equal to 0 and the prediction errors larger than 0 are increased by one. In the Left Shifted type, M is equal to -1 and the prediction errors smaller than 0 are reduced by one. In addition, in the L&R Shifted type, Left Shifted and Right Shifted types are used simultaneously.

Table 1. PSNR (dB) and payload capacities (bits) of the test images of the UWaterloo database [13] with different block size ($N \times N$)

Shift type		Right shifted $N = 1$	Right Shifted $N = 4$	L&R shifted $N = 1$	L&R shifted $N = 4$	Ratio
Mandrill	PSNR	51.13	51.1	48.3	48.24	2.12
	Payload	7630	6137	15088	12031	
Lena	PSNR	51.12	51.16	48.58	48.49	2.58
	Payload	28932	22616	52664	42334	
Peppers	PSNR	51.13	51.17	48.49	48.43	2.70
	Payload	21158	17810	40929	34775	
Zelda	PSNR	51.16	51.13	48.61	48.48	2.75
	Payload	27120	19977	53474	39201	
Goldhill	PSNR	51.26	51.34	48.43	48.36	2.63
	Payload	17431	13063	34492	25948	
Barbara	PSNR	51.15	51.19	48.51	48.40	2.33
	Payload	22409	15680	43682	31167	
Boat	PSNR	51.12	51.06	48.59	48.49	2.60
	Payload	27522	21735	52543	41591	

The experimental results of the L&R Shifted type exhibit that the PSNR of all marked images is above 48 dB.

Although in the description of the method only 4×4 blocks were used for embedding and detecting data, other block sizes could be easily used. Fig. 3 shows how the performance of the proposed scheme varies with different block sizes: 1×1, 2×2, 3×3, 4×4, and 8×8. As shown in the figure, the performance of the proposed scheme is increased when the block size is the smallest one (1×1). The smaller the block size, the larger the amount of the embedding information and the more computation time is required. Since in most of image data hiding applications the computation time is not critical, using 1×1 blocks to achieve high capacity and PSNR is desirable. In Table 1, the "Ratio" is the ratio of computation time for embedding process using 1×1 blocks to embedding time using 4×4 blocks. In H.264/AVC, using 4×4 blocks is more effective because the frames are divided into 4x4 blocks for coding and decoding and, furthermore, the execution time is a critical issue in video coding.

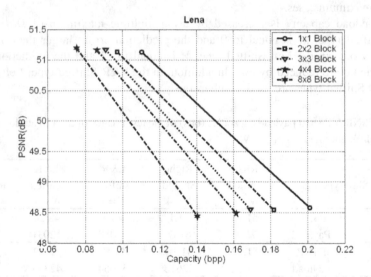

Fig. 3. Comparison between the embedding capacity in bpp and distortion in PSNR with different block sizes for the Lena image

Fig. 4 illustrates the performance comparison of the SIPE with the methods reported in [4], [6], and [7] for the Lena image in terms of PSNR (dB) and payload (bpp: bits per pixel). As shown in Fig 4, the SIPE scheme provides high enough bound of the PSNR (above 48dB) with a quite large data embedding capacity, indicating a fairly better performance of the SIPE method.

It is worth to mentioning that the selection of the prediction mode is an important step. Fig. 5 demonstrates the effects of the selection mode on capacity and distortion. As an example for the Lena image, vertical prediction has higher capacity with lower distortion.

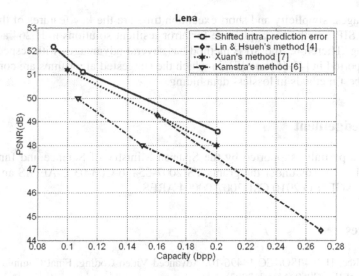

Fig. 4. Comparison between the embedding capacity (bpp) and distortion (PSNR) for the Lena image

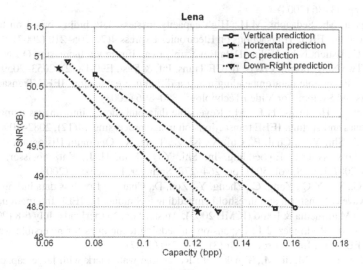

Fig. 5. Comparison of embedding capacity versus distortion with different prediction modes for the Lena image

5 Conclusion

This paper presents a novel high-capacity reversible data hiding algorithm, called shifted intra prediction error (SIPE), which is based on shift differences between the cover image pixels and their predictions. Large capacity of embedded data (15-120 kbits for a 512×512 greyscale image), PSNR above 48 dB, applicability to almost all types of images, simplicity and short execution time are the key features of this algorithm. The SIPE method is applicable for error resilient solutions in H.264 advanced video coding. Therefore, the SIPE method has several advantages with respect to the methods reported in [4], [6] and [7], in which the suggested algorithms are considered among the best methods in lossless data hiding.

Acknowledgement

This work is partially supported by the Spanish Ministry of Science and Innovation and the FEDER funds under the grants TSI2007-65406-C03-03 E-AEGIS and CONSOLIDER-INGENIO 2010 CSD2007-00004 ARES.

References

1. ITU-T Rec. H.264/ISO/IEC 14496-10, Advanced Video Coding, Final Committee Draft, Document JVTG050 (March 2003)
2. Shi, Y.Q., Ni, Z., Zou, D., Liang, C., Xuan, G.: Lossless data hiding: Fundamentals, algorithms and applications. In: Proc. IEEE Int. Symp. Circuits Syst., Vancouver, BC, Canada, vol. II, pp. 33–36 (2004)
3. Fallahpour, M., Sedaaghi, M.H.: High capacity lossless data hiding based on histogram modification. IEICE Transactions on Electronics Express 4(7), 205–210 (2007)
4. Lin, C.C., Hsueh, N.L.: Hiding Data Reversibly in an Image via Increasing Differences between Two Neighboring Pixels. IEICE Trans. Inf. & Syst. E90-D(12), 2053–2059 (2007)
5. Tian, J.: Reversible data embedding using a difference expansion. IEEE Transactions on Circuits and Systems for Video Technology, 890–896 (2003)
6. Kamstra, L., Heijmans, H.J.A.M.: Reversible data embedding into images using wavelet techniques and sorting. IEEE transactions on image processing 14(12), 2082–2090 (2005)
7. Xuan, G., Shi, Y.Q., Chai, P., Cui, X., Ni, Z., Tong, X.: Optimum Histogram Pair Based Image Lossless Data Embedding. In: Shi, Y.Q., Kim, H.-J., Katzenbeisser, S. (eds.) IWDW 2007. LNCS, vol. 5041, pp. 264–278. Springer, Heidelberg (2008)
8. Xuan, G., Shi, Y.Q., Yang, C., Zheng, Y., Zou, D., Chai, P.: Lossless data hiding using integer wavelet transform and threshold embedding technique. In: IEEE International Conference on Multimedia & Expo (ICME 2005), Amsterdam, Netherlands, July 6-8 (2005)
9. Thodi, D.M., Rodriguez, J.J.: Expansion embedding techniques for reversible watermarking. IEEE Trans. Image Process. 16(3), 723–730 (2007)
10. Kuribayashi, M., Morii, M., Tanaka, H.: Reversible watermark with large capacity based on the prediction error expansion. IEICE Trans. Fundamentals E91-A(7), 1780–1790 (2008)
11. Richardson, I.E.G.: H.264 and MPEG-4 Video Compression, pp. 120–145. Wiley, Chichester (2003)
12. Sullivan, G.J., Wiegand, T.: Rate-distortion optimization for video compression. IEEE Signal Process. Mag. 15, 74–90 (1998)
13. Waterloo Repertoire GreySet2 (October 27, 2008),
    http://links.uwaterloo.ca/greyset2.base.html

Scalability Evaluation of Blind Spread-Spectrum Image Watermarking

Peter Meerwald and Andreas Uhl

Dept. of Computer Sciences, University of Salzburg,
Jakob-Haringer-Str. 2, A-5020 Salzburg, Austria
{pmeerw,uhl}@cosy.sbg.ac.at

Abstract. In this paper, we investigate the scalability aspect of blind watermark detection under combined quality and resolution adaptation of JPEG2000 and JPEG coded bitstreams. We develop two multi-channel watermarking schemes with blind detection, based on additive spread-spectrum watermarking: one employs the DCT domain, the other the DWT domain. We obtain watermark scalability by combining detection results from multiple channels modeled by Generalized Gaussian distributions. Both schemes achieve incremental improvement of detection reliability as more data of a scalable bitstream becomes available.

1   Introduction

Watermarking has been proposed as a technology to ensure copyright protection by embedding an imperceptible, yet detectable signal in digital multimedia content such as images or video [1]. Watermarks are designed to be detectable, even when the multimedia content is altered during transmission and provide a level of protection after presentation – an advantage over cryptographic methods [2].

With the advent of mobile devices capable of wireless transmission and ubiquitous presentation of multimedia content, scalable image coding is more and more employed to allow adaptation of a single multimedia stream to varying transmission and presentation characteristics. A scalable image bitstream can be adapted to fit different resolution and quality presentation demands.

The JPEG2000 standard for image coding already addresses scalability by relying on a wavelet transformation and embedded, rate-distortion optimal coding [3]. The previous standard, JPEG [4], provides only limited support for sequential and progressive quality scalability (Annex F and G, resp.) and resolution scalability (Annex J), which is rarely implemented.

Streaming and scalable multimedia transmission poses challenges as well as potentials for watermarking methods [5], but has received little attention so far. An explicit notion of scalability first appears in the work of Piper et al. [6]. They evaluate the robustness of different coefficient selection methods with regards to quality and resolution scalability in the context of the basic spread-spectrum scheme proposed by Cox et al. [7]. Later, Piper et al. [8] combine resolution and quality scalability and argue that both goals can be achieved by exploiting the

H.J. Kim, S. Katzenbeisser, and A.T.S. Ho (Eds.): IWDW 2008, LNCS 5450, pp. 61–75, 2009.

human visual system (HVS) characteristics appropriately in order to maximize the watermark energy in the low-frequency components of the images which are typically best preserved by image coding methods. However, only non-blind watermarking was considered so far [9,8] . In this case, the original host signal can be used to completely suppress the interference of the host noise during watermark detection, thus a relatively small number of coefficients suffices for reliable detection.

In this paper we revisit the scalable watermarking problem. We aim for blind scalable watermark detection and design two schemes where the watermark information is embedded into multiple, diverse host signal components which are roughly aligned with resolution or quality enhancement layer components of a scalable bitstream. As the scalable bitstream of the watermarked image is transmitted, more and more host signal channels become available. We propose a multi-channel watermark detector which combines the detection results of the independent watermarks that are embedded in the different channels. The detection reliability should increase as more channels are transmitted. This allows to explicitly investigate the scalability of the watermark.

We propose to split the host image into independent subbands obtained using the DWT and $8 \times 8-$block DCT and model the different characteristics of the resulting host signal channels separately using Generalized Gaussian distributions (GGDs) [10]. We investigate the impact of scalable JPEG2000 bitstream adaptation and JPEG coding on the global detection performance and present results for blind spread-spectrum watermarking.

In section 2, we discuss the application scenario for scalable watermarking, then turn to the multi-channel watermark detection problem in section 3. Based on this foundation, we design two blind spread-spectrum watermarking schemes with scalable detection in section 4. In section 5, we present experimental results and offer concluding remarks in section 6.

2   Application Scenario

In this paper we study the scenario that a watermark is embedded in the host image during content creation to denote the copyright owner before it is distributed. A single bit watermarking method can be employed to this end where the seed used to generate the watermark identifies the copyright owner. Before distribution, the watermarked content is compressed. Primarily, we are interested in JPEG2000 as a vehicle for image coding and bitstream adaptation. Depending on the capabilities of the content consumer's presentation device, the compressed bitstream may undergo adaptation before or during transmission in order to save bandwidth or processing time on the presentation device. adaptation of the coded bitstream might reduce the quality and/or resolution of the coded image, yet the watermark shall remain detectable [9]. See Figure 1 for an overview of the scenario. For comparison, we also examine JPEG coding. However, we do not use JPEG's scalability features but rather simulate adaptation by coding separate bitstreams for the required resolution and quality settings since JPEG's hierarchical scalability (Annex J) shows poor performance.

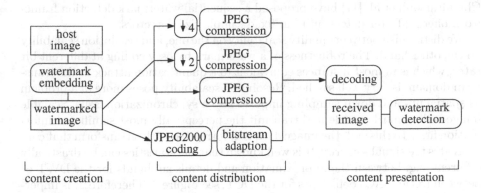

Fig. 1. The application scenario

Alternatively, in a different scenario, the watermark could also be embedded in the bitstream either at the distribution stage, during transmission or before presentation of the content. In all of these cases, the secret watermark key must be exposed to areas which are not under the control of the content creator. Hence, the problem of watermark key distribution must be addressed in technical ways or by assuming trust between the involved entities [11].

The presentation device tries to make a fast decision on the presence or absence of the watermark. If possible, the watermark should be detectable from the base layer. In addition, we expect an increase in detection reliability as more data is transmitted, improving the quality and resolution of the image. A blind watermark detector that can incrementally provide more reliable detection results when decoding scalable image data has not been discussed in the literature so far but seems highly desirable for resource-constraint multimedia clients, such as mobile devices.

2.1 Scalable Watermarking

Lu et al. [12] claim that a watermark is scalable if it is detectable at low quality or low resolution layers. A number of watermarking methods have been proposed which allow for progressive detection [13,14,15,16,8]. However, they either are non-blind [16,8], or do not consider resolution scalability [13,14,15]. The impact of fully scalable image coding as enabled by JPEG2000 has not been investigated as far.

Piper et al. [8] refine Lu et al.'s definition and put forward two properties for scalable watermarking along with numeric measures: detectability and graceful improvement. The detectability property states that a watermark shall be detectable in any version of the scaled content which is of acceptable quality. Graceful improvement refers to the desirable property that as increased portions of the content data become available, the watermark detection shall become more reliable. Note that detection reliability may also be traded for faster detection [15], i.e. watermark detection utilizing a minimum number of host coefficients.

Chandramouli et al. [17] have proposed a sequential watermark detection framework offering faster detection than fixed sample size detectors.

We distinguish between quality scalability on the one, and resolution scalability on the other hand. The robustness of a watermark to lossy coding at different bit rates, which is in most instances equivalent to a quantization attack in the transform domain, is very well studied. Resolution scalability poses more of a problem as spatial up- and down-sampling imposes also a synchronization issue. Scalable image coding must encode and transmit the perceptually most significant information first. To this end, the image data is decorrelated with a transform that concentrates the signal's energy. It is well known that the statistics can be drastically different, e.g. between the approximation and details subbands for the DWT or between DC and AC coefficients for the DCT, see Figure 2. Therefore, it is imperative to accurately model the host signal statistics per channel for effective blind watermark detection which we address in the next section relying on the GGD.

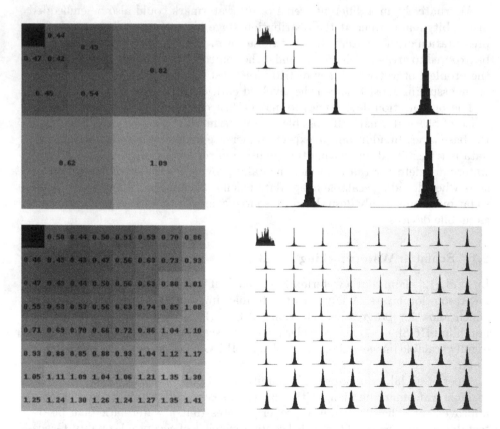

Fig. 2. GGD shape parameter per DWT subband (top) and DCT frequency subband (bottom); histograms of the subband coefficients are shown right (Lena image)

3   Watermark Detection

We review optimal blind detection of an additive spread-spectrum watermark
with the assumption that the transform domain host signal coefficients can be
modeled with i.i.d Generalized Gaussian distributions [18,10]. The next section
extends the results to the multi-channel case.

For blind watermarking, i.e. when detection is performed without reference
to the unwatermarked host signal, the host energy interferes with the watermark signal. Modeling of the host signal $\mathbf{x}$ is crucial for watermarking detection
performance. The GGD given by

$$p(\mathbf{x}) = A\exp(-|\beta\mathbf{x}|^c), \quad -\infty < x < \infty \tag{1}$$

where $\beta = \frac{1}{\sigma_x}\sqrt{\frac{\Gamma(3/c)}{\Gamma(1/c)}}$ and $A = \frac{\beta c}{2\Gamma(1/c)}$ has been successfully used in image
coding to model subband and DCT coefficients of natural images [19]. The shape
parameter c is typically in the range of 0.5 to 0.8 for DCT and DWT coefficients,
see Figure 2. The watermark detection problem on a received signal $\mathbf{y}$ can be
formulated as a hypothesis test

$$\begin{aligned} H_0 &: \ y[k] = x[k] \\ H_1 &: \ y[k] = x[k] + \alpha w[k] \end{aligned} \tag{2}$$

where $\mathbf{x}$ represents the original signal, α denotes the watermark embedding
strength and $\mathbf{w}$ is the pseudo-random bipolar watermark sequence generated
from a secret key identifying the copyright owner. The optimal decision rule is

$$l(\mathbf{y}) = \frac{p(\mathbf{y}|H_1)}{p(\mathbf{y}|H_0)} = \frac{\prod_{k=1}^{N}\exp\left(-|\beta(y[k]-\alpha w[k])|^c\right)}{\prod_{k=1}^{N}\exp\left(-|\beta y[k]|^c\right)} \overset{H_1}{\underset{H_0}{\gtrless}} T, \tag{3}$$

where $l(\mathbf{y})$ is the likelihood function with $\mathbf{x}$ modeled by a GGD and T is a
decision threshold, usually set according to the Neyman-Pearson criterion. The
detection statistic is then given by the log-likelihood ratio

$$L(\mathbf{y}) = \sum_{k=1}^{N} \beta^c(|y[k]|^c - |y[k] - \alpha w[k]|^c) \tag{4}$$

for which the PDFs under hypothesis H_1 and H_0 are approximately Gaussian
with the same variance

$$\sigma^2_{L(\mathbf{y})|H_1} = \sigma^2_{L(\mathbf{y})|H_0} = \frac{1}{4}\sum_{k=1}^{N} \beta^{2c}(|y[k]+\alpha|^c - |y[k]-\alpha|^c)^2 \tag{5}$$

and mean

$$\mu_{L(\mathbf{y})|H_0} = \sum_{k=1}^{N} \beta^c(|y[k]|^c - \frac{1}{2}\sum_{k=1}^{N} \beta^c(|y[k]+\alpha|^c + |y[k]-\alpha|^c), \tag{6}$$

where $\mu_{L(\mathbf{y})|H_1} = -\mu_{L(\mathbf{y})|H_0}$ (see [18] for details). The probability of missing the watermark, P_m, is then given by

$$P_m = \frac{1}{2}\operatorname{erfc}\left(\frac{\mu_{L(\mathbf{y})|H_1} - T}{\sqrt{2\sigma_{L(\mathbf{y})}^2}}\right) \tag{7}$$

for a detection threshold T which is set to achieve a desired false-alarm rate denoted by P_{fa},

$$T = \sqrt{2\sigma_{L(\mathbf{y})}^2}\operatorname{erfc}^{-1}(2P_{fa}) - \mu_{L(\mathbf{y})|H_0}, \tag{8}$$

where $\operatorname{erfc}(\cdot)$ is the complement of the error function [20].

3.1   Multi-channel Detection

So far, we have only addressed the detection problem for one channel. However, in case we mark more than one channel, we have to discuss how to combine the detector responses and how to determine a suitable global detection threshold. We will consider the straightforward approach of simply summing up the detector responses of each channel (i.e. subband and/or frequency band), normalized to unit variance. In order to derive a model for the global detection statistic, we assume that the detector responses $L(\mathbf{y}_i)$, $1 \leq i \leq K$ for each of the K channels are independent. Further, the watermark sequences are independent as well. This assumption allows to exploit the reproductivity property of the Normal distribution, namely that the sum of Normal random variables is again normally distributed.

Formally, if we have K random variables $L(\mathbf{y}_1), ..., L(\mathbf{y}_K)$ which all follow Normal distributions with standard deviation $\sigma_1, ..., \sigma_K$ and mean $\mu_1, ..., \mu_K$ (under H_0), we obtain a global detection statistic

$$L_{global}(\mathbf{y}) = \sum_{i=1}^{K} \frac{L(\mathbf{y}_i) - \mu_i}{\sigma_i} \tag{9}$$

which again follows a Normal distribution with variance K. We can then determine a threshold $T_{global} = \sqrt{2}\operatorname{erfc}^{-1}(2P_{fa})$ for the global detector response in order to decide on the presence or absence of the watermark.

4   Two Watermarking Schemes

Based on the multi-channel detection strategy outlined in the previous section, we now formulate a DCT-domain as well as a DWT-domain watermarking scheme. Our DCT-domain scheme is very similar to the method discussed in [18] and serves as a reference. Wavelet-domain watermarking algorithms can exploit the inherently hierarchical structure of the transform [21]. Especially, when watermark detection is integrated with image decoding from an embedded bit-stream (such as JPEG2000), progressive watermarking detection can be easily achieved [22,23].

The embedding strength α is determined for each channel such that a fixed document-to-watermark ratio (DWR) is maintained for each channel. More sophisticated watermark energy allocation strategies have been proposed [24], but are not adopted here to keep the scheme simple. Furthermore, perceptual shaping of the watermark is not used.

4.1 DCT-Domain Reference Scheme

For the DCT-domain watermarking method, a 8×8-block DCT is computed on the host image. From each transform-domain block, the frequency bands 3 to 20 in zig-zag order are extracted and concatenated to construct the marking space. These bands correspond to the low- and mid-frequency range commonly used as marking space [18]. An independent, bipolar watermark $\mathbf{w}$ is added to the coefficients of each frequency band (denoted by $\mathbf{x}$), $y[k] = x[k] + \alpha w[k]$, where $1 \leq k \leq N$ with $N = \frac{W \cdot H}{64}$ and W,H correspond to the width and height of the image. Thus we have 18 channels with N coefficients each.

4.2 DWT-Domain Scheme

The DWT-domain scheme decomposes the host image using a $J-$level pyramidal DWT to obtain $3 \cdot J$ detail subbands and the approximation subband. The approximation subband is further decorrelated with a $8 \times 8-$block DCT. As above, an independent, bipolar watermark $\mathbf{w}$ is embedded in each of the separate signal components obtained by the transformations: we have $3 \cdot J$ channels relating to the detail subbands with $N = \frac{W \cdot H}{2^{2j}}$ coefficients, where $1 \leq j \leq J$ is the decomposition level of the subband, and 18 channels with $N = \frac{W \cdot H}{2^{2J+6}}$ coefficients, each relating to one concatenated DCT frequency band derived from the approximation subband.

4.3 Watermark Detection

For watermark detection, first the GGD parameters β and c (the shape parameter) are computed for each channel using Maximum Likelihood Estimation (MLE), e.g. with the Newton-Raphson algorithm given in [25]. Next, the variance and mean of the detection statistic are determined per channel invoking Eq. 5 and 6, respectively. Eq. 9 permits to combine the individual per-channel detection results to obtain the global detector response and fix the global detection threshold. Note that the channels are independent and no order is imposed by the watermark detector, thus it can be applied bottom-up (i.e. integrated in a scalable decoder) or top-down (i.e. detector decomposes received image). The experimental results presented in the next section relate to the bottom-up case: base layer image data is incrementally augmented with resolution as well as quality enhancement data and we observe the combined watermark detection performance.

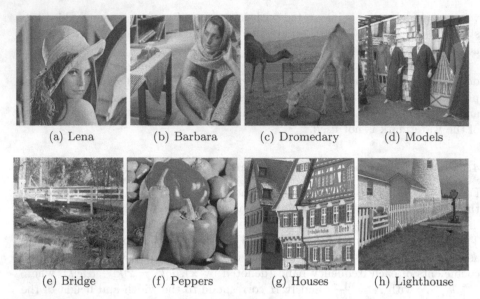

(a) Lena   (b) Barbara   (c) Dromedary   (d) Models

(e) Bridge   (f) Peppers   (g) Houses   (h) Lighthouse

Fig. 3. Test images

5   Experimental Results

Experimental results are reported on eight 512×512 gray-scale images, including six common test images and two images taken with a popular digital camera, see Figure 3.

We employ our DCT- and DWT-domain watermarking schemes presented in section 4. The document-to-watermark ratio (DWR) for each channel is fixed to 15 dB and 20 dB, for the DCT- and DWT-scheme, respectively, see Table 1 for the resulting PSNR. A two-level wavelet decomposition with biorthogonal 9/7 filters is used by the DWT algorithm. In the interest of reproducible research, the source code and image data is available at http://www.wavelab.at/sources.

Table 1. Average PSNR in dB after embedding as well as JPEG and JPEG2000 compression for the DWT and DCT scheme

Image	Embedding		JPEG $Q = 90$		JPEG $Q = 30$		JPEG2000 2.0 bpp		JPEG2000 0.3 bpp	
	DWT	DCT	DWT	DCT	DWT	DCT	DWT	DCT	DWT	DCT
Lena	42.34	42.81	38.54	38.67	33.72	33.87	40.35	40.68	34.17	34.49
Barbara	39.98	40.61	37.24	37.39	29.82	29.91	38.40	38.68	28.82	28.88
Dromedary	46.25	46.09	40.30	40.16	33.85	33.89	43.94	43.81	33.88	33.95
Models	39.88	39.89	38.05	37.95	32.60	32.69	39.27	39.30	31.10	31.36
Bridge	39.79	38.60	35.61	35.08	27.83	27.74	34.91	34.54	25.24	25.26
Peppers	41.19	42.40	36.89	37.24	32.94	33.17	39.02	39.78	33.32	33.75
Houses	36.86	35.22	34.63	33.52	28.87	27.81	34.63	33.57	23.95	23.96
Lighthouse	40.00	37.44	36.41	35.04	29.84	29.53	37.29	35.70	28.27	28.28

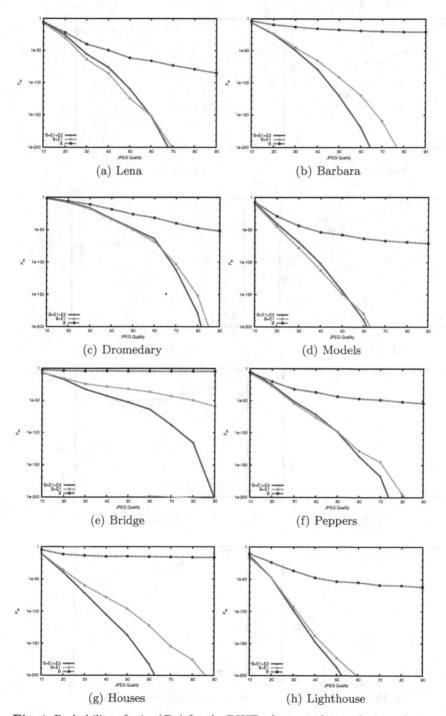

Fig. 4. Probability of miss (P_m) for the DWT scheme under resolution adaptation and JPEG compression

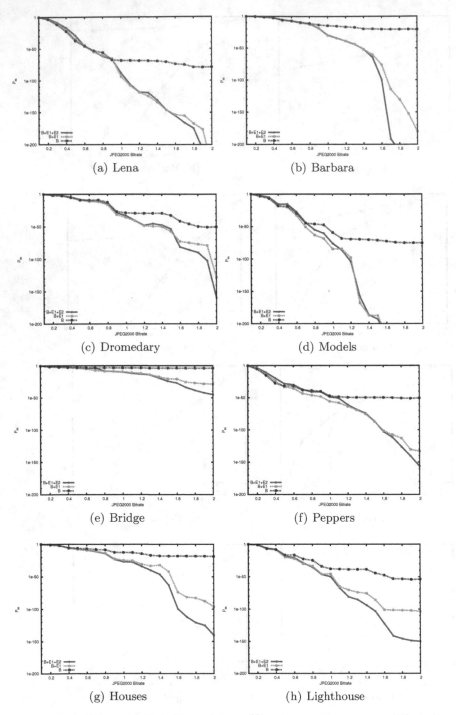

Fig. 5. Probability of miss (P_m) for the DWT scheme under resolution adaptation and JPEG2000 compression

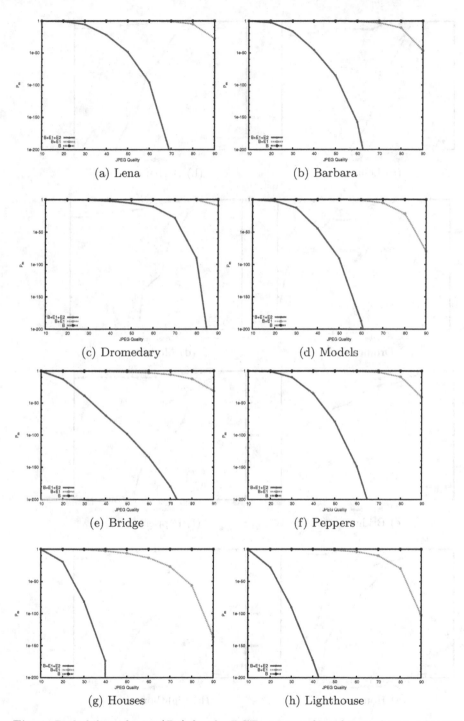

Fig. 6. Probability of miss (P_m) for the DCT watermark under resolution adaptation and JPEG compression

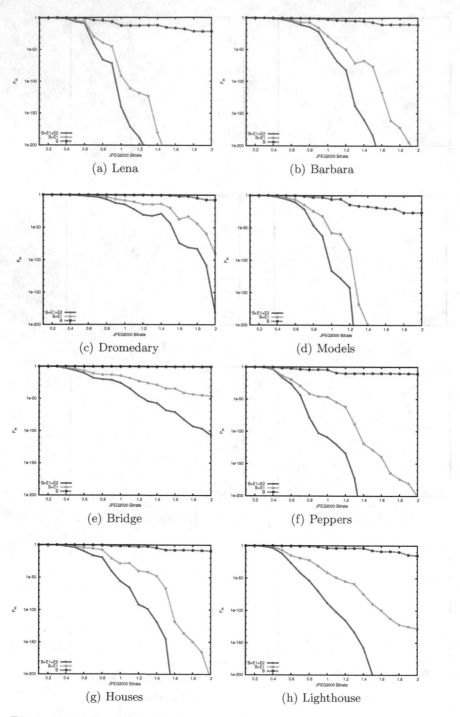

Fig. 7. Probability of miss (P_m) for the DCT watermark under resolution adaptation and JPEG2000 compression

We evaluate the performance of our watermarking schemes in the context of the application scenario depicted in Figure 1. We rely on the Kakadu 6.0 JPEG2000 implementation for constructing scalable bitstreams with quality layers ranging from 0.1 to 2.0 bits per pixel (bpp) from the watermarked test images. In the case of JPEG, we choose to simulate scalable bitstream formation for comparison: the watermarked image is downsampled with a bilinear filter and then compressed with a specific JPEG quality factor Q. Note that the methods described in JPEG Annex F, G, J are not used.

Watermark detection performance is evaluated in terms of the probability of missing the watermark (P_m), see Eq. 7, given a false-alarm rate (P_{fa}) of 10^{-6}. For each test image, we have generated 1000 copies with unique watermark seeds which have been subjected to JPEG and JPEG2000 coding. Watermark detection is performed after decoding the quality- and resolution adapted bitstream. A base layer (denoted B), one sixteenth the size of the full resolution image, and two resolution enhancement layers (denoted $E1$ and $E2$), each doubling the resolution, are examined at different quality layers. In case of the DCT watermark detector, the received image is upsampled to its original size before detection. For each setting, the 1000 detection results are used to estimate the mean and variance of the global detection statistic, in order to compute P_m.

In Figures 4 and 5, we observe the detection performance of our DWT watermarking scheme. The probability of missing the watermark (P_m) is plotted against a varying JPEG quality factor $(Q = 10, ..., 90)$ and JPEG2000 bit rates ranging from 0.1 to 2.0 bits per pixel (bpp). Combining the detection response of the base layer with the first resolution layer result significantly boosts the detection performance, unless the decoded image has very poor quality $(Q \leq 20$ for JPEG, bit rate ≤ 1.0 bpp for JPEG2000). Only for the Barbara, Bridge, Houses and Lighthouse images, the second resolution layer contributes to an improvement of the detection result at high bit rates. As expected, the detection reliability increases also with improved quality; the effect is more pronounced for higher resolutions.

The DCT scheme fails to detect the watermark solely from the base resolution layer when using JPEG coding, see Figure 6. The first resolution aids in detection only for high quality images $(Q \geq 70)$. However, with the second enhancement layer, reliable detection is achieved in all cases. Note that the DCT scheme outperforms the DWT detector for full-resolution images. In Figure 7 we observe that the DCT scheme's base-layer detection fares better with JPEG2000 coding. Both resolution layers improve the detection results. Note that the upsampling operation before detection must carefully match the downsampling due to JPEG2000's wavelet decomposition.

6   Conclusions

We have proposed two additive spread-spectrum watermarking schemes with blind, scalable detection and evaluated their detection performance in the context of scalable JPEG2000 and JPEG coding. Both schemes fulfill the properties of a scalable watermark set out by Piper et al., i.e protection of the base layer and graceful improvement as more image data is transmitted, to some extent. However, the DCT scheme fails to protect the base layer in the JPEG coding experiment and the DWT scheme does not benefit from the second resolution layer except for high bit rates. A more sophisticated watermark energy allocation strategy together with perceptual shaping might improve the detection performance and lift these deficiencies; we have not studied the impact of the selection of frequency bands on detection performance [26,15].

The proposed multi-channel modeling and detection approach enables investigation of blind, scalable watermark detection which we believe will become increasingly important as scalable codec gain widespread use. Further work will incorporate temporal scalability for video watermarking applications and assess protection of scalable video.

Acknowledgments

Supported by Austrian Science Fund project FWF-P19159-N13.

References

1. Cox, I.J., Miller, M., Bloom, J., Fridrich, J., Kalker, T.: Digital Watermarking and Steganography. Morgan Kaufmann, San Francisco (2007)
2. Uhl, A., Pommer, A.: Image and Video Encryption. In: From Digital Rights Management to Secured Personal Communication. Advances in Information Security, vol. 15. Springer, Heidelberg (2005)
3. ISO/IEC 15444-1: Information technology – JPEG2000 image coding system, Part 1: Core coding system (December 2000)
4. ISO 10918-1/ITU-T T.81: Digital compression and coding of continuous-tone still images (September 1992)
5. Lin, E.T., Podilchuk, C.I., Kalker, T., Delp, E.J.: Streaming video and rate scalable compression: what are the challenges for watermarking? Journal of Electronic Imaging 13(1), 198–208 (2004)
6. Piper, A., Safavi-Naini, R., Mertins, A.: Coefficient selection method for scalable spread spectrum watermarking. In: Kalker, T., Cox, I., Ro, Y.M. (eds.) IWDW 2003. LNCS, vol. 2939, pp. 235–246. Springer, Heidelberg (2004)
7. Cox, I.J., Kilian, J., Leighton, T., Shamoon, T.: Secure spread spectrum watermarking for multimedia. In: Proceedings of the IEEE International Conference on Image Processing, ICIP 1997, Santa Barbara, USA, October 1997, pp. 1673–1687 (1997)
8. Piper, A., Safavi-Naini, R., Mertins, A.: Resolution and quality scalable spread spectrum image watermarking. In: Proceeding of the 7th Workshop on Multimedia and Security, MMSEC 2005, New York, USA, August 2005, pp. 79–90 (2005)
9. Bae, T.M., Kang, S.J., Ro, Y.M.: Watermarking system for QoS aware content adaptation. In: Cox, I., Kalker, T., Lee, H.-K. (eds.) IWDW 2004. LNCS, vol. 3304, pp. 77–88. Springer, Heidelberg (2005)

10. Voloshynovskiy, S., Deguillaume, F., Pun, T.: Optimal adaptive diversity watermarking with state channel estimation. In: Proceedings of SPIE, Security and Watermarking of Multimedia Contents III, San Jose, USA, January 2001, vol. 4314 (2001)

11. Zeng, W., Land, J., Zhuang, X.: Security architectures and analysis for content adaptation. In: Proceedings of SPIE, Security, Steganography, and Watermarking of Multimedia Contents VII, San Jose, USA, January 2005, pp. 84–95 (2005)

12. Lu, W., Safavi-Naini, R., Uehara, T., Li, W.: A scalable and oblivious digital watermarking for images. In: Proceedings of the International Conference on Signal Processing, ICSP 2004, August 2004, pp. 2338–2341 (2004)

13. Chen, T.P.C., Chen, T.: Progressive image watermarking. In: Proceedings of the IEEE International Conference on Multimedia and Expo, ICME 2000, New York, USA, July 2000, pp. 1025–1028 (2000)

14. Biswal, B.B., Ramakrishnan, K.R., Srinivasan, S.H.: Progressive watermarking of digital images. In: Proceedings of the IEEE Pacific-Rim Conference on Multimedia, Beijing, China (October 2001)

15. Tefas, A., Pitas, I.: Robust spatial image watermarking using progressive detection. In: Proceedings of the International Conference on Acoustics, Speech and Signal Processing, ICASSP 2001, Salt Lake City, USA, May 2001, pp. 1973–1976 (2001)

16. Seo, J.H., Park, H.B.: Data protection of multimedia contents using scalable digital watermarking. In: Proceedings of the 4th ACIS International Conference on Information Science, Sydney, Australia, July 2005, pp. 376–380 (2005)

17. Chandramouli, R., Memon, N.D.: On sequential watermark detection. IEEE Transactions on Signal Processing 51(4), 1034–1044 (2003)

18. Hernández, J., Amado, M., Pérez-González, F.: DCT-domain watermarking techniques for still images: Detector performance analysis and a new structure. IEEE Transactions on Image Processing 9(1), 55–68 (2000)

19. Birney, K., Fischer, T.: On the modeling of DCT and subband image data for compression. IEEE Transactions on Image Processing 4(2), 186–193 (1995)

20. Abramowitz, M., Stegun, I.: Handbook of Mathematical Functions with Formulas, Graphs, and Mathematical Tables. Dover, New York (1964)

21. Xia, X.G., Boncelet, C.G., Arce, G.R.: Wavelet transform based watermark for digital images. Optics Express 3(12), 497 (1998)

22. Su, P.C., Kuo, C.C.: An integrated approach to image watermarking and JPEG 2000 compression. Journal of VLSI Signal Processing 27(1), 35–53 (2001)

23. Simitopoulos, D., Boulgouris, N.V., Leontaris, A., Strintzis, M.G.: Scalable detection of perceptual watermarks in JPEG 2000 images. In: IFIP TC6/TC11 Fifth Joint Working Conference on Communications and Multimedia Security, CMS 2001, Darmstadt, Germany, pp. 93–102. Kluwer, Dordrecht (2001)

24. Kundur, D.: Water-filling for watermarking? In: Proceedings of the IEEE International Conference on Multimedia and Expo, ICME 2000, New York, USA, July 2000, pp. 1287–1290 (2000)

25. Do, M., Vetterli, M.: Wavelet-based texture retrieval using Generalized Gaussian density and Kullback-Leibler distance. IEEE Transactions on Image Processing 11(2), 146–158 (2002)

26. Wu, M., Yu, H., Gelman, A.: Multi-level data hiding for digital image and video. In: Proceedings of the SPIE, Multimedia Systems and Applications II, Boston, USA, September 1999, vol. 3845, pp. 10–21 (1999)

Run-Length and Edge Statistics Based Approach for Image Splicing Detection

Jing Dong[1], Wei Wang[1], Tieniu Tan[1], and Yun Q. Shi[2]

[1] National Laboratory of Pattern Recognition, Institute of Automation,
Chinese Academy of Sciences, P.O.Box 2728, Beijing
[2] Dept. of Electrical and Computer Engineering
New Jersey Institute of Technology, Newark NJ, USA
{jdong,wwang,tnt}@nlpr.ia.ac.cn,shi@nuit.edu

Abstract. In this paper, a simple but efficient approach for blind image splicing detection is proposed. Image splicing is a common and fundamental operation used for image forgery. The detection of image splicing is a preliminary but desirable study for image forensics. Passive detection approaches of image splicing are usually regarded as pattern recognition problems based on features which are sensitive to splicing. In the proposed approach, we analyze the discontinuity of image pixel correlation and coherency caused by splicing in terms of image run-length representation and sharp image characteristics. The statistical features extracted from image run-length representation and image edge statistics are used for splicing detection. The support vector machine (SVM) is used as the classifier. Our experimental results demonstrate that the two proposed features outperform existing ones both in detection accuracy and computational complexity.

Keywords: image splicing, run-length, edge detection, characteristic functions, support vector machine (SVM).

1 Introduction

Digital images are a powerful and widely used medium in our society. For example, newspapers and magazines depend on digital images to represent the news and information every day. However, the availability of powerful digital image processing tools and editing software packages, such as PhotoShop, also makes it possible to change the information represented by an image and create forgeries without leaving noticeable visual traces. Since digital images play a crucial role and have an important impact, the authenticity of images is significant in our social and daily life. How much we can believe in seeing is becoming an intractable problem [1]. The need of authenticity assurance and detection of image forgery (tampering) makes image forensics a very important research issue.

Generally speaking, existing image forgery detection approaches are described as active [2],[3],[4] and passive (blind) [5],[6] techniques. Active approaches are usually related to digital signature and watermarking. In these approaches, certain data (such as signature or proprietary information) for multimedia digital

H.J. Kim, S. Katzenbeisser, and A.T.S. Ho (Eds.): IWDW 2008, LNCS 5450, pp. 76–87, 2009.

rights protection and content tempering authentication are embedded into images. If the content of image has been changed, the embedded information will also be changed consequently. However, either signature-based or watermark-based methods require pre-processing (e.g. watermark embedding) to generate the labeled images for distribution. Unless all digital images are required to be embedded with watermarks before presented in the Internet, it will be unlikely to detect image alteration using active approaches. In contrast to active approaches, passive approaches look into the problem of image tampering from a different angle. These approaches analyze images without requiring prior information (such as embedded watermarks or signatures) and make blind decision about whether these images are tampered or not. Passive techniques are usually based on supervised learning by extracting certain features to distinguish the original images from tampered ones. The practicality and wider applicability of passive approaches make them a hot research topic.

Image splicing [7] is a common operation for generating a digital image forgery, defined as a simple cut-and-paste of image regions from one image onto the same or another image without performing post-processing such as matting and blending in image compositing. The vide availability of image processing software makes the creation of a tampered image using splicing operation particularly easy. The artificial region introduced by image splicing may be almost imperceptible by human eyes. The detection of image splicing is a preliminary but desirable study for image forensics. In this paper we focus on this issue. We present two kinds of simple but efficient statistical features for splicing detection in terms of image run-length representation and sharp image characteristics. The analysis of the performance of the proposed features are made in our experiments. Also, we analyze the comparison of the proposed features and related features proposed in the literature as well as their combinations to evaluate their performance on splicing detection in the terms of both detection accuracy and computational complexity.

The remainder of this paper is organized as follows. Section 2 contains an introduction of related work on splicing detection. Section 3 introduces our proposed two kinds of statistical features for splicing detection. In Section 4, we carry out a number of experiments and analyze on the performance of the proposed features both in detection accuracy and computational complexity. Discussions and conclusions are presented in Section 5.

2   Related Work

Ng et al. [8],[9] detect the presence of the abrupt discontinuities in an image or the absence of the optical low-pass property as a clue for identifying spliced images. For detecting the abrupt splicing discontinuity, a higher order moment spectrum, bicoherence, is used as features. An important property of bicoherence is its sensitivity to a phenomena called quadratic phase coupling (QPC), while the splicing discontinuity leads to a variant of quadratic phase coupling which induces a $\pm\frac{1}{2}\pi$ phase. However, the detection accuracy evaluated on the

Columbia Image Splicing Detection Evaluation Dataset [10] by using only bicoherence features is reported as 62%. To improve the detection performance, they designed a functional texture decomposition method to decompose an image into a gross-structure component and a fine-texture component. With the aid of the decomposition, the detection rate improves from 62% to 71%.

In [11], *Johnson and Farid* developed a technique of image splicing detection by detecting the inconsistency in lighting in an image. It is often difficult to match the lighting conditions from the individual photographs when creating a digital composite, and large inconsistencies in lighting direction may be obvious. Lighting inconsistencies can therefore be a useful tool for revealing traces of digital tampering. At least one reason for this is that image tampering, especially the manipulation of object or people in an image may require the creation or removal of shadows and lighting gradients. The direction of the light source can be estimated for different objects or people in an image, and the presence of inconsistencies in lighting can be used as evidence of digital tampering.

Hsu and Chang take advantages of geometric invariants and camera characteristic consistency to detect spliced images in [12]. They proposed an authentic vs. spliced image classification method by making use of geometric invariants in a semi-automatic manner. For a given image, they identify suspicious splicing areas, compute the geometric invariants from the pixels within each region, and then estimate the camera response function (CRF) from these geometric invariants. If there exists CRF abnormality, the image is regarded as a spliced one. However, this method needs to label the suspicious region of the images before making decisions. It maybe unrealistic for real applications.

In a series of papers [13],[14],[15], *Shi et al.* studied the splicing detection based on statistical features of characteristic functions within a pattern recognition framework. In [13], Hilbert-Huang transform (HHT) is utilized to generate features for splicing detection due to the high non-linear and non-stationary nature of image splicing operation. The moments of characteristic functions with wavelet decomposition is then combined with HHT features to distinguish the spliced images from the authentic ones. The support vector machine (SVM) is utilized as the classifier. A detection accuracy of 80.15% is reported. In the following work of [14], phase congruency has been introduced as features for splicing detection by making use of its sensitivity to sharp transitions such as lines, edges and corners caused by splicing operation. The moments of wavelet characteristic functions form the second part of features for splicing detection in their method. A 82% detection accuracy is reported on Columbia Image Splicing Detection Evaluation Dataset.

3   Proposed Approach for Image Splicing Detection

It is meaningless to talk about the authenticity of a random pixel image, as it has no meaningful characteristics. Creation of a natural-scene forgery image by splicing often introduces abrupt changes around certain objects or regions such as lines, edges and corners. These changes may be much sharper and rougher compared to regular lines, edges and corners due to the unskillful cut-and-paste operation. Meanwhile, splicing may also introduce inconsistency in image statistics by replacing or bringing extra image content to its original content. Hence, the features for splicing detection should capture these variations. In this section, we introduce our proposed features which consist of two parts, namely run-length based statistic moments which are extracted in terms of the global disturbance of correlation of image pixels caused by splicing, and image edge statistic moments which focus on local intensity discontinuity due to splicing.

3.1   Run-Length Based Statistic Moments

The motivation of using run-length based statistic moments for splicing detection is due to a recent study by *Shi et al.* [15]. It is reported that these approaches used in steganalysis can promisingly make progress in splicing detection applications if appropriately applied. The conclusion was demonstrated by their analytical research and extensive experiments. Steganalysis is a counter technique of image steganography [16]. Image steganography is the art and science of invisible communication, by concealing the existence of hidden messages in images. The secret messages are usually embedded in the image by means of slight alterations in image content, which could not be observed by human eyes. The goal of steganalysis is to detect these alteration, in another word, to detect if there is any secret message hidden in the image. Since image splicing definitely alters image content and brings extra information to its original version, it is reasonable to make use of effective features developed for steganalysis to splicing detection as both bring changes on image characteristics and cause some statistical artifacts. The run-length based statistic moments were first proposed in our previous work on blind image steganalysis in [17]. The run-length based features outperform the state-of-art steganalysis approaches as very effective features. Inspired by the conclusion in [15], we employ our proposed efficient run-length based statistic moments in [17] for splicing detection in this section.

The concept of run-length was proposed in the 1950s [18]. A run is defined as a string of consecutive pixels which have the same gray level intensity along a specific linear orientation θ (typically in $0°$, $45°$, $90°$, and $135°$). The length of the run is the number of repeating pixels in the run. For a given image, a run-length matrix $p(i, j)$ is defined as the number of runs with pixels of gray level i and run length j. For a run-length matrix $p_\theta(i, j)$, let M be the number of gray levels and N be the maximum run length. We can define the image run-length histogram (RLH) as a vector:

$$H_\theta(j) = \sum_{i=1}^{M} p_\theta(i, j). \qquad 1 < j < N \qquad (1)$$

This vector represents the sum distribution of the number of runs with run length j in the corresponding image. The length of the runs reflects the size of image structure and texture. For example, smooth image often contains more long runs while an image with finer details usually consists of much more short runs.

The image run-length representation reflects the information of image structure and texture. However, splicing operation, which creates a tampered image by putting several different image regions together in one image, will cause the discontinuity and incoherency on image structure and pixel correlation. It will also be reflected by a variance on their image run-length representation. Hence we take this observation as a clue for splicing detection.

The statistical moments of the characteristic function (denoted as CF) of a histogram are claimed to be very effective features to detect the slight modification of image artifacts [14],[13],[17],[19]. Here we utilize the multi-order moments of the characteristic function of image run-length histograms (in four directions) as features for splicing detection, defined as:

$$M_n = \sum_{j=1}^{L/2} f_j^n |F(f_j)| / \sum_{j=1}^{L/2} |F(f_j)|. \tag{2}$$

where F is the characteristic function of image run-length histogram H (i.e. the Discrete Fourier Transform (DFT) of H), $F(f_j)$ is the component of F at frequency f_j, and L is the DFT sequence length.

In our experiment, we extracted the first three moments of the characteristic functions of image run-length histograms in four directions as features for splicing detection. The 12-D feature vector consists of run-length based features for splicing detection. Fig.1 demonstrates the extraction of the proposed run-length based statistic moment features. The feature extraction procedure is simple and fast, which makes it suitable for splicing detection on large-scale analysis and real applications.

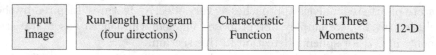

Fig. 1. Diagram of the extraction of the proposed run-length based statistic moment features for splicing detection

3.2   Edge Based Statistic Moments

In addition to the inconsistency of global pixel correlations caused by splicing, there is another change introduced by splicing operation: sharp image intensity variations. Simple copy-and-paste operations introduce extra edges, corners or blobs into images no matter they are visible or not. Such edges, corners or blobs are much sharper compared to natural ones of image content due to the blunt splicing. Hence, the detection of sharp image intensity variations may be served as significant cues for splicing. In [14], *Chen and Shi* focused on Fourier phase and made analysis on 2-D phase congruency to extract features for splicing detection. The presence of sharp edges in images dose cause variations in phase information. However, their method on 2-D phase congruency feature extraction is time consuming. For real applications, it requires more efficient methods. Since we notice that the 2-D phase congruency is usually regarded as a method for edge detection [20], here we also make attempts on analysis of related features extracted from edge, corner and blobs detection, in a relative simple but more efficient way, for image splicing detection.

The Sobel operator [21] is a well known first-order approach to edge detection. Technically, it is a discrete differentiation operator by computing an approximation of the gradient of the image intensity function. At each point in the image, the result of the Sobel operator is either the corresponding gradient vector or the norm of this vector. The Sobel operator is based on convolving the image with a small, separable, and integer valued filter in horizontal and vertical directions and is therefore relatively inexpensive in terms of computations.

Mathematically, the operator uses two 3×3 kernels which are convolved with the original image to calculate approximations of the derivatives for horizontal and vertical changes. If we define A as the source image, and G_x and G_y are two 2-D arrays which at each point contain the horizontal and vertical derivative approximations, the computations are as follow [21]:

$$
G_x = \begin{bmatrix} +1 & 0 & -1 \\ +2 & 0 & -2 \\ +1 & 0 & -1 \end{bmatrix} * A \quad and \quad G_y = \begin{bmatrix} +1 & +2 & +1 \\ 0 & 0 & 0 \\ -1 & -2 & -1 \end{bmatrix} * A \tag{3}
$$

where * denotes the 2-dimensional convolution operation. The x-coordinate is defined as increasing in the right-direction, and the y-coordinate is defined as increasing in the down-direction. At each point in the image, the resulting gradient approximations can be combined to give the gradient magnitude, using:

$$
G = \sqrt{G_x^2 + G_y^2} \tag{4}
$$

we can also calculate the gradient's direction:

$$
\Theta = \arctan(G_y/G_x) \tag{5}
$$

Besides Sobel operator, Laplacian of Gaussian (LoG)[22] is regarded as a common and effective detector for image corners and blobs. Since the Laplacian is very sensitive to noise, usually it was applied after the pre-smoothing of Gaussian filter. Given an input image $I(x, y)$,the LoG is defined as:

$$
\nabla^2 L(x, y, t) = L_{xx} + L_{yy} \tag{6}
$$

where $L(x, y, t) = g(x, y, t) * I(x, y)$ and $g(x, y, t) = e^{(x^2+y^2)/2t}/2t\pi$.

In the proposed approach, we take the 2-D array of image gradient's direction Θ from the result of Sobel operator and the result after LoG detector as the base of our proposed feature for splicing detection. Note the computation of the two edge detection operations is much more faster than phase congruency. Then we calculate the statistical moments of the two base features. We also extract the same features on their prediction-error image and wavelet reconstruction image as proposed in [14]. The prediction-error image I' is calculated by predicting each pixel gray-scale value in the original input image I using its neighboring pixels' gray-scale values. The prediction-error image also removes low frequency information and keeps high frequency information, which makes the splicing detection more efficient. To generate the reconstructed image $I_i(i = 1, 2, 3)$ from the input image I, we erase the information (set the wavelet coefficients to be zero) in sub-band LL_i of I after its Daubechies wavelet transform. Identical procedure is conducted for each reconstructed image $I'_i(i = 1, 2, 3)$. It is claimed that the splicing usually introduces the disturbance in high frequency components of spliced images, and the wavelet-based image by zeroing the approximation sub-band could enhance the difference between the authentic and spliced images. This conclusion has been proved to be reasonable by other's previous successful experience [13],[14]. Fig.2 demonstrates our proposed feature extraction for splicing detection based on sharp image intensity variations.

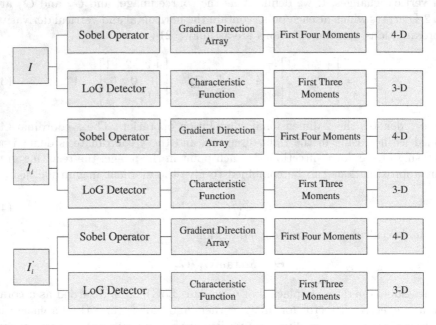

Fig. 2. Framework of edge based feature extraction from the reconstructed images

3.3   Feature Vector for Splicing Detection

Finally we obtain a 61-D feature vector for splicing detection, 12-D from run-length based statistic moments (denoted as RL features), and 49-D from edge detection statistic moments (denoted as SP features). As the feature extraction is implemented at image pixel level and based on simple filtering operation, the computation of the whole feature vector is fairly inexpensive. We will make further analysis on the performance and computation complexity of our proposed features for splicing detection in the following section.

(a) Some examples of authentic images

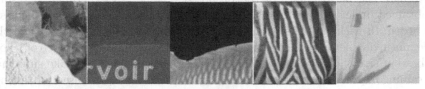

(b) Some examples of spliced images

Fig. 3. Image examples from the Columbia Image Splicing Detection Evaluation Dataset

4   Experimental Results

In this section, we present a set of experiments to demonstrate the effectiveness and efficiency of the proposed features.

4.1   Database Description

The only public available image database for splicing detection is provided by DVMM, Columbia University [10]. The Columbia Image Splicing Detection Evaluation Dataset has 933 authentic and 912 spliced image blocks of size 128 x 128 pixels. The authentic category refers to those images which are original without any splicing operation. This category consists of image blocks of an entirely homogenous textured or smooth region and those having an object boundary separating two textured regions, two smooth regions, or a textured regions and a smooth region. The location and the orientation of the boundaries are random. The spliced category has the same subcategories as the authentic one. For the spliced subcategories, splicing boundary is either straight or arbitrary object boundaries. The image blocks with arbitrary object boundaries are obtained from images with spliced objects. For the spliced subcategories with an entirely homogenous texture or smooth region, image blocks are obtained from the corresponding authentic subcategories by copying a vertical or a horizontal strip of 20 pixels wide from one location to another location within the same image block. More details about this database may be found in [10]. Example images from this database are shown in Fig.3.

4.2   Classifier

Support Vector Machine (SVM) is an optimal and efficient classifier which is commonly used for machine learning systems. Since our work in this paper only focuses on feature extraction rather than the design of classifier, we utilize the SVM^{light} as the classifier in our experiment and a non-linear kernel is chosen. All the experiments and comparisons are tested on the same database and the same classifier in this paper.

4.3 Detection Performance

To evaluate the performance of the proposed scheme, all experimental results are obtained under similar conditions, and the average rate of 5 repeating tests is recorded. In each run, the training samples were randomly selected from the whole image dataset to train the classifier. The training samples are 5/6 of whole database in size (i.e, 776 authentic and 760 spliced images) to make sure that the training model are well learned. The remaining images were used for testing.

The average detection rates of our proposed feature sets as well as the comparison with similar feature sets proposed in [13] are shown in Table 1, where the TP (true positive)represents the detection rate of spliced images, TN (ture negative)represents the detection rate of authentic images, and accuracy is the average detection rate. The corresponding ROC curves are shown in Fig.4 The denoted CF feature set and PC feature set are proposed in [13]. The CF feature set represents a 78-D feature vector which is calculated from the first three moments of the characteristic function of a three level DWT decomposition of test images as well as their prediction-error images. The PC feature set represents a 24-D feature vector extracted from four higher-order statistics of image 2-D phase congruency after image DWT LL_i sub-band zeroing reconstruction at level $i = 1, 2$ and 3. As the CF features are commonly used for image splicing detection and the PC features are also edge based features, we compared the performance of our proposed 12-D run-length based feature vector (denoted as RL feature set) and 49-D edge based feature vector (denoted as SP feature set) with these two feature sets both in detection accuracy and computational cost. The feature extraction time in Table 1 is computed under MATLAB7.0 code run time calculation.

From Table 1, we can see tat the detection accuracy of the SP feature set is 74.27%, higher than its similar feature set PC at 70.68%. However, the feature extraction of SP features takes only $\frac{1}{8}$ computational time of PC feature set. Since both of the two kinds of feature sets are extracted in terms of the sharper image characteristic caused by splicing operation, we can expect that these simple edge detectors are better and more efficient features for splicing detection. Moreover, although the detection results of PC feature set and RL feature set are close, their feature extraction cost differs by 200 times. We also list the detection rates achieved by applying different feature combinations in Table 1 to examine how effective the combination is for splicing detection. Although the detection performance of RL feature set and SP feature set are not as good as CF feature set in our experiments, the combination of them does improve the detection accuracy. Compared with the combination of CF and PC feature sets proposed in [13] which achieved a detection accuracy of 81.76%, the combination of CF and the proposed SP and RL feature sets could also achieve a detection

Table 1. Comparison of detection results and computation time between our proposed feature sets, similar feature sets in [13] and their combinations for splicing detection

Feature Set	TP	TN	Accuracy	Extraction Time (Second)
CF (78D)	78.06%	75.00%	76.22%	0.1056
PC (24D)	65.16%	76.32%	70.68%	5.2526
RL (12D)	65.81%	69.74%	69.75%	0.0245
SP (49D)	78.71%	69.74%	74.27%	0.6424
CF+PC (102D)	81.29%	82.24%	81.76%	5.3582
CF+RL (90D)	80.00%	75.00%	77.52%	0.1301
SP+RL(61D)	76.13%	76.32%	76.52%	0.6708
SP+RL+CF(139D)	83.87%	76.97%	80.46%	0.7480
ALL (163D)	83.23%	85.53%	84.36%	6.0251

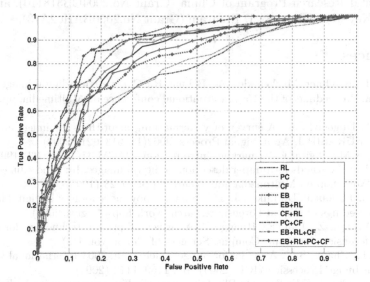

Fig. 4. Comparison of the several feature sets and their combinations for splicing detection

accuracy of 80.46% whereas the computational complexity is just about $\frac{1}{7}$ of the former. Also, when combining all the listed feature sets together, a detection rate as high as of 84.36% is achieved.

5   Conclusion

In this paper, a simple but efficient splicing detection scheme has been proposed. To capture the global inconsistency of pixel correlation caused by splicing, we generated a 12-D feature set based on the statistic moments of characteristic function of image run-length histograms. These features are fast to compute and prove to be valid for splicing detection. To detect the shaper edges introduced by splicing, we utilize a Sobel operator and LoG detector to obtain the local sharp image intensity variations. Since the two sets of features are very simple to extract, the computation complexity of the proposed shame is very low. Also, we have demonstrated the performance of our proposed edge based approach is better than phase congruency approach, while providing a much lower computational cost. Our study in this paper is also a verification of the validity of edge information for splicing detection. Surely other methods related to image intensity variation, such as edges, could be investigated for splicing detection. Also, the experimental results indicated that the combination of our proposed features with the state-of-art features further improve the detection accuracy for splicing detection without any significant extra costs.

Acknowledgments. This work is funded by research grants from the National Fundamental Research Program of China (Grant No.2004CB318110), and the National Key Technology R&D Program (Grant No.2006BAH02A13).

References

1. Gloe, T., Kirchner, M., Winkler, A., Behme, R.: Can we trust digital image forensics? In: Proceedings of the 15th international conference on Multimedia, pp. 78–86 (2007)
2. Rey, C., Dugelay, J.L.: A survey of watermarking algorithms for image authentication. EURASIP J. Appl. Signal Process. 2002(1), 613–621 (2002)
3. Yeung, N.M.: Digital watermarking introduction. CACM 41(7), 31–33 (1998)
4. Fridrich, J.: Methods for tamper detection in digital images. In: Proceedings of the ACM Workshop on Multimedia and Security, pp. 19–23 (1999)
5. Fridrich, J., Soukal, D., Lukas, J.: Detection of copy-move forgery in digital images. In: Proceedings of Digital Forensic research Workshop (August 2003)
6. Luo, W., Qu, Z., Pan, F., Huang, J.: A survey of passive technology for digital image forensics. In: Front. Comput. Sciences of China, vol. 1(2)
7. Ng, T.T., Chang, S.F.: A model for image splicing. In: 2004 International Conference on Image Processing (ICIP 2004), pp. 1169–1172 (2004)
8. Ng, T.T., Chang, S.F., Sun, Q.: Blind detection of photomontage using higher order statistics. In: IEEE International Symposium on Circuits and Systems (2004)
9. Ng, T.T., Chang, S.F.: Blind detection of photomontage using higher order statistics. In: ADVENT Technical Report 201-2004-1. Columbia University (June 8, 2004)
10. Ng, T.T., Chang, S., Sun, Q.: A data set of authentic and spliced image blocks. In: ADVENT Technical Report 203-2004-3. Columbia University (June 2004), http://www.ee.columbia.edu/trustfoto
11. Johnson, M.K., Farid, H.: Exposing digital forgeries by detecting inconsistencies in lighting. In: ACM Multimedia and Security Workshop (2005)
12. Hsu, Y., Chang, S.: Detecting image splicing using geometry invariants and camera characteristics consistency. In: IEEE ICME (July 2006)
13. Fu, D., Shi, Y.Q., Su, W.: Detection of image splicing based on hilbert-huang transform and moments of characteristic functions with wavelet decomposition. In: Shi, Y.Q., Jeon, B. (eds.) IWDW 2006. LNCS, vol. 4283, pp. 177–187. Springer, Heidelberg (2006)

14. Chen, W., Shi, Y.Q.: Image splicing detection using 2-d phase congruency and statistical moments of characteristic function. In: Imaging: Security, Steganography, and Watermarking of Multimedia Contents (January 2007)
15. Shi, Y.Q., Chen, C., Xuan, G.: Steganalysis versus splicing detection. In: Shi, Y.Q., Kim, H.-J., Katzenbeisser, S. (eds.) IWDW 2007. LNCS, vol. 5041, pp. 158–172. Springer, Heidelberg (2008)
16. Johnson, N.F., Jajodia, S.: Exploring steganography: Seeing the unseen. In: Computer, vol. 31, pp. 26–34. IEEE Computer Society, Los Alamitos (1998)
17. Dong, J., Tan, T.: Blind image steganalysis based on run-length histogram analysis. In: 2008 IEEE International Conference on Image Processing(ICIP 2008) (2008)
18. Galloway, M.M.: Texture analysis using gray level run lengths. In: Cornput. Graph. Image Proc., vol. 4, pp. 171–179 (1975)
19. Shi, Y.Q., Chen, C., Chen, W.: A natural image model approach to splicing detection. In: ACM Workshop on Multimedia and Security, ACM MMSEC 2007 (2007)
20. Kovesi, P.: Phase congruency: A low-level image invariant. Psych. Research 64, 136–148 (2000)
21. Sobel, I., Feldman, G.: A 3x3 isotropic gradient operator for image processing. In: Duda, R., Hart, P. (eds.) Pattern Classification and Scene Analysis, pp. 271–272. John Wiley and Sons, Chichester (1973)
22. Arfken, G.: Mathematical methods for physicists, 3rd edn. Academic Press, Orlando (1985)

Scale-Space Feature Based Image Watermarking in Contourlet Domain

Leida Li[1], Baolong Guo[1], Jeng-Shyang Pan[2], and Liu Yang[1]

[1] Institute of Intelligent Control and Image Engineering (ICIE), Xidian University,
Xi'an 710071, P.R. China
reader1104@hotmail.com
[2] Department of Electronic Engineering, Kaohsiung University of Applied Sciences,
Kaohsiung 807, Taiwan

Abstract. In scale-space feature based watermarking schemes, the watermark is usually embedded in spatial domain so that watermark robustness is not satisfactory. In this paper, a novel image watermarking scheme is presented by combining scale-space feature based watermark synchronization and nonsubsampled Contourlet transform (NSCT) based watermark embedding. Watermark synchronization is achieved based on the local circular regions, which can be generated using the scale-invariant feature transform (SIFT). In the encoder, the watermark is embedded into the NSCT coefficients in a content-based and rotation-invariant manner by odd-even quantization. In the decoder, the watermark can be extracted directly from the local regions using the proposed coefficient property detector (CPD). Simulation results and comparisons show that the proposed scheme can efficiently resist signal processing attacks, geometric attacks as well as some combined attacks.

1 Introduction

Digital watermarking is a promising way to protect the copyright of multimedia data [1]. Watermark robustness to traditional signal processing attacks has been addressed extensively, for example JPEG compression, median filtering, and added noise etc. However, many of the existing schemes are fragile to geometric attacks, such as rotation, scaling and translation (RST) [2,3]. Current countermeasures can be classified into invariant domain embedding [4,5,6,7,8], template based embedding [9], and image feature based embedding [10,11,12,13,14,15,16,17].

The basic idea of feature based embedding is to determine the position for both embedding and extraction by referring to intrinsic image features, i.e. watermark synchronization. Feature based schemes belong to the second generation watermarking [10], and they have become an active research field recently. This kind of scheme first extracts feature points from the image and then decomposes it into a set of disjointed local regions, such as triangles [11,12] or circles [13,14,15,16,17]. Then the watermark is embedded into the regions repeatedly. Tang *et al.* adopt the Mexican Hat wavelet scale interaction to extract feature points. Then local circular regions are generated and the watermark is embedded in Discrete Fourier

H.J. Kim, S. Katzenbeisser, and A.T.S. Ho (Eds.): IWDW 2008, LNCS 5450, pp. 88–102, 2009.

Transform (DFT) domain [13]. In [14], Seo *et al.* extract the feature points using the Harris-Laplace detector [18], and the local circular regions are determined using the characteristic scale. Then the watermark is embedded in spatial domain after partitioning the circular region. Similar methods can be found in [15,16]. As the watermark is embedded in spatial domain, watermark robustness is not satisfactory. More recently, Wang *et al.* present a new feature based watermarking scheme in DFT domain [17]. The feature point is also extracted using the Harris-Laplace detector. For each circular region, it is first zero-padded to form a block. Then the block is transformed into DFT domain and the watermark is embedded therein. At last, the block is zero-removed to replace the original region. This scheme outperforms the aforementioned methods in that the watermark is embedded in DFT domain.

By comparison, we know that the advantage of using scale-space feature points is that the extracted regions can always cover the same image content even when the image is scaled. However, as most of the schemes embed the watermark in spatial domain, watermark robustness is not satisfactory. Although scheme [17] is a transform domain scheme, it has the following inherent deficiencies. First, the loss of watermark energy is inevitable during watermark embedding. This is due to the zero-padding and zero-removing operations. In other words, the whole block contains watermark but only part of it is used to produce the watermarked image. As a result, the watermark cannot be extracted accurately even the watermarked image is free of attacks. Second, image rotation should be rectified before watermark detection. The rectification operation will introduce the interpolation error, which can be seen as an additional signal processing attack. This also affect the overall performance. In this paper, we present a novel feature based robust image watermarking scheme. It is based on the following two findings: (1) In scale-space feature based watermarking schemes, the watermark should be embedded in a content-based manner, whether it is embedded in spatial domain or transform domain. (2) To achieve good robustness, the watermark should be embedded in transform domain. Besides, compared with DFT or discrete cosine transform (DCT), the multiscale geometric analysis (MGA) tools are more appropriate for this kind of schemes. The proposed scheme consists of two phases, namely scale-invariant feature transform (SIFT) based watermark synchronization and nonsubsampled Contourlet transform (NSCT) domain odd-even quantization based embedding/extraction. Extensive simulations and comparisons demonstrate the advantages of our scheme.

2   Watermark Synchronization

In our scheme, watermark synchronization is achieved using the SIFT [19]. As a result, we will first introduce the scale space theory and how the SIFT keypoints can be extracted. Then we present the watermark synchronization procedure.

2.1   Image Scale Space and the SIFT

For an image $I(x,y)$, the scale space is defined as a function $L(x,y,\sigma)$, that is produced from the convolution of a variable-scale Gaussian $G(x,y,\sigma)$, with it.

$$L(x,y,\sigma) = G(x,y,\sigma) * I(x,y), \tag{1}$$

where

$$G(x,y,\sigma) = \frac{1}{2\pi\sigma^2} e^{-(x^2+y^2)/2\sigma^2}. \tag{2}$$

The SIFT feature points are detected using a staged filtering approach. The first step is to identify locations and scales by finding the scale-space extrema in the Difference-of-Gaussian (DoG) function $D(x,y,\sigma)$. The DoG images can be obtained by subtracting two nearly scale-space images separated by a constant multiplicative factor k.

$$D(x,y,\sigma) = L(x,y,k\sigma) - L(x,y,\sigma). \tag{3}$$

Fig.1 illustrates image scale space and how the DoG images are generated. The original image is incrementally convolved with Gaussians to produce images separated by a constant factor k in scale space shown on the left. Then adjacent images are subtracted to produce the DoG images shown on the right. To detect the local maxima and minima of $D(x,y,\sigma)$, each sample point is compared with its eight neighbors in the current image and nine neighbors in the scale above and below. It is selected only if it is larger than all of the 26 neighbors or smaller than all of them.

The second step is to determine the location and scale of each feature point accurately using a detailed model. In addition, candidate locations that have low contrast or are poorly localized along edges are removed. In order to achieve invariance to image rotation, a consistent orientation is then assigned to each feature point by finding the peak of the orientation histogram, which can be generated from the neighboring region around the keypoint.

The last step is to compute a descriptor for each feature point from the local image region, which is typically a 128 dimensional vector. This descriptor is highly distinctive so that it can be used for reliable feature point matching. Further details on the detection of the SIFT keypoints can be found in [19].

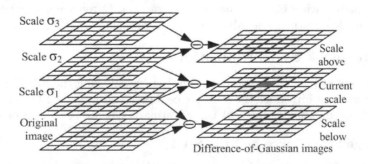

Fig. 1. Image scale space and the DoG images

2.2   SIFT Based Watermark Synchronization

The SIFT extracts feature points with their location, characteristic scale and orientation. A feature point is used to generate a circular region as follows.

$$(x - x_0)^2 + (y - y_0)^2 = (k\sigma)^2 \tag{4}$$

where (x_0, y_0) is the coordinate of the feature point, σ is the characteristic scale and k is a magnification factor which controls the size of the circular region. A feature point whose characteristic scale is small or large has a low probability of being redetected, because it is unstable when the image is scaled [16]. Besides, if the characteristic scale of a feature point is too small, the region will be too small for effective watermark embedding. On the contrary, if the characteristic scale is too large, the number of local regions will be too few for effective watermark detection. In this paper, a feature point is used for watermark embedding if its characteristic scale is between 4 and 8, which can produce about 6 to 10 local regions for common images.

In the encoder, the SIFT features are first extracted and the points with scales between 4 and 8 are selected. Then the feature point with the biggest DoG function value is first adopted to generate a circular region using (4). Afterwards, the feature point with the biggest DoG value among the remaining points is also used to generate a circular region. If the new region overlaps with the previous ones, this point is rejected. The above operation is done for all points, producing non-overlapped circular regions. In the above operations, the feature points with the biggest DoG value are used because they have the best stability. Finally, the descriptors of the used feature points are saved as a secret key.

In the decoder, the SIFT features are first extracted and their descriptors are matched to the descriptors in the secret key using the fast nearest neighbor algorithm [19]. Then the matched points are used to generate the circular regions. It should be noted that the range of characteristic scale used for watermark detection should be wider, because the image may have been distorted by scaling

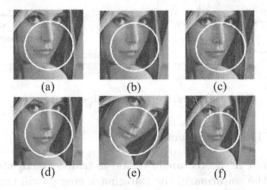

(a)   (b)   (c)

(d)   (e)   (f)

Fig. 2. Local circular regions extracted from: (a) Original image, (b) Median-filtered image, (c) Added Gaussian noise image, (d) 30% JPEG image, (e) 30-degree rotated image, (f) 0.8× scaled image

attacks. In this paper, the range is set to [2, 16] so that the features can be redetected even the image is distorted by 0.5× or 2× scaling.

The extracted regions using the above synchronization scheme are invariant to RST attacks. Fig.2 shows one of the regions extracted from image Lena and some distorted versions. It is observed that the region can be generated robustly whether the image is distorted by traditional signal processing attacks or geometric attacks.

3   Watermark Embedding Scheme

The key idea of the proposed scheme is that it combines scale-space feature based watermark synchronization with transform domain watermark embedding, aiming to achieve good robustness. Fig.3 shows the diagram of watermark embedding. For a digital image, the local circular regions are first generated. As several regions can be extracted, we embed the watermark into all the regions repeatedly. It should be noted that circular regions are used during watermark synchronization. However, due to the discrete property of digital image, the region we can use is actually a rectangle region. The rectangle region can be produced from the circular region by padding its border with zeros.

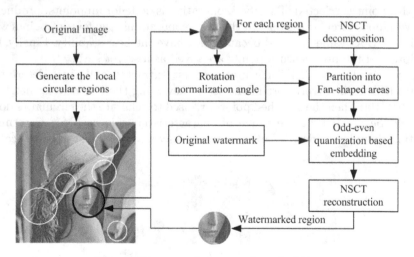

Fig. 3. Diagram of watermark embedding

3.1   Transform Domain Selection

It is well known that if the rectangle region is transformed into DFT or DCT domain to embed the watermark, the watermark energy will spread over all the pixels of the rectangle region. However, when we replace the original region with the watermarked one, the border has to be discarded. This will inevitably lose some watermark energy. The recent development in multiscale geometric analysis

(MGA) has brought us a promising solution, such as the discrete wavelet transform (DWT), the Contourlet transform etc. The MGA tools have a desirable property, namely they are localized in both spatial and transform domain. In this case, if we embed the watermark by modifying coefficients within the circular region, the watermark energy will only spread over the corresponding pixels within the circular region. In other words, the border area does not contain any watermark. In the proposed scheme, we embed the watermark in NSCT domain [20]. The NSCT is a flexible multiscale, multidirection, and shift-invariant version of Contourlet transform. We employ the NSCT in that it produces subband images with the same size with the original image. Besides, it is a redundant transform. As a result, it is a quite an efficient tool for information hiding.

3.2 Partition of the Circular Region

During embedding, a local region is first transformed into NSCT domain. Then the low frequency subband image is divided into fan-shaped sub-regions (SR), as shown in Fig.4. These SRs have the same area. Then the watermark is embedded into each SR by odd-even quantization.

It should be noted that when rotation occurs, the two regions extracted from the original and the rotated images differ in orientation, see Fig.2(a) and Fig.2(e). In order to achieve watermark robustness to image rotation, the partition should be done in a rotation invariant way. In other words, the start radius should keep constant relative to the image content. This can be done by first align the region to a standard orientation using rotation normalization [21]. However, in that case, the watermarked region has to be rotated back to the original orientation, which will inevitably introduce interpolation error. The interpolation error can be seen an signal processing attack. In order to avoid interpolation error, we determine the start radius by only using the rotation normalization angle, as shown in Fig.4.

If the watermark to be embedded is $w = \{w_1, w_2, \cdots, w_N\}$, $w_i \in \{0, 1\}$, then the angle of each SR is $2\pi/N$. In order to describe the SR, the coordinate of the circular region is first transformed into polar system as follows.

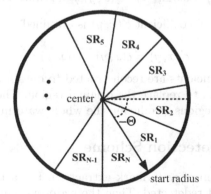

Fig. 4. Partition of the circular regions

$$\rho_{x,y} = \sqrt{x^2 + y^2}, \quad and \quad \theta_{x,y} = arctan(y/x). \tag{5}$$

Suppose that the rotation normalization angle of the local region is Θ, then each SR can be described as follows.

$$SR_i = \{(x,y)| -\Theta + (i-1)\frac{2\pi}{N} \le \theta_{x,y} < -\Theta + i\frac{2\pi}{N}\}, \tag{6}$$

where $i = 1, 2, \cdots, N$.

3.3   Odd-Even Quantization Based Embedding

In our scheme, the watermark is embedded using odd-even quantization. Besides, a watermark bit w_i is embedded into a SR with the same index, i.e. SR_i. In implementation, all coefficients in the same SR are quantized into odd or even coefficients according to the watermark bit. First, the coefficient $c(x,y)$ is assigned sign "0" or "1" using the quantization function.

$$Q(x,y) = \begin{cases} 0, & if \ \ k\Delta \le c(x,y) < (k+1)\Delta \ \ for \ \ k = 0, \pm2, \pm4, \cdots \\ 1, & if \ \ k\Delta \le c(x,y) < (k+1)\Delta \ \ for \ \ k = \pm1, \pm3, \pm5, \cdots \end{cases} \tag{7}$$

where Δ is the quantization step. In order to enhance watermark robustness, the modified NSCT coefficient should locate at the middle of the corresponding quantization interval [22]. Therefore, we modify the coefficients as follows.

The quantization noise is first computed as

$$r(x,y) = c(x,y) - \left\lfloor \frac{c(x,y)}{\Delta} \right\rfloor \cdot \Delta \tag{8}$$

where $\lfloor \cdot \rfloor$ is the floor operation. The amount of modification is determined by

$$u(x,y) = \begin{cases} -r(x,y) + 0.5\Delta, & if \ \ Q(x,y) = w_i \\ -r(x,y) + 1.5\Delta, & if \ \ Q(x,y) \ne w_i \ \ and \ \ r(x,y) > 0.5\Delta \\ -r(x,y) - 0.5\Delta, & if \ \ Q(x,y) \ne w_i \ \ and \ \ r(x,y) \le 0.5\Delta \end{cases} \tag{9}$$

Then the modified NSCT coefficient $c'(x,y)$ is obtained.

$$c'(x,y) = c(x,y) + u(x,y) \tag{10}$$

Then the modified coefficients are reconstructed to obtain the watermarked region, which is then used to replace the original region. The above operation is done circularly for all regions, producing the whole watermarked image.

4   Watermark Detection Scheme

Fig.5 shows the diagram of watermark extraction. For a test image, the local circular regions are first redetected. Then the watermark are detected based on the synchronized regions in NSCT domain.

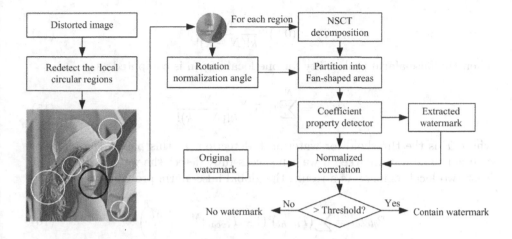

Fig. 5. Diagram of watermark extraction

4.1 Coefficient Property Detector

We present a coefficient property detector (CPD) to extract the watermark. For a circular region, it is first transformed into NSCT domain. Then it is divided into N fan-shaped SRs using (6). The CPD is designed as follows. The coefficients inside SR_i are first assigned number "0" or "1" using (7). The number of "0" coefficients is denoted by $NUM_{i,0}$, and the number of "1" coefficients is denoted by $NUM_{i,1}$. Then the watermark bits are extracted as follows.

$$w_i' = \begin{cases} 0, & if \;\; NUM_{i,0} > NUM_{i,1} \\ 1, & if \;\; NUM_{i,1} > NUM_{i,0} \end{cases} \qquad (11)$$

At last, the normalized Hamming similarity (NHS) is computed to evaluate the effectiveness of the proposed algorithm. The NHS between the embedded binary watermark w and the extracted one w' is defined as

$$NHS = 1 - HD(w, w')/N \qquad (12)$$

where $HD(\cdot, \cdot)$ denotes the number of bits different in the two binary strings.

4.2 False Alarm Analysis

The presence of the watermark can be claimed if NHS is above a threshold. In this paper, the threshold is determined based on the false-alarm probability. Suppose that each bit of the watermark is an independent variable. The probability of a k-bit match between N-bit extracted and original watermark is

$$P_k = \binom{N}{k} p^k (1-p)^{N-k} \qquad (13)$$

where p is the possibility that the extracted bit matches the original watermark bit. It is reasonable to assume $p = 0.5$, so that P_k can be rewritten as

$$P_k = (0.5)^N \frac{N!}{k!(N-k)!} \tag{14}$$

Then the false-alarm probability for one local region is computed as

$$P_{local} = \sum_{k=T}^{N} (0.5)^N \frac{N!}{k!(N-k)!} \tag{15}$$

where T is the threshold for watermark detection. In this paper, the image is claimed to be watermarked if we can successfully detect the watermark from at least two local regions. As a result, the global false-alarm probability is

$$P_{global} = \sum_{i=2}^{M} (P_{local})^i (1 - P_{local})^{M-i} \binom{M}{i} \tag{16}$$

where M is number of the local regions.

5   Experimental Results

Experiments have been conducted on various standard gray images with size 512×512, including Lena, Peppers, Baboon etc. In experiments, the binary watermark is 32-bit long. The watermark detection threshold is set to 24, which achieves a false-alarm probability as low as 1.8×10^{-4}.

5.1   Watermark Synchronization

The proposed scheme embeds the watermark into the local circular regions. As a result, the redetection of these regions is crucial for correct watermark extraction. Fig.6 shows an example of all matched regions between the image Lena and the attacked images, which have been distorted by a few signal processing attacks and geometric attacks as indicated in the captions of the figure. Note the sizes of Fig.6(e), Fig.6(f), Fig.6(h) have been adjusted for better display.

It can be seen from Fig.6 that enough number of regions can be detected to facilitate watermark detection, regardless of traditional signal processing attacks and geometric attacks.

5.2   Watermark Invisibility

As our scheme embeds the watermark by quantization, it is necessary to determine the optimal quantization step that can achieve a trade-off between watermark invisibility and robustness. Fig.7 shows the relation between the peak signal to noise ratio (PSNR) and the quantization step on Lena and Baboon.

It can be found that the PSNR decreases with increasing quantization steps. When $\Delta < 0.025$, the PSNR values are above 40dB. When $\Delta \in (0.03, 0.04)$,

Fig. 6. The local circular regions extracted from: (a) Original image. (b) Median-filtered image. (c) 20% JPEG image. (d) Added-Gaussian noise image. (e) 10-deg-rotated image. (f) 1.2× scaled image. (g) 40-pixel-translated image. (h) 30-deg-rotated and 0.8× scaled image.

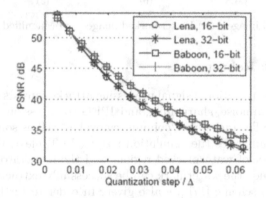

Fig. 7. The relation between PSNR and the quantization step

the PSNR is still higher than 35dB. However, we have found that the artifacts become visible around the watermarked regions. In this paper, we set the quantization step to 0.02, which achieves PSNR values higher than 40dB for all images. Fig.7 also shows that the length of the watermark has little influence on the invisibility. In our experiments, the watermark is 32-bit long.

Fig.8 shows an example of watermark embedding. The original images and the watermarked images are shown in Fig.8(a) and Fig.8(b), respectively. Fig.8(c) shows the corresponding residual images, which are magnified 100 times for better display. It is easily seen from Fig.8 that we insert the watermark so as not to be visible to the naked eyes. The PSNR values of the watermarked images are 44.039dB, 45.282dB, and 45.578dB, respectively.

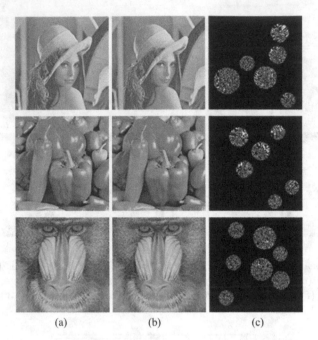

(a)   (b)   (c)

Fig. 8. (a) Original images. (b) Watermarked images. (c) Magnified residual images.

5.3   Watermark Robustness

In our experiments, traditional signal processing attacks consists of median filtering, added Gaussian noise, sharpening, and JPEG compression, while geometric attacks include RST attacks, row/column removal as well as some combined attacks. Table 1 summarizes the simulation results. In Table 1, the denominator denotes the number of watermarked regions and the numerator is the number of regions from which the watermark can be successfully extracted. For comparison, the results in scheme [17] are also given. In order to further demonstrate the effectiveness of the proposed scheme, Table 2 lists the watermark similarities for each of the distorted image. While several similarities can be obtained from different circular regions, only the maximum value and the mean value are given, which are denoted by NHS_{max} and NHS_{mean}, respectively.

Signal processing attacks: Traditional signal processing attacks act on the watermarking system by reducing the watermark energy. Our scheme can resist median filtering, added Gaussian noise, image sharpening, JPEG compression etc. Besides, the watermark can be detected even the quality factor of JPEG compression is up to 30. Our results are comparable with those in scheme [17].

Image rotation: Our scheme can resist rotation attacks and it can extract the watermark directly from the rotated image without any auxiliary operations. This is because that the partition of the circular region is rotation invariant so that the watermark is embedded in a rotation invariant pattern. In [17], image rotation has to be rectified by first obtaining the rotation angle using the

Table 1. Fraction of correctly detected watermark regions under different attacks

Attacks	Lena		Peppers		Baboon	
	Proposed	Wang[17]	Proposed	Wang[17]	Proposed	Wang[17]
Med-filter (3 × 3)	4/6	3/6	5/6	4/8	2/7	7/12
Sharpening (3 × 3)	6/6	3/6	6/6	5/8	4/7	6/12
Added Gaussian noise	3/6	2/6	4/6	4/8	5/7	4/12
JPEG compression 30	4/6	2/6	4/6	4/8	2/7	8/12
JPEG compression 50	4/6	4/6	6/6	6/8	6/7	8/12
JPEG compression 70	4/6	4/6	6/6	7/8	6/7	9/12
Med-filter (3 × 3) + JPEG 90	5/6	3/6	4/6	4/8	3/7	7/12
Sharpening (3 × 3) + JPEG 90	6/6	3/6	6/6	6/8	4/7	6/12
Remove 8 rows and 16 cols	4/6	5/6	4/6	6/8	5/7	7/12
Cropping 55%	2/6	4/6	2/6	5/8	1/7	6/12
Rotation 5 degree	5/6	4/6	6/6	5/8	4/7	5/12
Rotation 15 degree	4/6	3/6	6/6	4/8	4/7	4/12
Rotation 30 degree	4/6	2/6	5/6	2/8	2/7	4/12
Translation -x-10 and -y-10	4/6	5/6	4/6	5/8	6/7	10/12
Scaling 0.6×	1/6	1/6	4/6	3/8	1/7	2/12
Scaling 0.9×	4/6	3/6	4/6	3/8	2/7	5/12
Scaling 1.4×	3/6	1/6	3/6	2/8	2/7	1/12
Cropping 10% + JPEG 70	4/6	2/6	3/6	4/8	5/7	5/12
Rotation 5 + Scaling 0.9×	4/6	3/6	4/6	5/8	2/7	5/12 1
Translation x-10 and y-10 + Rotation 5 + Scaling 0.9×	3/6	2/6	3/6	1/8	2/7	3/12

DFT magnitude. The rectification process will inevitably introduce interpolation error, which can be seen as a signal processing attack before watermark detection. By comparison, our scheme can resist large rotations, such as 60 degree and 90 degree. Besides, the mean NHS values are all above 0.8.

Image scaling: Two factors are crucial for scale invariance of the watermark. First, the local regions should always cover the same content in an image before and after an attack. Second, the watermark should be embedded in a content-based manner. In our scheme, the SIFT based watermark synchronization meets the first requirement, which can be seen from Fig.2 and Fig.6. Embedding the watermark into the fan-shaped SRs meets the second requirement.

Image translation: During translation, some of the image content is missing. However, most of the local regions can be redetected. As a result, the watermark can be successfully detected if only at least two regions can survive the attack.

Other geometric attacks: Besides RST attacks, we also test some other geometric attacks, such as cropping and row/column removal. For cropping, scheme [17] performs a little better than our scheme. This is because they extract more

Table 2. The maximum and mean NHS values under different attacks

Attacks	Lena		Peppers		Baboon	
	NHS_{max}	NHS_{mean}	NHS_{max}	NHS_{mean}	NHS_{max}	NHS_{mean}
Added Gaussian noise	0.969	0.906	0.969	0.891	0.875	0.819
Med-filter (3 × 3)	1.000	0.969	1.000	0.906	0.938	0.860
JPEG compression 30	0.969	0.900	0.938	0.867	0.813	0.813
JPEG compression 50	1.000	0.969	1.000	0.932	0.969	0.860
JPEG compression 70	1.000	0.969	1.000	0.958	1.000	0.980
Med-filter 3 × 3 + JPEG 90	1.000	0.988	1.000	0.906	0.969	0.833
Rotation 10 degree	0.969	0.963	1.000	0.975	1.000	0.866
Rotation 30 degree	0.969	0.922	0.969	0.919	1.000	0.906
Rotation 45 degree	0.906	0.891	0.969	0.888	1.000	0.922
Rotation 60 degree	0.906	0.850	0.969	0.896	0.938	0.875
Rotation 90 degree	0.969	0.844	0.969	0.869	0.813	0.813
Rotation 10 degree + cropping	1.000	0.969	1.000	0.956	1.000	0.917
Rotation 15 degree + cropping	1.000	0.969	1.000	0.984	1.000	0.938
Rotation 30 degree + cropping	0.969	0.922	1.000	0.980	1.000	0.927
Rotation 45 degree + cropping	0.969	0.938	1.000	0.953	1.000	0.865
Rotation 60 degree + cropping	0.938	0.917	0.969	0.885	0.906	0.828
Scaling 0.6×	0.844	0.844	0.938	0.898	0.750	0.750
Scaling 0.8×	1.000	0.888	1.000	0.969	0.938	0.906
Scaling 1.2×	1.000	0.960	1.000	0.980	1.000	0.860
Scaling 1.4×	1.000	0.969	1.000	0.906	0.875	0.875
Translation -x-40 and -y-40	1.000	1.000	1.000	1.000	1.000	0.969
Rotation 10 deg + Scaling 0.9×	1.000	0.938	1.000	0.938	0.969	0.896
Rotation 30 deg + Scaling 0.9×	0.906	0.875	0.969	0.919	0.969	0.875

regions than we do, so that more regions are preserved. This can be improved by using smaller magnification factor during the circular region generation.

Combined attacks: Except for individual attack, we also conduct some attacks that combine two or more kind of attacks. The combined attacks include rotation plus cropping, rotation plus scaling, median filtering plus JPEG compression etc. From Table 1 and Table 2, we know that the watermark can be detected successfully from all the distorted images.

6   Conclusion

This paper presents a novel feature based robust image watermarking scheme for resisting signal processing attacks and geometric attacks simultaneously. The proposed scheme combines scale-space feature point based watermark synchro- nization with transform domain based watermark embedding. We have shown that the watermark should be embedded in a content-based manner in this kind of schemes. What is more, the multiscale geometric analysis tools are more suitable for watermark embedding due to its localizing property in both spatial and frequency domain. The proposed scheme achieves watermark synchronization using the scale-invariant feature transform. During embedding, the nonsubsampled Contourlet transform is employed by odd-even quantization. Besides, the watermark is embedded in a content-based and rotation-invariant manner. In the decoder, the watermark can be extracted using the proposed coefficient property detector (CPD) on the distorted images directly. The proposed scheme achieves good watermark invisibility and the PSNR values are all higher than 40dB. Extensive simulations and comparisons have demonstrated the superiority of our scheme.

Acknowledgements

This work is supported by National Natural Science Foundation of China (60802077), Ph.D. Programs Foundation of Ministry of Education of China (20060701004), and National High Technology Research and Development Program of China (2006AA01Z127).

References

1. Cox, I., Miller, M., Bloom, J.: Digital Watermarking. Morgan Kaufmann, San Mateo (2001)
2. Licks, V., Jordan, R.: Geometric attacks on image watermarking systems. IEEE Multimedia 12(3), 68–78 (2005)
3. Zheng, D., Liu, Y., Zhao, J., El-Saddik, A.: A survey of RST invariant image watermarking algorithms. ACM Comput. Surv. 39(2), 1–91 (2007)
4. O'Ruanaidh, J., Pun, T.: Rotation, scale and translation invariant digital image watermarking. Signal Process 66(3), 303–317 (1998)
5. Lin, C., Wu, M., Bloom, J., Cox, I., Miller, M., Lui, Y.: Rotation, scale, and translation resilient watermarking of images. IEEE Trans. Image Process. 10(5), 767–782 (2001)
6. Zheng, D., Zhao, J., El-Saddik, A.: RST-invariant digital image watermarking based on log-polar mapping and phase correlation. IEEE Trans. Circuits Syst. Video Technol. 13(8), 753–765 (2003)
7. Kim, H., Lee, H.: Invariant image watermark using Zernike moments. IEEE Trans. Circuits Syst. Video Technol. 13(8), 766–775 (2003)
8. Xin, Y., Liao, S., Pawlak, M.: Circularly orthogonal moments for geometrically robust image watermarking. Pattern Recognit. 40(12), 3740–3752 (2007)
9. Pereira, S., Pun, T.: Robust template matching for affine resistant image watermarks. IEEE Trans. Image Process. 9(6), 1123–1129 (2000)
10. Kutter, M., Bhattacharjee, S., Ebrahimi, T.: Towards second generation watermarking schemes. In: Proc. IEEE Int. Conf. Image Process., Kobe, Japan, vol. 1, pp. 320–323 (1999)
11. Bas, P., Chassery, J., Macq, B.: Geometrically invariant watermarking using feature points. IEEE Trans. Image Process. 11(9), 1014–1028 (2002)

12. Qi, X., Qi, J.: A robust content-based digital image watermarking scheme. Signal Process. 87(6), 1264–1280 (2007)
13. Tang, C., Hang, H.: A feature-based robust digital image watermarking scheme. IEEE Trans. Signal Process. 51(4), 950–959 (2003)
14. Seo, J., Yoo, C.: Localized image watermarking based on feature points of scale-space representation. Pattern Recognit. 37(7), 1365–1375 (2004)
15. Seo, J., Yoo, C.: Image watermarking based on invariant regions of scale-space representation. IEEE Trans. Signal Process. 54(4), 1537–1549 (2004)
16. Lee, H., Kim, H., Lee, H.: Robust image watermarking using local invariant features. Opt. Eng. 45(3), 037002(1-11) (2006)
17. Wang, X., Wu, J., Niu, P.: A new digital image watermarking algorithm resilient to desynchronization attacks. IEEE Trans. Inf. Forensics Security 2(4), 655–663 (2007)
18. Mikolajczyk, K., Schmid, C.: Scale and affine invariant interest point detectors. Int. J. Comput. Vis. 60(1), 63–86 (2004)
19. Lowe, D.: Distinctive image features from scale-invariant keypoints. Int. J. Comput. Vis. 60(2), 91–110 (2004)
20. Cunha, A., Zhou, J., Do, M.: The nonsubsampled Contourlet transform: theory, design, and applications. IEEE Trans. Image Process. 15(10), 3089–3101 (2006)
21. Alghoniemy, M., Tewfik, A.: Geometric invariance in image watermarking. IEEE Trans. Image Process. 13(2), 145–153 (2004)
22. Yu, G., Lu, C., Liao, H.: Mean-quantization-based fragile watermarking for image authentication. Opt. Eng. 40(7), 1396–1408 (2001)

A Practical Print-and-Scan Resilient Watermarking for High Resolution Images

Yongping Zhang[1], Xiangui Kang[2], and Philipp Zhang[3]

[1] Research Department, Hisilicon Technologies Co., Ltd, 100094 Beijing, China
zhangyongping79@huawei.com
[2] School of Infor. Sci. and Tech., Sun Yat-Sen University, 510275 Guangzhou, China
isskxg@mail.sysu.edu.cn
[3] Research Department, Hisilicon Technologies Co., Ltd, 750075 Plano, Texas, USA
pzhang@huawei.com

Abstract. Detecting the watermark from the rescanned image is still a challenging problem, especially for high resolution image. After printing and scanning, an image usually suffers global geometric distorts (such as rotation, scaling, translation and cropping), local random nonlinear geometric distortions which is simulated in Random Bending function of Stirmark, as well as nonlinear pixel value distortions. The combination of both of the latter distortions is called nonlinear distortions. Many watermarking techniques, including the pilot-based watermarking techniques, are robust against the global geometric distortions but sensitive to the nonlinear distortions. Local random nonlinear geometric distortions remain a tough problem for image watermarking. In the setting of print-and-scan process, it becomes severer as combining with nonlinear pixel value distortions. This may defeat a watermarking scheme, especially for the print-scanning of high resolution image. This paper proposes an effective pilot-based watermarking algorithm for the print-and-scan process. After careful analysis of the print-and-scan process, the pilot signal and the watermark are embedded and detected in the down-sampled low resolution image to deal with the combination of local random nonlinear geometric distortions and nonlinear pixel value distortions. Theoretical analysis and experimental results demonstrate the proposed algorithm is robust to the print-and-scan process and at low computational complexity. The major contribution of this paper is that we analyze the impact of nonlinear distortions in print-and-scan process and propose an effective watermarking scheme to conquer it. To our best knowledge, our work also first addresses the issues of print-and-scan resilient watermarking for high resolution images.

Keywords: Print-and-scan, watermark, non-linear distortions, robustness.

1 Introduction

The print-and-scan process is commonly used for image reproduction and distribution. Print-and-scan resilient data hiding provides a viable authentication mechanism via the multi-bit watermark hidden in a picture in the document. Document authentication of ID card, passport, driving license, image publication etc. is important today as the security concerns are higher than ever before. For example, the authentication of South

H.J. Kim, S. Katzenbeisser, and A.T.S. Ho (Eds.): IWDW 2008, LNCS 5450, pp. 103–112, 2009.

China Tiger Image drew extensively concerns in China, even in the whole world. But print-and-scan resilient data hiding has not been extensively researched. Many of them focus on detecting watermarks [1]. The rescanned image is usually changed both in geometric and pixel value, and in both linear, non-linear way. The watermark can be detected from the rescanned image only if the watermark can resist the combination of these distortions.

Many watermarking techniques robust to geometric distortions are presented. One category is to embed the watermark in the global geometric transform invariant domain [1][2]. The computational complexity of this technique is high, especially for high resolution images. Besides, the non-uniform log-polar mapping adopted in this technique might cause severe image fidelity problem.

Another category[3][4][5] is to embed a pilot multiple times in the image at different spatial locations. The autocorrelation function of the watermarked image will then yield a pattern of peaks corresponding to the embedded locations. Changes in this pattern of peaks can be used to estimate the global geometric distortions that the watermarked image has undergone. These methods have significant potential, but, similar to the above methods, do not consider the effect of the nonlinear distortions including nonlinear geometric distortions and nonlinear pixel value distortions. The watermark can be detected based on the pilot signals. But after printing-and-scanning, the nonlinear distortions destroy the periodicity of the pilot signal and the autocorrelation of the rescanned image will not yield a pattern of peaks.

After careful analysis of the print-and-scan process, our proposal employs a down-sampling operation in the watermark embedding and extraction process that simulate the watermarking embedding and extraction in the low-resolution, thus reduce the effect of nonlinear distortions and make sure that the pilot signal can be detected correctly. We can get the parameters of the global geometric distortions with the help of pilot signal and invert the rescanned image before applying the watermark extractor.

In this paper, we begin with the analysis of the print-and-scan process in Section 2. Then we propose a novel watermarking algorithm in Section 3. Experimental results are presented in Section 4. In Section 5, we make a summary.

2   Nonlinear Distortion Problems to Watermarking and Solutions

In this section, we first make a careful analysis on the print-and-scan process. Here we present how the print-and-scan process affects the watermark. Then we propose the following solutions.

2.1   Analysis on the Print-and-Scan Process

The process of print and scan is a complex composite of various attacks, which cause various distortions including geometric distortion and pixel value distortion.

The original image is transmitted into control system of the printer. Before Raster Image Processor (RIP) in the control system converts the original image into halftone image and corresponding electronic pulse level, the printer will make the Dots Per Inch (DPI) value of image equal to the physical DPI value of the printer. The electronic pulse level is converted into optical level and forms latent image on Organic Photo Conductor (OPC) drum of electrophotographic system. After development process, transferring process and fusing process, latent image is transformed into hardcopy output. In the scanning process, a pure light emitted to the hardcopy global. Charge Coupled Device (CCD) sensors receive the luminance of the light and convert it into the electronic signal. At last, the generated digital image generated is transmitted to PC via Graphic User Interface (GUI).[7]

According to the process of the print-and-scan process, we can find that both the printing process and scanning process introduce geometric distortions. The distortions during these two processes are different. Converting the electronic pulse into optical level and hardcopy process introduces the nonlinear geometric distortions. Consider an image I, and a nonlinear distorted version of this image, I'. Then we can write that

$$D(x, y) = I'(x, y) - I(x, y) \qquad (1)$$

where $D(x,y)$ represents the difference of pixel value between I and I'.

Let Δx and Δy represent the magnitude of nonlinear geometric distortion in the horizontal direction and in the vertical direction respectively. The position of pixel $I(x,y)$ is translated to the position $(x + \Delta x, y + \Delta y)$ in the rescanned image.

The nonlinear adjustment in the printer and scanner also introduce the distortion on the value of pixel. The nonlinear adjustment, which is called gamma tweaking, is performed in the printer to make sure the printed images appear the same as on a monitor:

$$B = I(x, y)^{\gamma_p} \qquad (2)$$

where B represents the actual light brightness. When the image is scanned, it must be compensated to make sure it looks fine to us when viewed on a monitor:

$$I'(x, y) = B^{1/\gamma_s} \qquad (3)$$

So,

$$I'(x, y) = I(x, y)^{\gamma_p/\gamma_s} \qquad (4)$$

Besides these distortions, there is some random noise in the print-and-scan process. And let N represent the random noise. Then we can get

$$D_{x,y} = I'(x,y) - I(x,y) = I(x - \Delta x, y - \Delta y)^{\gamma_p/\gamma_s} + N - I(x,y) \qquad (5)$$

When $\gamma_p > \gamma_s$, this nonlinear adjustment amplify the effect of nonlinear geometric distortion. And D is distributed in the whole image. It may become large enough to affect the watermark extracting.

There is one measure to deal with this problem if the resolution of image is low. We can scale the watermarked low resolution image up to a higher resolution before printing it out. And in the scanning process, the printed watermarked image is down-sampled to a low resolution as the original resolution. This operation can improve the robustness against the nonlinear distortions. Some experiments on Digimarc watermarking tool in Photoshop are performed. In our experiments, 100 images with 800-by-600(4:3) are utilized. The embedding strength is as large as possible under the limitation of invisibility and the PSNR value is 36.13dB averagely. It shows good performance when we up-scale the watermarked image to 4096-by-3076(4:3, which is the size of A4) in the printing process. The Bit Error Ratio (BER) of watermark extraction from the rescanned watermarked images is about 13.15%.

But unfortunately, with the development of the multimedia technologies, the resolutions of images become higher and higher in the application. Especially in the document printing, the driver of printer will convert a text to an image. The resolution of this image is very high and based on the size of paper printed on. Now the above measure does not work. For example, when the watermarked image whose resolution is 4096-by-3076 is printed on A4, we cannot up-scale the watermarked image because of the paper scope. Now the nonlinear distortions can destroy the watermark easily. Experiments on the Digimarc watermarking tool are performed on four images with 4096-by-3076 or 3076-by-4096. The watermarked images are printed with the original resolution on A4 paper. No watermark can be extracted from the rescanned watermarked images.

Another problem is that many existing print-and-scan resilient watermarking algorithms are complex and time consuming. Many complex methods are included to make the watermarking algorithm to be robust against the print-and-scan process. Computation efficiency is an important consideration in application.

There is still another measure proposed in this paper to deal with the above problems. We can adopt the method of embedding and extracting watermark at low resolution shown as follows to get better robustness against nonlinear distortions and higher computation efficiency.

2.2 Proposed Solution

Based on the above analysis and the experiments with Digimarc watermarking tool on the print-and-scan process, our solution is to add a down-sampling operation to achieve the low-resolution which is necessary in the watermark embedding and extraction process.

In watermark embedding, we first down-sample the original image with k and embed the watermark in the down-sampled image. Let $\bar{I}_w(\bar{x}, \bar{y})$ represent one pixel in the down-sampled watermarked image. We can obtain,

$$\bar{I}_w(\bar{x}, \bar{y}) = \frac{\sum_{i,j=1}^{k} I_w\left((\bar{x}-1)\times k + i, (\bar{y}-1)\times k + j\right)}{k^2} \tag{6}$$

Assume $I_w(x,y)$ to be the pixel of the watermarked image and $I'_w(x, y)$ to be the pixel of the rescanned watermarked image, where

$$\begin{cases} x = (\bar{x}-1)\times k + i \\ y = (\bar{y}-1)\times k + j \end{cases} \quad i, j \in [1, k] \tag{7}$$

In watermark extraction, the rescanned image is also down-sampled to $1/k$ of the original size. We can obtain,

$$\overline{I}'_w(\overline{x},\overline{y}) = \frac{\sum_{i,j=1}^{k} I'_w\left((\overline{x}-1)\times k+i,(\overline{y}-1)\times k+j\right)}{k^2} \tag{8}$$

The difference between Eq.(6) and Eq.(8), which is the effect on the image at low resolution from the nonlinear distortion, can be described as,

$$\overline{I}_w(\overline{x},\overline{y}) - \overline{I}'_w(\overline{x},\overline{y}) = \frac{\sum_{i,j}^{k} D_{i,j}}{k^2} = \frac{D}{k^2} \tag{9}$$

where $D = \sum_{i,j=1}^{k} D_{i,j}$ and $D_{i,j} = I_w(x,y) - I'_w(x,y) = I_w((\overline{x}-1)\times k+i,(\overline{y}-1)\times k+j) - I'_w((\overline{x}-1)\times k+i,(\overline{y}-1)\times k+j)$,

which is shown in Eq.(5). The impact of random noise N in Eq.(5) can be removed or lessened with the summation operation in Eq.(9). Most importantly, the nonlinear geometric distortion in a down-sampled rescanned image will be reduced to $1/k^2$ of the distortion in a rescanned image without down-sampling. If k is great enough, the effect of nonlinear geometric distortion almost may be ignored.

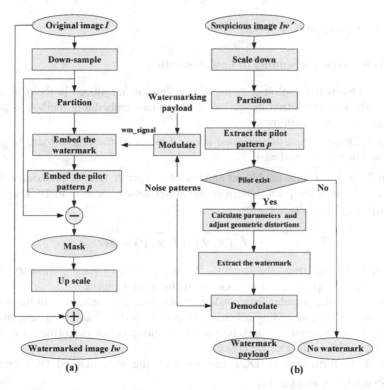

Fig. 1. Block diagram of the proposal scheme. (a) is watermark embedding process. (b) is watermark extraction process.

3 Proposed Scheme

The key idea of our proposed scheme is to embed and detect the watermark in low resolution. This will reduce the effect of nonlinear geometric distortion and ensure the survival of watermark from the print-and-scan process. The diagram of our watermarking scheme is illustrated in Fig. 1.

3.1 Watermark Generation

In our watermarking algorithm, there are two types of watermarking signal. One is pilot pattern p, which is a white noise pattern and embedded in certain blocks. Through the extraction of pilot patterns, the rescanned image can be resynchronized. The other is watermark payload $\{b_i\}$. The watermark payload is often chosen to be orthogonal noise-like patterns or spread spectrum modulated by a basis of n orthogonal noise-like patterns $\{u_i\}$ via,

$$w = \sum_{i=1}^{n} b_i u_i \tag{10}$$

where $b_i \in \{0,1\}$ or $b_i \in \{-1,1\}$ [8].

3.2 Watermark Embedding in Low Resolution

In order to resist the nonlinear geometric distortions introduced in the print-and-scan process, the low-resolution image based watermark embedding method is proposed. The main steps are as follows.

Step 1. Down-sample the original image I to $1/k$ of the original size to get $\bar{I}$. In our experiments, we use $k=10$.

Step 2. Divide $\bar{I}$ into the blocks and perform Discrete Cosine Transform (DCT).

Step 3. Embed the pilot pattern p and watermark pattern w in DCT coefficients f of different blocks (see Fig. 2).

$$f_w(\overline{x}, \overline{y}) = f(\overline{x}, \overline{y}) + \alpha w \tag{11}$$

where α is the watermarking strength. We use $\alpha = 0.6$ in our experiments. And here p is a white noise pattern and w is a set of white noise patterns.

Here we use the popular linear, additive watermarking scheme[8] in the watermark embedding. Other embedding schemes, such as linear[10][11] and nonlinear multiplication watermark scheme[12], may be used to achieve better performance in terms of robustness or invisibility of watermark.

Step 4. Perform inverse DCT transform on the watermarked DCT coefficients f_w of every block and get $\bar{I}_w$.

Step 5. Get the watermark mask as,

$$\textbf{Mask} = \bar{I}_w - \bar{I} \tag{12}$$

Step 6. Up-scale *Mask* with k to the original size and add it to the original image to get the watermarked image I_w.

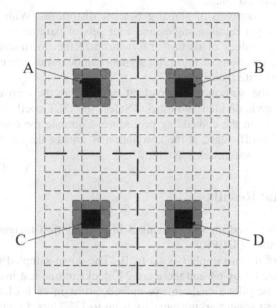

Fig. 2. The partition model. The pilot patterns are embedded in the black blocks. And the watermark patterns are embedded in the other blocks. The gray regions are exhaustive searching windows.

3.3 Watermark Extraction

The watermark extraction procedure includes three steps.

Step 1. Down-sample the rescanned image at different ratio and detect the positions of four pilot patterns by exhaustive searching in the certain windows (see Fig. 2).

Assume R_{img} to be the resolution of the original image and R_{scan} to be resolution of the scanner. It is well known that the difference between R_{img} and R_{scan} will introduce scaling between the original image and the rescanned one. And the scaling parameter is

$$P_{scale} = \frac{R_{scan}}{R_{img}} \tag{12}$$

In order to extract the pilot signal in the low resolution, we also down-sample the watermarked image with $1/k$. According to Section 3.2, the effect of nonlinear geometric distortions can be reduced. So the possible ratio of down-sample is

$$Rate = \frac{R_{img}}{k \cdot R_{scan}} \tag{13}$$

R_{scan} is certain in the extraction process. And although R_{img} is unknown to the detector, the possible value is limited in application. We will exhaustive searching in all possible value of Eq.13. We can get the searching result at which the correlation between the original pilot signal and the extracted one is highest. Down-sample the rescanned image with that result ratio and find the locations of four pilot signals in the down-sampled rescanned image.

Step 2. Get the parameters of global geometric distortions. With the help of four pilot signal, we can get rotation angles and four sets of parameters of translation[7]. We use the average values of them to resynchronize the down-sampled rescanned image. Because the positions of four pilot signals can be gotten correctly, we can get the exact orientation with their help.

Step 3. Decode the watermark payload bit by bit from the recovered image. The down-sampled image is divided into blocks. Select the DCT coefficients of blocks at the same positions as in the embedding procedure. Compute the correlation between u_i and the selected coefficients. If the correlation is greater than a threshold T, then indicate that the bit of watermark b_i is "1".

4 Experimental Results

We evaluate the proposed algorithm using 4 high resolution images from Standard High Precision Pictures (SHIPP).

The parameter of down-sample k is set to 10. The down-sampled image is divided into blocks with size 64-by-64 and there are 24 blocks in one test image. Except four blocks embedded the pilot signals in, we embed 64bits in each of the other twenty blocks. So the largest watermarking capacity is up to 1280 bits. In our application, we embed the same watermark in these twenty blocks and the watermarking capacity is 64 bits.

The PSNRs of the watermarked images are presented in Table 1. From this table, we can find that visual quality of the watermarked images is acceptable. And the watermarked images are shown in Fig. 3.

Table 1. PSNRs of the watermarked images

Image	Bride	Harbor	Wool	Bottle
Resolution	3072-by-4096	4096-by-3072	4096-by-3072	3072-by-4096
PSNR(dB)	36.07	33.49	33.80	33.25

Fig. 3. Watermarked images

We printed the watermarked image out by Ricoh Aficio 2032 and scanned the printed watermarked images by HP scanjet 8250, both common commercial products. We employ different models of Ricoh laser printer and HP scanner to evaluate the robustness of our proposed watermarking algorithm. The resolution of Aficio 2032 is set to 600 dpi. And the scanning resolution is set to 300 dpi and 600 dpi. The detection results are shown in Table 2. The BER of extrcted 64 bits is less than 1.6%. So we can conclude that the proposed algorithm is robust to print-and-scan process.

Table 2. The detection results

| | BER(%) | |
Image	Scanning Resolution = 300 dpi	Scanning Resolution = 600 dpi
Bride	0	0
Harbor	0	0
Wool	0	0
Bottle	1.6	0

We also evaluate the proposed algorithm's robustness against JPEG compression. The watermarked images are compressed under different JPEG quality factors. The quality factors are set to be from 50 to 90 and at intervals of 10. All watermarks can be extracted correctly without any error bit from compressed images.

In the experiments, the watermarking algorithm is implemented with CPU PIV 2.4G, RAM 2G and Matlab 6.5. It takes about 7.89s for the embedding, and takes 216.08s for the extraction despite the very high resolution of images, so it is efficient in computation.

5 Conclusions

In this paper, we analyze the impact of the combination of local random nonlinear geometric distortions and nonlinear pixel value distortions in print-and-scanning process, and a novel print-and-scan resilient image watermarking algorithm is proposed. The watermarks are embedded in the down-sampled low resolution image to reduce the effect of nonlinear distortions in printing process. And the pilot patterns are embedded to be resilient global geometric distortions in print-and-scan process. Theoretical analysis and the experimental results demonstrate that the proposed algorithm is robust to the print-and-scan process and its computational efficiency is high. Future work is to perform more testing on this algorithm by more test images.

Acknowledgement

This work was supported by NSFC (60403045), NSF of Guangdong (04009742), and NSF of Guangzhou (2006Z3-D3041).

References

1. Lin, C.Y., Wu, M., Bloom, J.A., Cox, I.J., Miller, M.L., Lui, Y.M.: Rotation, Scale and Translation Resilient Watermarking for Images. IEEE Transactions on Image Processing 10(5), 767–782 (2001)
2. O'Ruanaidh, J., Pun, T.: Rotation, Scale and Translation Invariant Spread Spectrum Digital Image Watermarking. Signal Processing 66(3), 303–317 (1998)
3. Su, P.C., Jay Kuo, C.C.: Synchronized Detection of The Block-Based Watermark with Invisible Grid Embedding. In: SPIE Photonics West, Security and Watermarking of Multimedia Contents III, San Jose, California (2001)
4. Kutter, M.: Watermarking Resisting to Translation, Rotation and Scaling. In: Proceedings of the SPIE: Multimedia systems and applications, Boston, USA, vol. 3528, pp. 423–431 (1998)
5. Voloshynovskiy, S., Deguillaume, F., Pun, T.: Multibit watermarking robust against local nonlinear geometrical distortions. In: IEEE International Conference on Image Processing 2001, Thessaloniki, Greece, pp. 999–1002 (2001)
6. Lin, C.Y., Chang, S.F.: Distortion Modeling and Invariant Extraction for Digital Image Print-and-Scan Process. In: Int. Symposium on Multimedia Information Processing (1999)
7. Yu, L., Niu, X., Sun, S.: Print-and-Scan Model and the Watermarking Countermeasure. Image and Vision Computing 23, 807–814 (2005)
8. Cox, I.J., Kilian, J., Leighton, T., Shamoon, T.: Secure Spread Spectrum Watermarking for Multimedia. IEEE Transactions on Image Processing 6(12), 1673–1687 (1997)
9. Solanki, K., Madhow, U., Manjunath, B.S., Chandrasekaran, S., El-Khalil, I.: Print and Scan Resilient Data Hiding in Images. IEEE Transactions on Information Forensics and Security 1(4), 464–478 (2006)
10. Cheng, Q., Huang, T.S.: Robust Optimum Detection of Transform Domain Multiplicative Watermarks. IEEE Transactions on Signal Processing 51(4), 906–924 (2003)
11. Kang, X., Zhong, X., Huang, J., Zeng, W.: An Efficient Print-Scanning Resilient Data Hiding Based on a Novel LPM. In: IEEE International Conference on Image Processing 2008, San Digeo, CA, USA (2008)
12. Liu, W., Dong, L., Zeng, W.: Optimum Detection for Spread-Spectrum Watermarking That Employs Self-Masking. IEEE Transactions on Information Forensics and Security 2(4), 457–460 (2007)
13. He, D., Sun, Q.: A Practical Print-Scan Resilient Watermarking Scheme. In: IEEE International Conference on Image Processing 2005, vol. 1, pp. 257–260 (2005)

Adaptive SVD-Based Digital Image Watermarking

Maliheh Shirvanian and Farah Torkamani Azar

Shahid Beheshti University
Evin, 1983963113 Tehran, Iran
maliheh.shirvanian@gmail.com, f-torkamani@cc.sbu.ac.ir

Abstract. Digital data utilization along with the increase popularity of the Internet has facilitated information sharing and distribution. However, such applications have also raised concern about copyright issues and unauthorized modification and distribution of digital data. Digital watermarking techniques which are proposed to solve these problems hide some information in digital media and extract it whenever needed to indicate the data owner. In this paper a new method of image watermarking based on singular value decomposition (SVD) of images is proposed which considers human visual system prior to embedding watermark by segmenting the original image into several blocks of different sizes, with more density in the edges of the image. In this way the original image quality is preserved in the watermarked image. Additional advantages of the proposed technique are large capacity of watermark embedding and robustness of the method against different types of image manipulation techniques.

1 Introduction

Digital data utilization along with the increase popularity of the Internet has facilitated information sharing and distribution. However, such applications have also raised concern about copyright issues and unauthorized modification and distribution of digital data such as digital image, audio and video. Digital watermarking is proposed to solve these problems. Digital watermarking is an adoption of traditional watermarking in digital applications and hides some information in digital media and extracts it whenever needed to indicate the data owner.

In general in a watermarking system, watermark which is usually some information related to the original data or the owner is produced and is embedded in the original data and then the watermarked data is saved or is distributed through computer networks. Considering the application of the watermarking system, the watermark can be removed or extracted from the media in conditions that are necessary.

Watermark can be applied to different sorts of media including digital text, audio, image and video. Because watermarking is chiefly applied to images, in this paper we focus on this field. In recent years several techniques for digital image watermarking is proposed. Each technique attempts to satisfy requirement of watermarking applications with regard to robustness, capacity and imperceptibility. Watermark is said to be robust if it survives common types of image manipulation including compression, rotation, scaling, cropping, filtering, noise, sharpening, blurring and etc. Capacity is the amount of watermarking information that can be embedded in the image. In most

H.J. Kim, S. Katzenbeisser, and A.T.S. Ho (Eds.): IWDW 2008, LNCS 5450, pp. 113–123, 2009.
© Springer-Verlag Berlin Heidelberg 2009

of the applications, distinguishing the original image and the watermarked image should be almost impossible, which implies watermarking imperceptibility. Every watermarking technique suggested for copyright protection should satisfy the robustness and imperceptibility requirements and the capacity should be high unless it reduces the ability of technique to satisfy the other two requirements.

The watermark can be applied to spatial domain [1 - 3]. An alternative to spatial domain watermarking is frequency domain watermarking [4 - 9]. There are also some other techniques which work on singular value decomposition of the image [10 - 14]. Watermark in spatial domain changes luminance of pixels directly, but in frequency domain some characteristics of transformation such as DCT and wavelet are changed. Some other methods compute singular value decomposition of the image matrix and embed watermark by altering singular value (SV) matrix.

Some previous works especially those in singular value decomposition domain are introduced later on. A new method of image watermarking based on singular value decomposition of image is presented which considers human visual system characteristic prior to embedding watermark by segmenting the original image into several blocks of different sizes, with more density in the edges of the image. In this way the original image quality is preserved in the watermarked image. Additional advantages of the proposed technique are large capacity of watermark embedding and robustness of the method against different types of image modification techniques.

2   An Overview on Previous Works

Watermarking techniques are categorized into spatial and frequency domain techniques. Among those in spatial domain a block based spatial watermarking method can be mentioned [11]. In this technique the image is subdivided into 8*8 blocks. Each block is classified to two zones based on the block luminance characteristics. Blocks are also categorized into A and B by a security categorization matrix. Each bit of watermark is spread over one block through the relationship between the categories mean values in each zone following an embedding rule.

The most famous method among frequency domain watermarking techniques is the Cox method [4] and many other techniques are based on this method. In the mentioned method, the Discrete Cosine Transform of the image is computed and some coefficients related to perceptually significant area of image are combined with the watermark according to an embedding equation. A scaling parameter is used to determine the extent of noise added by the watermark to the image.

Another group of techniques work on singular values (SVs) of image matrix. In E. Ganic method [10], watermark is applied to the original image in two layers. In the first layer the image is subdivided into 16*16 bocks and one SV of the watermark image is scaled by a variable scaling factor and added to the largest SV of the original image block. The result is a substitute for the largest SV of the block. In the second layer all SVs of the image are scaled using a constant scaling factor and combined with the SVs of the watermarked and the resulted values are substituted for the original image SVs.

In another technique [11], the watermark image matrix is scaled by a constant scaling factor and is added to the original image's SVs matrix. The image's SVs matrix is a diagonal matrix with singular values of the image in diagonal. Result of this combination is another matrix which its SVs are substituted for the SVs of the original Image.

Chandra has proposed three other methods [12]. The most popular method of them combines SVs of the original image with SVs of the watermark image by a constant scaling factor. The other one spread each bit of the watermark bit stream in one SV of the original image by increasing the SV of the original image by a constant value ,K , whenever the embedding bit is one and decreasing it by K whenever the watermark bit is zero. Chandra has presented another algorithm in which the SVs of the original image are altered whenever the watermark bit is one and are preserved whenever the embedding bit is zero. The alteration takes place by adding a constant value to the original image SVs.

Ganik and Eskicioglu [13] have proposed a SVD based watermarking using Discrete Wavelet Transform (DWT-SVD). In this scheme after decomposition of the host image into four subbands the SVD of the watermark image will be applied to the mentioned subbands. The beneficial point of this algorithm is its robustness since modification will be applied to all subbands. But this fact leads to quality degradation of watermarked image.

The last method introduced here is the one presented by Gorodetsky [14]. In the proposed technique the original image is segmented into 12*12 sub blocks. Then the SVD of each block and the second order norm of the SVs vector related to the block are calculated. The norm value is divided by a quantization step, called d. The result of the division is made odd in case the watermark bit is one and is made even in case the watermark bit is zero. The SV values are reconstructed with new values of the norm and are substituted for the SVs of the block. The same theory is used by Gorodetsky in another technique in which the SVs of each sub-block are divided by the quantization step and the results are made even and odd according to the watermark bits. Then the new SVs are reconstructed and are substituted for the original SVs of the block. A summary of SVD-based methods is given in Table 1 Appendix A.

3 Proposed Method

Previous SVD-based watermarking techniques treat different parts of the image the same, so the edges and the smooth areas of the image accept similar changes. Considering human visual system, which has more sensitivity to the edges, making similar changes to perceptually significant and insignificant area of the image will lead to noticeable alternation in smooth areas and as a result a quality degradation of the image.

Most of the techniques discussed here use constant scaling factor, to scale the watermark and embed it in the image. Considering the large range of watermark SVs, in a combination of SVs of the original image and the watermark image, in some cases the SV of the watermark image is lot smaller than the SV of the original image and in some cases it is lot larger and in some cases there is a little difference. So using a constant scaling factor may sometimes result a considerable change in original image or sometimes it may result a minute change that may cause the watermark to be destroyed or removed due to small attacks.

In addition in the previous techniques bit streams or binary images have been used as watermark in most cases. In these situations the watermark is more vulnerable to intentional or unintentional attacks and the capacity of the information which can be embedded in the image is reduced.

Image quality degradation due to previous techniques along with limited capacity of the watermark in the techniques have motivated the author to propose a simple image watermarking technique with unnoticeable image quality alteration. Additional advantages of the presented technique are large capacity and robustness of the method against different types of image modification techniques.

The presented method is an adaptive technique based on Singular Value Decomposition – SVD with variable scaling factor that use an image as a watermark and embed it in the original image. The watermarking process consists of segmentation of image into smaller blocks and embedding of SVs of the watermark in the original image blocks.

Considering less of human visual system sensitivity in the edges comparing to the smooth areas, the probability of distinguishing the watermarked image from the original image is reduced in those situations that the watermark alter the edges and so the quality of the image is degraded less. In the presented method the image is segmented into blocks with different sizes in the first step. The density of these blocks is more in perceptually insignificant areas comparing to perceptually significant areas. Segmentation of image is done based on Quadtree Decomposition method [15 - 16]. The whole image is considered as one large block and is segmented into four equal size sub-blocks in Quadtree method. The degree of activity, means the difference between lightest and darkest pixel of each sub-block, is calculated. If the activity is larger than a predefined threshold called T_A in a sub-block, the sub-block will be divided into four sub-blocks again. This process will continue on each sub-block until the degree of activity becomes less that T_A in each sub-block.

After segmenting the image into sub-blocks, SVDs of each sub-block is calculated. The SVD for square matrices was discovered independently by Beltrami in 1873 and Jordan in 1874. In recent years, it is massively used in image processing applications. Due to SVD every real matrix A can be decomposed into a product of 3 matrices $A = U\Sigma V^T$, where U and V are orthogonal matrices, $U^T U = I$, $V^T V = I$, and $\Sigma = $ diag $(\lambda_1, \lambda_2, ...)$. The diagonal entries of Σ are called the singular values (SVs) of A, the columns of U are called the left singular vectors of A, and the columns of V are called the right singular vectors of A. This decomposition is known as the Singular Value Decomposition (SVD) of A.

The main theoretical backgrounds of SVD in image watermarking are: (1) Singular values present luminance characteristic of the image themselves (2) Singular values of an image have very good stability. When a small change is added to singular values, great variance of image does not occur. (3) An image can be rebuilt just by a few highest singular values. In the presented method only the largest singular value of the sub-block is altered. Singular values have some other good characteristics as well which leads to more robustness of the watermarked image to common image manipulation such as compression, cropping and etc.

After finding the largest SV of the block, this amount is combined with a SV of the watermark image. Equation (1) shows the combination formula structure. In this equation λ_i is the largest SV of the ith block of the original image, λ_i^* is the largest SV of corresponding block in the watermarked image, λ^w_i is the ith SV of the watermark image and α_i is the variable scaling factor which will be discussed later.

$$\lambda_i^* = \lambda_i + \alpha_i \lambda_i^w \tag{1}$$

Scaling Factor is used to achieve better quality of the watermarked image. Influence of the watermark signal on the original image is controlled by this factor. Using constant scaling factor will lead to SVs of the watermark which have a wide range to be combined with largest SVs of the original image which have a limited range. This situation may lead to large influence of the watermark on the image, in the case the SV of the watermark is large and small influence in the case it is small. Using variable scaling factor which considers original image and watermark characteristics will solve this problem and leads to better watermarked image quality. The proposed scaling factor in the presented technique is defined by Equation (2) where m_i is the block size in which the watermark is going to be embedded, λ^w is the descending vector of watermark image SVs and $\|\lambda^w\|^2$ is the norm of it. b and c are constant parameters for achieving better quality of the watermarked image and are experimentally chosen off-line.

$$\alpha_i = c\frac{m_i}{\|\lambda^w\|^2} + b \tag{2}$$

By using Equations (1) and (2), SVs of the watermark image are placed in the range of zero and one. As the largest SV of the block has direct relation with the block size, the m_i controls the influence of the watermark based of the original image SVs. To have the second order norm of λ^w vector available during extraction, this amount is embedded in the first block of the image during embedding phase according to Equation (3) with a constant scaling factor α.

$$\lambda_1^* = \lambda_1 + \alpha\|\lambda^w\|^2 \tag{3}$$

Extraction of the watermark is done by the Equation (4) which is the inverse of the Equation (1). The second order of the watermark SV vector is done by the   Equation (5) which is the inverse of the Equation (2).

$$\lambda_i^w = \frac{(\lambda_i^* - \lambda_i)}{\alpha_i} \tag{4}$$

$$\|\lambda^w\|^2 = \frac{(\lambda_1^* - \lambda_1)}{\alpha} \tag{5}$$

Number of blocks in the image can be controlled with changing T_A which is the activity threshold and also with setting a maximum and minimum block size in Quadtree method. Therefore, if the number of blocks is more than the number of SVs of the watermark image, the watermark can be embedded redundantly in the image.

4   Experimental Results

The chosen image in the performed experiments is the famous cameraman image in 512x512 size and the watermark image is @ picture in 512x512 size consisting both edges and smooth areas. T_A which is the activity threshold in Quadtree method is 0.3, maximum block size is 32x32 and the minimum block size is 4x4. The constant "b" in the scaling factor formula is zero and the constant "c" is 6. In this experiment 3741 blocks are generated and therefore 3741 SVs of watermark image can be embedded. Figure 1 relates to the results of the experience for proposed method and shows the original image, watermark, watermarked image, extracted watermark, image blocks and difference between the original image and the watermarked image.

Fig. 1. Experimental result for the proposed method: a – Original image, b – Watermarked image, c – Image blocks, d – Watermarked image, e – Extracted watermark, f – Difference between original and watermarked image

Fig. 2. Experimental result for the E. Ganic method: a – Original image, b – Watermarked image, c – Image blocks, d – Watermarked image, e – Extracted watermark, f – Difference between original and watermarked image

Experiments are also performed for the Chandra [12] and Ganic [13] Methods. The image used here is cameraman photo, the watermark is the picture of @ and the constant scaling factor used here is set to 0.2.

Figure 2 shows results including watermarked image, extracted watermark and the difference between the original image and the watermark for the Chandra method. The amount of watermark that can be embedded in Chandra method is 512 singular values. The watermarked image in the Chandra method is much brighter than the original image and the image quality is obviously degraded due to usage of constant

Table 1. Calculating Amount of Error

Difference Distortion Metrics	Formula	Proposed Method	Chandra Method
Maximum Distortion	$MD = \max_{m,n} \left| I_{m,n} - \tilde{I}_{m,n} \right|$	25.712	56.468
Average Absolute Error	$AD = \dfrac{1}{MN} \sum_{m,n} \left| I_{m,n} - \tilde{I}_{m,n} \right|$	9.8082e-5	2.1541e-4
Mean Square Error	$MSE = \dfrac{1}{MN} \sum_{m,n} \left(I_{m,n} - \tilde{I}_{m,n} \right)^2$	13.113	1275.7
Signal to Noise Ratio	$SNR = \sum_{m,n} I_{m,n}^2 \Big/ \sum_{m,n} \left(I_{m,n} - \tilde{I}_{m,n} \right)^2$	5.4469e+5	5.5992e+3
Peak Signal to Noise Ratio	$PSNR = MN \max_{m,n} I_{m,n}^2 \Big/ \sum_{m,n} \left(I_{m,n} - \tilde{I}_{m,n} \right)^2$	1.0325e+6	1.0614e+4
Masked Peak Signal to Noise Ratio	$MPSNR = 10 \log_{10}^{\left(255^2 / MSE \right)}$	36.954	17.073

value in combining watermark and the image and also due to the fact that different parts of the image, edges and the smooth areas, are treated the same without paying attention to human visual system characteristics.

Calculation of mathematical parameters such as Signal to Noise Ratio, Mean Square Error and etc [17 - 18] shows the amount of error in the watermarked image compared to the original image. Table 1 shows result of calculating mentioned parameters in proposed method and the Chandra method.

To calculate robustness of the proposed method and the Chandra method against image manipulation, the watermarked image is compressed using JPEG up to 85% and the watermark is extracted from the compressed watermarked image. The extracted watermark for the proposed method and the Chandra method are shown in figure 3. The mean square error which shows difference between original watermark and the extracted one is 0.78883 for the proposed method and as the figure shows the watermark is extracted completely. The error for the Chandra method is 48.396 which is much larger than the error in the proposed method.

To calculate robustness of the technique against cropping, 100 pixels from right and bottom of the watermarked image are removed and the cropped watermarked

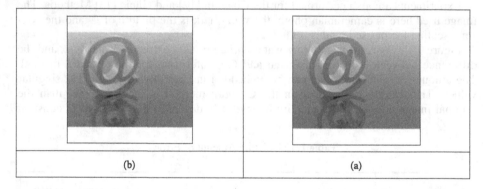

Fig. 3. Experimental result for the Chandra method: a – watermarked image, b – extracted watermark, c – difference between original and watermarked image

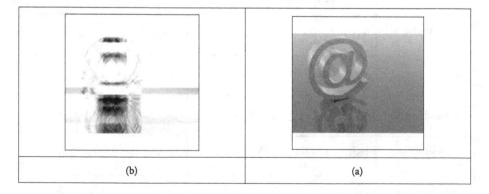

Fig. 4. Extracted watermark from compressed watermarked image: a – in proposed method, b – Chandra method, c – E. Ganic Method

image is used for watermark extraction. The same experiment is also performed for the Chandra method. The extracted watermark for proposed method and the Chandra method are shown in Figure 4. The mean square error which shows difference between original watermark and the extracted one is 6.2577 for the proposed method and as the figure shows the watermark is extracted with a little bit quality degradation. The error for the Chandra method is 13993 which is much larger than the calculated error for the proposed method.

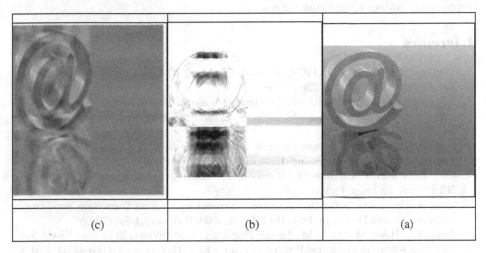

(c)	(b)	(a)

Fig. 5. Extracted watermark from cropped watermarked image a – in proposed method, b – in Chandra method, c – E. Ganic Method

Table 2. Results of image manipulation on extracted watermark

Action	MSE – Proposed Method	MSE – Chandra
JPEG – 85% Compression	0.78883	48.396
Cropping (Delete 100 Pixels from Right and Bottom)	6.2577	13993
Adding White Gaussian Noise – 5dB S/N	583.26	0.51119
Blurring	8.7043e+007	544.78
Flip - Vertical	1.412e+011	9.2968e+013
Flip – Horizontal	8.8074e+012	4.121e+011
Rotation - 15°	9.1119e+014	8.5629e+014
Histogram Normalization(Lena as Original Image, @ as Watermark)	31891	21.014

In addition to compression and cropping other common image manipulation are used to test the robustness of the technique. Mean square error calculation results are summarized in table 2 which shows difference between original watermark and the extracted one for the proposed method and the Chandra method. Results show good resistance of the proposed method against JPEG compression, cropping and white Gaussian noise.

5 Summary

In this paper an adaptive image watermarking method based on image singular value decomposition is presented. In this method the image is segmented into several blocks of different sizes, with more density in the edges of the image. In this way the original image quality is preserved in the watermarked image. Additional advantages of the technique are large capacity and robustness of the method against JPEG compression, cropping and adding white Gaussian noise.

References

1. Darmstaedter, V., Delaigle, J.F., Quisquater, J.J., Macq, B.: Low Cost Spatial Watermarking. Computers & Graphics 22(4) (1998)
2. Kimpan, S., Lasakul, A., Kimpan, C.: Adaptive Watermarking in Spatial Domain for Still Images. In: Proceedings of the International Conference on Information and Knowledge Engineering (2004)
3. Nikolaidis, N., Pitas, I.: Robust Image Watermarking in the Spatial Domain. In: European Association for Signal Processing (EURASIP), vol. 66(3) (1998)
4. Cox, I.J., Kilian, J., Leighton, T., Shamoon, T.: Secure Spread Spectrum for Multimedia. IEEE Trans. on Image Processing 6(12) (1997)
5. Guo, H., Georganas, N.D.: Digital Image Watermarking for Joint Ownership Verification without a Trusted Dealer. In: Proc. IEEE ICME 2003, Baltimore, USA (2003)
6. Jung, Y.J., Hahn, M., Ro, Y.M.: Spatial Frequency Band Division in Human Visual System Based-Watermarking. In: Petitcolas, F.A.P., Kim, H.-J. (eds.) IWDW 2002. LNCS, vol. 2613, pp. 224–234. Springer, Heidelberg (2003)
7. Yu, D., Sattar, F.: A New Blind Watermarking Technique Based on Independent Component Analysis. In: Petitcolas, F.A.P., Kim, H.-J. (eds.) IWDW 2002. LNCS, vol. 2613, pp. 51–63. Springer, Heidelberg (2003)
8. Guitart Pla, O., Lin, E.T., Delp, E.J.: A Wavelet Watermarking Algorithm Based on a Tree Structure. In: Proceedings of the SPIE International Conference on Security, Steganography, and Watermarking of Multimedia Contents VI, San Jose, vol. 5306 (2004)
9. Hu, A.T.S., Chow, A.K.K., Woon, J.: Robust Digital Image-in-Image Watermarking Algorithm Using the Fast Hadamard Transform. In: IEEE International Symposium on Circuits and Systems, Thailand, vol. 3 (2003)
10. Ganic, E., Zubair, N., Eskicioglu, A.M.: An Optimal Watermarking Scheme Based on Singular Value Decomposition. In: Proceedings of the IASTED International Conference on Communication, Network, and Information Security (CNIS 2003), Uniondale, NY (2003)
11. Liu, R.Z., Tan, T.N.: SVD-Based Watermarking Scheme for Protecting Rightful Ownership. IEEE Trans. on Multimedia 4(1) (2002)
12. Chandra, D.V.S.: Digital image watermarking using singular value decomposition. In: Proceedings of 45th IEEE Midwest Symposium on Circuits and Systems, Tulsa OK (2002)
13. Ganic, E., Eskicioglu, A.M.: Robust Embedding Of Visual Watermarks Using DWT-SVD. Journal of Electronic Imaging, 1–13 (2005)
14. Gorodetskii, V.I., Samoilov, V.I.: Simulation-Based Exploration of SVD-Based Technique for Hidden Communication by Image Steganography Channel. In: Proceedings of the 2nd International Workshop on Mathematical Methods, Models, and Architectures for Computer Network Security MMM–ACNS, Russia (2003)
15. Smith, J.R., Chang, S.F.: Quadtree Segmentation for Texture-Based Image Query. In: Proc. ACM Intern. Conf. Multimedia, San Francisco, CA (1994)

16. Remias, E., Sheikholeslami, G., Zhang, A.: Block-Oriented Image Decomposition and Retrieval in Image Database Systems. In: The 1996 International Workshop on Multi-media Database Management Systems, New York (1996)
17. Kutter, M., Petitcolas, F.A.P.: Fair Evaluation Methods for Image Watermarking Systems. Journal of Electronic Imaging 9(4) (2000)
18. Petitcolas, F.A.P., Anderson, R.J.: Evaluation of Copyright Marking Systems. In: IEEE Multimedia Systems 1999, Italy (1999)

Appendix A

Table 3. Summary of SVD-based watermarking techniques

	Description	Scaling Factor	Block No	Capacity	Advantages	Disadvantages
Emir Ganic et al	First Layer: image segmentation into 16x16 blocks, combination of the largest SV of the block with one of SVs of watermark with the formula $\lambda_{max}^w = \lambda_{max} + \alpha_i \lambda_{wi}$ Second Layer: Combination of SVs of the image with SVs of the watermark with the formula $\lambda_i^w = \lambda_i + \alpha \lambda_{wi}$	First Layer: Variable Scaling Factor $\alpha_i = c\frac{\lambda_{max}}{\lambda_{wi}}$ c = 0.05 Second Layer: Constant Scaling Factor α=0.1	First Layer: 16x16 Second Layer: N/A	1024 Singular Values	Robust to JPEG, Rotation, Blurring, Noise, Cropping and Rescaling	Image Quality Degradation due to double watermark insertion and non adaptive image segmentation
Ruizhen Liu	Computing SVD of original image and combining diagonal SV matrix (S) with watermark matrix (W) with the formula S+αW, substituting SVs of the resulted matrix with SVs of the original image	Constant scaling factor α=0.2	N/A	2500 integer or a 50x50 picture	Robust to JPEG, Rotation, Blurring, Cropping and Noise	Image Quality Degradation due to usage of constant scaling factor and ignoring image segmentation
Calindra 1	Combination of SVs of image with SVs of watermark with the formula $\lambda_{yi} = \lambda_{xi} + \alpha_i \lambda_{wi}$	Constant scaling factor α=0.2	N/A	4096 bits	Robust to JPEG	Image Quality Degradation due to usage of constant scaling factor and non adaptive image segmentation and small capacity
Calindra 2	Segmentation of the image into 8x8 blocks and embedding each bit of watermark in one block with one the following formulas: $\lambda_{yi} = \lambda_{xi} + \alpha \omega_b$ non − adaptive $\lambda_{yi} = \lambda_{xi} + \alpha \lambda_{xi} \omega_b$ adaptive	Constant scaling factor α=0.2 in adaptive scheme and α=0.2λ_{xi} in non adaptive scheme	8x8	4096 bits	Robust to JPEG	Image Quality Degradation due to usage of constant scaling factor and non adaptive image segmentation and small capacity
Calindra 3	Segmentation of the image into 8x8 blocks and embedding each bit of watermark in one block with the formula $\lambda_{yi} = \lambda_{xi} + K$ K is positive whenever watermark bit is one and negative whenever watermark bit is zero.	N/A	8x8	4096 bits	Robust to JPEG	Image Quality Degradation due to usage of constant scaling factor and non adaptive image segmentation and small capacity
Gorodetsky 1	Segmentation of the image into 12x12 blocks, choosing a quantization step d, and making reminder of largest SV divided by d even or odd based on watermark bit. Reminder would be even whenever the watermark bit is one and odd otherwise.	Three constant scaling factors: d(red)=46 d(green)=22 d(blue)=52	12x12	1920 0 bits in each color page	Robust to JPEG and high capacity	Constant scaling factor , non adaptive image segmentation and complex computation
Gorodetsky 2	Segmentation of the image into 12x12 blocks, choosing a quantization step d, and making reminder of second order norm of the watermark singular values vector divided by d even or odd based on watermark bit. Reminder would be even whenever the watermark bit is one and odd otherwise.	Three constant scaling factors: (red)=40 d(green)=24 d(blue)=48	12x12	1920 0 bit in each color page	Robust to JPEG and high capacity	Constant scaling factor and complex computation

Robust Audio Watermarking Based on Log-Polar Frequency Index

Rui Yang, Xiangui Kang, and Jiwu Huang

School of Information Science and Technology,
Sun Yat-sen University, Guangzhou, China, 510275
{isskxg,isshjw}@mail.sysu.edu.cn

Abstract. In this paper, we analyze the audio signal distortions introduced by pitch-scaling, random cropping and DA/AD conversion, and find a robust feature, average Fourier magnitude over the log-polar frequency index(AFM), which can resist these attacks. Theoretical analysis and extensive experiments demonstrate that AFM is an appropriate embedding region for robust audio watermarking. This is the first work on applying log-polar mapping to audio watermark. The usage of log-polar mapping in our work is basically different from the existing works in image watermarking. The log-polar mapping is only applied to the frequency index, not to the transform coefficients, which avoids the reconstruction distortion of inverse log-polar transform and reduces the computation cost. Comparison with the existing methods, the proposed AFM-based watermarking scheme has the outstanding performance on resisting pitch-scaling and random cropping, as well as very approving robustness to DA/AD conversion and TSM (Time-Scale Modification). The watermarked audio achieves high auditory quality. Experimental results show that the scheme is very robust to common audio signal processing and distortions introduced in Stirmark for Audio.

1 Introduction

Synchronization is important for detecting watermarks especially when the audio suffers from geometrical attacks [1]. Most of the existing audio watermarking schemes are often position-based. That is, the watermark bits are embedded into specific positions and later extracted from these positions. However, some audio processing operations, especially random cropping, time-scaling, pitch-scaling and DA/AD conversion, make watermarks shift in position, and thus fail the watermark extraction. Synchronization attacks are still challenging audio watermarking.

Up to date, there have been many works aiming at solving the synchronization problem. Haitsma et al. [2] proposed an algorithm by slightly modifying the magnitudes of the Fourier coefficients. It was reported to be robust to ±4% of TSM and DA/AD conversion. But the performance to resist cropping was not mentioned. Kirovski et al. [3] applied redundant-chip coding to spread spectrum watermarking. Only the central sample of each expanded chip was detected and

H.J. Kim, S. Katzenbeisser, and A.T.S. Ho (Eds.): IWDW 2008, LNCS 5450, pp. 124–138, 2009.

used for computing correlation. The authors claimed that the method could resist synchronization attacks. However, the experimental results on time-scaling and pitch-scaling were not presented in the papers. Tachibana et al. [4,5] calculated and manipulated the magnitudes of segmented areas in the time-frequency plane using short-time DFTs. The detector correlated the magnitudes with a pseudo-random array in the time-frequency plane. The method is robust against time-scaling and pitch-scaling up to ±8%. Li et al. [6] proposed an algorithm by embedding the watermark in the perceptually important localized regions. The algorithm is robust against certain synchronization attacks, such as ±15% of TSM. However, the method has its inherent limitation. It only works well on the audio with obvious rhythm. Xiang et al.[7] found that the relative relation in the number of samples among different bins in audio histogram and the audio mean were robust to the TSM attacks. A robust audio watermarking algorithm based on the two statistical features was proposed by using the histogram specification. The algorithm is robust against synchronization attacks, such as ±30% of TSM and random cropping. But it does not perform good to resist pitch-scaling.

However, most existing works focus on solving the time-scaling problem, and pitch-scaling hasn't attracted enough researchers' eyes. To the best of our knowledge, the most approving performance against pitch-scaling is only up to ±8%[5]. This doesn't fulfill the requirement of IFPI (International Federation of the Phonographic Industry) which is up to ±10%[8]. As pitch-scaling is a very common manipulation on audio, it is necessary to design robust audio watermarking method against pitch-scaling. Since the synchronization attack is always not single, the proposed method in this paper also considers on another two common attacks, random cropping and DA/AD conversion. Via analyzing the distortions introduced by pitch-scaling, random cropping and DA/AD conversion, we find a robust feature, average Fourier magnitude over the log-polar frequency index, which can resist these attacks. Although the concept "log-polar" has been introduced in robust image watermarking which is invariant to RST (rotation, scaling, and translation) by Ruanaidh et al.[9] a few years ago, log-polar mapping applied to audio watermarking hasn't been reported. The usage of log-polar mapping in our work is basically different from Ruanaidh's method. The log-polar mapping is only performed on the frequency index, not on the transform coefficients, which avoids the reconstruction distortion of inverse log-polar transform and reduces the computation cost.

The rest of this paper is organized as follows. In Section 2, we demonstrate *AFM* is very robust to synchronization attacks via theoretical analysis and extensive experiments. Based on selecting *AFM* as the embedding region, we present the embedding and detection strategies of our watermarking method in Section 3. A schematic explanation and experimental results are given in Section 4. Finally, the conclusions and discussions are shown in Section 5.

2   Motivation and Selection of Embedding Region

In this Section, we will focus on three types of synchronization attacks, random cropping, pith-scaling and DA/AD conversion. We review the degradations introduced by these three attacks, and analyze why the embedding features we select can resist the signal degradations.

2.1 Three Types of Synchronization Attacks

Random cropping drops some samples of the original audio. Here is a common example. D samples of the original signal $s = [s_1 \cdots s_k \cdots s_{k+D} \cdots s_N]$ are removed away, and the signal becomes $s' = [s_1 \cdots s_k, s_{k+D+1} \cdots s_N]$. So if the watermarking strategy is sample-based or segment-based, the watermark bits embedded in samples $[s_{k+1} \cdots s_{k+D}]$ will be lost.

Pitch scaling [10] is a very common processing to change the base frequency without changing the speed of audio signal. It may be described as follows:

$$f' = \beta \cdot f \tag{1}$$

where β stands for pitch-scaling factor. f and f' are the original frequency and the scaled frequency, respectively. We denote $S(f)$ as the original frequency magnitude, the pitch-shifted frequency magnitude becomes

$$S'(f') = S'(\beta \cdot f) = S(f) \tag{2}$$

Unlike pitch-scaling, DA/AD conversion still has not an accurate model although some works tried to solve it [11]. Some silence samples are often added at the beginning and the end of the original audio during DA/AD conversion, which makes watermark extraction in time domain difficult. Also DA/AD conversion introduces amplitude scaling and background noise. A successful DA/AD conversion should make sure that after these degradations the re-recorded audio is still tolerant to human auditory system (HAS). Frequency components play an important part in HAS, so perceptually similar audio means that their frequency components have strong correlation.

2.2 Selection of Embedding Region

To resist random cropping, the watermarking strategy must be global, so we perform global Fourier transform on the audio clip. As the length of audio clip varies after random cropping and DA/AD conversion, the frequency index must be normalized after global Fourier transform. Until now, we have taken considerations on resisting samples mismatch (in the time domain) introduced by synchronization attacks, but not on frequency scaling problem.

A powerful tool to deal with the scaling problem is log-polar transform. A logarithm could convert the scaling into the shifting in the logarithm axis. That is, taking the logarithm of Eq. (1), the equation is rewritten as follows:

$$\log_a f' = \log_a \beta + \log_a f \tag{3}$$

where $log_a(.)$ is the common logarithm of base a. From Eq. (3), we see that scaling by β is converted into shifting by $log_a \beta$.

As amplitude scaling is almost unavoidable during attacks, we select a correlation-based watermarking strategy which can resist amplitude scaling. Thus if we apply the logarithm on the normalized frequency axis, the correlation-based detection would still work well after pitch-scaling. Based on the above considerations, we propose a correlation-based audio watermarking method by applying discrete log-polar mapping on the normalized frequency index. The host feature is the average Fourier magnitude over the log-polar frequency index.

However, the log-polar mapping(LPM) in Eq. (3) cannot be directly implemented because the signal is in discrete form. In the following we design a LPM suitable for discrete-time signal. Given a signal $s = [s_1 \cdots s_N]$, we perform global Fourier transform on it, and get the Fourier magnitude $S(f), -N/2 \leq f \leq N/2$. After selecting part of the normalized frequency index $[d \cdot N, 2d \cdot N], 0 < d < 0.25$, we apply LPM on this frequency interval. All frequencies in the interval are mapped to a much smaller interval $[0, M-1]$, where M is a random number specified by users.

$$r = floor(\log_a \tfrac{f}{R}) + M/2, f \in [d \cdot N, 2d \cdot N]$$
$$a = 2^{1/M} \tag{4}$$
$$R = \sqrt{2}d \cdot N$$

The AFM is calculated as follows:

$$AFM(r) = \tfrac{1}{(f_2 - f_1)} \int_{f_1}^{f_2} S(f)df, 0 \leq r < M$$
$$f_1 = \min\{f | floor(\log_a \tfrac{f}{R}) + M/2 = r\} \tag{5}$$
$$f_2 = \max\{f | floor(\log_a \tfrac{f}{R}) + M/2 = r\}$$

2.3 Effectiveness of AFM

In the previous subsection, we have shown that why the AFM is selected as embedding region. How about the ability of this region to resist synchronization attacks? We randomly select a 30s audio clip, and apply random cropping (cropping 20%), DA/AD conversion (100cm between speaker and microphone) and pitch scaling (+5%) on it, respectively. Fig.1 (b) shows the AFM of the audio clip after random cropping 20%. Its shape is almost identical with the AFM of original audio in Fig.1 (a), as well as its amplitude. The correlation coefficient of AFM between the cropped audio and the original one is 0.9837, which means that AFM is stable after random cropping. The case of DA/AD conversion is illustrated in Fig.1 (c). As we see, the AFMs in Fig.1(a) and Fig.1(c) are with very similar shapes although with different amplitudes. The correlation coefficient between them, 0.8735, shows that they have very strong correlation. As shown in Fig.1 (d), the AFM of the audio clip after pitch-shift +5% is just the shifted version of Fig. 1(a). We use correlation function to align these two AFMs, and calculate the correlation coefficient (0.9120 in this work) between the aligned parts.

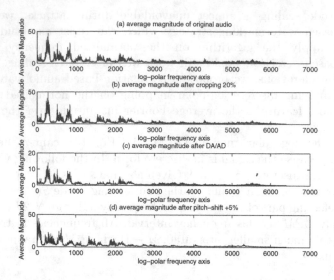

Fig. 1. *AFM* of the original audio (a), random cropping 20% one (b), DA/AD conversion one (c) and pitch scaling 5% one (d)

From Fig.1 we observe that *AFM* may be a robust feature for watermarking, but we still need extensive tests and theoretical analysis to confirm this observation. We pick 100 audio clips (each is 30s long) from ten music classes [12]. We randomly crop 5% (1.5s) and 20% (6s) of all 100 clips, and get 100 clips of 28.5s and another 100 clips of 24s. Fig.2(a) shows the correlation coefficients of *AFM* between the original ones and the attacked ones. As we see, all correlation coefficients are above 0.9 and most are above 0.95, no matter cropping 5% or cropping 20%. The mean values of these coefficients after cropping 5% and after cropping 20% are 0.9851 and 0.9944, respectively.

Let's go through why *AFM* is robust to random cropping. For random cropping, it means some samples lost in time domain, but in frequency domain it only introduces tiny fluctuation. In the following, we will give the theoretical proof that the distortion on the normalized frequency components introduced by random cropping is puny.

Support that S is the Fourier transform of $s = [s_1 \cdots s_k \cdots s_{k+D} \cdots s_N]$, and S' is the Fourier transform of $s' = [s_1 \cdots s_k \cdots s_{k+D} \cdots s_N]$, where s' is generated from s after cropping away D samples. We have

$$S(k) = \sum_{n=1}^{N} s(n)e^{-j2\pi(k-1)(\frac{n-1}{N})}, -\frac{N}{2} \leq k \leq \frac{N}{2}$$

$$S'(k') = \sum_{n=1}^{N'} s'(n)e^{-j2\pi(k'-1)(\frac{n-1}{N'})}, -\frac{N'}{2} \leq k' \leq \frac{N'}{2} \tag{6}$$

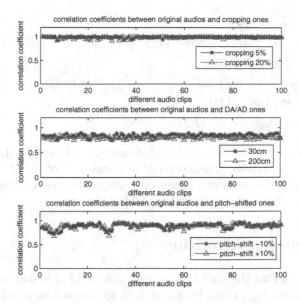

Fig. 2. Correlation coefficients of *AFM* between original audios and distorted ones, (a) random cropping, (b) DA/AD conversion, (c) pitch scaling

After the frequency index is normalized, the k-th element of S' is

$$S'(\tfrac{k}{N} \cdot N') = \sum_{n=1}^{N'} s'(n)e^{-j2\pi(\tfrac{k}{N} \cdot N'-1)(\tfrac{n-1}{N'})} = \sum_{n=1}^{N'} s'(n)e^{-j2\pi(\tfrac{kN'-N}{N})(\tfrac{n-1}{N'})}$$

$$kN' - N \approx N'(k-1)$$

$$\Rightarrow S'(\tfrac{k}{N} \cdot N') = \sum_{n=1}^{N'} s'(n)e^{-j2\pi(\tfrac{N'(k-1)}{N})(\tfrac{n-1}{N'})} \approx \sum_{n=1}^{N'} s'(n)e^{-j2\pi(k-1)(\tfrac{n-1}{N})} \quad (7)$$

$$\Rightarrow |S'(\tfrac{k}{N} \cdot N')| \approx |\sum_{n=1}^{N'} s'(n)e^{-j2\pi(k-1)(\tfrac{n-1}{N})}| \approx |S(k)|$$

The case of DA/AD conversion is also tested with the 100 audio clips. We output all 100 audio clips with speakers and capture them with a microphone at two distances, 30cm and 200cm. The correlation coefficients of *AFM* between the original ones and the re-recorded ones are shown in Fig.2(b). The mean values of these coefficients are 0.8412 and 0.7973 for 30cm-distance and 200cm-distance, respectively.

We also test the robustness of *AFM* against pitch-scaling with the 100 clips. We apply pitch-scaling -10% and +10% to 100 clips, respectively. The correlation coefficients of *AFM* between the original ones and the pitch-shifted ones are shown in Fig.2(c). We observe that most of these correlation coefficients are larger than 0.9, and the mean values are 0.8537 and 0.8456, respectively.

Here we will give the theoretical proof that *AFM* is stable to pitch-scaling. Denote the *AFM* of the original audio as $AFM(r)$, and the pitch-shifted one as $AFM'(r')$. We have

$$r = floor(\log_a \tfrac{f}{R}) + M/2$$
$$r' = floor(\log_a \tfrac{f'}{R}) + M/2 = floor(\log_a \tfrac{\beta \cdot f}{R}) + M/2 = r + floor(\log_a \beta)$$
$$AFM(r) = \tfrac{1}{(f_2 - f_1)} \int_{f_1}^{f_2} S(f)df$$
$$AFM'(r') = \tfrac{1}{f_2' - f_1'} \int_{f_1'}^{f_2'} S'(f')df' = \tfrac{1}{\beta(f_2 - f_1)} \int_{\beta \cdot f_1}^{\beta \cdot f_2} S(f)d(\beta \cdot f) \qquad (8)$$
$$= \tfrac{1}{(f_2 - f_1)} \int_{f_1}^{f_2} S(f)df = AFM(r)$$
$$AFM'(r + floor(\log_a \beta)) = AFM(r)$$

Via theoretical analysis and extensive experiments, we show that AFM is robust to synchronization attacks. The correlation coefficient of AFM between the original audio and the attacked one always approximates 1. This provides a solid foundation for our proposed watermarking method.

3   Watermarking Algorithm

Based on the point that the correlation coefficient of AFM is still close to 1 after synchronization attacks, a multi-bit watermark scheme aiming at solving random cropping, pitch-scaling and DA/AD conversion attacks is proposed. The basic idea is to embed watermark bits into the AFM and utilize correlation-based detection when extracting watermark bits.

3.1   Watermark Embedding

The detailed embedding procedures are shown in Fig. 3. The hidden bit stream consists of two parts, multi-bit information $m\{m_i; m_i \in \{-1, +1\}, 0 \leq i < L\}$ and

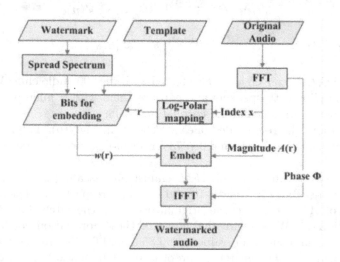

Fig. 3. Watermark embedding framework

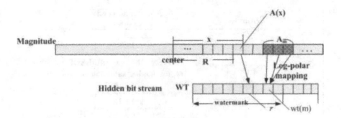

Fig. 4. Log-polar mapping used during watermark embedding

template $T\{T_n; T_n \in \{-1,+1\}, 0 \leq n < N_T\}$ generated by a key. With a N_p-bit bipolar PN sequence $p\{p_j; p_j \in \{-1,+1\}, 0 \leq j < N_p\}$, each m_i of m is direct sequence spread spectrum ($DSSS$) by p, where '1' is encoded as spread spectrum sequence $W_i\{w_{(i-1) \cdot N_p + j}; w_{(i-1) \cdot N_p + j} \in \{-1,+1\}, 0 \leq j < N_p\} = +1 \times p$, while '0' as $W_i = -1 \times p$. Then we obtain the spread-spectrum information bits $W\{w_i; w_i \in \{-1,+1\}, 0 \leq j < L \cdot N_p\}$. Spread-spectrum information bits W and template T are conjoined to form watermark $WT\{wt(m); 0 \leq m < M\}$, where $M = L \cdot N_p + N_T$.

Apply DFT on the original audio signal $s(n)(0 \leq n < N)$, and obtain the magnitude $A(x)(-N/2 \leq x \leq N/2)$ and the phase $\phi(x)(-N/2 \leq x \leq N/2)$. Performing discrete log-polar transform to the index x, we obtain the discrete log-polar coordinate r which corresponds to the index of watermark.

$$r = floor(\log_a \frac{|x|}{R}) + D_{offset}$$
$$R = f_n \cdot N \tag{9}$$
$$a^{-M/2} < \frac{|x|}{R} \leq a^{M/2}$$

where $a = 2^{1/M}$ is the base of logarithm, and f_n is a normalized frequency used as the embedding center. Modify the magnitude $A(x)(-N/2 \leq x \leq N/2)$ to embed the watermark with its corresponding index r.

$$A'(x) = A(x) \times (1 + \alpha \times WT(r)) \tag{10}$$

As shown in Fig.4, there are several magnitude coefficients to host a same watermark bit. Use the modified magnitude $A'(x)$ and the original phase $\phi(x)$ to construct watermarked audio signal $s'(n)$, and the embedding process ends.

$$s' = IDFT(A' \cdot e^{j\phi}) \tag{11}$$

3.2   Watermark Extraction

Fig.5 demonstrates the framework of watermark detection. First, apply DFT on the watermarked audio signal $s'(n)(0 \leq n < N')$, and obtain the magnitude $A'(x)(-N'/2 \leq x \leq N'/2)$. Then perform discrete log-polar transform to the

Fig. 5. Watermark detection framework

index x, and average the magnitude $A'(x)$ whose discrete log-polar coordinate are the same. Denote the average magnitude as $AFM(r)(0 \leq r < M')$.

$$r = floor(\log_a \frac{|x|}{R'}) + D_{offset}$$
$$R' = f_n \cdot N' \qquad\qquad (12)$$
$$a^{-M'/2} < \frac{|x|}{R'} \leq a^{M'/2}$$

where M' is greater than M (i.e. $M' = 2M$) because the track interval is chosen to be larger than the watermark embedding interval, $a = 2^{1/M}, D_{offset} = M'/2$.

Computing the correlation between the template T and the average magnitude AFM, we can determine the watermark positions. One way to track the location of watermark is to perform exhaustive shifting along the r axis and calculate the cross correlation between T and AFM, then find the maximum correlation to locate the match position. However the computation cost for this exhaustive searching is heavy. A fast way to search the maximum correlation value is using correlation theorem. Append T with zeros to the same size of AFM to obtain $g(m)(0 \leq m < M')$, the correlation between them is:

$$C(k) = AFM(m) * g(m) = IDFT(FAFM(u) \cdot G^*(u)), 0 \leq k \leq M' \qquad (13)$$

where $FAFM(u) = DFT(AFM(m))$, $G(u) = DFT(g(m))$. $G^*(u)$ is the conjugate of $G(u)$.

Pick N_p magnitude coefficients in order from AFM to a sequence W_i', which corresponds to the embedded spread spectrum sequence W_i from WT, and correlate with the original PN sequence p. If the correlation is larger than 0, the extracted bit is decided to be '1', else '0'. The watermark is retained and the extraction process ends.

4   Analysis and Experimental Results

4.1   Schematic Explanation

Here is the schematic explanation of log-polar mapping applying to the frequency index. We analyze the variety of the watermark during embedding and extraction. Fig.6 (a) is the original watermark signal used for embedding. If we embed the watermark signal into the Fourier magnitude of audio one-to-one and same-scale, the original watermark will suffer the same distortion as the audio signal. When synchronization attacks are applied to the watermarked audio, the Fourier magnitude of the audio will fluctuate and the frequency index will be scaled. The case of the watermark is the same. So after synchronization attacks the survival watermark is shown as Fig.6 (b). Obviously the original watermark and the survival watermark are not correlative. Correlation-based method would not extract watermark bits correctly.

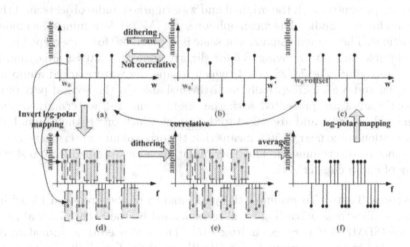

Fig. 6. Schematic explanation of robustness introduced by log-polar mapping

However, the case is different at all when applying log-polar mapping in watermark embedding and extraction. We apply inverse log-polar mapping to watermark index and generate embedding positions in DFT magnitude, as shown in Fig.6 (d). This means one watermark bit will be embedded in multiple magnitudes. After attacks there are some fluctuations in magnitudes, as shown in Fig.6 (e). And Fig.6 (f) is the average version from Fig.6 (e). Before watermark extraction we utilize log-polar mapping to retrieve the watermark index, and we can get Fig.6 (c). It is obvious that signals between Fig.6 (a) and (c) have strong correlation, and correlation-based extraction will be very effective. The watermark bits can be extracted from the survival watermark signal correctly.

4.2   Experimental Results

We test our algorithm with 15 audio clips from SQAM(Sound Quality Assessment Material)[14]. They are WAV format, mono, 44.1 kHz sampling frequency, 16-bit. For each audio clip, a watermark of 72 bits information is *DSSS* by a 32-bit *PN* sequence, and totally 2304 bits data are embedded into it. Watermark detection performance is evaluated by the bit error ratio (BER) of extracted watermarks.

Quality of Watermarked Audio. The objective quality is measured by SNR and ODG (Objective Difference Grade). The SNRs of 15 clips are all over 40dB which are far beyond 20dB requested by IFPI [8], and the average ODG performed by EAQUAL 0.1.3alpha [13] is -0.93. This means that the watermarked audio are similar to the original ones in objective test.

Subjective quality evaluation of the watermarking method has been done by asking 10 persons to listen the 15 audio clips. In the first part of the test, the persons are presented with the original and watermarked audio objects, and then give scores for each audio. The mean opinion score (MOS) determines the amount of distortions. The 5-point impairment scale is applied, 5.0 for imperceptible, 4.0 for perceptible but not annoying, 3.0 for slightly annoying, 2.0 for annoying, 1.0 for very annoying. The MOS is 4.4, which means the watermarked audio and the original audio are perceptually undistinguishable. In the second part of the test, participants are presented with the original and the watermarked audio clips in random order and are asked to determine which one is the original clip. Discrimination value near to 50% means that the original and watermarked audio clips cannot be discriminated. The discrimination value obtained from test with 15 pairs of audio clips is 52%.

Robustness Tests. To evaluate the performance of the proposed algorithm, we test its robustness according to all tests defined by the Secure Digital Music Initiative (SDMI) industry committee [15]. The audio editing and attacking tools adopted in our experiments are GlodWave v5.06, CoolEdit Pro v2.1, and Stirmark for Audio v0.2. The experimental conditions and test results under common audio signal processing, random cropping, time-scale modification, pitch scaling, DA/AD conversion and Stirmark for Audio [16] are listed in Tables 1-5. The *BER*s in the tables are the average results over all 15 clips.

From Table 1, we can see that our algorithm is robust enough to common audio signal processing, such as MP3 compression up to 32 kbps (22:1), low-pass filtering of 8kHz, noise, echo, amplitude scaling, etc.

Table 2 shows strong robustness to random cropping. In our experiments, even randomly cropping 60% of the watermarked audio, the performance does not degrade. As for jittering attacks, an evenly performed random cropping, the algorithm also shows strong robustness. The reason is that random cropping responding on Fourier amplitude spectrum is small fluctuation as stated in Section 2.

Table 1. Robustness Performance to Common Attacks

Attack Type	BER(%)	Attack Type	BER(%)
Requantization 16 → 32 → 16(bit)	0	Resample 44.1 → 8 → 44.1(kHz)	0
Low pass (8 kHz)	1.39	Normalize	0
Equalization	0	Volume (150%)	0
Denoise	0	Volume (10%)	0
Gaussian (30 dB)	0	MP3 (32 kbps)	0
Echo (100ms, 40%)	0	MP3 (48 kbps)	0

Table 2. Robustness Performance to cropping attacks

Attack Type	BER(%)	Attack Type	BER(%)
Cropping (5%)	0	Cropping (50%)	0
Cropping (10%)	0	Cropping (60%)	4.17
Cropping (20%)	0	Jittering (1/1500)	0
Cropping (30%)	0	Jittering (1/1000)	0
Cropping (40%)	0	Jittering (1/500)	2.78

Table 3. Robustness Performance to time-scaling and pitch-scaling

Pitch Shift (preserve tempo)	BER(%)	Time Stretch (preserve pitch)	BER(%)
80%	1.39	80%	0
85%	0	85%	0
90%	0	90%	0
98%	0	98%	0
110%	0	110%	0
115%	1.39	115%	2.78
120%	2.78	120%	4.17

Time stretch (preserve pitch) changes the duration of an audio signal without changing its pitch, while pitch shift (preserve tempo) modifies the base frequency without affecting the speed. In our test dataset, the proposed algorithm shows strong robustness to these attacks up to 20%, far beyond the 4% requested by the SDMI [15]. The test results under time-scaling and pitch-scaling attacks from -20% to +20% are listed in Table 3.

To evaluate the robustness to DA/AD conversion, we design two test environments similar with that in [11], as shown in Fig.7. Each DA/AD conversion test consists of speakers and a microphone. We set four test points with different distances: 30cm, 100cm, 200cm and 300cm. We have not set test points beyond 300cm due to the limitation of our experimental environment. And 30cm-300cm can cover many re-record situations in real life. As shown in Table 4, the watermarks are successfully extracted from the recorded version with no error bits.

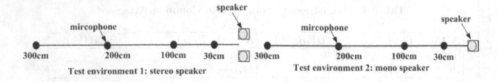

Fig. 7. test environments of DA/AD conversion

Table 4. Robustness Performance to DA/AD conversion

Stereo Speaker	BER(%)	Mono Speaker	BER(%)
30cm	0	30cm	0
100cm	0	100cm	0
200cm	0	200cm	0
300cm	0	300cm	0

Stirmark for Audio [16] is a standard robustness evaluation tool for audio watermarking techniques. All listed operations are performed by default parameters. From Table 5, we can observe that the proposed algorithm shows strong robustness to those distortions. The only attack that introduced large *BER* is "copysample" which has a strong impact on the fidelity of the audio so that the attacked clip almost does not resemble the original one. There are two kinds of attack "addfftnoise" and "resample" having bugs in Stirmark for Audio v0.2, the audio after applying these two attacks became silence. So we implemented these two attacks using Matlab 7.0 with the same parameters instead.

For easy to compare the performance on resisting synchronization attacks between our method and the existing methods, we list the reported results in the literatures in Table 6. We observe that our method has approving performance to all synchronization attacks, especially to pitch-scaling and random cropping.

Table 5. Robustness performance to the attacks in StirMark Benchmark for Audio v0.2

Attack Type	BER(%)	Attack Type	BER(%)
addbrumm_100	0	dynoise	0
addbrumm_10100	0	echo	0
addfftnoise*	0	exchange	0
addnoise_100	0	extrastereo_30	0
addnoise_900	0	ExtraStereo_70	0
addsinus	0	fft_hlpass	5.56
amplify	0	fft_invert	0
compressor	0	fft_real_reverse	0
copysample	failed	resample*	0
cutsamples	0	rc_highpass	2.78

Table 6. Comparison of performance to synchronization attacks

Synchronization Attack	best existing method and performance	our method
Pitch-scaling	Tachibana[5], up to ±8%	**up to ±20%**
Random cropping	Xiang[7], up to 20%	**up to 60%**
DA/AD conversion	Steinebach[11], 5cm to 400cm	30cm to 300cm
TSM	Xiang[7], up to ±20%	up to ±15%

5 Discussions and Conclusions

As we see, the biggest advantage of our method is the outstanding performance on resisting pitch-scaling (up to ±20%) and random cropping (up to 60%). At the same time it has approving robustness to DA/AD conversion, which make it contribute to digital right protection in broadcasting.

On the other hand, we can see that there are some inconveniences of the scheme due to so many parameters. These parameters need to preset for good performance. From Fig.1, we observe that not the whole range of AFM is suitable for embedding watermarks. The amplitudes of the part near the highest index (corresponding to the high frequency components) are small, and this part is sensitive to low-pass filtering. So we may set the f_n in Eq.(9) into the range $[0.1, 0.2]$, which means that the embedding region corresponds to the low and middle frequency components.

Given an audio clip, the length of the output AFM is $M = L \cdot N_p + N_T$, which depends on the bits of the watermark L, the length of PN sequence N_P, and the length of the template N_T. The value of M could not be too large, because it represents the number of discrete intervals that the frequency indices are mapped to, and we must make sure that each interval contains at least one frequency index. From Eq.(9), we know the length of the selected frequency indexes is $\frac{f_n \cdot N}{\sqrt{2}}$. To ensure that the embedded watermarks robust to +10% pitch-scaling, we must set $M < \frac{f_n \cdot N}{\sqrt{2}}$. Often we set the average repeat embedding time $\frac{f_n \cdot N}{\sqrt{2}M} \geq 5$. It means that each original watermark bit is embedded into the audio more than one time due to applying log-polar mapping on the normalized frequency index. And the repeat embedding times of each bit are different. The bits embedded into the higher frequency repeat more times. The choice of N_T will affect the performance of synchronization. Theoretically a larger N_T is better. But large N_T means to decrease hidden bits, so we often set $\frac{M}{6} \leq N_T \leq \frac{M}{3}$.

Based on the above parameter settings, we can achieve both strong robustness and transparency, but what is the capacity in this case? Given an audio clip of t seconds with sampling frequency F_s , the bits of watermark we can embed into the clip are $L = \frac{M-N_T}{N_p} = \frac{\frac{5}{6} \cdot M}{N_p} = \frac{\frac{5}{6} \cdot \frac{0.2N}{5\sqrt{2}}}{N_p} = \frac{t \cdot F_s}{30\sqrt{2}N_p}$. We can obtain the capacity as $C = \frac{L}{t} = \frac{F_s}{30\sqrt{2}N_p}$. As we see, the capacity depends on only two factors, the sampling frequency F_s and the length of PN sequence for $DSSS$. From our experiments, $N_p = 32$ is often enough. So for the audio with 44.1kHz sampling frequency, the capacity can achieve $C = 44100/(30\sqrt{2} \cdot 32) \approx 32.5$ bits/s.

Acknowledgements

This work was supported by NSFC (90604008, 60633030, 60403045), and 973 Program (2006CB303104). The authors would like to appreciate the anonymous reviewers for their constructive comments. Their suggestions will be very helpful for our future works.

References

1. Kim, H.J.: Audio watermarking techniques. In: Proc. of Pacific Rim Workshop on Digital Steganography, pp. 1–17 (2005)
2. Haitsma, J., Kalker, T., Bruekers, F.: Audio watermarking for monitoring and copy protection. In: Proc. of the 8th ACM Multimedia Workshop, Los Angeles, pp. 119–122 (2000)
3. Kirovski, D., Malvar, H.S.: Spread-spectrum watermarking of audio signals. IEEE Trans. on Signal Processing 51, 1020–1033 (2003)
4. Tachibana, R., Shimizu, S., Nakamura, T., Kobayashi, S.: An audio watermarking method robust against time- and frequency-fluctuation. In: Proc. of SPIE on Security and Watermarking of Multimedian Contents III, San Jose, USA, vol. 4314, pp. 104–115 (2001)
5. Tachibana, R.: Improving audio watermark robustness using stretched patterns against geometric distortion. In: Chen, Y.-C., Chang, L.-W., Hsu, C.-T. (eds.) PCM 2002. LNCS, vol. 2532, pp. 647–654. Springer, Heidelberg (2002)
6. Li, W., Xue, X., Lu, P.: Localized audio watermarking technique robust against time-scale modification. IEEE Transactions on Multimedia 8, 60–69 (2006)
7. Xiang, S., Huang, J.: Histogram-Based Audio Watermarking Against Time-Scale Modifications and Cropping Attacks. IEEE Transactions on Multimedia 9(7), 1357–1372 (2007)
8. Katzenbeisser, S., Petitcolas, F.A.P.: Information Hiding Techniques for Steganography and Digital Watermarking. Artech House, Inc. (2000)
9. O'Ruanaidh, J.J.K., Pun, T.: Rotation, scale and translation invariant spread spectrum digital image watermarking. Signal Process. 66(3), 303–317 (1998)
10. Arfib, F.K.D., Zolzer, U.: DAFX - Digital Audio Effects, pp. 232–292. J.Wiley & Sons, Chichester (2002)
11. Steinebach, M., Lang, A., Dittmann, J., Neubauer, C.: Audio watermarking quality evaluation: robustness to DA/AD processes. In: Proc. of IEEE Int. Conf. on Information Technology: Coding and Computing, pp. 100–103 (2001)
12. http://music.ucsd.edu/~sdubnov/
13. http://www.mp3-tech.org/programmer/sources/eaqual.tgz
14. Sound Quality Assessment Material,
    http://sound.media.mit.edu/mpeg4/audio/sqam/
15. SDMI, SDMI Phase II Screening Technology Version 1.0 (February 2000), http://www.sdmi.org/download/FRWG00022401-Ph2_CFPv1.0.pdf
16. Steinebach, M., Petitcolas, F.A.P., Raynal, F., Dittmann, J., Fontaine, C., Seibel, S., Fates, N., Ferr, L.C.: StirMark benchmark: audio watermarking attacks. In: Proc. of International Conference on Information Technology: Coding and Computing, pp. 49–54 (2001)

Adaptive Threshold Based Robust Watermark Detection Method

Cagatay Karabat

Tubitak Uekae (Turkish National Research Institute Of Electronics And Cryptography),
Kocaeli, Turkey
cagatay@uekae.tubitak.gov.tr

Abstract. In this paper, we propose a novel blind watermark detector based on channel estimation and adaptive threshold scheme to enhance the robustness of watermarking system. There are channel estimation based blind detectors in the literature. These detectors use reference watermarks to estimate degradations and information watermarks to send the hidden data. As in all estimation problems, however, unreliable estimates may deteriorate detection capability of the system. We employ adaptive threshold scheme to remedy these problems. First, we estimate the raw reliability scores of the watermark channels by using reference watermarks. Then, we transform the raw reliability scores by employing min-max method and logistic function respectively. Hence, we get the channel parameters of the watermark channels. Finally, the information watermark is recovered by using the extracted watermarks and their corresponding channel parameters. The simulations demonstrate that the proposed detector is more robust than existing detectors against various degradations and malicious attacks.

1 Introduction

Nowadays, Internet and digital multimedia products are widely used everywhere. This raises, however, security concerns since the digital multimedia products are highly vulnerable to the illegal copying, distribution, manipulation, and other types of malicious attacks. To remedy these problems, the digital watermarking systems, that the information to be hidden is carried by the watermark signal which is transmitted over the host signal, have been proposed in the literature.

The digital watermarking process can be modeled as a communication task, in which the watermark information is transmitted over the watermark channels within the host signal. The detection techniques employed at the receiver side is very important issue concerning with the watermarking systems. The watermark detectors which are called non-blind such as used in [1] require the host signal for the watermark detection. However, there are some other detectors, that are called blind, can detect the watermark bits without exploit the host signal features such as used in [2]-[6]. Especially, in case of malicious attacks and channel distortions, i.e. filtering, noise, cropping and compression etc., the performance of the detection schemes is very crucial to properly recover the hiding information in these systems.

In the literature, diversity and attack characterization based blind detector is proposed by Kundur and Hatzinakos [6]. In this method, they both embed information and reference watermarks. They called the region as a watermark channel where sin-

H.J. Kim, S. Katzenbeisser, and A.T.S. Ho (Eds.): IWDW 2008, LNCS 5450, pp. 139–151, 2009.
© Springer-Verlag Berlin Heidelberg 2009

gle information and single reference watermark sequence are embedded alternatively. In this channel, the information watermark is employed to carry specific information and the reference watermark is used to estimate the bit error rate (BER) of the corresponding channel. Finally, the receiver linearly combines the extracted information watermarks to recover the embedded information watermark. The linear combination weights are determined based on the BERs that are estimated using the reference watermarks. It was shown that this method has the higher detection performance than the majority rule based detector that is proposed in [5].

They, however, assume that all the watermark estimates are reliable; hence, they do not employ any threshold scheme. Therefore, the unreliable estimates deteriorate the watermark detection capability of the system against severe channel distortions and attacks. In this work, we propose a new blind detector that is called adaptive threshold based watermark detector (ATWD). In the proposed scheme, we, first, estimate the raw reliability scores of the watermark channels using reference watermarks. Then, we map these scores to the [0, 1] employing the min-max method [7]. Following that, we adaptively threshold them employing the logistic function and hence get the channel parameters of the watermark channels. Finally, we determine the embedded information watermark by using the estimated information watermarks and their corresponding channel parameters. The simulation results show the superior performance of the proposed detector. It improves the detection capability of the watermarking system.

The rest of this paper is organized as follows. In section 2, we described the quantization based watermarking system. Next, we present the detection methods in section 3. In section 4, we express the proposed detection method. The simulation results are given in section 5. Finally, we give some concluding remarks in section 6.

2 Quantization Based Watermarking System

In this section, we briefly describe the quantization based digital watermarking system that is proposed in [5] and [6]. In this system, we use random binary watermark sequences $w(i) \in \{\pm 1\}$ and $r(i) \in \{\pm 1\}$, where $1 \le i \le N$ for information and reference watermarks respectively.

2.1 Watermark Embedding Stage

In subsequent analysis, $x[m,n]$ and $y[m,n]$ denotes the host image and watermarked image in spatial domain respectively. Also, $\mathbf{X}_{o,l}[u,v]$ and $\mathbf{Y}_{o,l}[u,v]$ denotes the oth frequency orientation at the lth resolution level of the host image and the watermarked image in DWT domain respectively. In this notation, $o \in \{h,v,d\}$ expresses the horizontal, vertical and diagonal image details respectively, $l \in \{1,2,...,L\}$ is the resolution level and $[u,v]$ is the particular spatial location index at t resolution level l.

First, the host image $x[m,n]$ is transformed into the DWT domain by performing the L-level DWT decomposition. Thus, we get $3L$ detail images and an approximation image at the coarsest level. Next, we sort the detail coefficients at the spatial location $[u,v]$ and at the resolution level l in the ascending order such that:

$$\mathbf{X}_{o1,l}[u,v] \le \mathbf{X}_{o2,l}[u,v] \le \mathbf{X}_{o3,l}[u,v] \qquad (1)$$

where $o1,o2,o3 \in \{h,v,d\}$ and $o1 \ne o2, o2 \ne o3, o1 \ne o3$. We divide the range of values between the minimum and the maximum detail coefficient into the bins of width Δ.

$$\Delta = \frac{\mathbf{X}_{o3,l}[u,v] - \mathbf{X}_{o1,l}[u,v]}{2Q-1} \qquad (2)$$

We quantize the median value of the detail coefficients at the lth resolution level to embed one watermark bit as illustrated in Fig. 1. The value of quantization parameter Q is determined by making a trade-off between perceptual transparency and robustness. Finally, we compute the L-level inverse DWT and construct watermarked image in spatial domain.

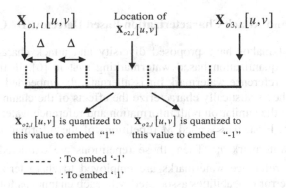

Fig. 1. Embedding watermark bits using quantization based method [5], [6].

2.2 Watermark Extraction Stage

The watermarked image $y[m,n]$ is transform into DWT domain by performing the L-level DWT decomposition at the receiver. Since we employ blind watermarking system we do not need the original image itself for the watermark extraction process. Then, we sort the detail coefficients in ascending order as follows:

$$\mathbf{Y}_{o1,l}[u,v] \le \mathbf{Y}_{o2,l}[u,v] \le \mathbf{Y}_{o3,l}[u,v] \qquad (3)$$

where $o1,o2,o3 \in \{h,v,d\}$ and $o1 \ne o2, o2 \ne o3, o1 \ne o3$. The watermark bit value is determined by using the relative position of $\mathbf{Y}_{o2,l}[u,v]$ and the same value of Q that is assigned in the watermark embedding process. Finally, we find the closest quantized value to $\mathbf{Y}_{o2,l}[u,v]$ and convert it to its associated binary value.

3 Watermark Detection Methods

In this section, we address the detection methods devised for the quantization based watermarking system. We briefly describe the majority rule based watermark detector [5] and diversity and attack characterization based watermark detector [6].

3.1 The Majority Rule Based Detector (MRD)

The quantization based watermarking system employing the majority rule based detection scheme to recover the information watermark is presented in [5]. In this detection method, Kundur and Hatzinakos do not use reference watermarks to characterize the intentional and/or unintentional degradations in the watermarked image. They embed the information watermarks into the host signal multiple times, $i.e.$ M times, at the transmitter and then M watermark estimates are extracted at the receiver. The most common bit value among the ith bits of the M watermark estimates is assigned as the ith bit of the recovered information watermark by the majority rule based detector. Therefore, all the bits of the embedded information watermark sequence are recovered.

3.2 Diversity and Attack Characterization Based Detector (DACD)

Kundur and Hatzinakos have proposed diversity and attack characterization based detector for the quantization based watermarking system [6]. In this method, each information and reference watermark bit is alternatively embedded in each localized regions. Thus, they statistically characterize the effects of the channel distortions that is assumed to be the similar on both information and reference watermark in the same localized region. First, they embed M repetitions of the information watermark w_k and reference watermark r_k. Then, these repetitions are extracted at the receiver. Following that, reference watermarks are employed to characterize the watermark channels by bit error probabilities associated with each channel as follows:

$$p_{Ek} = \frac{1}{N}\sum_{i=1}^{N} r_k(i) \oplus \hat{r}_k(i) \tag{4}$$

where $\oplus$ denotes the exclusive or (XOR) operation, $\hat{r}_k$ denotes the kth extracted reference watermark repetition and N denotes the watermark length. Finally, the extracted information watermark repetitions are linearly weighed and added to recover the information watermark as follows:

$$\hat{w}(i) = \text{sgn}\left[\sum_{k=1}^{M} \alpha_k \hat{w}_k(i)\right] \tag{5}$$

where $i=1,2,..,N$ and the linear combination coefficients α_k's are calculated:

$$\alpha_k = \frac{\log\left(\frac{1-p_{Ek}}{p_{Ek}}\right)}{\sum_{j=1}^{M}\log\left(\frac{1-p_{Ej}}{p_{Ej}}\right)} \tag{6}$$

The information and reference watermark bits are the elements of the set $\{\pm 1\}$ rather than the set $\{0, 1\}$ as in [6]. Thus, we employ sgn (.) function instead of $round$ (.) function in order to recover the information watermark bits.

4 The Proposed Adaptive Threshold Based Watermark Detection Method (ATWD)

In this section, we present the proposed adaptive threshold based watermark detection method (ATWD). The goal is to improve the detection capability of the watermarking system in case of channel distortions and attacks. In this method, we estimate the reliability of the watermark channels, normalize them using min-max method and threshold them adaptively by employing the logistic function. Finally, we use the channel parameters and estimated watermarks to recover the embedded watermark.

First, we calculate the raw reliability score of the kth watermark channel as follows:

$$R_k = \sum_{i=1}^{N} S_k(i) \tag{7}$$

where

$$S_k(i) = \begin{cases} 1, & r(i) = \hat{r}_k(i) \\ 0, & \text{otherwise} \end{cases} \tag{8}$$

where $k=1,2,..,M$, and M is the number of embedded reference watermarks, R_k denotes the raw reliability score of the kth watermark channel, $r(i)$ is the ith bit of the reference watermark, $\hat{r}_k(i)$ is the ith bit of the extracted reference watermark from the kth watermark channel, N is the length of extracted reference watermark.

Then, we map the raw reliability scores to the $[0, 1]$ range by using the Min-Max (MM) method [7] as follows:

$$N_{MM}^k = \frac{R_k - min(R)}{max(R) - min(R)} \tag{9}$$

where $R = [R_1, R_2, ..., R_M]$ and N_{MM}^k denotes the normalized reliability score of the reference watermark extracted from the kth watermark channel. The quantities $max(R)$ and $min(R)$ denotes the maximum and minimum value of the reliability parameters respectively.

Finally, we transform the normalized reliability scores of the watermark channels in order to calculate the channel parameters of watermark channels by using the logistic function [7], $L(.)$, whose input is N_{MM}^k, as follows:

$$L\left(N_{MM}^k\right) = \frac{1}{1 + A \cdot e^{-B \cdot N_{MM}^K}} \tag{10}$$

where $L\left(N_{MM}^{k}\right)$ is the channel parameter of the kth watermark channel and $k=1,2,..,M$. In addition, the constants A and B are calculated as follows:

$$A = \frac{1}{\delta} - 1 \text{ and } B = \frac{\ln A}{c} \tag{11}$$

where c is the inflection point (point where the derivative of the logistic function changes its sign) and $\delta = L(0)$.

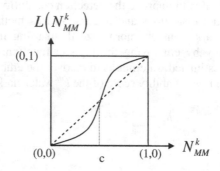

Fig. 2. The adaptive threshold scheme based on Logistic function

In all estimation problems, some of the estimates might be unreliable and these may deteriorate the detection performance of the system. In our method, we threshold the reliabilities of the watermark channel adaptively. Due to the nature of the DWT domain and the quantization based watermarking system, the reliabilities of the watermark channels are equal when there is no channel distortion and attack. In this case, we set the channel parameter to one in our experiments. Finally, we recover the ith bit of the information watermark sequence by using the watermark channel parameters λ_{k} and i^{th} bits of the information watermark estimates $\hat{w}_{k}$ as follows:

$$\hat{w}(i) = \text{sgn}\left[\sum_{k=1}^{M} \frac{\hat{w}_{k}(i)}{1 + A \cdot e^{-B \cdot N_{MM}^{K}}}\right] = \text{sgn}\left[\sum_{k=1}^{M} L\left(N_{MM}^{K}\right)\hat{w}_{k}(i)\right] \tag{12}$$

where $i=1,2,..,N$ and $sgn(.)$ is the *sign* function. Finally, we recover the embedded information watermark bits.

5   Simulation Results and Discussions

In all experiments, randomly generated information and reference watermarks of size 256 bit from the set $\{-1, 1\}$ are embedded into the test images of size $512x512$. In the quantization based watermarking system, DWT has been performed with the number of resolution level $L = 4$ and 10-point Daubechies filter. In addition, we set $c = 0.9$ and $\delta = 10^{-3}$ by experimentally (various values of c and δ gives similar results) in the simulations. First, we evaluate the robustness of the detectors by using the normal-ized correlation coefficient that is calculated as follows [6]:

$$c(w,\hat{w}) = \frac{\sum_{i=1}^{N} w(i)\hat{w}(i)}{\sqrt{\sum_{i=1}^{N} w^2(i)}\sqrt{\sum_{n=1}^{N} \hat{w}^2(i)}} \tag{13}$$

where w denotes the embedded watermark and $\hat{w}$ denotes the recovered watermark.

We calculate "*Peak Signal to Noise Ratio*" (PSNR) and "*Watermark to Document Ratio*" (WDR) to assess the perceptual quality of the watermarked image. The PSNR is a measure of the degradation in the host image introduced by the embedded water-marks as well as by the other factors. On the other hand, the WDR is the measure of ratio of watermark energy to the host image energy. Their formulas are given as follows:

$$PSNR = 10\log\frac{\mathbf{x}_{peak}^2}{\dfrac{1}{MN}\sum_{m=1}^{M}\sum_{n-1}^{N}(\mathbf{x}[m,n]-\mathbf{y}[m,n])^2} \tag{14}$$

$$WDR = 10\log\frac{\sum_{i=1}^{M}\sum_{j=1}^{N}(\mathbf{x}[m,n]-\mathbf{y}[m,n])^2}{\sum_{i=1}^{N}\sum_{j=1}^{N}\mathbf{x}^2[m,n]} \tag{15}$$

where $\mathbf{x}[m,n]$ and $\mathbf{y}[m,n]$ denotes host and watermarked image respectively, M and N denotes the size of image along both dimensions, and $\mathbf{x}_{peak}^2$ is the maximum pixel luminance value in the host image.

Table 1. WDR and PSNR values with respect to various quantization parameter values

Quantization Parameter (Q)	WDR [dB]	PSNR [dB]
1	-20,0147	25,6803
2	-29,9845	35,1969
3	-34,3355	39,3697
4	-37,2580	42,3996
5	-39,3664	44,8551
6	-41,02229	46,2361

In the watermark embedding stage, the quantization parameter (Q) has to be chosen so that the watermark power is maximized, while the perceptual quality is kept above the minimum acceptable level. The watermarked image should have PSNR value of 38 dB to satisfy the perceptual quality [8]. In simulations shown in Fig. 3-Fig. 10, we set the quantization parameter to $Q = 4$ and the corresponding PSNR value is 42.39 dB and WDR value is -37,25 dB as shown in Table 1.

The simulations shown in Fig. 3 and Fig. 4, the watermarked image is exposed to mean and median filtering attack with varying filter sizes respectively. The simulation results demonstrate that the proposed ATWD detector is more robust these attacks in comparison to other detectors at the same filter size. Since the proposed detector increases the correlation coefficient of the recovered information watermark sequences.

Moreover, we investigate the effects of Low-Pass Filtering on the correlation coefficient of the recovered information watermark sequences. The formula of the low-pass filter is given as follows:

$$h[m,n] = \frac{\mu_f^{\sqrt{m^2+n^2}}}{K} \tag{16}$$

where $K = \sum_{m=1}^{M}\sum_{n=1}^{N} h[m,n]$, μ_f denotes the filter parameter. In simulations shown in Fig. 5-Fig. 8, the watermarked image is exposed to the low-pass filtering attack at various filter sizes and with various filter parameters. As the value of the filter parameter increases, the performances of the detectors also decrease. Besides, filter size directly affects the correlation coefficient between the embedded and the recovered information watermarks. The proposed ATWD detector is more robust than the other aforementioned detectors at the same filter size. Besides, it increases the correlation coefficient at the same filter parameter.

Furthermore, we study the influence of AWGN on the performance of the detectors. The watermarked image is degraded by adding AWGN at various Signal-to-Noise Ratios (SNR) by using the below formula.

$$SNR = 10\log_{10}\left(\frac{\frac{1}{MN}\sum_{m=1}^{M}\sum_{n=1}^{N}(\mathbf{y}[m,n])^2 - \left(\frac{1}{MN}\sum_{m=1}^{M}\sum_{n=1}^{N}(\mathbf{y}[m,n])\right)^2}{\frac{1}{MN}\sum_{m=1}^{M}\sum_{n=1}^{N}(\mathbf{x}[m,n])^2 - \left(\frac{1}{MN}\sum_{m=1}^{M}\sum_{n=1}^{N}(\mathbf{x}[m,n])\right)^2}\right) \tag{17}$$

where $\mathbf{x}[m,n]$ and $\mathbf{y}[m,n]$ denotes host and watermarked image respectively, M and N denotes the size of image along both dimensions. The Fig. 9 point outs the corresponding correlation coefficients of the recovered watermarks at various SNR values.

We, also, assess the detection performance of the system against lossy data compression in Fig. 10. The watermarked image is exposed to the JPEG compression with varying quality factors. We can state that the proposed ATWD is more robust than the other detectors at the same JPEG quality factor. In addition, the performance of the proposed ATWD is increased with the increasing JPEG quality factor.

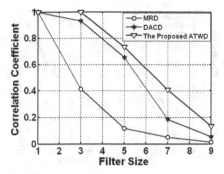

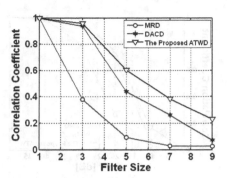

Fig. 3. Performances of the detectors against mean filtering attack

Fig. 4. Performances of the detectors against median filtering attack

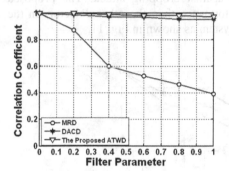

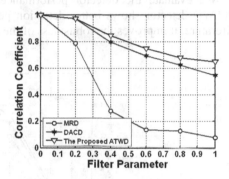

Fig. 5. Performances of the detectors against *3x3* low-pass filtering attack

Fig. 6. Performances of the detectors against *5x5* low-pass filtering attack

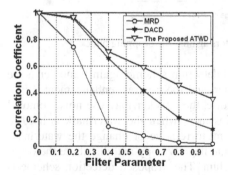

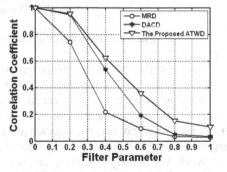

Fig. 7. Performances of the detectors against *7x7* low-pass filtering attack

Fig. 8. Performances of the detectors against *9x9* low-pass filtering attack

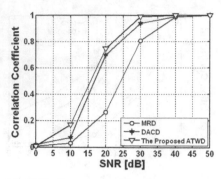

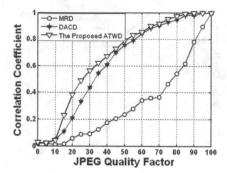

Fig. 9. Performances of the detectors against AWGN attack

Fig. 10. Performances of the detectors against JPEG compression attack

We evaluate the detector performances against cropping attack. The proposed ATWD detector shows better detection performance than the other detectors in the literature. It improves the robustness of the system as shown in Fig. 11.

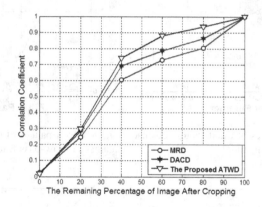

Fig. 11. Performances of the detectors against the cropping attack

We assess the robustness of the proposed detection scheme against another noise removal filters, i.e. wiener and Gaussian filter, and the contrast enhancement. Since the watermarks have lower power like the noise, the attacker may use noise removal filters to remove and/or change them. For instance; when the watermark is used to trace the illegal copies; the attacker may want to remove or change the watermark embedded into the original work. Thus, he can illegally copy and distribute the original work and nobody can trace and accuse him. The proposed detection scheme is more robust than the other detection schemes employed in the quantization based watermarking system against aforementioned attacks as shown in Table 2.

We also use the BER as a performance metric to compare the performance of the detectors. It is the probability that an information bit is decoded erroneously during the watermark detection process. In other words, it is the ratio of the number of bits

Table 2. Comparison of the detection schemes according to the correlation coefficient values of the extracted watermarks

ATTACK	THE PROPOSED ADAPTIVE THRESHOLDING BASED DETECTOR	DIVERSITY AND ATTACK CHARAC-TERIZATION BASED DETECTOR	MAJORITY RULE BASED DETECTOR
Contrast Enhancement	0.62375	0.58593	0.57031
Wiener filter of size 3x3	0.9962	0.9918	0.0333
Wiener filter of size 5x5	0.8750	0.8396	0.0229
Wiener filter of size 7x7	0.5980	0.5188	0.0214
Wiener filter of size 9x9	0.4576	0.3688	0.0208
Gaussian filter of size 3x3	1.0000	1.0000	0.9945
Gaussian filter of size 5x5	1.0000	1.0000	0.9938
Gaussian filter of size 7x7	1.0000	1.0000	0.9938
Gaussian filter of size 9x9	1.0000	1.0000	0.9938

received in error to the total number of received bits. In Fig. 12 – Fig. 15, we evaluate robustness of the detectors against mean filtering, median filtering, AWGN and JPEG compression attacks with respect to various quantization parameters used in the embedding process. Furthermore, we can see the effects of quantization parameter over the BER of the extracted watermark sequence in these simulations. As we mentioned previously, the watermark power increases with the decreasing quantization parameter value as shown in Table 1 and the simulations shown in Fig. 12 - Fig.15. In these experiments, the proposed ATWD detector shows superior performance in comparison to the other detectors. It improves the detection capability of the watermarking system and enhances the robustness of the quantization based watermarking system.

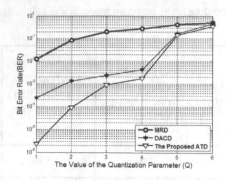

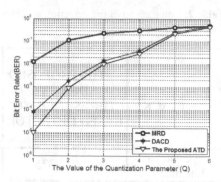

Fig. 12. Performances of detectors against *3x3* mean filtering attack

Fig. 13. Performances of detectors against *3x3* median filtering attack

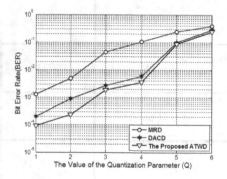

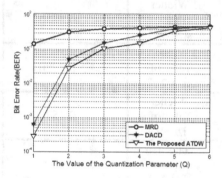

Fig. 14. Performances of detectors against AWGN at 30 dB SNR

Fig. 15. Performances of detectors against JPEG compression with quality factor 30

6 Conclusion

In this paper, we devise and implement a new blind detection method for quantization based watermarking system. We consider the importance of a watermark detection process which makes use of information concerning the adversary's actions to optimally estimate the watermark. By optimal we mean that the probability of bit error rate for watermark detection is minimized. Moreover, the correlation coefficient between embedded and the extracted watermark sequences is maximized. We employ basic communications theory principles, i.e. diversity, channel estimation, reliability estimation and threshold to enhance the robustness of the digital watermarking system. The main contribution of the proposed detector is to employ an adaptive threshold scheme and decrease the effects of the unreliable watermark estimates which deteriorate the performance of the system. Thus, these make the proposed detector more robust and reliable against common channel distortions and attacks such as AWGN, filtering, lossy data compression and cropping. We can conclude from the simulation results that the proposed ATWD detector increases the correlation coefficient and decreases the bit error rate between the embedded and the extracted information watermark sequences. Hence, it is more suitable for many real time applications.

References

1. Cox, I.J., Kilian, J., Leighton, F.T., Shamoon, T.: Secure Spread Spectrum Watermarking for Multimedia. IEEE Trans. on Image Processing 6, 1673–1687 (1997)
2. Hernandez, J.R., Amado, M., Perez-Gonzalez, F.: DCT-Domain Watermarking Techniques for Still Images: Detector Performance Analysis and a New Structure. IEEE Trans. on Image Processing 9, 55–68 (2000)
3. Perez-Gonzalez, F., Amado, M., Hernandez, J.R.: Performance Analysis of Existing and New Methods for Data Hiding with Known-Host Information in Additive Channels. IEEE Trans. on Signal Processing 51, 960–980 (2003)
4. Briassouli, A., Tsakalides, P.: Hidden Messages in Heavy Tails: DCT-Domain Watermark Detection Using Alpha-Stable Models. IEEE Trans. on Multimedia 7, 700–715 (2005)
5. Kundur, D., Hatzinakos, D.: Digital Watermarking using Multiresolution Wavelet Decomposition. In: Proc. IEEE Int. Conf. On Acoustics, Speech and Signal Processing, Seattle, Washington, May 1998, vol. 5, pp. 2969–2972 (1998)
6. Kundur, D., Hatzinakos, D.: Diversity and Attack Characterization for Improved Robust Watermarking. IEEE Trans. on Image Processing 49, 2383–2396 (2001)
7. Huber, P.J.: Robust Statistics. Wiley, Chichester (1981)
8. Petitcolas, F.A.P., Anderson, R.J.: Evaluation of copyright marking systems. In: Proc. IEEE Multimedia Systems 1999, Florence, Italy, June 7-11, vol. 1, pp. 574–579 (1999)

A Digital Forgery Image Detection Algorithm Based on Wavelet Homomorphic Filtering

Jiangbin Zheng and Miao Liu

School of Computer, Northwestern Polytechnical University,
P.O. Box 756, Northwestern Polytechnical University, Xi'an 710072, China
Zhengjb@nwpu.edu.cn, nwpu_lm@sina.com

Abstract. A novel forgery image detection algorithm is proposed to recognize some traces of artificial blur operation that is one of common ways to forge a digital image. Firstly, a wavelet homonorphic filtering is applied to enhance the high frequency edges after the blurring process. Secondly, the natural edges are eroded by mathematical morphology method, and then the enhanced artificial blur edges are preserved. Finally, the forgery image regions are localized by the region labeling method. Experimental results demonstrate the proposed method can detect forgery area accurately and reduce the detecting errors when some artificial blur operations are used to create a forgery image.

Keywords: Wavelet Homomorphic Filtering; Digital Forgery Image Detection; Blur Edges.

1 Introduction

The development of sophisticated photo-editing software and technology to manipulate digital image, has made it remarkably easy to manipulate and alter digital images and leave no visual clues of having been tampered with [1]. There are many images in various news, messages, articles, and *etc* in Internet, and to prove the authenticity and integrity of digital image becomes increasingly important. At present, the research on detection of image authenticity is mainly focus on digital watermarking, digital authentication, *etc.*, and the blind detecting techniques, which are used to distinguish a forgery image without any prior artificial embedded information, are becoming arresting.

Several researchers have proposed some detecting algorithms based on the method used in forgery images. Farid and Popescu describe how re-sampling (e.g. scaling or rotating) introduces periodic correlations, and describe how these correlations can be automatically detected in any portion of an image [1, 2]. They also proposed an algorithm based on main component analysis, which solves the problems of high complexity and bad robustness for forgery images which copy and paste in the same image [3]. Luo Weiqi *et al.* proposed an efficient and robust method for detecting and localizing region duplication forgery image [4]. Zhou Linna *et al.* proposed a new blur edge detection scheme based on the edge processing and analysis using edge preserving smoothing filtering and mathematical morphology, and then use it to detect forgery region [5]. Zhang Chi *et al.* proposed a detecting method based on

H.J. Kim, S. Katzenbeisser, and A.T.S. Ho (Eds.): IWDW 2008, LNCS 5450, pp. 152–160, 2009.

weighted local entropy, which is used to detect the traces of feather operation in forgery images [6].

Region synthetic forgery, in which a part of a digital image is copied and then pasted to another image, is a common method used in forgery images. In order to reduce the degree of discontinuity or to remove unwanted defects caused by copy and paste operations on the fake images, forgers usually apply retouching methods such as blurring, desalting, shading. Of the retouching skills, blurring is a very common process in digital image manipulation [5, 7]. Therefore, if we could trace such a process the existence of digital image tampering could be exposed. Zhou Linna et al. firstly proposed a detecting algorithm based on homonorphic filtering and mathematical morphology, it can localize the forgery region effectively [8]. Zhang Xinming et al. proposed a wavelet homonorphic filtering method based on automatic zooming property in spatial and frequency domain of wavelet transform [11]. In contrast with the traditional Homonorphic filtering, the proposed method can maintain original appearance of image and enhance the high frequency component and local contrast on the basis of. In this paper, a novel forgery image detecting algorithm is proposed based on wavelet homonorphic filtering. Because this newly proposed algorithm enhances the blurring edges effectively and applies region labeling, it reduces the detecting errors.

(a) original image 1   (b) original image 2   (c) forgery image

Fig. 1. The forgery image is obtained by copying the part of yacht and spray in image (b) and pasting it into image (a)

2 The Principle of Detecting Algorithm

Considering the making of digital image frauds, blurring is a very common process in digital image manipulation, it could be used to reduce the degree of discontinuity or to remove unwanted defects, ultimately, it is used to generate plausible digital image forensics. Hence if an additional blurring process applied on an image is detectable, possible fraud can be exposed even in a credible image. After applying artificial blurring operation, the gray range of artificial blur edge is reduced and the detail can not be distinguished. On the other hand, the dynamic range of regular edge without process is great, and has obvious gray difference [8]. Therefore, if we could enhance the artificial blur edge, and erode the poorer regular edge using mathematical morphology, the forgery regions could be detected.

To enhance artificial blur edge, using the linear transform can not be successful. Because extending the gray level of image uniformly can improve contrast of different object, but image could have a great dynamic range and also enhance the regular edge. Hence this process will lead to wrong result. On the other hand, if the gray level is compressed, the detail and gray level of blurring regions is not clear, this will bring on being unable to set threshold. Therefore, a detecting method, which utilizes homonorphic filtering and mathematical morphology and combines frequency filter and gray transform to detect the forgery image after blurring operation, is proposed in [8]. The method is based on that the regular part and blurring part of image are corresponding to illumination component and reflectance component respectively [8]. However, it has some detecting errors.

In this paper, wavelet homonorphic filtering is adopted, in which decomposing coefficients of different resolution are processed using a high-pass filter to enhance artificial blurring edge of high frequency reflectance component. And then erosion operation of mathematical morphology and region labeling method are applied to localize the forgery regions. The proposed algorithm reduces the detecting errors. It is composed of the following steps:

1) After the filter function is created, the wavelet homonorphic filtering is applied to the detected image in order to enhance the contrast in wavelet domain. This process amplifies the artificial blurring edge in the less affected regular edge.

2) Erosion operation is performed to the image after wavelet homonorphic filtering with the structure elements SE, to extract the edge after artificial blur operation and to erode the regular edge without enhancement.

3) 8-adjacent connection region labeling method is performed in eroded image to localize the forgery regions.

3   Improved Design of Wavelet Homonorphic Filter Function

The traditional homonorphic filtering processed the image as a whole, and can maintain its original appearance. But it does not consider spatial-frequency local characteristic of image, and the enhancement of high frequency component is not satisfying. On the other hand, the blurring region after average of adjacent region is corresponding to high frequency reflectance component. Therefore, its enhancement performance of blurring edges of traditional homonorphic filtering is unsatisfactory.

Wavelet transform can decompose the signal into several components, including one low frequency component and several high frequency components, each of which captures information present at a given scale. In the theory of filter, wavelet transform of image is two directional filtering operations and sub-sampling by a factor of two. Since the scaling and wavelet function are separable, image decomposition can break down into one-dimensional decomposition on rows and columns respectively. After applying wavelet transform of Jth level to image, we can obtain $3J$ detail images (LH_j, HL_j and HH_j, j=1,2,…,J) and one coarse approximation LL_J [9, 10]. In [11], decomposing coefficients in different resolution are processed using a high-pass filter to attenuate low frequency component and to amplify the high frequency component. Since the characteristic in spatial and frequency domain of wavelet transform, this process preserves the whole appearance of image. Fig.2 shows flowchart of the wavelet homonorphic filtering. Where, LL, LH, HL and HH are the four different components after wavelet transform, H_LL, H_LH, H_HL and H_HH denote coefficients of the decomposing filter.

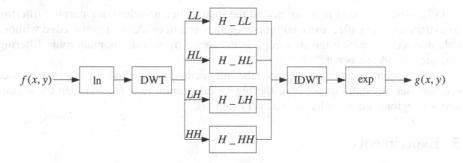

Fig. 2. The process of wavelet Homomorphic filtering

The exponential homonorphic filter function is defined as follows:

$$H(u,v) = (\gamma_H - \gamma_L)[1 - e^{-c(\frac{D(u,v)}{D_0})^2}] + \gamma_L \tag{1}$$

where, D_0 is cut-off frequency, $D(u,v) = \sqrt{u^2 + v^2}$, c is a constant which is used to control the slope of filter function, $c \in [\gamma_L, \gamma_H]$, and $\gamma_H > 1$, $\gamma_L < 1$.

In this paper, the exponential homonorphic filter function is improved, to use it in wavelet homonorphic filtering algorithm. The improved filter function is defined as follows:

$$H(j, w_h, w_v) = (\gamma_H - \gamma_L)[1 - e^{-(\frac{\sqrt{w_h^2 + w_v^2}}{2^j k_c})^2}] + \gamma_L \tag{2}$$

where, j denotes the level of wavelet decomposition, k_c is cut-off coefficient. w_h and w_v are horizontal and vertical weighted coefficient respectively. $w_h = 0$ and $w_v = 1$ for LH_j component, $w_h = 1$ and $w_v = 0$ for HL_j component, $w_h = 1$ and $w_v = 1$ for HH_j component. This improvement satisfies the frequency domain characteristics of wavelet coefficient in different level, and ensures the relationship between frequency characteristics and filter parameter of traditional high-pass filter.

The coefficients of LL_J component can be adjusted in linear or non-linear to modulate the non-uniformity of image based on its characteristic. The linear adjustment mode is adopted in this paper, it is described as follows:

$$H_{LL_J} = \gamma_L(k(x-m)+m) \tag{3}$$

where, the expression $k(x-m)+m$ performs linear adjustment to LL_J component, x is the wavelet coefficients of LL_J component, m is mean of these coefficients, k is the contrast regulation parameter ($0 \le k \le 1$).

4 Localizing Forgery Regions

In our approach, localization of forgery region is done in two phases, they are described as follows:

1) Erosion operation is performed to the image after wavelet homonorphic filtering to extract the edge after artificial blur operation and to erode the regular edge without enhancement, in which the first edge is enhanced by wavelet homonorphic filtering and the second one is not.

2) 8-adjacent connection region labeling method is applied to the image after erosion. Set threshold to preserve the bigger connection regions, and then these connection regions are considered as forgery regions.

5 Experiments

In order to testify the performance of different detecting algorithm, a database of forgery image which is classified by the method used in forgery image is established. There are more than 4000 forgery images presently, and the images used in this paper

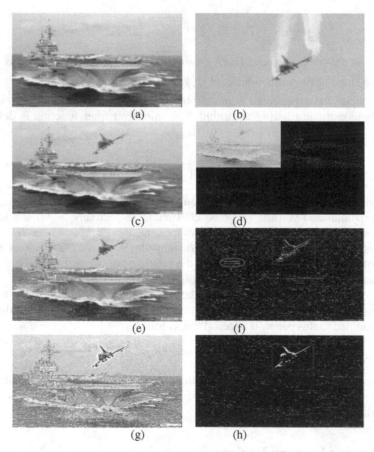

Fig. 3. The experiment results show the proposed algorithm reduces the detecting errors. (a) original image 1, (b) original image 2, (c) forgery image, (d) wavelet transform result of gray component of forgery image, (e) the processed image after traditional homonorphic filtering, (f) the detecting result of [8], (g) the processed image after wavelet homonorphic filtering, (h) the detecting result of proposed algorithm.

belong to this database. In order to testify the effect of proposed algorithm, a lot of experiments are performed using the images in this database. Some results are shown in Fig.3 and Fig.4, in which, the detecting performance of proposed algorithm will be compared with the method in [8].

Fig.3 (a) and (b) are two original images, Fig.3 (c) is forgery image, and Fig.3 (d) is wavelet transform result of gray component of forgery image. The detecting results of different algorithms are showed in Fig.3 (e)-(h). In experiments, adopts biorthogonal wavelet 'bior4.4', the parameters used in exponential function of wavelet

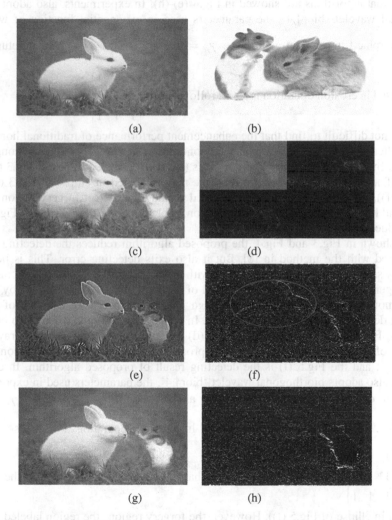

(a)   (b)

(c)   (d)

(e)   (f)

(g)   (h)

Fig. 4. The experiment results show the proposed algorithm reduces the detecting errors. (a) original image 1, (b) original image 2, (c) forgery image, (d) wavelet transform result of gray component of forgery image, (e) the processed image after traditional homonorphic filtering, (f) the detecting result of [8], (g) the processed image after wavelet homonorphic filtering, (h) the detecting result of proposed algorithm.

homonorphic filter are set as follows: $\gamma_H = 2.2$, $\gamma_L = 0.9$, $k_c = \dfrac{1}{8}$. The structure

element used in erosion operation is set as follow: $SE = \begin{bmatrix} 1 & 1 & 1 \\ 1 & 1 & 1 \\ 1 & 1 & 1 \end{bmatrix}$.

Fig.4 (a) and (b) are two original images, Fig.4 (c) is forgery image, and Fig.4 (d) is wavelet transform result of gray component of forgery image. The detecting results of different algorithms are showed in Fig.4 (e)-(h). In experiments, also adopts bior-thogonal wavelet 'bior4.4', the parameters used in exponential function of wavelet homonorphic filter are set as follows: $\gamma_H = 3.1$, $\gamma_L = 1.3$, $k_c = \dfrac{1}{6}$. The structure ele-

ment used in erosion operation is set as follow: $SE = \begin{bmatrix} 1 & 1 & 1 \\ 1 & 1 & 1 \\ 1 & 1 & 1 \end{bmatrix}$.

It is not difficult to find that the enhancement performance of traditional homonor-phic filtering is worse than wavelet homonorphic filtering. Therefore, although the method in [8] can detect the forgery regions (the region labeled in box of Fig.3 (f) and Fig.4 (f)), there exists detecting errors (the region labeled in ellipse of Fig.3 (f) and Fig.4 (f)). On the other hand, the proposed method can detect forgery regions after artificial blurring operation better (the region labeled in box of Fig.3 (h) and Fig.4 (h)) and reduce detecting errors.

As shown in Fig.3 and Fig.4, the proposed algorithm reduces the detecting errors compared with the method in [8]. But it also exits detecting error. This is because there is some regular edges are similar with artificial blur edge in dynamic range in the forgery image. So when the gray level of artificial blur edge is extended by wave-let homonorphic filtering, the regular edge is also enhanced. An example of image exiting detecting errors is shown in Fig.5. In which, Fig.5 (a) and (b) are two original images, Fig.5 (c) is forgery image, Fig.5 (d) is wavelet transform result of gray com-ponent of forgery image. Fig.5 (e) is the processed image after wavelet homonorphic filtering , and the Fig.5 (f) is the detecting result of proposed algorithm. In experi-ments, also adopts biorthogonal wavelet 'bior4.4', the parameters used in exponential function of wavelet homonorphic filter are set as follows: $\gamma_H = 2.5$, $\gamma_L = 1.2$,

$k_c = \dfrac{1}{11}$. The structure element used in erosion operation is set as follow:

$SE = \begin{bmatrix} 1 & 1 & 1 \\ 1 & 1 & 1 \\ 1 & 1 & 1 \end{bmatrix}$. It is not difficult to see that there exists detecting errors (the region

labeled in ellipse of Fig.5 (f)). However, the forgery region (the region labeled in box of Fig.5 (f)) is also detected. So we can conclude that although the proposed algo-rithm exits detecting errors, there is no undetected forgery region.

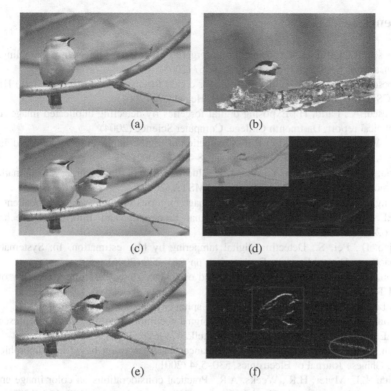

Fig. 5. The experiment results show the proposed algorithm reduces the detecting errors. (a) original image 1, (b) original image 2, (c) forgery image, (d) wavelet transform result of gray component of forgery image, (e) the processed image after wavelet homonorphic filtering, (f) the detecting result of proposed algorithm.

6   Conclusion

In this paper, a digital forgery image detection algorithm based on wavelet homomor-phic filtering is proposed. The algorithm adopts wavelet homonorphic filtering, in which decomposing coefficients of different resolution are processed using a high-pass filter to enhance artificial blurring edge. And then erosion operation of mathe-matical morphology is applied to extract the edge after artificial blur operation and to erode the regular edge without enhancement, in order to localize the forgery regions. Compared with the method using traditional homomorphic filtering, the proposed algorithm implements compression of gray dynamic range of the whole image better and enhance the blurring edge effectively, hence it reduces the detecting errors.

Acknowledgement

This work is supported by Research Foundation for the Doctoral Program of Higher Education of China (No.2008-2010) and NPU Foundation for Fundamental Research (No. 2007-2009).

References

1. Popescu, A., Farid, H.: Exposing digital forgeries by detecting traces of re-sampling. IEEE Transactions on Signal Processing, 758–767 (2005)
2. Popescu, A., Farid, H.: Statistical tools for digital forensics. In: Fridrich, J. (ed.) IH 2004. LNCS, vol. 3200, pp. 128–147. Springer, Heidelberg (2004)
3. Popescu, A., Farid, H.: Exposing digital forgeries by detecting duplicated image regions. Technical report, Dartmouth College, Computer Science (2004)
4. Luo, W., Huang, J., Qiu, G.: Robust detection of region-duplication forgery in digital image. J. Chinese Journal of Computers, 1998–2007 (2007)
5. Zhou, L., Wang, D., Guo, Y., Zhang, J.: Blue detection of digital forgery using mathematical morphology. Technical report, KES AMSTA (2007)
6. Zhang, C., Zhang, H.: Detecting digital image forgeries through weighted local entroy. In: IEEE International Symposium on Signal Processing and Information Technology, pp. 62–67 (2007)
7. Hsiao, D., Pei, S.: Detecting digital tampering by blur estimation. In: Systematic Approaches to Digital Forensic Engineering, pp. 264–278 (2005)
8. Zhou, L.: Study of digital forensics based on image content, Beijing University of Posts and Telecommunications (2007)
9. Daubechies, I.: Ten lectures on wavelet. Capital city press, Philadephia (1992)
10. Mallat, S.: A theory for multiresolution signal decomposition: the wavelet representation. IEEE Trans. on Pattern Anal. Machine Intell., 674–693 (1989)
11. Shen, L., Zhang, X.: Image contrast enhancement by wavelet based homomorphic filtering. Chinese Journal of Electronics, 530–534 (2001)
12. Voicu, L.I., Myler, H.R., Weeks, A.R.: Practical considerations on color image enhancement using homomorphic filtering. Journal of Electronic Imaging, 108–113 (1997)
13. Ng, T.T., Chang, S.F.: A model for image splicing. In: IEEE International Conference on Image Processing, pp. 1169–1172 (2004)

Blind Detection of Digital Forgery Image Based on the Local Entropy of the Gradient

Zhe Li and Jiang-bin Zheng

MailBox 886, Department of Computer Science and Engineering,
School of Computer, Northwestern Polytechnical University, Dong-Xiang Road,
Dong-Da Jie-Dao-Ban, Chang-An District, Xi'an, 710129, China
nwpulizhe@gmail.com, zhengjb@nwpu.edu.cn

Abstract. A novel method based on the local entropy of the gradient is proposed to detect the forged digital images. The method can discover some traces of artificial feather operation, which is necessary to create a smooth transition between a forged image region and its surroundings. The local entropy of the gradient is used to determine the forged region, and the credibility is computed to show the reality level of the image. Results of experiments on several forged images demonstrate the effectiveness of the algorithm.

Keywords: forged digital images, feather operation, the gradient, the local entropy, the credibility.

1 Introduction

Digital photography and powerful image editing software make it very easy today to create believable forgeries of digital images even for a non-specialist. However, the digital forgeries may cause some troubles or even great economic loss to our social and living, thus, the need for reliable of digital image contents is quickly increasing. Recently, several different methods for detecting digital forgeries have been proposed. Johnson et al. proposed a method [1] based on the inconsistencies in lighting, and assumed nearly Lambertian surface for both the forged and original areas. Understandingly it might not work when the object does not have a compatible surface, when pictures of both the original and forged objects were taken under approximately same lighting conditions. Popescu et al. developed several methods [2] for identifying digital forgeries by detecting traces of re-sampling, which might produce less reliable results for processed images stored in the JPEG format. Min wu et al. [3] [4]in Maryland University analyzed and detected several operations on digital images including space filtering, double JPEG compression, re-sampling and adjusting intensity, and these methods based on the relativity of DCT coefficients, spectrum analysis and signal/noise ratio prediction. Weihong Wang and Hany Farid developed a method [5] for identifying digital forgeries based on EM algorithm to detect traces of double mpeg compression. Chen, M et al. [6] provided a unified framework for identifying the source digital camera from its images and for revealing digitally altered images using photo-response non-uniformity noise (PRUN), which is a unique stochastic fingerprint

H.J. Kim, S. Katzenbeisser, and A.T.S. Ho (Eds.): IWDW 2008, LNCS 5450, pp. 161–169, 2009.

of imaging sensors. Obviously, it is a complex problem with no universally applicable solution to detect digital forged images. This paper proposes that the traces of artificial feather operation can be detected through comparing the local gradient entropy and the gradient, and it reasonable to assume that artificial feather operation is always used to create a smooth transition between a forged image region and its surroundings. What is more, the results of this algorithm are shown by the form of credibility, which is reasonable and understandable.

2   The Feather Operation Model

The feather operation is necessary when forging digital images, and it always incidents to the copy and paste operations, in helping the pasted object to blend smoothly and unobtrusively with its surroundings. The paper [7] models the feather operation, and assumes that the grayscale value of transition region is given by:

$$f = (1 - \frac{i}{2r})f_1 + \frac{i}{2r}f_2 \tag{1}$$

Where f_1 and f_2 are grayscale values of two original images, r is feather radius, and i is the row number or the column number. For example, if $r = 3$, then $i = 0, 1, 2, 3, 4, 5, 6$, which made $i/2r$ change from 0 to 1. If there are obvious differences of luminance between two original images, this operation will create a smooth transition to remove the splicing traces.

3   The Gradient of Color Images

The gradient is commonly used to detect the edge in gray images, and for a RGB color image, the usual method for detecting edge is to compute the gradient of R, G, B channels respectively, and then regard the combination of the three gradients as the final gradient of the color image. In consideration of the dependence among the three channels, the gradients using the method noted above are always undesirable. Thus, we can denote one pixel in color images as a vector, and outreach the concept of gradient from scalar function to the vector function.

Usually, an image is defined as an integer-valued matrix $\mathbf{f}$ within the range of [0, 255], and we convert it to a real-valued one $\mathbf{f}'$ with the domain [0, 1]. Where, $\mathbf{f}(x, y)$ is a color pixel and denoted as a vector.

$$\mathbf{f}'(x, y) = \frac{1}{255}\mathbf{f}(x, y) = \frac{1}{255}\begin{bmatrix} f_1(x, y) \\ f_2(x, y) \\ f_3(x, y) \end{bmatrix} \tag{2}$$

Let $\mathbf{u}$ and $\mathbf{v}$ denote two pixels of the color image, and $\mathbf{r}, \mathbf{g}, \mathbf{b}$ denote the unit vector in RGB color space. Then the dot multiply operation can be written as:

$$g_{xx} = \mathbf{u} \bullet \mathbf{u} = \mathbf{u}^T \mathbf{u} = \left|\frac{\partial \mathbf{R}}{\partial x}\right|^2 + \left|\frac{\partial \mathbf{G}}{\partial x}\right|^2 + \left|\frac{\partial \mathbf{B}}{\partial x}\right|^2 \tag{3}$$

$$g_{yy} = \mathbf{v} \bullet \mathbf{v} = \mathbf{v}^T \mathbf{v} = \left|\frac{\partial \mathbf{R}}{\partial y}\right|^2 + \left|\frac{\partial \mathbf{G}}{\partial y}\right|^2 + \left|\frac{\partial \mathbf{B}}{\partial y}\right|^2 \tag{4}$$

$$g_{xy} = \mathbf{u} \bullet \mathbf{v} = \mathbf{u}^T \mathbf{v} = \frac{\partial \mathbf{R}}{\partial x}\frac{\partial \mathbf{R}}{\partial y} + \frac{\partial \mathbf{G}}{\partial x}\frac{\partial \mathbf{G}}{\partial y} + \frac{\partial \mathbf{B}}{\partial x}\frac{\partial \mathbf{B}}{\partial y} \tag{5}$$

Where

$$\mathbf{u} = \frac{\partial \mathbf{R}}{\partial x}\mathbf{r} + \frac{\partial \mathbf{G}}{\partial x}\mathbf{g} + \frac{\partial \mathbf{B}}{\partial x}\mathbf{b} \tag{6}$$

$$\mathbf{v} = \frac{\partial \mathbf{R}}{\partial y}\mathbf{r} + \frac{\partial \mathbf{G}}{\partial y}\mathbf{g} + \frac{\partial \mathbf{B}}{\partial y}\mathbf{b} \tag{7}$$

Then the direction angle $\theta(x, y)$ and value of the gradient $F_\theta(x, y)$ defined as follow:

$$\theta(x, y) = \frac{1}{2}\arctan\left[\frac{2g_{xy}}{g_{xx} - g_{yy}}\right] \tag{8}$$

$$F_\theta(x, y) = \sqrt{\frac{1}{2}\left[(g_{xx} + g_{yy}) + (g_{xx} - g_{yy})\cos 2\theta + 2g_{xy}\sin 2\theta\right]} \tag{9}$$

Where, θ and F_θ are images with the same size as the input color image.

4   The Local Gradient Entropy

Shannon [8] defined the entropy of an n-state system as follow:

$$H = -\sum_{s=1}^{n} p_s \log p_s \tag{10}$$

Where p_s is the probability of occurrence of the event s and

$$\sum_{s=1}^{n} p_s = 1, \ 0 < p_s < 1 \tag{11}$$

Following Shannon's definition of entropy Pun [9] defined the entropy of the image as follow:

$$H = -\sum_{s=0}^{L-1} p_s \log p_s \tag{12}$$

Where

$$p_s = \frac{n_s}{M \times N} \tag{13}$$

Where p_s is the probability of grayscale value s that appears in the image, L is the number of discrete gray levels, n_s is the number of pixels in the image having gray level s, $M \times N$ is the image size.

Following Pun's definition of image entropy we can defined the local gradient entropy. Firstly, the gradient of color image F_θ can be gained using the method in section 3, which is a real-valued matrix with the domain [0, 1]. We convert it to an integer-valued one F'_θ with the domain [0, 255], which will be a gray image.

$$F'_\theta(x, y) = \text{int} \ (\ 255 \bullet F_\theta(x, y) \ ) = \begin{bmatrix} \text{int} \ (\ 255 \times F_{\theta 1}(x, y) \ ) \\ \text{int} \ (\ 255 \times F_{\theta 2}(x, y) \ ) \\ \text{int} \ (\ 255 \times F_{\theta 3}(x, y) \ ) \end{bmatrix} \tag{14}$$

We give a small $M_i \times N_i$ neighborhood Ψ_i in the gradient image F'_θ, and the local entropy of Ψ_i can be defined as:

$$H(\Psi_i) = -\sum_{j=0}^{L-1} p_j \log p_j \tag{15}$$

Where

$$p_j = \frac{n_j}{M_i \times N_i} \tag{16}$$

Where p_j is the probability of grayscale value j that appears in the neighborhood Ψ_i, L is the number of discrete gray levels, n_j is the number of pixels which have the same gray level j in the neighborhood, and $H(\Psi_i)$ is the local entropy of neighborhood Ψ_i.

Entropy can well express the indeterminacy of information. In one image, the local entropy of a region will be lager, when the region includes lots of details and the grayscales change tempestuous. Otherwise, it will be smaller.

5   Detecting of Forgery Images

The feather operation can remove the splicing traces, and create a smooth transition between a forged image region and its surroundings. Therefore, the local gradient entropy of the feathered region will be much smaller than that of other regions.

Let (m,n) denotes the coordinate of a pixel point q in the gradient image. Centered with q, we define a neighborhood region Ω, $\Omega = \{(i, j) | |i - m| \leq W, |j - n| \leq W\}$. It is considered that the feather radius is usually to be 1, 2, or 3 in forging images, and 3 is the most commonly used. Thus, it is reasonable that W should not set to be too large, and the experimental results will get better as W is closer to the feather radius.

In this paper, $W = 3$ □and   the region is 7×7. Then we can compute the local entropy of the gradient image. Denote the gradient of pixel q as G_q, and the local entropy of gradient as E_q. Thus, we can find out the suspicious pixels and the forgery ones.

$$\begin{cases} G_q > T_1, \ E_q < T_2 + \varepsilon & \text{mark } q \text{ as a suspicious pixel} \\ G_q > T_1, \ E_q < T_2 & \text{mark } q \text{ as a forgery pixel} \end{cases} \tag{17}$$

$$T_1 = \frac{1}{M \times N} \sum G_q$$

$$T_2 = \frac{1}{M \times N} \sum E_q \tag{18}$$

$$\varepsilon = \frac{T_2}{5}$$

Where, T_1, T_2, ε are thresholds. It is better for T_1 to be smaller and T_2 to be larger, because the important edge and local entropy will not be likely lost. However, too small T_1 or too large T_2 will likely mistake the nature edge as forgery ones. Thus in this paper they are be chosen as follows, which can make good experimental results. More self-adapted thresholds will be attempted in later research.

Where q is each pixel of the image, and $M \times N$ is the size of the image.

Let num_1 and num_2 denote the number of the forgery pixels and the suspicious pixels, then the credibility of the image can be written as follows:

$$credibility = 1 - \frac{num_1}{num_2} \tag{19}$$

Finally, the forgery pixels are marked with 255 and others with 0, and the result will be presented by a binary image. We process the binary image using operations erosion and dilation, to remove the isolated points and make the density points conjoined. If a connected region can be described by the write points, we regard it as the suspicious region.

The steps of the algorithm can be summarized as follows:

1) Outreach the concept of gradient from scalar function to the vector function and compute the gradient of the color image defined by formula (8) and (9).
2) Compute the local entropy of the gradient value image defined by formula (15) and (16).
3) Find out the suspicious pixels and the forgery ones, by comparing the gradient with T_1 and the local entropy with T_2, shown by formula (17) and (18), then compute the credibility of the image defined by formula (19).
4) Mark the suspicious region by using operations erosion and dilation.

6   Results

In order to make experiment conveniently, the subject group has built an image database based on the different forged methods, which including nearly 4000 forged images. Several images have been chosen to examine the effectiveness of the proposed algorithm from the database, in which artificial feather operation was used. They were forged by using all kinds of methods, such as circumrotating, minifying, or magnifying the insert part and so on. Therefore, the experiment results are representative.

Several experiment results are shown in figure 1, 2 and 3, where we show the original images, the forgery ones, and the experimental results.

In figure 1, (a) and (b) are two original images, and we get image (c) by moving the dog from image (b) to image (a), and using the artificial feather operation. Image (d) is the gradient value, and the local entropy of gradient is shown in image (e). Finally, we show the traces of feather operation in image (f), which are located by the points, the gradient of which is larger than T_1 and the local entropy of which are smaller than T_2, where T_1 and T_2 are available from (16). Therefore, the dog is considered as the forgery region, and the credibility of image (c) is 0.5635 by using the method described in this paper.

(a)   (b)   (c)

(d)   (e)   (f)

Fig. 1. One of the experimental results. (a) and (b) are original images, (c) is the forgery image, (d) is the gradient of (c), (e) is the local entropy image of (d), (f)is the final experimental result of (c) and the traces of feather operation are shown by write points.

In order to show the sturdiness of the approach, we make experiment on the three images in figure 2 respectively. (a) and (b) are two original images, and we get image (c) by moving the tower from image (b) to image (a), and using the artificial feather operation. The experimental results are respectively shown in image (d) (e) (f), in which the traces of feather operation are presented by write points. In image (d) and (e), the white points we get are scattered and can not build up a connected region, however, in (f) the tower was marked suspicious region, and we regard it as forgery region. The credibility of image (a) is 0.9135, (b) is 0.8993, and (c) is 0.5392. What is clear, the credibility of forgery image (c) is much lower.

Fig. 2. One of the experiment results. (a) and (b) are two original images; (c) is a forgery by moving the tower from (b) to (a) and using feather operations. (d) (e) and (f) are respectively the experimental results of (a) (b) and (c), and the traces of feather operation are marked by white points.

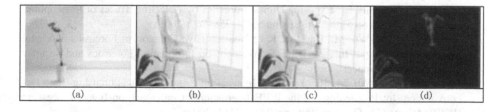

Fig. 3. One of the false experiment results. (a) and (b) are two original images; (c) is a forgery by moving the plant from (a) to (b) and using feather operations. (d) is the experimental results of (c) using the method in this paper, and the traces of feather operation are marked by white points.

It is a pity that there still are false detections of forgery images, although it is very few. In figure 3, (a) and (b) are two original images, and we get image (c) by moving the plant from image (a) to image (b), and using the artificial feather operation. The experimental result is shown in image (d), in which the traces of feather operation are presented by write points. However, it is clear that the detection result is false, and the bottom left of image (d) should not be the traces of feather operation. The reason for the false detection is that the nature image has blur regions, and the problem will be improved in our later research.

7 Conclusion

In this paper, we propose a method based on the local entropy of the gradient image, to compute the credibility of an image and to detect the forgery region. The experimental results show that the method is reliable in discovering the traces of feather operation. What is more, the results of this algorithm are shown by the form of credibility, which are more reasonable and understandable than those binary results. However, it might not work when either the feather operation was not used in forging the images or the nature image had blur regions, and it also need to improve in the later research. We expect that the method described here will make it increasingly stronger to detect the digital forgeries.

Acknowledgments

The letter is supported by Northwestern Polytechnical University Fundamental Research Grant.

References

1. Johnson, M.K., Farid, H.: Exposing Digital Forgeries by Detecting Inconsistencies in Lighting. In: Proc. ACM Multimedia and Security Workshop, New York, pp. 1–9 (2005)
2. Popescu, A.C., Farid, H.: Exposing Digital Forgeries by Detecting Traces of Resampling. IEEE Transactions on Signal Processing 53(2), 758–767 (2005)
3. Swaminathan, A., Wu, M., Ray Liu, K.J.: Non-Intrusive Component Forensics of Visual Sensors Using Output Images. IEEE Transactions on Information Forensics and Security (March 2007) (to appear)
4. Swaminathan, A., Wu, M., Ray Liu, K.J.: Image Tampering Identification using Blind Deconvolution. In: Proceedings of the IEEE International Conference on Image Processing (ICIP), Atlanta,GA, October 2006, pp. 2311–2314 (2006)
5. Wang, W., Farid, H.: Exopsing Digital Forgeries in Video by Detecting Double MPEG Compression, pp. 37–47. ACM, New York (2006)
6. Chen, M., Fridrich, J., Goljan, M., Lukas, J.: Determining Image Origin Integrity Using Sensor Noise. IEEE Transactions on information forensics and security 3(1), 74–90 (2008)

7. Zhang, C., Zhang, H.: Detecting Digital Image Forgeries Through Weighted Local Entropy. In: IEEE International Symposium on Signal Processing and Information Technology, pp. 62–67 (2007)
8. Shannon, C.E.: A mathematical theory of communication. ACM SIGMOBILE Mobile Computing and Communications Review 5(3-55) (January 2001)
9. Pun, T.: A new method for grey-level picture thresholding using the entropy of the histogram. Signal Processing 2, 223–237 (1980)

Exposure Time Change Attack on Image Watermarking Systems

Kaituo Li[1,*], Dan Zhang[2], and De-ren Chen[2]

[1] College of Software Technology, Zhejiang University, Hangzhou 310027, China
lee.kaituo@gmail.com
[2] College of Computer Science and Technology, Zhejiang University,
Hangzhou 310027, China
zhangdanny@nbip.net, drc@zju.edu.cn

Abstract. We propose an effective attack based on exposure time change and image fusion, called ETC attack. First, the ETC attack simulates a set of images with different exposure time using the watermarked image. Then it applies an image fusion method to blend the multiple images mentioned above and thus generates a new image similar to the original one. In this paper, we describe the rationale of the attack. To illustrate the effect of the ETC attack, we present the results of the ETC attack against one watermarking scheme Iterative Watermarking Embedding. It can be shown that the ETC attack has an important impact on watermark decoding and detection, while not severely reducing the quality of the image.

1 Introduction

Digital watermarking technology is the process of watermark host analysis, medium pre-processing, embedded information selection and embedding pattern design. It tries to seek an optimum scheme while keeping balance between transparency and robustness. Transparency refers to the imperceptibility of watermark. Robustness means that the inserted watermark is hard to remove. Along with the watermarking technologies, digital watermark attack appears. In fact, the watermark embedding can be regarded as efforts in defense of host image from the watermark attack. The watermark attack aims at breaking the practicality of the watermark and keeping the host work worthwhile at the same time, like removing the watermark or invalidating the extraction algorithm.

In fact there are two main types of attacks: those who attempt to remove the watermark and those who just prevent the detector from detecting them [1]. Attacks in the first category usually try to estimate the original non-watermarked cover-signal, considering the watermark as noise with given statistic. For instance, Maes presented a twin peaks attack to fixed depth image watermarks [2]. The attack is based on histogram analysis. If watermarked images with twin peaks of histogram are observed, we could reconstruct the histogram of the original image and estimate the watermark. Attacks in the second category, such as

* This research is supported partly by the National Basic Research Program of China (973 Program) under Grant No. 2006CB303104.

H.J. Kim, S. Katzenbeisser, and A.T.S. Ho (Eds.): IWDW 2008, LNCS 5450, pp. 170–183, 2009.
© Springer-Verlag Berlin Heidelberg 2009

cropping, horizontal reflection and deletion of lines or columns, try to change the content of watermarked images to make detectors cannot find watermark.

In this article, we propose an attack, called *exposure time change attack* (ETC), which aims at decreasing the cross-correlation score between the original watermark and the decoded watermark. ETC is a new breed of attack against generic image watermarking systems, that is, the ETC attack is not limited to a particular image watermarking algorithm. In order to launch the attack successfully, the adversary does not need to know the details of the watermarking algorithms and the watermarking keys. Also, the ETC attack requires no detectors. Only one copy of watermarked image is needed to break the watermark. In this paper, to illustrate the effect of the ETC attack, we present the results of the ETC attack against one watermarking scheme Iterative Watermark Embedding (IWE). The ETC attack can also be launched for many other watermarking technologies, such as Cox's spread spectrum watermarking [3] and Xia's watermarking [4]. It can be shown that the attack has an important impact on watermark decoding and detection, while not severely reducing the quality of the image. The strategy of an ETC attack contains two steps:

- *Step 1*: simulate a set of images with different exposure time using the watermarked image;
- *Step 2*: apply an image fusion method to blend the multiple images mentioned above.

This paper is organized as follows. The watermarking algorithm IWE is briefly reviewed in Sec. 2. The proposed ETC attack algorithm and the experimental results of the ETC attack against IWE are discussed in detail in Sec. 3 and 4. Discussion is presented in Sec. 5.

2   Reviewing the IWE

In this section, the watermarking scheme IWE presented by [5] is reviewed. The Single Watermarking Embedding (SWE) is introduced at first, since IWE is based on SWE.

2.1   Reviewing the SWE

The sequence that is used to embed a watermark bit sequence comes from the first P percent AC coefficients of the JPEG compressed domain. The sequence is called the host vector. For each of the 8×8 blocks in a JPEG compressed image, the first AC coefficient in zigzag order is extracted and ordered to form the first segment of the vector [5]. Next, the second AC coefficient in zigzag order is extracted and appended to the first segment of the vector and so on, until P percent of the AC coefficients are extracted. The host vector Y is defined as $Y = [y_1, y_2, \ldots, y_M]$ with length M. Watermark sequence $W = [w_1, w_2, \ldots, w_N]$ is a binary sequence with length N, where $M \gg N$. To embed the N bit watermark into M coefficients, the host vector Y is divided into N sub-vectors of length $P = \lfloor M/N \rfloor$. One bit watermark is inserted into every sub-vector. SWE uses two keys. The first key $D = [d_1, d_2, \ldots, d_N | d_i \in R^+, 1 \le i \le N]$ is a set of N pseudo-random positive real numbers. The second key is $K = [k_1, k_2, \ldots, k_M]$ with every key k_i being zero-mean Gaussian. Using these two keys, SWE embeds W into Y, and decodes watermark. The watermarked host vector Y is denoted by Y'. Based on above definitions, the basic embedding is shown as follows:

$$Y_i' = Y_i + \alpha_i K_i \tag{1}$$

$$\alpha_i = \begin{cases} \dfrac{d_i \cdot round(\langle Y_i, K_i \rangle / d_i) - \langle Y_i, K_i \rangle}{\|K_i\|_2^2}, \text{ for case 1,} \\ \dfrac{d_i \cdot [round(\langle Y_i, K_i \rangle / d_i) + 1] - \langle Y_i, K_i \rangle}{\|K_i\|_2^2}, \text{ for case 2,} \\ \dfrac{d_i \cdot [round(\langle Y_i, K_i \rangle / d_i) - 1] - \langle Y_i, K_i \rangle}{\|K_i\|_2^2}, \text{ for case 3,} \end{cases} \tag{2}$$

Case 1: $\lfloor round(\langle Y_i, K_i \rangle / d_i) \rfloor_2 = w_i$
Case 2: $\lfloor round(\langle Y_i, K_i \rangle / d_i) \rfloor_2 \ne w_i$ and $\langle Y_i, K_i \rangle \ge d_i \cdot round(\langle X_i, K_i \rangle / d_i)$
Case 3: $\lfloor round(\langle Y_i, K_i \rangle / d_i) \rfloor_2 \ne w_i$ and $\langle Y_i, K_i \rangle < d_i \cdot round(\langle X_i, K_i \rangle / d_i)$
The blind watermark decoding and detection is defined as follows:

$$w_i' = \lfloor round(\langle Y_i', K_i \rangle) / d_i \rfloor_2 \tag{3}$$

The cross-correlation score s_1 is used to test the watermark is present or not:

$$s_1 = \frac{\sum_{i=1}^{N}(2 \cdot w_i - 1) \cdot (2 \cdot w_i' - 1)}{\sqrt{\sum_{i=1}^{N}(2 \cdot w_i - 1)^2 \cdot \sum_{i=1}^{N}(2 \cdot w_i' - 1)^2}} \tag{4}$$

2.2 Reviewing the IWE

Under JPEG recompression attack, SWE cannot decode and detect the watermark correctly [5]. IWE is proposed to prevent the removal of the watermark in the JPEG recompression attack. Also, SWE is adopted as one part of IWE. Before SWE, Y_i is added to a random noise sub-vector N_i of length P. Each element of the random noise is uniformly distributed in $[-q/2, q/2]$ where q is the JPEG quantization step size (accounting for both the quantization matrix and the scaling factor). Then SWE is performed. After that, a trial JPEG re-quantization and watermark decoding is performed on the watermarked sub-vector. If computed s_1 is below the threshold value or the number of iterations exceeds a pre-defined threshold T, the iteration is forced to terminate. Otherwise, another iteration is carried out. Please refer to Fig. 5.1 in [5] for more details.

2.3 The Robustness of IWE

IWE is tested under Stirmark 4.0 attacks [5]. Experiments show that IWE is robust to JPEG compression attacks, small median filtering attacks, watermark embedding attacks, rotation-and-cropping attacks, rotation-and-scaling attacks, latest small random distortion attacks. IWE is not robust to affine transformation attacks, Gaussian filtering attacks, large noise attacks, rescaling attacks, lines removal attacks, large rotation attacks, large median filtering attacks. But most of the images under attacks in the second category are highly distorted. Thus, IWE is robust to attacks that do not degrade image quality severely.

3   ETC Phase 1: Changing Exposure Time

3.1   Introduction

Exposure time is the duration for which the light sensing element (CCD or film) is exposed to light from the scene. Exposure time is determined by camera shutter speed. The exposure of an image X is defined as [6]:

$$X = Et \tag{5}$$

where E is the irradiance that the light sensing element receives and t is the exposure time. Exposure time can have a dramatic impact on the image pixel values: if the scene luminance is constant, longer exposure time can make the image brighter; shorter exposure time can make the image darker. Varying exposure time can be used to adjust image brightness.

We find that many image watermarking systems can not derive the correct decoded watermark when the watermarked image is exposed differently. This is the main reason why we choose exposure time change as the watermark attack method. In addition, by making some adjustments to the exposure time of the watermarked images, we could simulate the physically-based photorealistic images. That is, we could capture the contents of these simulated images using our cameras. Thus, the resultant images are likely to preserve good image quality and look natural.

3.2   The Relationship between the Image Exposure Time Change and Change of Pixel Value

To illustrate the relationship between the image exposure time change and change of pixel value, we need to know the relationship between the image pixel exposure x and its intensity p. In [7], the relationship between the pixel value p (with a range between 0 and 255 corresponding to 8-bit quantization) and the exposure level x in Jm^{-2} is expressed using the Eq. (6):

$$p = \phi(x) = \frac{255}{1 + e^{-0.75 \cdot \log_{10} x}}, \tag{6}$$

If we change exposure time from t_1 to t_2 and E is constant, the pixel value p_1 in one location turns to p_2. According to Eq. (5), we get

$$x_1 = Et_1, \tag{7}$$

$$x_2 = Et_2, \tag{8}$$

Thus,

$$p_1 = \phi(x_1) = \frac{255}{1 + e^{-0.75 \cdot \log_{10} x_1}}, \tag{9}$$

$$p_2 = \phi(x_2) = \frac{255}{1 + e^{-0.75 \cdot \log_{10} x_2}} = \frac{255}{1 + e^{-0.75 \cdot \log_{10}(Et_2)}} = \frac{255}{1 + e^{-0.75 \cdot \log_{10}(x_1 t_2/t_1)}}, \tag{10}$$

From Eq. (9),

$$x_1 = 10^{-4/3 \cdot \ln(\frac{255}{p_1} - 1)}, \tag{11}$$

with $p_1 \neq 0$ and $p_1 \neq 255$. According to Eq. (11), Eq. (10) can be expressed in the form:

$$p_2 = \frac{255}{1 + e^{-0.75 \cdot \log_{10}(10^{-4/3 \cdot \ln(\frac{255}{p_1} - 1)} \cdot t_2/t_1)}} = \frac{255}{1 + e^{\ln(\frac{255}{p_1} - 1) - \frac{3}{4} \log_{10}(t_2/t_1)}}. \tag{12}$$

By using Eq. (12), we could figure out the new pixel value when the camera exposure time is changed. On the other hand, when a pixel value is 0 or 255, it approaches a limit, either too dark or too bright. If p_1 approximates to 0, $\ln(255/p_1 - 1)$ is close to $+\infty$, the effect of exposure change is slight; if p_1 approximates to 255, $\ln(255/p_1 - 1)$ is close to $-\infty$, the effect of exposure change is also slight. So under these two conditions, we do not change the pixel value. If $p_1 = 0$, we assume $p_2 = 0$; if $p_1 = 255$, we assume $p_2 = 255$.

3.3   The Effect of Exposure Time Change

Scaling Factor (SF) is used to scale the quantization matrix recommended in the JPEG standard. The 'Lena' JPEG image with SF =1 as shown in Fig. 1a is used in our experiments. The image size is 512-by-512 pixels. A 40-by-40 pixels binary logo called 'Fu' which is a Chinese character shown in Fig. 1b is used as a decodable watermark. The watermarked image through IWE and the corresponding decoded watermark are shown in Fig. 1c and Fig. 1d respectively. The exposure time of Fig. 1c is increased 1000 times. The resultant watermarked image and the corresponding decoded watermark are shown in Fig. 1e and Fig. 1f respectively. Then the exposure time of Fig. 1c is shortened 1000 times. The resultant watermarked image and the corresponding decoded watermark are shown in Fig. 1g and Fig. 1h respectively. Finally, Figs. 2 and 3 illustrate the influence of exposure time change of the watermarked image on normalized cross-correlation score s_1 and error rate during watermark decoding and detection. The error rate is used to refer to the percentage of incorrect pixels of the decoded watermark image. For any pixel of the decoded watermark image, if it is different from the pixel of the same location of the original watermark image, it is an incorrect pixel. Fig. 2 shows how s_1 drops and how error rate increases with the increase of the exposure time of the watermarked image. Fig. 3 shows how s_1 drops and how error rate increases with the decrease of the exposure time of the watermarked

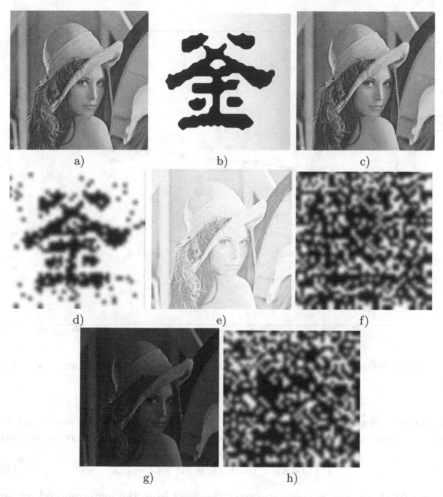

Fig. 1. ETC attack phase 1: a) Original host image; b) Original watermark; c) Watermarked image; d) Extracted watermark; e) Watermarked image with longer exposure time; f) Extracted watermark from e); g) Watermarked image with shorter exposure time; h) Extracted watermark from g)

image. As expected the larger extent of the exposure time change of the watermarked image, the harder for us to get the correct watermark during watermark decoding and detection.

3.4   Probability Analysis of the Results of Exposure Time Change

The aim of ETC attack is to decrease the cross-correlation score between the original watermark and the decoded watermark. It is expected that the

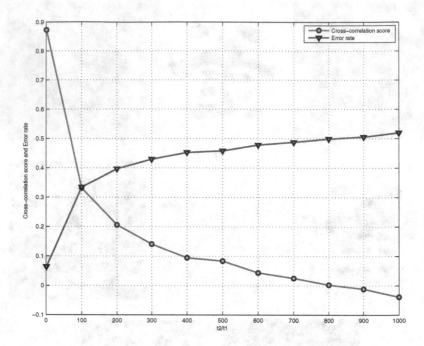

Fig. 2. The influence of longer exposure time on cross-correlation score s_1 and error rate

correlation between watermark decoded from the watermarked image and watermark decoded from the attacked image is weak. By probability term, we want

$$P\left(\mathrm{OWM}(\vec{i}) = \mathrm{ETCWM}(\vec{i})\right) = 0.5 \tag{13}$$

where OWM represents the watermark decoded from the watermarked image, ETCWM represents the watermark decoded from the attacked image, and $\vec{i}$ denotes the coordinate of watermarks.

For $P\left(\mathrm{OWM}(\vec{i}) = \mathrm{ETCWM}(\vec{i})\right) = 1$, it is obvious that the ETC is of no effect to the watermark decoding. For $P\left(\mathrm{OWM}(\vec{i}) = \mathrm{ETCWM}(\vec{i})\right) = 0$, we derive the opposite image of OWM, where 1 corresponds to 0 in OWM and 0 corresponds to 1 in OWM. By reversing the opposite image, we could get OWM. When $P\left(\mathrm{OWM}(\vec{i}) = \mathrm{ETCWM}(\vec{i})\right) = 0.5$, the attack effect is best. Neither detectors or human eyes could find the watermark in that case.

4   ETC Phase 2: Image Fusion

Changing exposure time causes the disparity between the original image and differently-exposed images, which are either darker or brighter than original

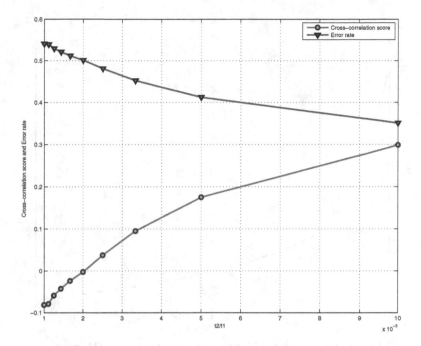

Fig. 3. The influence of shorter exposure time on cross-correlation score s_1 and error rate

image. In what follows, the disparity is larger as watermark detection becomes harder. The hope is that after a large exposure time is changed such that watermark detection becomes nearly impossible, the image after exposure time change can be turned to approximate to original image as closely as possible. We propose to use an image fusion method based on weighted average to combine differently-exposed images into a single composite image that is very similar to the original image for visual perception.

4.1   SSIM

We apply SSIM index [8] for measuring the disparity between the original image and differently-exposed images instead of PSNR or MSE because SSIM performs better with the qualitative visual appearance. When SSIM = 1, the disparity is non-existent. The two compared images look exactly the same. The smaller the SSIM index, the larger the disparity between the two compared images. The influence of exposure time change on SSIM index is shown in Fig. 4 and Fig. 5. Notably, the increase in exposure time change diminishes the similarity between the watermarked image and the original image. Note that the SSIM index between the original image and the watermarked image is 0.94. In comparison with Fig. 2 and Fig. 3, the problem now is how to keep balance between the similarity and the attack effect.

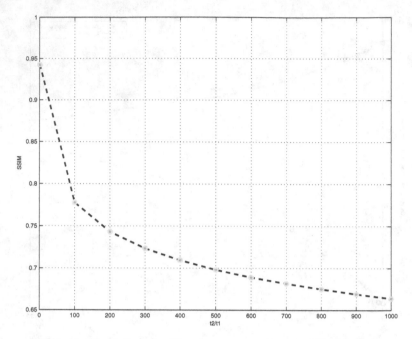

Fig. 4. The influence of longer exposure time on SSIM index

4.2   The Classification of Image Fusion Methods

Image fusion is the process of combining multiple images to offer an enhanced picture, which contains more desired features than input images. For a recent survey of image fusion research advances and challenges, the reader is referred to [9].

There are several types of image fusion methods. We distinguish between three main types based on their combination algorithms:

- *Average or weighted average method.* The resulting image pixel values are the arithmetic mean or the weighted average of corresponding pixel values of multiple images in the spatial domain. In general, it leads to a stabilization of the fusion result, whereas it introduces the problem of contrast reduction [10].
- *Image pyramids method* [11], [12]. A method first generates a sequence of image pyramids from each image. Then for each level, combination algorithm yields the composite image. At last, an inverse pyramid transform of the composite image is used to create the fused image.
- *Wavelet method* [13], [14]. A method similar to the image pyramids method. Instead, wavelet transform is used for the decomposition of the input images.

4.3   Fusion Algorithm

We have used three breeds of popular methods in image fusion mentioned above to combine differently-exposed images. Although the composite images constructed using the image pyramids method and the wavelet method are very

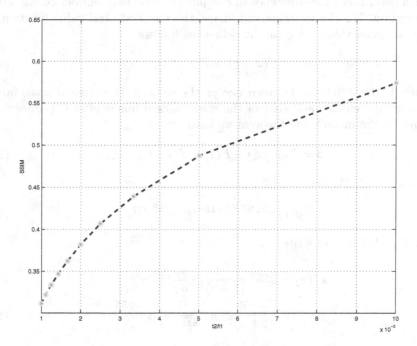

Fig. 5. The influence of shorter exposure time on SSIM index

similar to the original image, the attack effect is relatively poor when compared with the weighted average method. In addition, the image pyramids method and the wavelet method are more time-consuming than the weighted average method. Thus, We choose the weighted average approach to fuse images with different exposure time, which achieves better attack effect and is computationally efficient.

First we get a set of differently exposed images using method introduced in Sec. 3. Then each image is assigned to a weight. If there are k differently-exposed images $(x_1, x_2, \ldots, x_k)$ and their corresponding weights are $(\lambda_1, \lambda_2, \ldots, \lambda_k)$, the fused image is

$$x^{wave} = \frac{\sum_{i=1}^{k} \lambda_i x_i}{\sum_{i=1}^{k} \lambda_i}. \tag{14}$$

According to Eq. (12), when we change exposure time of one image from t_1 to t_2, for this change, the changed pixel value can be seen as a function of original pixel value. For the original image, the changed pixel value function F of original pixel value x is defined as

$$F(x) = \frac{255}{1 + e^{\ln(\frac{255}{x} - 1) - \frac{3}{4} \log_{10}(t_2/t_1)}}. \tag{15}$$

When we insert the watermark into the image to obtain the watermarked image either in the spatial domain or in the transform domain, we specify a parameter

ξ which determines the difference of one pixel's value between before and after the insertion. For the watermarked image, the changed pixel value function F of original pixel value $x + \xi$ can be defined as follows

$$F(x + \xi) = F(x) + \beta, \tag{16}$$

where β is the difference between one pixel's value of the original image after exposure time change and that of the watermarked image after exposure time change. By the mean value theorem we have

$$\beta = F(x + \xi) - F(x) = F'(x + \xi') \cdot \xi, \tag{17}$$

with $0 < \xi' < \xi$. We set

$$y(x) = e^{\ln(\frac{255}{x} - 1) - \frac{3}{4} \log_{10}(t2/t1)}. \tag{18}$$

Using Eq. (18) we can write

$$
\begin{aligned}
F'(x) &= 255 \cdot \frac{-1}{(1 + y)^2} \cdot \frac{dy}{dx}(x) \\
&= \frac{-255}{(1 + y)^2} \cdot y \cdot \frac{1}{\frac{255}{x} - 1} \cdot \frac{-255}{x^2} \\
&= 255^2 \cdot \frac{y}{(1 + y)^2} \cdot \frac{1}{(255 - x) \cdot x},
\end{aligned}
\tag{19}
$$

hence

$$\beta = 255^2 \cdot \frac{e^{\ln(\frac{255}{x + \xi'} - 1) - \frac{3}{4} \log_{10}(t2/t1)}}{(1 + e^{\ln(\frac{255}{x + \xi'} - 1) - \frac{3}{4} \log_{10}(t2/t1)})^2} \cdot \frac{1}{(255 - x - \xi') \cdot (x + \xi')} \cdot \xi. \tag{20}$$

Using Eq. (14) and (16), the pixel value $x_{p,q}^{wave}$ of the resultant fused image in the position (p, q) can be calculated as follows

$$
\begin{aligned}
x_{p,q}^{wave} &= \frac{\sum_{i=1}^{k} \lambda_i F_{i,p,q}(x + \xi)}{\sum_{i=1}^{k} \lambda_i} \\
&= \frac{\sum_{i=1}^{k} \lambda_i (F_{i,p,q}(x) + \beta_{i,p,q})}{\sum_{i=1}^{k} \lambda_i} \\
&= \frac{\sum_{i=1}^{k} \lambda_i F_{i,p,q}(x)}{\sum_{i=1}^{k} \lambda_i} + \frac{\sum_{i=1}^{k} \lambda_i \beta_{i,p,q}}{\sum_{i=1}^{k} \lambda_i},
\end{aligned}
\tag{21}
$$

where $F_{i,p,q}(x + \xi)$ is the pixel value of the i-th watermarked image after exposure time change in the position (p, q). Our purpose is that the resultant fused image is similar to the original image, while the watermark is nearly removed. The hope is to select the optimum weights to achieve our purposes. That is, after choosing optimum weight λ_i for each image to be fused, $\frac{\sum_{i=1}^{k} \lambda_i F_{i,p,q}(x)}{\sum_{i=1}^{k} \lambda_i}$ approaches to x

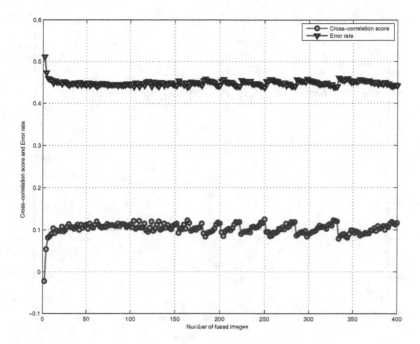

Fig. 6. The influence of the number of fused images on cross-correlation score s_1 and error rate

and $\frac{\sum_{i=1}^{k} \lambda_i \beta_{i,p,q}}{\sum_{i=1}^{k} \lambda_i}$ tends toward 0. We are currently investigating a method for selecting the optimum weights. In this paper, each image is assigned to a weight that depends on the extent of the exposure time change. The larger extent of the exposure time changes as compared to the watermarked image, the larger the weight.

4.4 Effect of Image Fusion

In our experiment, we choose images from a database of 2000 images with different exposure time as compared to the watermarked image. Fig. 6 illustrates how the cross-correlation score s_1 and error rate change as the number of fused image increases. Note that the error rates approximate to 0.5, which is a satisfactory result. The reasons can be seen in Sec. 3.4. Fig. 7 illustrates how the number of fused image influences the SSIM index. It can be observed that the attack effect of ETC phase 1 is preserved in the fusion process and the fused images are similar to the original image. The fused image from 400 images with different exposure time is shown in Fig. 8a. It can be seen that the fused image achieves good perceptual quality. Using the watermark decoding algorithm, we extract the watermark image from the fused image as shown in Fig. 8b. Clearly, the watermark cannot be successfully detected.

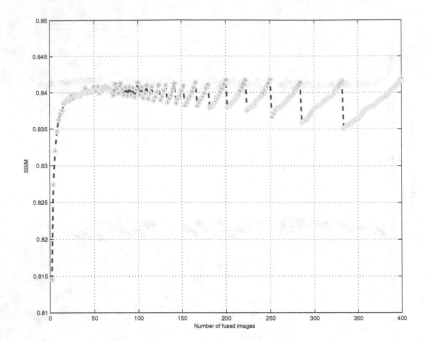

Fig. 7. The influence of the number of fused images on SSIM index

a)   b)

Fig. 8. ETC attack phase 2: a) Fused image; b) Decoded watermark from a).

5   Conclusion

In this paper, a new watermark attack paradigm ETC has been analyzed in detail. We find that many image watermarking systems can not derive the correct decoded watermark when the watermarked image is exposed differently. Both theoretical and experimental analyses have been given to support the proposed attack. The results show that our attack is successful with a reasonable image quality.

References

1. Kirovski, D., Petitcolas, F.: Blind pattern matching attack on watermarking systems. IEEE Transactions on Signal Processing 51, 1045–1053 (2003)
2. Maes, M.: Twin peaks: the histogram attack to fixed depth image watermarks. In: Aucsmith, D. (ed.) IH 1998. LNCS, vol. 1525, pp. 290–305. Springer, Heidelberg (1998)
3. Cox, I., Kilian, J., Leighton, F., Shamoon, T.: Secure spread spectrum watermarking for multimedia. IEEE Transactions on Image Processing 6, 1673–1687 (1997)
4. Xia, X., Boncelet, C., Arce, G.: A multiresolution watermark for digital images. In: Proceedings of the International Conference on Image Processing, vol. 1 (1997)
5. Wong, P.: Image watermarking and data hiding techniques. PhD thesis, Department of Electronic and Computer Engineering. The Hong Kong University of Science and Technology, Hong Kong, China (2003),
   http://www.ece.ust.hk/~eepeter/paperpdf/Peter.PhD.Thesis.pdf
6. Debevec, P., Malik, J.: Recovering high dynamic range radiance maps from photographs. In: Proceedings of the 24th Annual Conference on Computer Graphics and Interactive Techniques, pp. 369–378 (1997)
7. Bhukhanwala, S., Ramabadran, T.: Automated global enhancement of digitized photographs. IEEE Transactions on Consumer Electronics 40, 1–10 (1994)
8. Wang, Z., Bovik, A., Sheikh, H., Simoncelli, E.: Image quality assessment: from error visibility to structural similarity. IEEE Transactions on Image Processing 13, 600–612 (2004)
9. Ardeshir Goshtasby, A., Nikolov, S.: Image fusion: advances in the state of the art. Information Fusion 8, 114–118 (2007)
10. Piella, G.: A general framework for multiresolution image fusion: from pixels to regions. Information Fusion 4, 259–280 (2003)
11. Burt, P.: The pyramid as a structure for efficient computation. Multiresolution Image Processing and Analysis 35 (1984)
12. Sims, S., Phillips, M.: Target signature consistency of image data fusion alternatives. Optical Engineering 36, 743 (1997)
13. Zhang, Z., Blum, R.: A categorization of multiscale-decomposition-based image fusion schemes with a performance study for a digital camera application. Proceedings of the IEEE 87, 1315–1326 (1999)
14. Rockinger, O., Fechner, T.: Pixel-level image fusion: the case of image sequences. In: Proceedings of the SPIE, vol. 3374, pp. 378–388 (1998)

Steganalysis Based on Difference Image

Yifeng Sun[1,2], Fenlin Liu[1], Bin Liu[1], and Ping Wang[1]

[1] Zhengzhou Information Science and Technology Institute,
Zhengzhou, China, 450002
[2] Electronic Technology Institute, Zhengzhou, China, 450004
yfsun001@163.com

Abstract. This paper presents a performance comparison between Harmsen's method and a family of Steganalysis Method based on Difference Image (SMDI), in order to promote steganalysis theory. The theoretical analysis begins from the commonness of SMDI, and is focused on the assessment of the statistical change degree between cover and stego, showing that SMDI outperforms the Harmsen's method. The paper also analyzes that the improvement owes to two aspects: the larger variance of stego-noise difference and the correlation of adjacent pixels utilized by SMDI. A new detector, which can use the correlation of adjacent pixels in all directions, is proposed by generalizing the definition of difference. Experiments show the better performance of the new detector.

Keywords: Steganalysis, Steganography, Difference Image, Additive Noise, Gaussian Mixture Model.

1 Introduction

The aim of steganography is to hide information imperceptibly into a cover, so that the presence of hidden data cannot be diagnosed. Steganalysis aims to expose the presence of hidden data. Most steganalytical algorithms fall into two major types: embedding specific [1-4] and universal [5-18]. Universal (blind) steganalysis attempts to detect the presence of an embedded message independent of the embedding algorithm and, ideally, the image format. This paper mainly discusses universal steganalysis. The basis of universal steganalysis is that there exists difference between cover and stego. The key issue for steganalyzers is discovering some statistical features that can embody the difference between cover and stego.

In the state of the art, features are mainly extracted from the histogram of image pixels (or transform coefficients) [6-11] and the co-occurrence matrix of adjacent image pixels (or transform coefficients) [12-18]. The literatures [6-8] are all based on histogram moments. Harmsen et al. [9] proposed an important steganalysis method which exploits the change of the histogram characteristic function center of mass (HCF COM) after embedding. The additive stego-noise model in the method accurately describes the difference of the probability density function (PDF) between cover image and stego image. Xuan et al. [10] extended HCF COM to multiple-order characteristic function (CF) moments. Wang et al. [11] proved that the CF moments are more sensitive features than PDF moments. In the co-occurrence matrix based

H.J. Kim, S. Katzenbeisser, and A.T.S. Ho (Eds.): IWDW 2008, LNCS 5450, pp. 184–198, 2009.
© Springer-Verlag Berlin Heidelberg 2009

steganalysis methods, some literatures [12-14] directly used the part or total elements of co-occurrence matrix as features. Ker [15] proposed the adjacency HCF COM features which are statistical analysis of co-occurrence matrix. In [16], the marginal moments of 2-D CF are essentially the extensions of adjacency HCF COM. Liu et al. [17] extended co-occurrence matrix to the histogram of the nearest neighbors (NNH) and used the entropy of NNH as features.

Recently, some steganalysis methods [3,24-26] utilized difference image data to attack steganography. They can be classified to a family of Steganalysis Method based on Difference image (SMDI). Zhang [3] first utilized the difference image to attack LSB steganography. Based on pixel difference, He and Huang [24] proposed a method of estimating secret message length for stochastic modulation steganography [27]. Chen [25] gave a universal steganalysis method which used the part of difference histogram as features. Liu [26] proposed another universal steganalysis method based on statistics of the partial differentiation and total differentiation.

There are questions one may ask: Is better the universal method in SMDI family than the other methods? Is there a mathematical explanation for the superiority of method? This paper investigates the above questions. We begin from the commonness of SMDI, and analyze the SMDI performance by measuring the statistical change degree between cover and stego. We prove SMDI outperforms the Harmsen's method [9], due to the larger variance of stego-noise difference and the correlation of adjacent pixels utilized by SMDI. Furthermore, we give a new detector by generalizing the definition of difference, which can use the correlation of adjacent pixels in all directions. Experiments show the better performance of our detector in detecting spread-spectrum image steganography (SSIS) [18] and LSB matching [15,19].

2 The Commonness in SMDI

The methods [3,24-26] in SMDI family are all based on the distribution model of stego-image's pixel difference which is related to that of stego-noise difference. It has been the commonness of SMDI family.

Denote the intensity value of the image I at the position (i,j) as $I(i,j)$, and the pixel difference at the position (i,j) is defined by

$$D(i,j)= I(i,j)- I(i+1, j) \ or \ D(i,j)= I(i,j)- I(i, j+1). \tag{1}$$

$D=\{D(i,j)\}$ can be called as difference image[3]. Denote cover image signal as random variable I_c, and stego image signal as I_s. Let $P_c(x)$, $P_s(x)$ be the PDF of I_c, I_s respectively. Correspondingly, denote cover difference image signal as random variable D_c, and stego difference image signal as D_s. Let $q_c(x)$, $q_s(x)$ be the PDF of D_c, D_s respectively. From (1),

$$D_c= I_{c,1}- I_{c,2}, \ D_s= I_{s,1}- I_{s,2}. \tag{2}$$

$I_{c,1}$, $I_{s,1}$ refers to the cover and stego signal at the position (i,j) respectively, and $I_{c,2}$, $I_{s,2}$ refers to the cover and stego signal at the position $(i+1, j)$ or $(i, j+1)$.

Data hiding in the spatial domain, such as spread-spectrum image steganography (SSIS) [18] and LSB matching [15,19], can be modeled as adding stego-noise signal Z to cover image signal I_c [9]. Thus

$$I_s = I_c + Z. \tag{3}$$

The PDF of Z is denoted as $P_z(x)$. Z is independent of I_c. The literature [24] uses the pixel difference to describe data hiding.

$$D_s = (I_{c,1} + Z_1) - (I_{c,2} + Z_2) = D_C + (Z_1 - Z_2). \tag{4}$$

Z_1 and Z_2 are the stego-noise signal corresponding to $I_{c,1}$ and $I_{c,2}$. Z_1, Z_2 and Z are independent and identically distributed. Here Z_1-Z_2 is also named as stego-noise difference signal.

$$Var(Z_1 - Z_2) = 2 \cdot Var(Z) \tag{5}$$

where Var refers to the variance of random variable.

$$q_s(x) = q_c(x) * q_z(x) = \int q_c(x) q_z(x-s)ds. \tag{6}$$

where $q_z(x)$ be the PDF of Z_1-Z_2. The expression (5) shows that the intensity of stego-noise difference signal is twice as large as that of stego-noise signal. The expression (6) shows the distribution of stego difference image signal is the convolution of the distribution of cover difference image signal and that of stego-noise difference signal, and reflects the statistical change between cover and stego. The steganalysis methods in SMDI family, such as [24-26], are all based on (5) and (6). From (5) and (6), universal steganalysis method attempts to extract features. Chen [25] used some special part of difference histogram as features. Liu [26] used characteristic function moments of the partial differentiation and total differentiation as features.

3   Performance Analysis of SMDI

3.1   PDF of Difference Image Signal

Here we first analyze the relationship of the PDF of cover image signal and that of cover difference image signal. It will be used during performance analysis. It is generally believed that $q_c(x)$ is Generalized Gaussian Distribution (GGD)[20]. But the

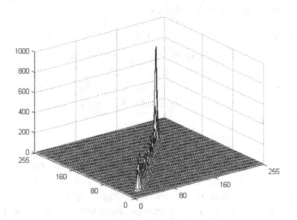

Fig. 1. Two dimensional histogram of a gray image

model cannot show the relationship between $P_c(x)$ and $q_c(x)$. Because D_c is derived from I_c in (2), there is some relation between $P_c(x)$ and $q_c(x)$. To get the relationship, we build the model of the joint probability density of $I_{c,1}$ and $I_{c,2}$.

Denote the joint probability density of $I_{c,1}$ and $I_{c,2}$ as $f_c(x_1, x_2)$. $f_c(x_1, x_2)$ usually has multiple peaks shape. For example, Fig.1 shows the two dimensional histogram of horizontal adjacent pixels in a gray image. Multiple peaks occur around its main diagonal in Fig.1. Finite mixture models, such as Gaussian Mixture Model (GMM), usually well approximate the complicated and unknown distributional shape [21]. Here GMM is used as the model of $f_c(x_1, x_2)$.

$$f_c(x_1,x_2) = \sum_i a_i \cdot f_{c,i}(x_1,x_2).$$ (7)

$$f_{c,i}(x_1,x_2) = \frac{1}{\sqrt{2\pi}\sigma_i^2\sqrt{1-\rho_i^2}} \exp\{\frac{-1}{2(1-\rho_i^2)}[\frac{(x_1-u_i)^2}{\sigma_i^2} - 2\rho_i\frac{(x_1-u_i)(x_2-u_i)}{\sigma_i^2} + \frac{(x_2-u_i)^2}{\sigma_i^2}]\}.$$ (8)

$$\sum_i a_i = 1 \quad 0 \le a_i \le 1.$$ (9)

$f_c(x_1, x_2)$ is mixture of several child two dimensional Gauss distribution $f_{c,i}(x_1, x_2)$. Every child distribution $f_{c,i}(x_1, x_2)$ is weighed by a_i. The two margin distributions of $f_{c,i}(x_1, x_2)$ have the same mean u_i and variance σ_i. The reason is that the local area of image can be regarded as stationary process, especially for two adjacent pixels. The correlation coefficient ρ_i reflects the correlation between the two adjacent pixel signals. $0 \le \rho_i \le 1$. The bigger ρ_i is, the stronger the correlation is, and the peak is more flat and its energy concentrates in diagonal.

The relation between $q_c(x)$ and $p_c(x)$ can be obtained by the integral of $f_c(x_1, x_2)$. Let the value of $I_{c,1}$ be x_1, and the value of $I_{c,2}$ be x_2. The difference value is $x=x_1-x_2$, then

$$q_c(x) = \int_{-\infty}^{\infty} f_c(x_1, x_1 - x)dx_1 = \sum_i a_i q_{c,i}(x).$$ (10)

$$q_{c,i}(x) = \int_{-\infty}^{\infty} f_{c,i}(x_1, x_1 - x)dx_1 = \frac{1}{\sqrt{2\pi}(\sqrt{2(1-\rho_i)}\sigma_i)} \exp\{-\frac{x^2}{2(\sqrt{2(1-\rho_i)}\sigma_i)^2}\}.$$ (11)

Let $\sigma_i' = \sqrt{2(1-\rho_i)}\sigma_i$,

$$q_{c,i}(x) = \frac{1}{\sqrt{2\pi}\sigma_i'} \exp\{-\frac{x^2}{2\sigma_i'^2}\}.$$ (12)

$p_c(x)$ is the margin distribution of $f_c(x_1, x_2)$

$$p_c(x) = \int_{-\infty}^{\infty} f_c(x_1, x)dx_1 = \sum_i a_i p_{c,i}(x).$$ (13)

$$p_{c,i}(x) = \int_{-\infty}^{\infty} f_{c,i}(x_1, x)dx_1 = \frac{1}{\sqrt{2\pi}\sigma_i} \exp\{-\frac{(x-u_i)^2}{2\sigma_i^2}\}.$$ (14)

From (10)-(14), both $q_c(x)$ and $p_c(x)$ are GMM. But the mean of $p_{c,i}(x)$ is different resulting in the multiple peaks shape of the mixture $p_c(x)$, while each $q_{c,i}(x)$ has the same mean zero and the mixture $q_c(x)$ is still one peak. The variance σ_i' of $q_{c,i}(x)$ and σ_i of $p_{c,i}(x)$ have the relationship: $\sigma_i' \leq \sigma_i$ if $\rho_i \geq 0.5$. It shows that the shape of $q_{c,i}(x)$ is more peaky than that of $p_{c,i}(x)$ because of the correlation of adjacent pixels.

3.2 Two Improvements Compared with Harmsen's Method

In this section, the paper gives a performance comparison between the universal steganalysis method in SMDI family and Harmsen's method [9]. The performance of universal steganalysis is influenced by features. The change degree of feature can be used as performance measurement [11]. But different algorithms, such as [25, 26], in SMDI family have different features. It makes the performance comparison difficult. Without loss of generality, we adopt the 'general' statistical change between cover and stego as performance measurement. The more the degree of change is, the better the performance is.

According to Harmsen's additive noise method [9],

$$p_s(x) = \int p_c(x) p_z(x - s) ds . \tag{15}$$

Characteristic function (CF) is Fourier transform of PDF. Apply Fourier transform to (15) and get

$$\left| \Phi_{p,s}(t) \right|^2 = \left| \Phi_{p,c}(t) \right|^2 \cdot \left| \Phi_{p,z}(t) \right|^2 \tag{16}$$

where $\Phi_{p,c}(t)$, $\Phi_{p,s}(t)$, and $\Phi_{p,z}(t)$ is the CF of $p_c(x)$, $p_s(x)$, and $p_z(x)$ respectively. $\left| \Phi_{p,c}(t) \right|^2$ and $\left| \Phi_{p,s}(t) \right|^2$ are the amplitude spectrum square of $q_c(x)$ and $q_s(x)$. Correspondingly, we can get (17) from (6) for SMDI.

$$\left| \Phi_{q,s}(t) \right|^2 = \left| \Phi_{q,c}(t) \right|^2 \cdot \left| \Phi_{q,z}(t) \right|^2 \tag{17}$$

where $\Phi_{q,c}(t)$, $\Phi_{q,s}(t)$ and $\Phi_{q,z}(t)$ is the CF of $q_c(x)$, $q_s(x)$ and $q_z(x)$ respectively. $\left| \Phi_{q,c}(t) \right|^2$ and $\left| \Phi_{q,s}(t) \right|^2$ are the amplitude spectrum square of $q_c(x)$ and $q_s(x)$. From (16) and (17), the embedding equals to a filter. Both (16) and (17) can capture the general statistical change of cover and stego. The statistical change is embodied in every frequency point t of the amplitude spectrum square. But which is more informative in terms of distinguishing cover and stego? SMDI outperforms Harmsen's method in two aspects.

Firstly, we discuss the normalized statistical change between cover and stego. Here the normalized statistical change is the ratio of statistic between cover and stego. At every frequency point t , the normalized statistical change of Harmsen's method can be described by

$$\left| \Phi_{p,s}(t) \right|^2 / \left| \Phi_{p,c}(t) \right|^2 = \left| \Phi_{p,z}(t) \right|^2 , \tag{18}$$

and the normalized statistical change of SMDI can be described by

$$\left|\Phi_{q,s}(t)\right|^2 / \left|\Phi_{q,c}(t)\right|^2 = \left|\Phi_{q,z}(t)\right|^2 . \tag{19}$$

In the case of SSIS [18], Z is distributed as Gaussian $N(0,\sigma^2)$. $Z_1 - Z_2$ is distributed as Gaussian $N(0,2\sigma^2)$. Thus

$$\left|\Phi_{p,z}(t)\right|^2 = \left|\int_{-\infty}^{+\infty} p_z(x)e^{jtx}dx\right|^2 = e^{-\sigma^2 t^2} \leq 1. \tag{20}$$

$$\left|\Phi_{q,z}(t)\right|^2 = \left|\int_{-\infty}^{+\infty} q_z(x)e^{jtx}dx\right|^2 = e^{-2\sigma^2 t^2} \leq 1. \tag{21}$$

In terms of discriminating between cover and stego, the smaller the ratio in (18) or (19), the more informative it is. Because $e^{-2\sigma^2 t^2} \leq e^{-\sigma^2 t^2}$, SMDI is better. The improvement owes to stego-noise difference. The stego-noise difference Z_1-Z_2 has the larger variance $2\sigma^2$. It is equal to enlarging the intensity of stego-noise. The larger the intensity of stego-noise is, the easier the detection is.

In contrast to the normalized statistical change, the 'absolute' statistical change between cover and stego refers to the difference of statistic between cover and stego. Here at every frequency point t, the 'absolute' statistical change of Harmsen's method is defined as

$$\left|\Phi_{p,c}(t)\right|^2 - \left|\Phi_{p,s}(t)\right|^2 = \left|\Phi_{p,c}(t)\right|^2 \cdot (1 - \left|\Phi_{p,z}(t)\right|^2), \tag{22}$$

and the 'absolute' statistical change of SMDI is defined as

$$\left|\Phi_{q,c}(t)\right|^2 - \left|\Phi_{q,s}(t)\right|^2 = \left|\Phi_{q,c}(t)\right|^2 \cdot (1 - \left|\Phi_{q,z}(t)\right|^2). \tag{23}$$

Proposition1: The amplitude spectrum square $\left|\Phi_{q,c}(t)\right|^2$ of $q_c(x)$ is larger than the amplitude spectrum square $\left|\Phi_{p,c}(t)\right|^2$ of $p_c(x)$ if each $\rho_i \geq 0.5$ where ρ_i is the correlation coefficient in child distribution $f_{c,i}(x_1,x_2)$.

Proof:

$$\Phi_{p,c}(t) = \int_{-\infty}^{\infty} p_c(x) \cdot \exp(jtx)dx = \sum_i a_i \exp(ju_i t - \frac{\sigma_i^2 t^2}{2}). \tag{24}$$

$$\left|\Phi_{p,c}(t)\right|^2 = \sum_i a_i^2 \cdot \exp(-\sigma_i^2 t^2) + \sum_{\substack{i,j \\ i \neq j}} 2a_i a_j \cdot \exp(-\frac{\sigma_i^2 + \sigma_j^2}{2}t^2) \cdot \cos(u_i - u_j)t^{\,1}. \tag{25}$$

where $u_i - u_j \geq 0$.

$$\Phi_{q,c}(t) = \int_{-\infty}^{\infty} q_c(x) \cdot \exp(jtx)dx = \sum_i a_i \exp(-\frac{\sigma_i^2 t^2}{2}). \tag{26}$$

The cosine function results to vibrating characteristic. It is shown in Fig.2(b).

$$\left|\Phi_{q,c}(t)\right|^2 = \sum_i a_i^2 \cdot \exp(-\sigma_i'^2 t^2) + \sum_{\substack{i,j \\ i \neq j}} 2a_i a_j \cdot \exp(-\frac{\sigma_i'^2 + \sigma_j'^2}{2} t^2) . \tag{27}$$

$$\left|\Phi_{q,c}(t)\right|^2 - \left|\Phi_{p,c}(t)\right|^2 = \sum_i a_i^2 \cdot \left[\exp(-\sigma_i'^2 t^2) - \exp(-\sigma_i^2 t^2)\right]$$

$$+ \sum_{\substack{i,j \\ i \neq j}} 2a_i a_j \cdot \left[\exp(-\frac{\sigma_i'^2 + \sigma_j'^2}{2} t^2) - \exp(-\frac{\sigma_i^2 + \sigma_j^2}{2} t^2) \cdot \cos(u_i - u_j)t \right] \tag{28}$$

If $\rho_i \geq 0.5$, $\sigma_i' \leq \sigma_i$,

$$\exp(-\sigma_i'^2 t^2) - \exp(-\sigma_i^2 t^2) \geq 0, \tag{29}$$

$$\exp(-\frac{\sigma_i'^2 + \sigma_j'^2}{2} t^2) - \exp(-\frac{\sigma_i^2 + \sigma_j^2}{2} t^2) \cdot \cos(u_i - u_j)t \geq 0. \tag{30}$$

So $\left|\Phi_{q,c}(t)\right|^2 \geq \left|\Phi_{p,c}(t)\right|^2$ if each $\rho_i \geq 0.5$.   □

Proposition2: In the case of SSIS, given $\left|\Phi_{q,c}(t)\right|^2$, $\left|\Phi_{p,c}(t)\right|^2$, $\left|\Phi_{q,s}(t)\right|^2$ and $\left|\Phi_{p,s}(t)\right|^2$, the 'absolute' statistical change $\left|\Phi_{q,c}(t)\right|^2 - \left|\Phi_{q,s}(t)\right|^2 \geq \left|\Phi_{p,c}(t)\right|^2 - \left|\Phi_{p,s}(t)\right|^2 \geq 0$ at every frequency point t if each $\rho_i \geq 0.5$ where ρ_i is the correlation coefficient in child distribution $f_{c,i}(x_1, x_2)$.

Proof: In the case of SSIS,

$$\left|\Phi_{p,c}(t)\right|^2 - \left|\Phi_{p,s}(t)\right|^2 = \left|\Phi_{p,c}(t)\right|^2 \cdot (1 - e^{-\sigma^2 t^2}) . \tag{31}$$

$$\left|\Phi_{q,c}(t)\right|^2 - \left|\Phi_{q,s}(t)\right|^2 = \left|\Phi_{q,c}(t)\right|^2 \cdot (1 - e^{-2\sigma^2 t^2}) . \tag{32}$$

$$1 - \exp(-2\sigma^2 t^2) \geq 1 - \exp(-\sigma^2 t^2) \geq 0, \tag{33}$$

From Propostion1, if each $\rho_i \geq 0.5$, $\left|\Phi_{q,c}(t)\right|^2 \geq \left|\Phi_{p,c}(t)\right|^2 \geq 0$, and so

$$\left|\Phi_{q,c}(t)\right|^2 - \left|\Phi_{q,s}(t)\right|^2 \geq \left|\Phi_{p,c}(t)\right|^2 - \left|\Phi_{p,s}(t)\right|^2 \geq 0 \qquad □$$

Proposition2 shows the 'absolute' statistical change of SMDI is larger than that of Harmsen's method under the condition of each $\rho_i \geq 0.5$. The condition is usually satisfied because of the strong correlation of adjacent pixels. The correlation embodied by the PDF of pixel difference is the second reason for SMDI's better performance.

As a conclusion, SMDI outperforms Harmsen's method because SMDI utilizes the larger variance of stego-noise difference and the correlation of adjacent pixels. Fig.2 shows that the amplitude spectrum square of difference histogram is more informative in terms of distinguishing cover and stego than that of histogram.

(a) gray image

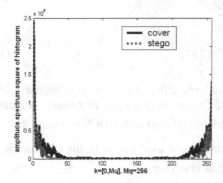

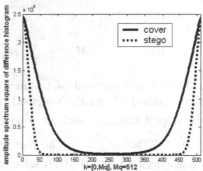

(b) amplitude spectrum square of histogram

(c) amplitude spectrum square of difference histogram

Fig. 2. amplitude spectrum square of histogram and difference histogram before and after SSIS embedding with max embedding rate for a gray image. The stego-noise variance $\sigma^2 = 4$.

4   New Steganalysis Method Based on Generalized Difference Image

From section 3.2, SMDI utilizing the correlation of adjacent pixels is one reason for better performance. But from the definition of difference, the expression (1), only the horizontal correlation or the vertical correlation is utilized. If we generalize the definition of difference as following, the correlation in other direction can all be used.

The generalized difference at the position (i,j) is defined by

$$D^{r,\theta}(i, j) = I(i, j) - I(i + r\cos\theta, j + r\sin\theta) . \tag{34}$$

where $I(i,j)$ and $I(i+r\cos\theta, j+r\sin\theta)$ represents two adjacent pixels which have correlation characteristic. The parameter r is the step, and θ is the direction. $D^{r,\theta} = \{D^{r,\theta}(i, j)\}$ is the generalized difference image. The difference image in [3, 24-26] is the special case of the generalized one. When r, θ varies, the PDF of difference image $D^{r,\theta}$ can reflect correlation in different direction and distance.

Based on (17), we use Difference Characteristic Function (DCF) moments [26] as features. The histogram is used as the estimation of PDF. Let $\{h_q^{r,\theta}(m)\}_{m=0}^{M_q-1}$ be M_q-bin histogram of $D^{r,\theta}$. Difference Characteristic Function (DCF) is denoted as $\{\Phi_q^{r,\theta}(k)\}_{k=0}^{M_q-1}$ which is calculated by

$$\Phi_q^{r,\theta}(k) = \sum_{m=0}^{M_q-1} h_q^{r,\theta}(m)\exp(\frac{j2\pi nk}{K}). \tag{35}$$

The n order DCF moments $\hat{M}_q^{n,r,\theta}$ are calculated by

$$\hat{M}_q^{n,r,\theta} = \frac{M_q^{n,r,\theta}}{M_q^{0,r,\theta}}, \tag{36}$$

$$M_q^{n,r,\theta} = \sum_{k=0}^{M_q/2-1} \left|\Phi_q^{r,\theta}(k)\right| \cdot k^n. \tag{37}$$

The detector is composed of features and classifier. In order to compare with Harmsen's method [9] and Ker's method [15] which only use the first order moment, we only adopt 4 features which are four directional first order DCF moments $\hat{M}_q^{1,1,0}$, $\hat{M}_q^{1,1,\pi/2}$, $\hat{M}_q^{1,\sqrt{2},\pi/4}$ and $\hat{M}_q^{1,\sqrt{2},-\pi/4}$. The four features are used as the inputs of classifier. Fisher linear discriminator (FLD) is selected as classifier.

5 Experiments

In the experiments, the cover database consists of 3000 images which were downloaded from USDA NPCS Photo Gallery [22]. The images are as very high resolution TIF files (mostly 2100×1500) and appear to be scanned from a variety of paper and film sources. The database is difficult to detect steganography [8]. In the experiments, we only discuss the problem of detecting steganography that is performed in the image spatial domain, such as SSIS [19] and LSB matching [15,19]. The reason is that it is believed that the best (most sensitive) features are obtained when they are calculated directly in the embedding domain [8]. For testing, the cover images were resampled to 614×418 and converted to grayscale(The tool used is Advanced Batch Converter 3.8.20, and the selected interpolation filter is bilinear). They are regarded as seeds to generate 2 groups of 3000 stego images of LSB matching and SSIS. The max embedding rate is adopted. The variance of Gaussian stego-noise σ^2 is 4 in SSIS.

5.1 Feature Evaluation

It has been proved that SMDI is better than Harmsen's method in terms of distinguishing cover and stego in section 3.2. By experiments, we show the excellent performance of DCF moment $\hat{M}_q^{n,1,0}$, comparing to the HCF moment [9,10] and the adjacency CF (ACF) moment which is the extension of Ker's adjacency HCF COM [15].

Denote M_p-bin image histogram as $\{h_p(m)\}_{m=0}^{M_p-1}$, and the corresponding CF as $\{\phi_p(k)\}_{k=0}^{M_p-1}$. The n order HCF moment $\hat{M}_p^n$ is calculated by:

$$\hat{M}_p^n = (\sum_{k=0}^{M_p/2-1} |\Phi_p(k)| \cdot k^n) / \sum_{k=0}^{M_p/2-1} |\Phi_p(k)| . \tag{38}$$

The two dimensional adjacency histogram $\{h_p(m,n)\}$ is essentially $M_p \times M_p$ co-occurrence matrix [15], and the corresponding CF $\{H_p(k,l)\}$ can be calculated by two dimensional discrete Fourier transform of $\{h_p(m,n)\}$. The n order ACF moment $\hat{M}_A^n$ is calculated by:

$$\hat{M}_A^n = \{ \sum_{k,l=0}^{M_p/2-1} |H_p(k,l)| \cdot (k+l)^n \} / \sum_{k,l=0}^{M_p/2-1} |H_p(k,l)| . \tag{39}$$

The impact of feature on classification accuracy is determined by the feature distribution. Several criteria such as the Bhattacharyya distance from the pattern recognition and machine learning literature may be used to evaluate the usefulness of a feature in discriminating between classes [23]. The Bhattacharyya distance is defined as the following:

$$B(p_0, p_1) = -\log \int \sqrt{p_0(x)p_1(x)} . \tag{40}$$

where x is a feature, and $p_0(x)$ and $p_1(x)$ are the feature PDFs under Class 0 and Class 1, respectively. The larger the $B(p_0,p_1)$ is for a feature, the better that feature is for classification. When $p_0=p_1$, $B(p_0,p_1) =0$ and the feature is useless. In practice, p_0 and p_1 are often unavailable, and we use their histogram estimation from train features and compute the empirical Bhattacharyya distance in experiments. The Bhattacharyya distance of DCF moments $\hat{M}_q^{n,1,0}$ is compared with that of HCF moments $\hat{M}_p^n$ and ACF moments $\hat{M}_A^n$ in Fig. 3 and Fig. 4. Fig. 3 show the case of distinguishing cover

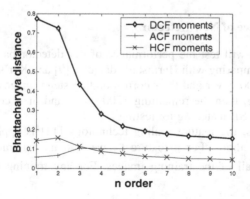

Fig. 3. Empirical Bhattacharyya distance for DCF moment $\hat{M}_q^{n,1,0}$, ACF moment $\hat{M}_A^n$ and HCF moment $\hat{M}_p^n$. Data are gathered from the cover database and the stego database of SSIS where the additive noise variance $\sigma^2 = 4$ with max embedding rate.

and SSIS stego, and Fig. 4 show the case of distinguishing cover and LSB matching stego. The two figures show that DCF moment feature is a better feature than ACF moment and HCF moment since the empirical Bhattacharyya distance of DCF moments is larger than others. On the other hand, the Bhattacharyya distance of the first order DCF moment $\hat{M}_q^{1,1,0}$ is 0.774 in Fig. 3, while it is 0.214 in Fig.4. It shows that DCF moments are better in detecting SSIS than detecting LSB matching. The reason may be that the stego-noise variance of LSB matching is lower than that of SSIS. It validates the conclusion that the larger the intensity of stego-noise is, the easier it is to detect. So the idea of enlarging the variance of stego-noise is useful to improve the performance of steganalysis.

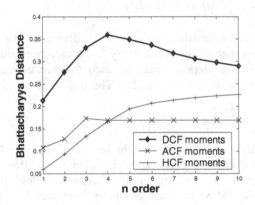

Fig. 4. Empirical Bhattacharyya distance for DCF moment $\hat{M}_q^{n,1,0}$, ACF moment $\hat{M}_A^n$ and HCF moment $\hat{M}_p^n$. Data are gathered from the cover database and the stego database of LSB matching with max embedding rate.

5.2   The Performance of New Detector

In the following, we will test the performance of our detector based on generalized difference image, comparing with Harmsen's detector [9] and Ker's detector [15]. We randomly choose 1500 cover and their corresponding stego images of SSIS and LSB matching for training, then the remaining 1500 cover and their corresponding stego images of SSIS and LSB matching for testing.

In the experiments, the feature scaling technology [11] is applied for the above three steganalysis methods. For a feature f , we find its maximum value f_{max} and minimum f_{min} from all of the training images. For any training or test image, the feature f is scaled as

$$\tilde{f} = \frac{f - f_{min}}{f_{max} - f_{min}}.$$

(41)

For all of the train images, $\tilde{f} \in [0,1]$; For most test images, most of $\tilde{f}$ will also be between 0 and 1. This scale step prevents features with large numerical ranges from dominating those with numerical ranges.

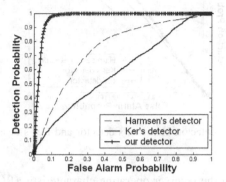

Fig. 5. ROC curves of our detector, Harmsen's detector and Ker's detector for SSIS with max embedding rate. The noise variance $\sigma^2 = 4$.

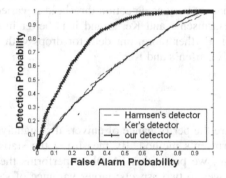

Fig. 6. ROC curves of our detector, Harmsen's detector and Ker's detector for SSIS with 50% embedding rate. The noise variance $\sigma^2 = 4$.

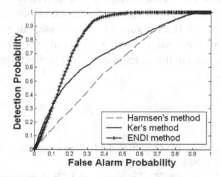

Fig. 7. ROC curves of our detector, Harmsen's detector and Ker's detector for LSB matching

with max embedding rate

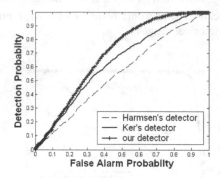

Fig. 8. ROC curves of our detector, Harmsen's detector and Ker's detector for LSB matching with 50% embedding rate

Fig.5 and Fig.6 show the receiver operating characteristic (ROC) curves for SSIS, and Fig.7 and Fig.8 show the ROC curves for LSB matching. The ROC curves show how the detection probability (the fraction of the stego images that are correctly classified) and the false alarm probability (the fraction of the cover images that are misclassified as as stego images) vary as detection threshold is varied. From figures, our detector is better than Harmsen's and Ker's, and it is better in detecting SSIS than detecting LSB matching. Further more, our detector drops with the embedding rate, but it is still better than Harmsen's and Ker's.

6 Conclusion

In this paper, we compare the performance of universal steganalysis method in SMDI family with that of Harmsen's method. By measuring the statistical change degree between cover and stego, we prove that SMDI outperforms the Harmsen's method [9]. The improvement owes to two aspects: larger variance of stego-noise difference and the correlation of adjacent pixels utilized by SMDI. Based on the second reason for better performance, we propose a new detector by generalizing the definition of difference image, which takes full advantage of the correlation of adjacent pixels in all directions. Experiments show the better performance of our detector.

On the other hand, we think the mixture of stego-noise, such as stego-noise difference, provides new idea in the design of steganalysis algorithm. The larger the intensity of stego-noise is, the easier the detection is. In the next step, the corresponding study should be strengthened.

References

1. Fridrich, J., Goljan, M., Du, R.: Detecting LSB Steganography in Color and Gray-scale Images. IEEE Multimedia 8(4), 22–28 (2001)
2. Dumitrescu, S., Wu, X., Wang, Z.: Detection of LSB Steganography via Sample Pair Analysis. IEEE Trans. on Signal Processing 51(7), 1995–2007 (2003)

3. Zhang, T., Ping, X.: A New Approach to Reliable Detection of LSB Steganography in Natural Images. Signal Process. 83(10), 2085–2093 (2003)
4. Luo, X., Wang, Q., Yang, C., Liu, F.: Detection of LTSB Steganography Based on Quartic Equation. In: Proceedings of ICACT 2006, pp. 20–22 (2006)
5. Avcıba, I., Memon, N., Sankur, B.: Steganalysis of Watermarking Techniques Using Image Quality Metrics. In: Proceedings of Security and Watermarking of Multimedia Contents II. SPIE, vol. 4314, pp. 523–531 (2000)
6. Farid, H.: Detecting Hidden Messages Using Higher-order Statistical Models. In: Proceedings of IEEE ICIP 2002, pp. 905–908 (2002)
7. Holotyak, T., Fridrich, J., Voloshynovskiy, S.: Blind Statistical Steganalysis of Additive Steganography Using Wavelet Higher Order Statistics. In: Dittmann, J., Katzenbeisser, S., Uhl, A. (eds.) CMS 2005. LNCS, vol. 3677, pp. 273–274. Springer, Heidelberg (2005)
8. Goljan, M., Fridrich, J., Holotyak, T.: New Blind Steganalysis and its Implications. In: Proceedings of Security, Steganography, and Watermarking of Multimedia Contents. SPIE, pp. 1–13 (2006)
9. Harmsen, J., Pearlman, W.: Steganalysis of Additive Noise Modelable Information Hiding. In: Proceedings of. Security, Steganography, and Watermarking of Multimedia Contents VI. SPIE, pp. 131–142 (2003)
10. Xuan, G., Shi, Y., Gao, J., et al.: Steganalysis based on Multiple Features Formed by Statistical Moments of Wavelet Characteristic Function. In: Proceedings of Information Hiding Workshop, Barcelona, Spain, pp. 262–277 (2005)
11. Wang, Y., Moulin, P.: Optimized Feature Extraction for Learning-based Image Steganalysis. IEEE Trans. Information Forensics and Security 2(1), 31–45 (2007)
12. Sullivan, K., Madhow, U., Chandrasekaran, S., et al.: Steganalysis for Markov Cover Data with Applications to Images. IEEE Trans. Information Forensics and Security 1(2), 275–287 (2006)
13. Xuan, G., Cui, X., Shi, Y., et al.: JPEG Steganalysis based on Classwise Non-principal Components Analysis and Mutil-directional Markov Model. In: Proceedings of IEEE ICME 2007, pp. 903–906 (2007)
14. Shi, Y., Chen, C., Chen, W.: A Markov Process Based Approach to Effective Attacking JPEG Steganography. In: Camenisch, J.L., Collberg, C.S., Johnson, N.F., Sallee, P. (eds.) IH 2006. LNCS, vol. 4437, pp. 249–264. Springer, Heidelberg (2007)
15. Ker, A.: Steganalysis of LSB Matching In Grayscale Images. IEEE Signal processing letters 12(6), 441–444 (2005)
16. Chen, C., Shi, Y., Chen, W., et al.: Statistical Moments based Universal Steganalysis Using JPEG 2-D Array and 2-D Characteristic Function. In: Proceedings of IEEE ICIP 2006, pp. 105–108 (2006)
17. Liu, Q., Sung, A.H., Chen, Z., Xu, J.: Feature Mining and Pattern Classification for Steganalysis of LSB Matching Steganography in Grayscale Images. Pattern Recognition 41, 56–66 (2008)
18. Marvel, L.M., Boncelet, C.G., Retter, C.T.: Spread Spectrum Image Steganography. IEEE Trans. on Image Processing 8(8), 1075–1083 (1999)
19. Ker, A.: Improved Detection of LSB Steganography in Grayscale Images. In: Fridrich, J. (ed.) IH 2004. LNCS, vol. 3200, pp. 97–115. Springer, Heidelberg (2004)
20. Huang, J., Mumford, D.: Statistics of Natural Images and Models. In: Proceedings of IEEE Conference Computer Vision and Pattern Recognition, pp. 541–547 (1999)
21. McLachlan, G., David, P.: Finite Mixture Models. Wiley, New York (2001)
22. USDA NPCS Photo Gallery, http://photogallery.nrcs.nsda.gov

23. Ben-Bassat, M.: Use of Distance Measures, Information Measures and Error Bounds on Feature Evaluation. In: Krishaiah, P.R., Kanal, L.N. (eds.) Handbook of statistics: Classification, Pattern Recognition and Reduction of Dimensionality, pp. 773–791. North-Holland, Amsterdam (1987)

24. He, J., Huang, J.: A New Approach to Estimating Hidden Message Length in Stochastic Modulation Steganography. In: Barni, M., Cox, I., Kalker, T., Kim, H.-J. (eds.) IWDW 2005. LNCS, vol. 3710, pp. 1–14. Springer, Heidelberg (2005)

25. Chen, D., Wang, Y.: Universal Steganalysis Algorithm for Additive Noise Modelable Steganography in Spacial Domain. Journal of Southeast University (Natural Science Edition) 37(suppl.), 48–52 (2007)

26. Liu, Z., Pan, X., Shi, L., et al.: Effective Steganalysis Based on Statistical Moments of Differential Characteristic Function. In: Proceedings of IEEE ICCIS 2006, pp. 1195–1198 (2006)

27. Fridrich, J., Goljan, M.: Digital Image Steganography Using Stochastic Modulation. In: Proceedings of Security and Watermarking of Multimedia Contents V, vol. 5020, pp. 191–202. SPIE, San Jose (2003)

A Novel Steganographic Algorithm Resisting Targeted Steganalytic Attacks on LSB Matching

Arijit Sur, Piyush Goel, and Jayanta Mukhopadhyay

Department of Computer Science and Engineering,
Indian Institute of Technology, Kharagpur
{arijits,piyush,jay}@cse.iitkgp.ernet.in

Abstract. In this paper, a new steganographic algorithm is proposed where embedding is done by swapping two randomly chosen pixels in a gray scale cover image. Proposed steganographic scheme inherently preserves first order image statistics such as image histogram and thus remains undetectable against histogram based spatial domain steganalytic attacks. Experimental results show that the proposed scheme clearly outperforms LSB matching and its improved version against existing targeted attacks. The proposed scheme also remains almost undetectable against wavelet absolute moment based blind attack at low embedding rates for never compressed image dataset.

1 Introduction

Steganography is the art of hiding information in an innocent looking cover objects and thus visual and statistical undetectability is one of the major concerns in the steganography. From early days of image steganography, LSB replacement has been a very useful concept where least significant bit plane of the image is replaced by pseudorandom sequence of secret data. However it is observed that there happenes some structural asymmetry due to the LSB replacement and these are exploited to mount different steganalytic attacks[1, 2, 3]. However, the LSB matching which is a counterpart of LSB substitution does not simply overwrite the LSB plane. Rather in this technique randomly selected pixel values are increased or decreased based on a pseudorandom key if its LSB does not match the secret message bit to be embedded. By adopting this meance, the LSB matching method eliminates the problem of structural asymmetry and hence it remains undetectable against aforesaid attacks [1, 2, 3].

Steganalytic attacks on LSB matching scheme are broadly categorized into two classes, namely targeted attacks and blind attacks. There are few targeted attacks are proposed in the recent literature. Some of them [4,5] are especially suitable for color and compressed images and not perform well on never compressed gray scale images. Most of the efficient targeted steganalytic attacks [6,7,8] are based on first order image statistics. These attacks can be easily prevented by image histogram. This fact motivated us to propose an embedding scheme which inherently preserves the image histogram. In blind attacks [9,10,11,12], features are extracted from noise residual. These features are used

H.J. Kim, S. Katzenbeisser, and A.T.S. Ho (Eds.): IWDW 2008, LNCS 5450, pp. 199–208, 2009.

to train a supervised learning classifier which is used to distinguish between cover and stego images. In recent literature, wavelet absolute moment based blind attack [12] is most efficient. From the above discussion, it can be concluded that there are some limitations in the LSB matching algorithm and those can be successfully utilized to mount attacks on LSB matching scheme. In this paper, we introduce a novel embedding algorithm called *Pixel Swapping based Steganographic Algorithm (PSSA)* which is based on the simple idea of swapping the pixels in spatial domain for embedding. This similar kind of idea was first proposed for DCT domain embedding in [13]. To the best of our knowledge this idea of pixel swapping for embedding is not proposed before in the spatial domain image steganography. The one of the important characteristics of the proposed scheme is that this method inherently restors image histogram and thus can resist histogram based targeted attacks. We evaluate our proposed scheme against several recent targeted steganalysis attacks which can detect LSB matching and its improved version [17]. These attacks include Ker's calibrated HCF and HCF of Adjacency Histogram based attacks [6], Jun Zhang et al 's high frequency noise based attack [7] and Fangjun Huang et al 's targeted attack based on features from two least significant bit planes [8]. We found that proposed *PSSA* algorithm is completely undetectable for most of the embedding configurations while the competing algorithm, LSB matching and its improved version [17] are detectable at the same embedding rates. For the rest of the paper, we use the notation **LSBM** to refer to the LSB matching and **ILSBM** to refer improved version of LSB matching. The rest of the paper is organized as follows. The proposed encoding and decoding algorithm and a statistical test showing zero *KL Divergence* between cover and stego histograms are described in section 2; in sec 3 steganalytic security of the proposed scheme is presented. Discussion on experimental results are presented in section 4.

2   The Proposed Scheme

2.1   Embedding Algorithm

Let I be the gray scale cover image of size $M \times N$. In the proposed scheme, at first a pseudo random sequence without repetition (let it be denoted as $\chi[1 \ldots (M \times N)]$) is generated with a shared secret seed. Then first two consecutive elements of the pseudorandom sequence (χ)are taken, (say α and β). Suppose that image pixels corresponding to α and β locations are I_α and I_β. In our embedding rule, if I_α is greater than I_β, the embedded bit is taken as 1 on the other hand if I_β is greater than I_α the embedded bit is taken as 0. If I_α and I_β are same or absolute difference between I_α and I_β is greater than or equal to an prescribed threshold value (let it be denoted as δ) then this pair is not used for embedding. Now if present secret bit is 1 and present I_β is greater than I_α then we just swap I_α and I_β pixels. Similarly swapping is done if present secret bit is 0 and present I_α is greater than I_β. Then we take next two consecutive pixels from χ and continue this process until all secret bits are being embedded. Step by step encoding algorithm is stated below.

Algorithm. *Pixel Swapping based Steganographic Algorithm (PSSA)*
Input: *Cover Image I, Secret Bit Sequence S*, present secret bit is represented by $S_{present}$
Input Parameters: *Shared secret seed for generating pseudorandom sequence,*
Threshold (δ)
Output: *Stego Image I_s*

1. Generate a pseudorandom sequence ($\chi[1 \ldots (M \times N)]$) without repetition using given shared secret seed where $M \times N$ is the size of cover image I.
2. Take first two consecutive elements (say α and β) from χ as two locations of the Image I.
3. Determine the image locations with respect to α and β as (x_α, y_α) and (x_β, y_β) respectively as follows
$$x_\alpha = \lfloor \tfrac{\alpha}{N} \rfloor \quad ; \quad y_\alpha = mod(\alpha, \ N)$$
$$x_\beta = \lfloor \tfrac{\beta}{N} \rfloor \quad ; \quad y_\beta = mod(\beta, \ N)$$
4. **if** $(I(x_\alpha, y_\alpha) = I(x_\beta, y_\beta)$ **OR** $(I(x_\alpha, y_\alpha) - I(x_\beta, y_\beta) \geq \delta)$
     this pair is *not suitable* for embedding.
     Go for next pair.
   else
        if ($S_{present} = 0$ and $I(x_\alpha, y_\alpha) > I(x_\beta, y_\beta)$)
           swap $I(x_\alpha, y_\alpha)$ and $I(x_\beta, y_\beta)$
        if ($S_{present} = 1$ and $I(x_\alpha, y_\alpha) < I(x_\beta, y_\beta)$)
           swap $I(x_\alpha, y_\alpha)$ and $I(x_\beta, y_\beta)$
   *end of **else** loop*
5. Go for next pair and continue until S or χ are all exhausted.
6. Using above steps elements of S embedded into I to get stego image I_s.

End *Pixel Swapping based Steganographic Algorithm (PSSA)*

2.2 Extraction Algorithm

The extraction algorithm is also very simple. With the help of shared secret seed, pseudorandom sequence (without repetition), identical as in encoder, can be again generated and then just extract the secret bits for two consecutive numbers in the pseudorandom sequence as 1 when I_α is greater than I_β and 0 when I_β is greater than I_α. For termination of extracting scheme a special string is used which acts as end of secret bits. Bit stuffing is used to distinguish between secret sequence and terminator string.

Steganographic security was first described using an information theoretic model by Christian Cahin in [14] where $\epsilon - secure$ is defined as the relative entropy (also called Kullback-Leibler or K-L divergence) between the cover and stego distributions is less than or equal to ϵ. KL Divergence between two probability distributions P_{Q_0} and P_{Q_1} (let it be denoted as $KLD(P_{Q_0}||P_{Q_1})$) is defined as follows

$$KLD(P_{Q_0}||P_{Q_1}) = \sum_{q \in \zeta} P_{Q_0} \log \frac{P_{Q_0}}{P_{Q_1}}. \tag{1}$$

where ζ is the probability space of measurement for probability distributions P_{Q_0} and P_{Q_1}. From this information theoretic model of steganographic security it can be shown that proposed scheme is completely secured with respect to first order image statistics such as histogram. As histograms of cover and stego images are completly identical the $KL\ Divergence$ between the cover image histogram and stego image histogram is $zero$ for the proposed embedding scheme.

3   Experimental Results

In this section a comprehensive set of experiments and their results are given to demonstrate the applicability of the proposed algorithm. First, the results for the embedding capacity and PSNR between stego and cover images are presented for some standard images under different threshold values (Section 3.1). Next, in Section 3.2, we present the detection results of our scheme using recent targeted steganalysis methods against $LSBM$ method for a dataset (comprising of several thousand never compressed natural images). We also compare the detection results of our method with those of $LSBM$ and $ILSBM$ [17] in sec 3.3, and demostrate that our method clearly outperforms $LSBM$ and $ILSBM$.

3.1   Embedding Capacity

In Table 1, we list the number of bits that can be hidden in several standard images using $PSSA$ with different embedding parameters. δ denotes a threshold value up to which a change for a single bit embedding can be tolerated. It is very easy to understand that there is certain tradeoff between payload and additive noise in choosing the value of delta. For large value of δ, payload will be increased at the cost of large additive noise due to embedding. On the other hand for a relatively small value of δ, additive noise will decreased but payload also become significantly less. For the proposed scheme, theoretically, the maximum payload can be 0.5 bit per pixels (bpp). This is possible only when, for all chosen pixel pairs, pixel values are not equal but their difference is less than the prescribed threshold (δ). Clearly in practice there are very rare occasions when such conditions occour.

In table 2, we list the PSNR between cover and stego images using proposed $PSSA$ algorithm where several standard images are used as cover images with

Table 1. Embedding Rate of Some Standard Images at Different Values of Thresholds δ

Threshold	Embedding Rate (bpp) for Some Standard Gray Scale Images							
(δ)	Lena	Peeper	Crowd	Goldhill	airplane	man	boats	harbour
2	0.0788	0.0658	0.0776	0.0698	0.1175	0.0762	0.0962	0.1083
4	0.2055	0.1821	0.1788	0.1845	0.2489	0.1853	0.2332	0.2024
8	0.3359	0.3334	0.2653	0.3157	0.3379	0.3012	0.3406	0.2823
16	0.4094	0.4280	0.3350	0.4038	0.3824	0.3840	0.3976	0.3406

Table 2. PSNR between Cover and Stego Image for Some Standard Images at Different Values of Thresholds δ

Threshold	PSNR (dB) for Some Standard Gray Scale Images							
(δ)	Lena	Peeper	Crowd	Goldhill	airplane	man	boats	harbour
2	59.1243	59.9665	59.2242	59.6943	57.4465	59.2747	58.2862	57.8033
4	48.7034	49.0399	49.9556	49.1390	48.5293	49.3331	48.3396	49.8707
8	41.5736	40.9760	43.1278	41.5187	42.8779	42.1101	42.2959	43.4235
16	36.9359	36.1825	37.4890	36.4235	38.7642	36.6847	37.8625	38.1206

different threshold values (δ). It can be shown from the table 2, that the PSNR of cover and stego images are strictly decreasing with increase of threshold value (δ) and even with very high threshold value ($\delta = 16$), $PSNR$ is dropped not less than $36dB$.

3.2   Security against Steganalytic Attacks

The steganographic security of proposed scheme is evaluated against the targeted attacks especially suitable for $LSBM$ scheme. We have used those attacks because the proposed attacks can be thought of a novel expansion of the $LSBM$ method especially from view point of additive noise spatial domain steganography. Since two least significant bit planes are modified quite often in $LSBM$ method, proposed algorithm can easily be described by $LSBM$ method when threshold value $\delta \leq 4$, except that almost double amount of additive noise will be added using proposed one. One of the main aims of our proposed algorithm is the improvement over $LSBM$ algorithm such that the most of the targeted attacks on $LSBM$ algorithm will fail.

The security of the proposed scheme is evaluated against following targeted attacks:

1. Ker's HCF COM based attacks using calibration by down sampling Image [6]
2. Ker's HCF COM based attacks using Adjacency Histogram [6]
3. Jun Zhang et al 's targeted attack [7]
4. Fangjun Huang et al 's targeted attack [8]

For testing the performance of the $PSSA$ algorithm we conducted experiments on a data set of two thousand test images which were divided into two equal sets of one thousand cover images and one thousand stego images. We have used 1000 cover images from an uncompressed image database, UCID [15] and generate 1000 stego images with proposed algorithm. For comparison with $LSBM$ another set of 1000 stego images are generated using $LSBM$ embedding scheme. All the images in UCID are high resolution TIFF files with the size 512×384 or 384×512 and appear to be cut from a variety of uncompressed digital camera images. Before testing, the images were converted to grayscale using the $rgb2gray$ function of Matlab 7.0 directly.

Table 3. Area under ROC and Detection Accuracy of Proposed Algorithm at Different Values of Thresholds δ

Steganalytic Attacks	Threshold	Emb Rate	Area ROC	P_{detect}
HCF COM Calibration [6]	2	0.0900	0.127353	0.5000
	4	0.1823	0.127353	0.5000
	8	0.2687	0.127353	0.5000
	16	0.3422	0.127353	0.5000
HCF COM Adjacency [6]	2	0.0900	0.025830	0.5250
	4	0.1823	0.033197	0.5400
	8	0.2687	0.034013	0.5300
	16	0.3422	0.036183	0.5350
Jun Zhang et al 's targeted attack [7]	2	0.0900	0.066116	0.5000
	4	0.1823	0.066116	0.5000
	8	0.2687	0.066116	0.5000
	16	0.3422	0.066116	0.5000
Fangjun Huang et al 's targeted attack [8]	2	0.0900	0.126114	0.5140
	4	0.1823	0.146517	0.5150
	8	0.2687	0.149884	0.5240
	16	0.3422	0.155668	0.5310

Area under the Receiver Operating Characteristic Curve (ROC) and the Detection accuracy (P_{detect})[16] which is computed using equations 2 and 3 have been used as the evaluation metrics.

$$P_{detect} = 1 - P_{error} \tag{2}$$

$$P_{error} = \frac{1}{2} \times P_{FP} + \frac{1}{2} \times P_{FN} \tag{3}$$

where P_{FP}, P_{FN} are the probabilities of false positive and false negative respectively. A value of $P_{detect} = 0.5$ shows that the classification is as good as random guessing and $P_{detect} = 1.0$ shows a classification with 100% accuracy. The area under ROC and detection accuracy (P_{detect}) of $PSSA$ Algorithm at different threshold values are shown in table 3. It can be observed from table 3 that the area under ROC and P_{detect} is same and negligibly small for any embedding rate against HCF COM Calibration and Jun Zhang et al 's targeted attack. This result indicates that the detection accuracy for proposed $PSSA$ algorithm is as good as random guessing against those attacks. For HCF COM Adjacency and Fangjun Huang et al 's targeted attack, the detection accuracy is also very near to the random guessing.

3.3   Comparison with Existing Method(s)

In Figures $(1, 2, 3, 4)$ we present a comparison of our steganographic scheme, $PSSA$, to $LSBM$ and $ILSBM$ [17]. To enable fair comparison, we must use the same hiding rate for the competing scheme. From these results, it can be

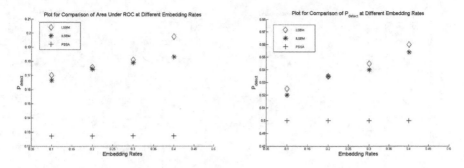

Fig. 1. Comparison of Area under ROC and P_{detect} against HCF Calibration Attack

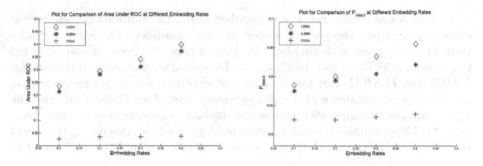

Fig. 2. Comparison of Area under ROC and P_{detect} against HCF Adjacency Attack

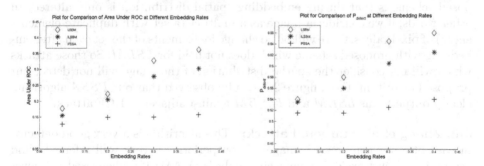

Fig. 3. Comparison of Area under ROC and P_{detect} against Fangjun Huang et al 's targeted attack

easily figured out that for any embedding rate proposed scheme clearly outperforms $LSBM$ and $ILSBM$ schemes against mentioned targeted attacks. Some intuitive explanation of the results against different attacks are given below:

HCF COM with Calibration by Downsampling: This attack is based on image histogram, since the image histogram is completely restored by the proposed algorithm, this attack can't detect the existence of stego bits even

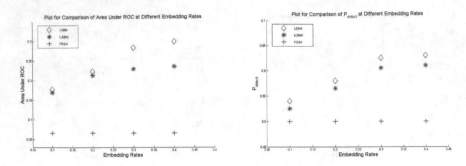

Fig. 4. Comparison of Area under ROC and P_{detect} against Jun Zhang et al 's targeted attack

at very high embedding rates. As described in [6], the calibration technique is efficient only when the secret message is fully embedded. Experiment results from figure 1 agrees with this fact. In spite of poor detection, it can be noted that Area Under ROC and the P_{detect} are increasing with increase of payload for $LSBM$ and $ILSBM$. But for the proposed algorithm results suggest detection accuracy as camparable with random guessing. Also Area Under ROC and the P_{detect} are not changing with increase of payload. Also note that at embedding rate of $0.40bpp$, almost all suitable pixels pairs are used for embedding. In spite of that, proposed method with $0.40bpp$ embedding rate is completely undetectable.

HCF COM of Adjacency Histogram: This is slightly rigorous attacks which consider second order image statistics. One of the major characteristics of proposed scheme is that during embedding spatial distribution is only altered. In other words, if we consider a image as a multiset (a set with duplicate elements entry) of pixel values, there will be no change for elements of the set even after embedding with proposed scheme which does not hold for $LSBM$. So those attacks which will not consider the spatial distribution of the image will not detect our proposed algorithm. From figure 2 it can be observed that our $PSSA$ algorithm clearly outperforms $LSBM$ and $ILSBM$ against adjacency HCF attack.

Jun Zhang et al 's targeted attack: This algorithm is a very good detector in case of never compressed images. It outperforms a very well referred blind attack based on wavelet moment steganalysis (WAM)[12] when used for never compressed images. The weakness of this attack is that it only deals with features using first order image statistics i.e. image histogram and thus can not detect our proposed algorithm.

Fangjun Huang et al 's targeted attack: This attack also fail because of probably the same reason which we have mentioned in case of HCF COM of Adjacency Histogram attack.

From figures 3 and 4, it can be observed that $PSSA$ scheme performs much better than $LSBM$ and $ILSBM$ against both Jun Zhang et al 's targeted attack and Fangjun Huang et al 's targeted attack.

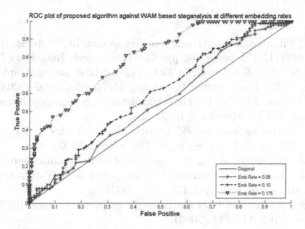

Fig. 5. ROC plot of proposed algorithm against WAM based steganalysis at different embedding rates

3.4 Security against Blind Steganalysis

Proposed algorithm perform poorer than $LSBM$ against any blind steganalyzer based on additive noise as it adds almost double additive noise during embedding than that of $LSBM$. Wavelet absolute moment steganalyzer (WAM) [12] is one of such steganalysis tool. The performance of our algorithm will be inferior than $LSBM$ against WAM based steganalysis. But, the limitation of WAM is that it is not very good detector for stego images among never compressed image dataset. In Figure 5, we have shown experimentally that proposed algorithm is almost undetectable against WAM at some lower embedding rate for never compress image dataset.

4   Discussion and Conclusion

In this paper, we have proposed a spatial domain embedding scheme which is based on a simple concept of pixel swapping. Experimental results show that against some well referred targeted steganalysis attacks, the proposed algorithm clearly outperforms $LSBM$ and $ILSBM$ algorithms which are considered one of the most effective steganographic schemes in recent past. But there are some obvious limitations in proposed scheme. Firstly, the maximum embedding rate is 0.5 bpp. This is not very significant issue because most of the spatial domain steganographic sachems are completely detectable with such high embedding rate. The most important limitation of our scheme is that it adds almost double additive noise during embedding than that of $LSBM$. The additive noise due to embedding in our scheme can be reduced if number of swapping per embedding can be reduced and this is the future direction of our research.

References

1. Westfeld, A., Piftzmann, A.: Attacks on Steganographic Systems. In: Pfitzmann, A. (ed.) IH 1999. LNCS, vol. 1768, pp. 61–76. Springer, Heidelberg (2000)
2. Dumitrescu, S., Wu, X., Wang, Z.: Detection of LSB steganography via sample pair analysis. IEEE Trans. Signal Processing 51(7), 1995–2007 (2003)
3. Fridrich, J., Goljan, M., Du, R.: Detecting LSB steganography in color and gray-scale images. IEEE Multimedia 8(4), 22–28 (2001)
4. Westfeld, A.: Detecting low embedding rates. In: Petitcolas, F.A.P. (ed.) IH 2002. LNCS, vol. 2578, pp. 324–339. Springer, Heidelberg (2003)
5. Harmsen, J.J., Pearlman, W.A.: Steganalysis of additive noise modelable information hiding. In: Proc. SPIE, Security, Steganography, and Watermarking of Multimedia Contents V, vol. 5020, pp. 131–142 (2003)
6. Ker, A.D.: Steganalysis of LSB matching in grayscale images. IEEE Signal processing letters 12(6), 441–444 (2005)
7. Zhang, J., Cox, I.J., Doerr, G.: Steganalysis for LSB Matching in Images with High-frequency Noise. In: Proc. IEEE 9th Workshop on Multimedia Signal Processing, MMSP 2007 (October 1-3, October 19, September 16) pp. 385–388 (2007)
8. Huang, F., Li, B., Huang, J.: Attack LSB Matching Steganography by Counting Alteration Rate of the Number of Neighbourhood Gray Levels. In: Proc. IEEE International Conference on Image Processing, ICIP 2007, vol. 1, pp. 1401–1404 (2007)
9. Fridrich, J., Soukal, D., Goljan, M.: Maximum likelihood estimation of length of secret message embedded using $\pm k$ steganography in spatial domain. In: Proc. SPIE, Security, Steganography, and Watermarking of Multimedia Contents VII, vol. 5681, pp. 595–606 (2005)
10. Holotyak, T., Fridrich, J., Soukal, D.: Stochastic approach to secret message length estimation in $\pm k$ embedding steganography. In: Proc. SPIE, Security, Steganography, and Watermarking of Multimedia contents VII, vol. 5681, pp. 673–684 (2005)
11. Holotyak, T., Fridrich, J., Voloshynovskiy, S.: Blind statistical steganalysis of additive steganography using wavelet higher order statistics. In: Dittmann, J., Katzenbeisser, S., Uhl, A. (eds.) CMS 2005. LNCS, vol. 3677, pp. 273–274. Springer, Heidelberg (2005)
12. Goljan, M., Fridrich, J., Holotyak, T.: New blind steganalysis and its implications. In: Proceedings of SPIE Security, Steganography, and Watermarking of Multimedia Contents VIII, ser., January 2006, vol. 6072, pp. 1–13 (2006)
13. Hetzl, S., Mutzel, P.: A graph theoretic approach to steganography. In: Dittmann, J., Katzenbeisser, S., Uhl, A. (eds.) CMS 2005. LNCS, vol. 3677, pp. 119–128. Springer, Heidelberg (2005)
14. Cachin, C.: An information theoretic model for steganography. In: Aucsmith, D. (ed.) IH 1998. LNCS, vol. 1525, pp. 306–318. Springer, Heidelberg (1998)
15. Schaefer, G., Stich, M.: UCID - An Uncompressed Colour Image Database. In: Proc. SPIE, Storage and Retrieval Methods and Applications for Multimedia, vol. 5307, pp. 472–480 (2004)
16. Solanki, K., Sarkar, A., Manjunath, B.S.: YASS: Yet another steganographic scheme that resists blind steganalysis. In: Furon, T., Cayre, F., Doërr, G., Bas, P. (eds.) IH 2007. LNCS, vol. 4567, pp. 16–31. Springer, Heidelberg (2008)
17. Mielikainen, J.: LSB matching revisited. IEEE Signal Processing Letters 13(5), 285–287 (2006)

A Novel Approach for JPEG Steganography

Vasiliy Sachnev, Hyoung Joong Kim, Rongyue Zhang, and Yong Soo Choi

Center of Information Security Technologies,
Graduate School of Information Security and Management,
Korea University, Seoul 136-701, Korea
bassvasys@hotmail.com, khj-@korea.ac.kr

Abstract. In this paper we present a novel steganography method for JPEG images. The proposed data hiding technique bases on removing and inserting AC coefficients 1 or -1. The candidates for removing and inserting are searched among all non rounded AC coefficients after quantization. The proposed search technique flags the most appropriate coefficients, which provide less degradation after modification. The proposed data hiding technique based on the modification the flagged coefficients significantly decreases distortion after data hiding and, as result, shows the less detectability for steganalysis. The proposed method was tested with steganalysis method proposed by T. Pevny and J.Fridrich. The experimental results show that the proposed method has less detectability compared to the existed widely used steganography methods as F5 and OutGuess.

1 Introduction

The steganography is the information security related tools provided the undetectable communication. In general, the steganography is the set of the specific data hiding technologies, which hide existence of the embedded message. As result, the modified after data hiding stegos become invisible for detection among non modified covers. The adversary who tries to interrupt communication can not detect the modified stegos among covers.

The most popular object for steganography methods becomes digital images. They widely used in modern human society as common way for communication. Among all image's formats JPEG is the most popular. Thus, developing the JPEG steganography methods becomes the main, most important research direction in the steganography area.

The one of the first steganography method for JPEG images was JSteg. It utilizes very popular LSB substitution technique in order to hide data to the quantized DCT coefficients. JSteg significantly distorts the histogram of the DCT coefficients. As result, this method can be easily detected by estimating the shape of the histogram of modified DCT coefficients. The next generation of the JPEG steganography methods tried to remove drawback of the JSteg and keep the histogram of the DCT coefficients just slightly modified [12], [16].

Provos [12] keeps histogram of the DCT coefficients by compensating distortion caused after data hiding. He divided the set of the DCT coefficients into two

H.J. Kim, S. Katzenbeisser, and A.T.S. Ho (Eds.): IWDW 2008, LNCS 5450, pp. 209–217, 2009.

disjoint subsets. The first subset is used for hiding data. The second subset is used for compensating distortions after data hiding to the first subset. As result, the histogram of the DCT coefficients after data hiding has the same shape as original histogram. Methods presented in [2] and [10] use a similar approach.

Another way to decrease detectability of the steganography methods is reducing total distortion caused after data hiding. Normally one DCT coefficient is used for hiding one bit of data. Westfeld [16] suggested to use matrix encoding for hiding data to DCT coefficients. Matrix encoding allows hide data with hither embedding efficiency. Proposed scheme enables to hide more than 1 bit by changing one coefficient. Thus, distortion after data hiding is significantly decreased.

Solanki et. al. [15] utilized the robust watermarking scheme for steganography purposes. They embed data to image in spatial domain by using robust against JPEG compression method. Their scheme provides less degradation of the features of DCT coefficients, and,as result, less detectability.

Most of the steganography algorithms base on manipulation of the magnitudes of DCT coefficients. The proposed technique hide data by inserting or removing 1 or -1. Inserting or removing 1 or -1 does not affect image significantly, the distortion is the same as the magnitude's changing. However, proposed approach has a big choice of candidates for inserting and removing. Method chooses the best combination of 1 and -1 in the set of non rounded quantized DCT coefficients which causes minimum distortion. Thus, proposed method achieves desirable data hiding capacity with less distortion and low detectability.

The paper is organized as follows. The section 2 describes the proposed algorithm based on inserting/removing 1 and -1 in detail. The encoder and decoder are presented in the section 3. The section 4 provides the experimental results. Section 5 concludes the paper.

2   Proposed Algorithm

The sequence of the DCT coefficients presents the specific information about image. Most of the steganography methods use this sequence as basic for hiding necessary information by keeping or slightly modifying sequence. The exception is methods proposed by Westfeld [16]. His methods change the sequence of coefficients by removing non necessary 1 and -1 in order to get proper sequence for data hiding. Definitely, such kind of basic's changes distort original features of the DCT coefficients, and, as result, provides high level of detectability for steganalysis. The proposed scheme allows removing and inserting 1 and -1. Thus, proposed scheme changes the sequence of the DCT coefficients and does not modified the features of the DCT coefficients significantly.

The main idea of the removing and inserting coefficients bases on the searching candidates among non rounded DCT coefficients after quantization. In general, the rounded quantized coefficients -1, 0, and 1 are the unions of the non rounded quantized coefficients (-1.5; -0.5], (-0.5; 0.5], and [0.5; 1.5), respectively. Thus, the candidates for discarding can be any non quantized coefficients (-1.5; -0.5] or

[0.5; 1.5). Definitely, the non quantized coefficients with magnitudes closed to 0.5 or to -0.5 are the best candidates for discarding. Similarly the best candidates for inserting 1 or -1 are non rounded quantized coefficients around 0.5 or -0.5, respectively. The distribution of the non rounded quantized coefficients shows that the proposed strategy allows a lot of different combination of removing and inserting. Such as big choice provides high flexibility, allows to choose the most appropriate combination of inserted and removed coefficients, and keeps the features of the DCT coefficients just slightly modified.

In order to keep compression ratio of the modified JPEG image inside of available ranges the coefficients belonging to the (-0.25; 0.25) should be removed from set of candidates. Reminded part of candidates is sorted according to the distortion effect caused after changing coefficients.

Distortion effect of each coefficient depends on magnitude of the corresponding quantization factor from the quantization table. Modification of the coefficient with large quantization factor provides higher distortion for the 8 * 8 block of image and visa versa.

In general, the distortion effect also depends on magnitude change of the inserted or removed coefficients. Magnitude's change M_c is computed as difference between magnitude of the original non rounded quantized DCT coefficient c_{nR}^q and rounded quantized coefficient after inserting or removing C_R^q.

The distortion effect D for each modified coefficient is computed as follows:

$$D_{(i,j)} = |C_{R\ (i,j)}^q - c_{nR\ (i,j)}^q| \cdot Q_{(i,j)}. \tag{1}$$

where (i, j) is the location of the modified coefficient; $Q_{(i,j)}$ is the corresponding quantization factor from quantization table.

Proposed strategy enable to find the most appropriate coefficient in the group of coefficients, if the data hiding algorithm allows to choose one coefficient among others.

The proposed data hiding technique is the kind of group embedding where group of coefficients presents one bit of information. The changing of anyone of the coefficient from the group inverses the corresponding bit of information. The candidates for inserting or removing belonging to the group are used for modifying the corresponding bit of information.

In the proposed method each DCT coefficient presents one bit according to the F4 algorithm proposed by Westfeld (see 2).

$$b_{(i,j)} = \begin{cases} c_{(i,j)} \ mod \ 2 & \text{if } c_{(i,j)} > 0, \\ c_{(i,j)} - 1 \ mod \ 2 & \text{if } c_{(i,j)} < 0 \end{cases} \tag{2}$$

where $b_{(i,j)}$ is the corresponding bit of the coefficient $c_{(i,j)}$.

The proposed method utilized the sequence S of the non rounded quantized DCT coefficients c_{nR}^q and sequence of the corresponding bits B as basis for data hiding. The sequence S includes all nonzero coefficients $c_{nZ} \in (c_{nR}^q > 1.5; c_{nR}^q < -1.5$, candidates for inserting $c_i \in (-1.5; -0.5] \cup [0.5; 1.5)$ and candidates for removing $c_r \in (-0.5; -0.25] \cup [0.25; 0.5)$. The sequence B covers sets c_{nZ} and c_r.

The basis idea of the proposed data hiding strategy is the modifying sequence B by excluding or inserting corresponding bits from B after removing or inserting coefficients from sets c_r or c_i, respectively.

The sequence S is divided by group of coefficients according to the necessary payload size and length of the sequence B. Thus, the length of the group of coefficients is computed as follows:

$$L_g = \lfloor \frac{|B|}{C} \rfloor. \tag{3}$$

where $|B|$ is the length of the sequence B; C is the necessary payload size.

The information bit of the group b_g is computed as follows:

$$b_g = \bigoplus_i^{L_g} B_i^g. \tag{4}$$

where B^g is the set of the corresponding bits from B for the group of coefficients.

The main requirement for the proposed method is keeping the number of rounded non zeros DCT coefficient in the group unchanged and equal to L_g.

The group G of the coefficients covers L_g number of the coefficients from sets c_{nZ} and c_r in the sequence S (see Figure 1). Thus, group G occupies coefficients from set S belonged to the sets c_{nZ}, c_r, and c_i. The set of the covered by group G coefficients is being modified after inserting or removing candidates. As result, the set of corresponding bits B^q and the information bit b_g become modified (see Figure 1). In order to hide necessary bit of data the proposed method should follow some specific requirements.

The example of the inserting and discarding shows the hiding bit 1 to the sequence S (see Figure 1). The inserting can be done in two different ways. First, by replacing 0.28 to the 1 and, second, by replacing -0.31 to the -1. In order to keep the main requirement of the proposed method the inserting coefficient 0.28 to the group discards coefficients -0.31 and -0.85. The coefficient -0.85 was involved to the equation for getting the information bit b_g. Thus, the resulted information bit b_g may or may not be complimented. The set of the corresponding bits B^q and resulted information bit b_g are modified according to the new inserted coefficient 1 (replacing 0.28 after inserting) and skipped coefficient -0.85. The replacing 0.28 to the 1 compliments the information bit b_g. The second way (inserting -0.31 to the group) does not complement the information bit, so this way is unacceptable case for data hiding in this certain situation.

The requirement for inserting is that the corresponding bit of the inserted coefficient should be different from the corresponding bit of the removed coefficient (note that this coefficient was removed from examined group, and was not discarded to 0 see Figure 1). In case of the first way the corresponding bit of the inserted 0.28 was 1 and the corresponding bit of the removed -0.85 was 0.

The similar conclusion is obtained after analyzing the removing coefficients (see Figure 1). The requirement for removing is that the corresponding bit of the removed coefficient should be different from the corresponding bit of the inserted

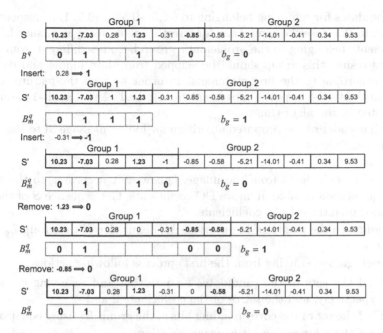

Fig. 1. Example of inserting and removing candidates from the group

coefficient (inserted to the group). In case of the first way the corresponding bit of the removed 1.23 was 1 and the corresponding bit of the inserted -0.58 was 0. Thus, the proposed requirements skip the unacceptable cases.

The proposed method has a choice between inserting 0.28 or removing 1.23 for described example. Assume that corresponding quality factors for coefficient 0.28 and 1.23 are 22 and 14, respectively. According to the equation 1 the distortion effects for the coefficients 0.28 and 1.23 are $|1 - 0.28| \cdot 22 = 15.84$ and $|0 - 1.23| \cdot 14 = 17.22$. Thus, the inserting 0.28 will cause less distortion than removing 1.23. When the group size is larger than 4 as in the described example the proposed method has the big choice for inserting or removing and can choose the candidate with the minimum distortion factor.

3   Encoder and Decoder

The core of encoder of the proposed algorithm bases on the algorithm described in previous section. In case when capacity is large and the group length is small, situations when, there is no candidates from sets c_i and c_r, or, the modification of the existed candidates violates the requirements for data hiding, are possible. For these certain cases data hiding is not possible and proposed algorithm should be modified. First, in case when there are no candidates for inserting or removing, the tested group should be excluded from data hiding process. Second, in case when inserting or removing violates the main requirements of the proposed data hiding method, the candidates from set c_r should move to 2 or -2 if the candidates for removing belonging to (-1.5; -0.5] or [0.5; 1.5), respectively. Definitely, in this case the candidates for inserting should stay unchanged. Thus, all coefficients belonging to the problematic group becomes different from 1 and -1 and, as result, this group should be skipped from data hiding and data recovering according to the first statement. In order to keep the features of the DCT coefficients the opposite movement from 2 and -2 to 1 and -1 should be implemented in the algorithm.

The full encoder of the proposed algorithm for hiding message M to the image is designed as follows.

1) Get DCT coefficients from the image. Quantize it according to the necessary quantization table. Remove DC coefficients. Get sequence S of the non rounded quantized DCT coefficients.
2) According to the length of the message M, define size of the group L_g using equation 3.
3) For each group (starting from the first) process following actions:
   3.1) Find the sets of the candidates for inserting and removing c_i and c_r, respectively, and the set c_2 of the coefficients 2 and -2.
   3.2) Find the set of the corresponding bits of the group B^q using equations 2.
   3.3) Find the information bit b_g using equation 4.
   3.4) If the information bit is different that corresponding bit from message M:
      3.4.a) Check all candidates for inserting, removing, and movement from sets c_i, c_r, and c_2, respectively. Choose the candidates, which can compliment the information bit b_g and meet the main requirements of the proposed algorithm. If there are no such candidates, modify existed candidates from set c_r to 2 or -2 according to its sign and switch to the next group. In this case the current group should be omitted from the data hiding process.
      3.4.b) Compute distortion effect D of the chosen candidates by using equation 1.
      3.4.c) Choose candidates with smallest distortion effect. Modify it and switch to the next group.
3) Recover order of the DCT coefficients. Add DC coefficients. Round all coefficients. Get JPEG image.

The decoder of the proposed method is organized as follows:

1) See Step 1 and 2 in encoder. Round all coefficients from the set S.
2) For each group (starting from the first) process following actions:
   2.1) Find set of the corresponding bits B^q.
   2.2) If there are no coefficients 1, -1 in the group, skip this group and move to the next.
   2.2) Compute the information bit (or hidden bit) of the group b_g by using equation 4 and move to the next group.

4   Experimental Results

Proposed method and examined existed algorithms (F5 and OutGuess) was tested by powerful steganalysis algorithm proposed by T. Pevny and J. Fridrich [11]. It uses 274 different features of the DCT coefficients and allows deeply investigate artificial changes in the tested images. The union of the 274 features from the original and modified images uses for making model in the support vector machine (SVM). The resulted model is used for testing the modified images.

The set of the 1174 test images distributed by CorelDraw was used in our experiments. The proposed and examined method were tested for 6 different capacities (1%, 3%, 5%, 10%, 15%, 20% of the full length of the DCT coefficients) and for 3 most popular quality factors (50, 75, 90). We used 3 models adapted to different quality factors and tested the 18 sets of test images for each examined capacity and quality factor. Each set of test images had 587 cover and 587 stego images. The result shows the accuracy of the steganalysis for each set of the stego images. Finally all results were separated into 3 sets according to the utilized models and quality factors, namely 50, 75, 90 (see Table 1 and Figures 2, 3).

Table 1. Steganalysis efficiency for examined methods and different capacities

Capacities (%)	Quality factors								
	50			75			90		
	Methods			Methods			Methods		
	Proposed	F5	OutGuess	Proposed	F5	OutGuess	Proposed	F5	OutGuess
1	51.36	52.89	69.01	50.97	52.88	70.71	49.53	52.94	67.54
3	55.69	62.24	93.3	54.41	65.36	94.9	58.31	66.92	93.18
5	60.45	74.57	98.55	63.29	79.79	99.06	70.35	79.71	98.97
10	75.59	94.64	100	80.13	96.0	100	90.28	97.86	99.91
15	87.84	98.80	100	94.82	98.98	100	98.72	99.48	100
20	93.96	99.65	100	98.21	99.74	100	99.74	99.82	100

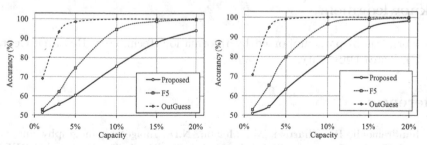

Fig. 2. Steganalysis accuracy for sets of images with quality factor 50 (left) and 75 (right)

The results show high detectability for OutGuess for all examined capacities and quality factors. F5 is more robust against steganalysis and for low (1%) and high (20%) capacities provides almost the same detectability. In case of low

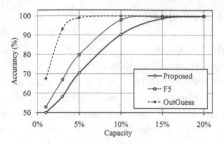

Fig. 3. Steganalysis accuracy for sets of images with quality factor 90

capacities (1 - 3 %) steganalysis can not distinguish the stego images among cover images, and provides the accuracy close to 50 %. For middle capacities (3 - 10 %) the proposed method significantly outperforms the examined methods. The difference between accuracies for F5 and proposed algorithm in case of capacity 5 % and quality factor 75 is 15 %. For capacities closed to the 20% the accuracy for proposed and examined methods is around 100%.

5   Conclusion

This paper presented the novel steganography approach based on inserting and removing DCT coefficients 1 and -1. The proposed data hiding strategy enable to choose one coefficient among the set of candidates in order to hide data to the group. In addition, the proposed scheme evaluates the distortion effect of each candidate's modification and chooses the best candidate with less distortion. The experimental results shows that the proposed scheme provides the decidable performance compared to the existed methods. The proposed idea can be used as basis for further improving the steganography methods.

Acknowledgment

This work was in part supported by Information Technology Research Center (ITRC), Korea University.

References

1. Chandramouli, R., Kharrazi, M., Memon, N.D.: Image steganography and steganalysis: Concepts and practice. In: Kalker, T., Cox, I., Ro, Y.M. (eds.) IWDW 2003. LNCS, vol. 2939, pp. 35–49. Springer, Heidelberg (2004)
2. Eggers, J., Bauml, R., Girod, B.: A communications approach to steganography. In: SPIE, Electronic Imaging, Security, and Watermarking of Multimedia Contents, San Jose, CA (2002)
3. Farid, H., Lyu, S.: Steganalysis using higher-order image statistics. IEEE Transactions on Information Forensics and Security, 111–119 (2006)

4. Farid, H., Siwei, L.: Detecting hidden messages using higher-order statistics and support vector machines. In: Petitcolas, F.A.P. (ed.) IH 2002. LNCS, vol. 2578, pp. 340–354. Springer, Heidelberg (2003)
5. Fridrich, J.: Minimizing the embedding impact in steganography. In: Proceedings ACM Multimedia and Security Workshop, Geneva, Switzerland, pp. 2–10 (2006)
6. Fridrich, J.: Feature-based steganalysis for JPEG images and its implications for future design of steganographic schemes. In: Fridrich, J. (ed.) IH 2004. LNCS, vol. 3200, pp. 67–81. Springer, Heidelberg (2004)
7. Fridrich, J., Filler, T.: Practical methods for minimizing embedding impact in steganography. In: Proceedings SPIE Electronic Imaging, Security, Steganography, and Watermarking of Multimedia Contents IX, San Jose, CA, vol. 6505, pp. 02–03 (2007)
8. Hetzl, S., Mutzel, P.: A graphtheoretic approach to steganography. In: Dittmann, J., Katzenbeisser, S., Uhl, A. (eds.) CMS 2005. LNCS, vol. 3677, pp. 119–128. Springer, Heidelberg (2005)
9. Lee, K., Westfeld, A.: Generalized category attackimproving histogram-based attack on JPEG LSB embedding. In: Furon, T., Cayre, F., Doërr, G., Bas, P. (eds.) IH 2007. LNCS, vol. 4567, pp. 378–391. Springer, Heidelberg (2008)
10. Noda, H., Niimi, M., Kawaguchi, E.: Application of QIM with dead zone for histogram preserving JPEG steeganography. In: Processing ICIP, Genova, Italy (2005)
11. Pevny, T., Fridrich, J.: Merging Markov and DCT features for multi-class JPEG steganalysis. SPIE (2007)
12. Provos, N.: Defending against statistical steganalysis. In: 10th USENIX Security Symposium, Washington, DC (2001)
13. Sallee, P.: Model-based steganography. In: Kalker, T., Cox, I., Ro, Y.M. (eds.) IWDW 2003. LNCS, vol. 2939, pp. 154–167. Springer, Heidelberg (2004)
14. Sallee, P.: Model-based methods for steganography and steganalysis. International Journal of Image Graphics, 167–190 (2005)
15. Solanki, K., Sakar, A., Manjunath, B.S.: YASS: Yet another steganographic scheme that resists blind steganalysis. In: Furon, T., Cayre, F., Doërr, G., Bas, P. (eds.) IH 2007. LNCS, vol. 4567, pp. 16–31. Springer, Heidelberg (2008)
16. Westfeld, A.: High capacity despite better steganalysis (F5a steganographic algorithm). In: Moskowitz, I.S. (ed.) IH 2001. LNCS, vol. 2137, pp. 289–302. Springer, Heidelberg (2001)
17. Xuan, G., Shi, Y.Q., Gao, J., Zou, D., Yang, C., Zhang, Z., Chai, P., Chen, C., Chen, W.: Steganalysis based on multiple features formed by statistical moments of wavelet characteristic function. In: Barni, M., Herrera-Joancomartí, J., Katzenbeisser, S., Pérez-González, F. (eds.) IH 2005. LNCS, vol. 3727, pp. 262–277. Springer, Heidelberg (2005)

A High Capacity Steganographic Algorithm in Color Images

Ting Li, Yao Zhao, Rongrong Ni, and Lifang Yu

Institute of Information Science, Beijing Jiaotong University, China
hdlt57@163.com, yzhao@center.njtu.edu.cn,
rrni@bjtu.edu.cn, yulifang@gmail.com

Abstract. In this paper, we propose a steganographic method which is conducted on the quantized discrete cosine transform (DCT) coefficients of YUV. Three coefficients at the same frequency in YUV components are selected respectively to compose a triplet as the message carrier, then decide the embedding capacity in each triplet according to the property of the triplet. The modulus 2 or modulus 4 arithmetic operations are applied to the valid triplet to embed 1 bit or 2bits secret messages. When modification is required for data embedding, the random number is produced to decide the flipping direction and the improved shortest route modification scheme (SRM) is applied to reduce distortion. To compare with other existing methods, the proposed method has high embedding capacity while approximately preserving the histogram of quantized DCT coefficients, and the spatial blockiness of stego image has no obvious changes, so the method can resist against histogram-based and blockiness-based steganalysis.

Keywords: Steganography, High capacity, JPEG, Color image, Modulus arithmetic.

1 Introduction

Steganography is the art and science of hiding secret data into innocuous-looking cover objects, such as digital still images, audio files or video files. Once the data have been embedded, it may be transferred across insecure channels or posted in public places. Steganography provides good security and becomes an extremely powerful security tool when combined with encryption.

The JPEG compression using the discrete cosine transform (DCT) is still the most common compression standard for still images, though the JPEG2000 standard has already been released. Getting less attention, JPEG images are therefore the most suitable cover images for steganography[1]. Recently, many steganographic techniques using JPEG images have been invented with high embedding capacity while preserving the marginal statistics of the cover coefficients, such as J-Steg[2], F5[3] and OutGuess[4]. It is well known that J-Steg method is detectable using the chi-square attack (Westfeld, 2001) since it is based on flipping the least significant bits (LSBs). F5 employs the technique

H.J. Kim, S. Katzenbeisser, and A.T.S. Ho (Eds.): IWDW 2008, LNCS 5450, pp. 218–228, 2009.
© Springer-Verlag Berlin Heidelberg 2009

of matrix encoding to hold secret information in LSB of quantized DCT coefficients (qDCTCs). Rather than simply flipping LSBs to encode the message bits, F5 increments and decrements coefficient values in order to maintain coefficient histograms unaltered. Thus, F5 cannot be detected by the chi-square attack. However, it can be detected by a specific technique proposed by Fridrich et al in 2003. In OutGuess method, the LSB flipping is applied to a part of usable coefficients. The rest part is used to make the stego image's histogram of quantized DCT coefficients match the histogram of cover image. Note that OutGuess can preserve only the global histogram for all frequencies, and cannot preserve histograms for individual frequencies.

Sallee proposed a model-based steganographic scheme that treats a cover medium as a random variable that obeys some parametric distribution such as Cauchy or Gaussian[5]. Mod4 is a blind steganographic method which applied modulus 4 arithmetic operations to the valid GQC (group of 2×2 spatially adjacent quantized DCT coefficients) as to embed a pair of bits, and the shortest route modification scheme is applied to reduce distortion as compared to the ordinary direct modification scheme[6]. YASS[7] is a newly proposed advanced steganographic method. Instead of embedding data directly in JPEG coefficients, YASS hides information in the quantized DCT coefficients of randomly chosen 8×8 host blocks, whose locations may not coincide with the 8×8 grids used in JPEG compression. Thus it have security against previously proposed blind steganalysis[8,9]. But it's obvious that the embedding efficiency of YASS is low for the use of erasure and error correction codes. Along with the evolution of steganographic method, many steganalysis methods are invented[10,11].

In this paper, a DCT-based steganographic method for color images is proposed. This method has high embedding capacity, good stego image quality and robustness against normal steganalysis. The rest of this paper is organized as follows. In section 2, details of steganographic method are presented. Section 3 shows the experimental results to compare the proposed method with some existing steganographic methods in DCT domain. Finally, conclusions are addressed in Section 4.

2   Steganographic Method

An image is denoted by a matrix of pixels. Each pixel is composed of three fundamental colors: red, green and blue. The RGB color model is an additive model in which red, green and blue are combined in various ways to reproduce other colors. Compared to the RGB model, the YUV model defines a color space in terms of one luminance (brightness) and two chrominance (color) components. YUV emulates human perception of color in a better way than the standard RGB model used in computer graphic hardware. So we conduct the proposed method on YUV coefficients, details of embedding process will be specified as follows.

2.1   Embedding Algorithm

The block diagram of embedding process is shown in Fig. 1.

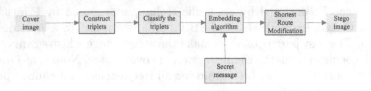

Fig. 1. Embedding diagram

Step1. Construct triplets

Select quantized DCT coefficients with the same frequency index from Y, U, V component respectively to construct a triplet $T(y, u, v)$, where y, u, v is the quantized DCT coefficients (qDCTCs). The number of triplets in an $H \times W$ image is $H \times W$.

Step2. Classify the triplets

All triplets are classified to three classes according to different properties of each triplet.

A triplet $T(y, u, v)$ is characterized as the first class $C1$ if it satisfies Eq.(1):

$$|\{t : t \in T(y, u, v) \cap (t > 1 \cup t < -1)\}| \geq 2 \tag{1}$$

Where $|\{t\}|$ denotes the cardinality of the set $\{t\}$.

Classified to the second class $C2$ if it satisfies:

$$|\{t : t \in T(y, u, v) \cap (t > 1 \cup t < -1)\}| = 1 \tag{2}$$

Otherwise, a triplet is characterized as the third class $C3$ if it satisfies:

$$|\{t : t \in T(y, u, v) \cap (t > 1 \cup t < -1)\}| = 0 \tag{3}$$

In the next step, embedding capacity in each triplet will be decided by its property.

Step3. Modification procedure

For any triplet $T(y, u, v)$, if it belongs to class $C1$, then 2bits secret message can be embedded in it; if it belongs to $C2$, then 1bit can be embedded; or else, there has no message hidden in it.

Suppose $m-bit$ secret message $wm \in \{0, 1, \cdots 2^m - 1\}$ is to be embedded in the triplet $T(y, u, v)$. Firstly, compute the sum of elements in T, i.e. $sum = y + u + v$. Then modify relevant coefficients so as to enforce $mod(sum, 2^m) = wm$. Here, we use improved SRM[6] method to modify the coefficients. The details are discussed in Section 2.2.

Step4. Compress the watermarked coefficients as a JPEG image.

2.2   Modification Method

Modification of quantized DCT coefficients is the most important part of our method. The advantage of our modification method is to make fewer changes to coefficients after stego.

The embedding scheme constrained by modification method is illustrated in Table 2, where we show a representative example of embedding 2bits to one triplet.

To ease the discussion, we define the following sets for one triplet T:

$$P := \{p : p \in T, p > 1\} \tag{4}$$

$$N := \{n : n \in T, p < -1\} \tag{5}$$

And relabeling the elements in P and N so that

$$p_1 \geq p_2 \geq p_3 \tag{6}$$

$$|n_1| \geq |n_2| \geq |n_3| \tag{7}$$

Obviously, some p_i or n_i might not exist, depending on the class that triplet belongs to, and $|P|+|N| \leq 3$, where $|P|$ indicates the number of positive coefficients in one triplet and $|N|$ denotes the number of negative coefficients.

add records values to be added to the coefficients to achieve $mod(sum, 2^m) = wm$, similarly, sub records values to be subtracted from the coefficients. For example when $wm = 00$, add and sub are calculated as in Table 1.

Table 1. calculate add and sub for $wm = 00$

$mod(sum, 2^m)$	add	sub
00	0	0
01	3	1
10	2	2
11	1	3

Then, we modify the coefficients in the triplet T based on the following rules:

(1) In order to maintain histograms of coefficients unaltered distinctly, random number $flag \in \{0, 1\}$ is produced to decide the flipping direction. If the random number is 1, unless specified otherwise, a certain value is added to the coefficients. Similarly, if random number is 0, then subtraction has the first priority.

(2) Always modify the coefficients with larger magnitudes first, and ensure each coefficients undergo as fewer changes as possible.

(3) Coefficients after stego cannot become to -1, 0 or 1, otherwise, triplets cannot be classified correctly in extraction process.

The shortest route modification (SRM)[6] always conducts add operation on positive coefficients and conducts subtract on negative coefficients, so the number of coefficients nearby 0 will be decreased. In order to avoid this shortage,

Table 2. Modification method for stego 2bits in one triplet

add	sub	$|P|$	$|N|$	adding $n_1\,n_2\,n_3\,p_1\,p_2\quad p_3$	subing $n_1\,n_2\,n_3\,p_1\,p_2\quad p_3$
0	0	any	any	no change	no change
		>0	any	p_1+1	no change
1	3	0	2	n_1+1 while $n_1<-2$	n_1-2,n_2-1 while $n_1=-2$
		0	3	n_1+1 while $n_1<-2$	n_1-1,n_2-1,n_3-1 while $n_1=-2$
		2or3	any	p_1+1,p_2+1	no change
2	2	1	≥ 1	n_1+1,p_1+1 while $n_1<-2$ p_1+2 while $n_1=-2$	no change
		0	2or3	n_1+1,n_2+1 while $n_2<-2$	n_1-1,n_2-1 while $n_2=-2$
		any	>0	no change	n_1-1
3	1	2	0	p_1+2,p_2+1 while $p_1=2$	p_1-1 while $p_1>2$
		3	0	p_1+1,p_2+1,p_3+1 while $p_1=2$	p_1-1 while $p_1>2$

an improved shortest route modification scheme is proposed to use a random number $flag$ to decide the flipping direction, Table 2 shows the modification(s) route for embedding 2bits in one triplet T ($flag=1$):

(1) If $add=sub=0$, no changes are made.

(2) If $add=1$ (i.e. $sub=3$), there are three cases to consider (Because a triplet carrying 2bits secret message belongs to class $C1$, there are two coefficients at least whose absolute values are larger than 1, i.e.$|P|+|N|\geq 2$):

Case1:$|P|>0$. p_1 exist at least. Let $p_1'=p_1+1$.

Case2:$|P|=0$ and $|N|=2$. n_1,n_2 exist.

If $n_1<-2$, let $n_1'=n_1+1$.

If $n_1=-2$, add operation has priority because of $flag=1$. However, n_1 becomes to -1 after adding 1, which is not allowed in rule (3). So subtracting operation is conducted instead of adding, that is, let $n_1'=n_1-2$ and $n_2'=n_2-1$ instead of $n_1'=n_1+1$.

Case3: $|P|=0$ and $|N|=3$. n_1,n_2,n_3 exist.

If $n_1<-2$, let $n_1'=n_1+1$.

If $n_1=-2$, similar to Case 2, adding 1 to n_1 is not allowed. Thus, let $n_1'=n_1-1,n_2'=n_2-1$and $n_3'=n_3-1$.

(3) If $add=sub=2$, there are three cases to consider:

Case1:$|P|=2$ or 3. p_1,p_2 exist at least. Let $p_1'=p_1+1$ and $p_2'=p_2+1$.

Case2:$|P|=1$ and $|N|\geq 1$. p_1, n_1 exist.

If $n_1<-2$, let $p_1'=p_1+1$ and $n_1'=n_1+1$.

If $n_1=-2$, let $p_1'=p_1+2$.

Case3:$|P| = 0$ and $|N| = 2$ or 3. n_1, n_2 exist at least.

If $n_2 < -2$, let $n'_1 = n_1 + 1$ and $n'_2 = n_2 + 1$.

If $n_2 = -2$, similar to Case2 in (2), let $n'_1 = n_1 - 1$ and $n'_2 = n_2 - 1$.

(4) If $sub = 1$ (i.e. $add = 3$), similar to (2), there are three cases to consider(In order to ensure fewer changes to coefficients, subtract 1 from coefficients is given the first priority):

Case1:$|N| > 0$. n_1 exist at least. Let $n'_1 = n_1 - 1$.

Case2:$|N| = 0$ and $|P| = 2$. p_1,p_2 exist.

If $p_1 > 2$, let $p'_1 = p_1 - 1$.

If $p_1 = 2$, p_1 becomes 1 after subtracting 1, which is not allowed. So adding operation is conducted on p_1 instead of subtracting, let $p'_1 = p_1 + 2$ and $p'_2 = p_2 + 1$.

Case3:$|N| = 0$ and $|P| = 3$. p_1,p_2,p_3 exist.

If $p_1 > 2$, let $p'_1 = p_1 - 1$.

If $p_1 = 2$, similar to Case2, let $p'_1 = p_1 + 1$, $p'_2 = p_2 + 1$ and $p'_3 = p_3 + 1$.

Similarly, modification(s) route for embedding 1bit in the triplet T ($flag = 1$) is illustrated as follows:

(1) If $add = sub = 0$, no changes are made.

(2) If $add = 1$ (i.e. $sub = 1$), there are two cases:

Case1:$|P| > 0$, adding 1 to p_1 is the shortest addition route, let $p'_1 = p_1 + 1$.

Case2:$|P| = 0$(i.e. $|N| > 0$).

If $n_1 < -2$, let $n'_1 = n_1 + 1$.

If $n_1 = -2$, let $n'_1 = n_1 - 1$.

The extensions to other conditions such as $flag = 0$ could be derived and we can adjust parameters to get balance between image quality and embedding capacity, for example just embed secret message in class $C1$.

2.3   Extraction Algorithm

The 8×8 blocks of qDCTCs are decompressed and extracted from the JPEG data stream, then extraction process is conducted on these coefficients. The block diagram of extraction is shown in Fig.2.

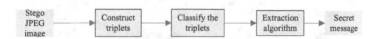

Fig. 2. Extraction diagram

The extraction process can be carried out by reversing the embedding procedure. After classify all triplets, the number of bits that should be extracted from every triplet is attained. Subsequently, extraction of the embedding message is completed by calculating a series of summations of qDCTCs in triplets and applying modulus 2 or 4 arithmetic.

3   Experimental Results

To compare the proposed method with existing methods in terms of embedding capacity, image quality and robustness against steganalysis, we use 14 standard 256×256 color images as the cover image. The JPEG quality factor is set to 80.

3.1   Embedding Capacity and Image Quality

The embedding capacity denotes the average number of bits that can be embedded per pixel into a given cover image. It depends on both the stego-algorithm and the cover image content. Usually, we use bpc (bit per non-zero DCT coefficients) denotes the embedding capacity. In our tests a random message of variable length is used so that the arbitrary embedding capacity can be calculated.

Figure 3 shows the cover image and stego image with different embedding capacity. The human eyes are unable to detect small variations in colors.

To quantify the invisibility of the embedded secret message, we compute PSNR for stego images produced by Mod4[6] method and our method with different embedding capacity.

Figure 4 shows the average embedding capacity and PSNR (Calculated by StirMark[12]) by our method and Mod4 method.

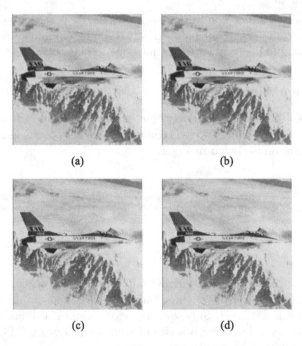

(a)   (b)

(c)   (d)

Fig. 3. Stego images of different embedding capacity (a) Cover image (b) stego image at 0.43 bpc (c) stego image at 0.27 bpc (d) stego image at 0.13 bpc

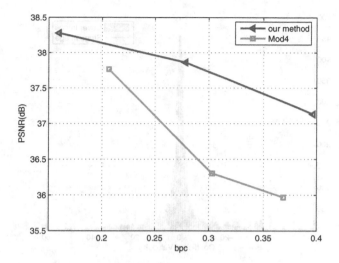

Fig. 4. Results of embedding experiments

3.2 Robustness against Steganalysis

For JPEG steganalysis, histogram distributions and blockiness changes between cover image and stego image are two important factors, so histogram-based and blockiness-based methods are basic steganalysis methods for JPEG images.

Fig. 5 shows two histograms of components on frequency (1,2) for Airplane: histogram of cover image and stego image by our method(bpc=0.4). The line denotes the stego image, and bars denote the cover image. We can see that the proposed method maintains the component-wise distribution for the (1,2)-AC coefficients of Y component. In fact, the other significant components (e.g. mode (2, 1), (2, 2) of Y, U or V component) also exhibit similar behaviors, and the same observations are obtained for other images as well. Actually, because of different direction of flipping and no change of coefficients in the bins labeled by -1, 0 or 1, a spike could hardly be seen and the general shape of the distribution (i.e., Laplacian) is maintained for each histogram. Therefore, we conclude that the proposed method in this paper shows no abnormality with respect to the histogram-based steganalysis methods.

Fig. 6 shows blockiness of cover image and stego image respectively. Blockiness of an image is defined as follows[13]:

$$B = \sum_{i=1}^{\lfloor (M-1)/8 \rfloor} \sum_{j=1}^{N} |g_{8i,j} - g_{8i+1,j}| + \sum_{j=1}^{\lfloor (N-1)/8 \rfloor} \sum_{i=1}^{M} |g_{i,8j} - g_{i,8j+1}| \qquad (8)$$

where $g_{i,j}$ are pixels' values in a $M \times N$ grayscale image and $\lfloor x \rfloor$ denotes the integer part of x.

In Figure 6, the blue beeline denotes B_0 in sign of blockiness of cover image, the red line (marked with triangle ◁) denotes the blockiness of stego image by

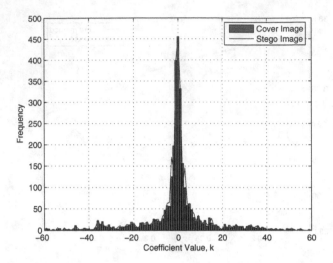

Fig. 5. Airplane-(1, 2)-mode AC qDCTCs distribution

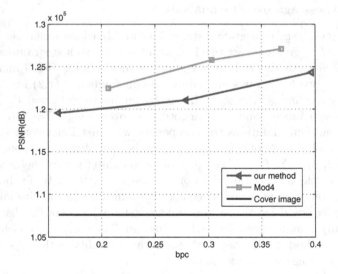

Fig. 6. Comparison of blockiness

our method signed by B_1, and green line (marked with square $\diamond$) denotes the blockiness of stego image by mod4 method signed by B_2. Using $a = B_1/B_0$ measures blockiness changes between cover image and stego image. We can see that a is close to 1 even when embedding capacity ups to 0.4bpc, so blockiness-based steganalysis methods are useless for the proposed method is concluded. Fig. 6 also reveals that B_1 is smaller than B_2, that is, our method makes fewer changes to cover image than Mod4 method.

4   Conclusion

A DCT-based steganographic method for color JPEG images is proposed. Different secret bits can be hidden into different triplets according to their properties. When modifications are conducted on a triplet, flipping direction of coefficients is considered to preserve the distributions of histogram, meanwhile, the improved shortest route modification(s) is employed to ensure that the expected number of modifications is minimal. Experiments are conducted to compare the proposed method with other existing steganographic methods, especially Mod4 method, in terms of embedding capacity (bpc), image quality (PSNR and blockiness) and robustness against steganalysis.

Acknowledgements. This work was supported in part by National Natural Science Foundation of China (No. 60776794, No. 90604032, No.60702013), 973 program (No. 2006CB303104), 863 program (No. 2007AA01Z175), Beijing NSF(No.4073038) and Specialized Research Foundation of BJTU (No. 2006XM008, No. 2005SZ005).

References

1. Noda, H., Niimi, M., Kawaguchi, E.: High-performance JPEG steganography using quantization index modulation in DCT domain. Pattern Recognition Letters 27(5), 455–461 (2006)
2. Upham, D.: (1997),
   http://ftp.funet.fi/pub/crypt/cypherpunks/steganography/jsteg/
3. Westfeld, A.: F5-A steganographic algorithm: high capacity despite better steganalysis. In: Moskowitz, I.S. (ed.) IH 2001. LNCS, vol. 2137, pp. 289–302. Springer, Heidelberg (2001)
4. Provos, N.: Defending against statistical steganalysis. In: Proceedings of the 10th USENIX Security Symposium, pp. 323–336 (2001)
5. Sallee, P.: Model-based methods for steganography and steganalysis. International Journal of Image and Graphics 5(1), 167–190 (2005)
6. Wong, K., Qi, X., Tanaka, K.: A DCT-Based Mod4 Steganographic Method. Signal Processing 87, 1251–1263 (2007)
7. Solanki, K., Sarkar, A., Manjunath, B.S.: YASS:Yet another steganographic scheme that resists blind steganalysis. In: Furon, T., Cayre, F., Doërr, G., Bas, P. (eds.) IH 2007. LNCS, vol. 4567, pp. 16–31. Springer, Heidelberg (2008)
8. Shi, Y.Q., Chen, C., Chen, W.: A Markov process based approach to effective attacking JPEG steganography. In: Camenisch, J.L., Collberg, C.S., Johnson, N.F., Sallee, P. (eds.) IH 2006. LNCS, vol. 4437, pp. 249–264. Springer, Heidelberg (2007)
9. Pevny, T., Fridrich, J.: Merging Markov and DCT features for multi-class JPEG steganalysis. In: Proc. of SPIE, San Jose, CA (2007)
10. Fridrich, J., Goljan, M., Hogea, D.: Steganalysis of JPEG images: breaking the F5 algorithm. In: Petitcolas, F.A.P. (ed.) IH 2002. LNCS, vol. 2578, pp. 310–323. Springer, Heidelberg (2003)

11. Fridrich, J., Goljan, M., Du, R.: Detecting LSB steganography in color and gray-scale images. IEEE Multimedia 8(4), 22–28 (2001)
12. Petitcolas, F., Anderson, R.J., Kuhn, M.G.: Attacks on copyright marking systems. In: Aucsmith, D. (ed.) IH 1998. LNCS, vol. 1525, pp. 219–239. Springer, Heidelberg (1998), http://www.cl.cam.ac.uk/~fapp2/papers/ih98-attacks/
13. Fridrich, J., Goljan, M., Hogea, D.: Attacking the OutGuess. In: Proceedings ACM Workshop on Multimedia and Security 2002, France, pp. 3–6 (2002)

A Novel Method for Block Size Forensics Based on Morphological Operations

Weiqi Luo[1], Jiwu Huang[1], and Guoping Qiu[2]

[1] Guangdong Key Lab. of Information Security Technology
Sun Yat-Sen University, Guangdong, China, 510275
[2] School of Computer Science, University of Nottingham, NG 8, 1BB, UK
isshjw@mail.sysu.edu.cn

Abstract. Passive forensics analysis aims to find out how multimedia data is acquired and processed without relying on pre-embedded or pre-registered information. Since most existing compression schemes for digital images are based on block processing, one of the fundamental steps for subsequent forensics analysis is to detect the presence of block artifacts and estimate the block size for a given image. In this paper, we propose a novel method for blind block size estimation. A 2×2 cross-differential filter is first applied to detect all possible block artifact boundaries, morphological operations are then used to remove the boundary effects caused by the edges of the actual image contents, and finally maximum-likelihood estimation (MLE) is employed to estimate the block size. The experimental results evaluated on over 1300 nature images show the effectiveness of our proposed method. Compared with existing gradient-based detection method, our method achieves over 39% accuracy improvement on average.

Keywords: Block Artifacts Detection, Digital Forensics, Morphological Operations.

1 Introduction

Recent development in digital multimedia processing methods and related software tools such as *Photoshop* has made it increasingly easy for ordinary users to alter (tamper) the contents of multimedia data without leaving obvious traces. As a result, verifying the trustworthiness of multimedia data has become ever more challenging.

Traditional approaches to multimedia security need additional processing at the time of data creation. For instance, watermark based methods need to embed an imperceptible digital watermark in advance in order to facilitate tampering detection at a later time. However, in many real forensics scenarios, there is no additional information such as digital watermark and/or digital signature can be used. Thus those active approaches which need such information would fail in such situations.

Some passive methods [1,2,3,4,5,6] have been reported recently to provide forensics information on how multimedia data is acquired and processed. The

H.J. Kim, S. Katzenbeisser, and A.T.S. Ho (Eds.): IWDW 2008, LNCS 5450, pp. 229–239, 2009.
© Springer-Verlag Berlin Heidelberg 2009

passive methods exploit inherent traces left by various modules in the imaging device or software system to exposure digital tampering without relying on any additional information. The artifacts left by the past operations performed on the multimedia signals can serve as an intrinsic feature for use in forensics analysis. By extracting and analyzing those inherent features, we can deal with many forensics problems related to source identification and forgery detection.

The block-based processing is widely employed in many well-known encoding mechanisms, and one of the inherent patterns for such data is the block artifact. To our knowledge, most existing works on block artifacts are proposed with the purposes of image restoration and enhancement, only a few are for forensics purposes [7,8,9,10]. It is well known that the block artifact is bad for the quality of the image. However, from the point of view of digital forensics, the block artifact is a useful feature for analyzing the content of the multimedia signal. For example, the authors in [7] proposed a method to determine whether a BMP image has been JPEG compressed previously by detecting the block artifact effects. It has also been shown that JPEG image splicing [8] and MPEG recompression [9] can be exposed based on the block artifacts. In most prior literatures, it is always assumed that the image is compressed by a known scheme such as JPEG, MPEG with a fixed block size of 8×8 or 16×16. However, in many forensics cases, there are many other options of the block size in different source coders, for instance, some vector quantization encoders employ block size as small as 2×2. JPEG 2000 has the option of tiling the image with any block size. Also, the block needs not be a regular square shape. Therefore, given a BMP image without any knowledge of prior processing, is it possible to detect the presence of block artifacts and estimate the block size? This is a crucial question for further forensics analysis such as source coder identification and quantization parameters estimation, because the inaccurate block size estimation would lead to invalid subsequent analysis.

One relevant work that we are aware of is reported in [10]. To estimate the block size, the method first obtains the gradient separately along each dimension, and then averages resulting data along the orthogonal direction and obtain a 1-D average values both in horizontal and vertical directions. By estimating the period of the 1-D signal using maximum-likelihood estimation (MLE), the block size can be estimated. From our analysis and large experiments, we find that detecting the block boundaries by observing the image gradient is very sensitive to the image contents such as edges. How to eliminate the effect of the image content is the key of the detection algorithm. In this paper, we first design a 2×2 cross-differential filter to search the possible positions that are along the block boundaries, and then employ morphological operations to eliminate the effect of the content edges, and finally estimate the block size using MLE. Compared with the gradient-based method [10], our proposed method can achieve over 39% accuracy improvement on average.

The rest of the paper is arranged as follows. Section 2 describes the estimation methodology, including the filter design for locating possible boundaries, block boundaries detection based on morphological operations and block size estimation using MLE. Section 3 shows the experimental results and discussion. The concluding remarks and future works are discussed in the Section 4.

2 Methodology

Block artifacts appear as artificial discontinuities in an image. Since the block-processing in commonly used compression standard is performed regularly on each non-overlapping block with a fixed size, we have following two observations:

- The block artifact boundary just appears in two directions, *i.e.*, horizontal and vertical. Any boundaries occur in other directions should be regarded as false block boundaries.
- For nature images, the lengths of the boundaries along block artifacts should be much longer than those content edges in horizontal and vertical directions.

Based on these observations, we first design a filter to locate all the possible pixels along the block boundaries in an image, and then employ morphological operations to remove the effect of boundaries that are in other directions and short edges along the block artifact directions. The details of our proposed method are described as follows.

2.1 Filter Design for Locating Possible Boundary

In prior literatures, *e.g.* [10,11], image gradient is commonly used for measuring the discontinuities in an image. However, based on large experiments, we find that the gradient-based measurement is very sensitive to the image content such as edges, and thus it is not a good way to detect the block artifact boundaries introduced by compression schemes such as JPEG and MPEG. To overcome these weakness, we first design a 2×2 cross-differential filter to eliminate the effect of the actual image contents, and then obtain a binary image from the filtered image in each dimension for subsequent block boundary detection.

Given an image $f(i,j)$, where $1 \le i \le M, 1 \le j \le N$, we first divide it into 2×2 overlapping small blocks. As illustrated in Fig.1, we consider all the adjacent block pairs in horizontal and vertical directions, respectively.

Taking horizontal filter for example[1], for each coordinate (i,j) in the image f, we compute the two cross-differences $\alpha(i,j)$ and $\beta(i,j)$ as follows.

$$\alpha(i,j) = |f(i-1,j-1) + f(i,j) - f(i,j-1) - f(i-1,j)|$$

$$\beta(i,j) = |f(i-1,j) + f(i,j+1) - f(i,j) - f(i-1,j+1)|$$

Then we obtain a binary image $f_V(i,j)$ which can indicate the vertical discontinuities in the image f by comparing the two differences α and β,

---

[1] In the paper, we just show the process of the block size estimation in horizontal direction. We can repeat the process similarly in vertical direction.

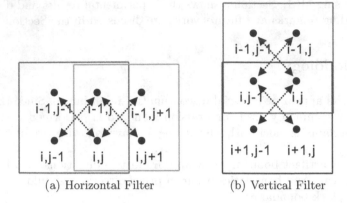

(a) Horizontal Filter   (b) Vertical Filter

Fig. 1. Filters in two directions

$$f_V(i,j) = \begin{cases} 1 & \text{if } \alpha(i,j) < \beta(i,j) \ \& \ 2 \leq i \leq M, 2 \leq j \leq N-1 , \\ 0 & \text{Others.} \end{cases}$$

Fig. 2 shows a test example using the horizontal filter. Fig.2 (a) is a JPEG Lena image in a relative high quality (PSNR=38db) without any obvious visual block effects, Fig.2 (b) is the binary image f_V after filtering. If zoom in, it can be observed that most 8×8 block boundary introduced by compression as well as the edges in image content has been detected.

We note that most edges in the image content are not along the vertical or horizontal direction, like the edges of the hat, hair and the shoulder. In next subsection, a structure element is designed and the erosion operation is employed to remove such edges and preserve the long vertical (horizontal) boundary which most likely comes from the block artifacts.

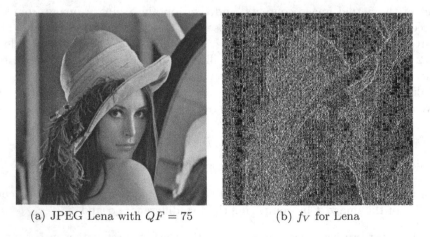

(a) JPEG Lena with $QF = 75$ (b) f_V for Lena

Fig. 2. Vertical discontinuities detection using horizontal filter

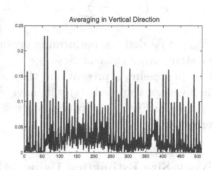

(a) Erosion Version of f_V with Se_V (b) Averaging Values in Vertical Direction

Fig. 3. Erosion operation on the filtered Lena image (Fig.2 (b)) and the average values in vertical direction

2.2 Block Boundary Detection Based on Morphological Operations

Mathematical morphology (MM) is a set-theoretical approach for digital signal analysis, and it is widely applied to digital image for the processing of geometrical structures such as boundary extraction, region filling, extraction of connected components and skeletons by locally comparing with a so-called structuring element. Erosion and dilation are the two basic operations of MM. In our algorithm, we employ the erosion operation to detect block artifact.

As mentioned earlier, the artificial boundaries introduced by the block-based processing are arranged regularly in an image. Therefore, after locating all the possible boundaries using our proposed filter and obtaining a binary image f_V, we design a structuring element combining with the erosion operation to eliminate the effect of the edges in the image, such as short edges along vertical (horizontal) direction and/or the bevel edges.

Two structuring elements Se_V, Se_H of size 5×5 are designed as follows.

$$Se_V = \begin{bmatrix} 0 & 0 & 1 & 0 & 0 \\ 0 & 0 & 1 & 0 & 0 \\ 0 & 0 & 1 & 0 & 0 \\ 0 & 0 & 1 & 0 & 0 \\ 0 & 0 & 1 & 0 & 0 \end{bmatrix} \quad Se_H = \begin{bmatrix} 0 & 0 & 0 & 0 & 0 \\ 0 & 0 & 0 & 0 & 0 \\ 1 & 1 & 1 & 1 & 1 \\ 0 & 0 & 0 & 0 & 0 \\ 0 & 0 & 0 & 0 & 0 \end{bmatrix}$$

where Se_V, Se_H are the structuring elements for short boundary removal in vertical and horizontal directions respectively.

As shown in Fig.3 (a), most bevel edges in Fig.2 (b) have been removed effectively after erosion operation. The remaining boundaries are most likely coming from the block artifacts. To estimate the horizontal block size, we average the binary values as shown in Fig.3 (a) in vertical direction.

$$A_V(j) = \frac{1}{M} \sum_{i=1}^{M} E_{Sev} \circ f_V(i,j) \qquad j = \{1, 2, \ldots N\}$$

where $E_{Sev} \circ f_V$ denotes performing erosion operation on the binary image f_V with the structuring element Sev.

If block processing is present and the block size is B_H in horizontal direction, then the average values $A_V(i)$ will have nearly periodic peaks at multiples of B_H as shown in Fig.3 (b). In next step, we want to estimate the period $\hat{B}_H$ of the average values.

2.3   Block Size Estimation Using MLE

The process of extracting the periodic peak signal (outline) from the average values $A_V(i)$ is similar to the scheme used in [10]. We first perform a median filter with the size 3 on the signal $A_V(i)$ and obtain its median version $M_V(i)$ as follows.

$$M_V(i) = median\{A_V(i-1), A_V(i), A_V(i+1)\}$$

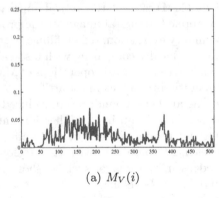

(a) $M_V(i)$

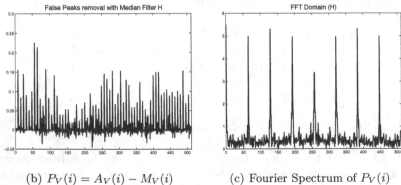

(b) $P_V(i) = A_V(i) - M_V(i)$ (c) Fourier Spectrum of $P_V(i)$

Fig. 4. Illustration of periodic peak signal extraction from A_V

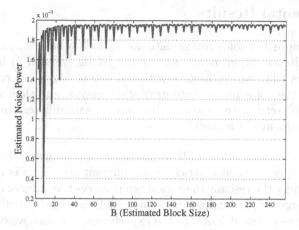

Fig. 5. Noise power $\hat{\sigma}^2(B_H)$ for the periodic signal $P_V(i)$ (Fig.4 (b))

And then we subtract M_V from A_V itself to obtain the approximate outline of the signal A_V.

$$P_V(i) = A_V(i) - M_V(i)$$

In this case, the outline of signal A_V is defined as the average values that are located at the peaks, namely $\{A_V(i)|i = kB_H, k \in Z\}$, and the period of the outline P_V is the estimated horizontal block size $\hat{B}_H$.

Fig.4 (a) is the median version of the average values $A_V(i)$; Fig.4 (b) is the outline of $A_V(i)$. By observing the spectrum of P_V in Fourier domain as shown in Fig.4 (c), we can conclude that the period of P_V is 8, since there are 8 impulses in its Fourier spectrum.

To determine the period from the signal P_V automatically, we use the MLE scheme employed in [12,10]. Suppose P_V consists of a periodic signal s plus an *i.i.d* Gaussian noise n with mean zero.

$$P_V(i) = s(i) + n(i) \qquad i \in \{1, 2, \dots N\}$$

where s is a periodic repetition of a signal q with the period B_H:

$$s(i) = q(i \ mod \ B_H)$$

To estimate the period B_H from the signal P_V, we can maximize the conditional probability density function $p(P_V|s, \sigma^2, B_H)$, where σ^2 is the variance of the noise $n(i)$, B_H is the period of $s(i)$, by minimizing the estimated noise variance $\hat{\sigma}^2(B_H)$ as a function of B_H.

$$\hat{B}_H = \arg\min_{B_H} \hat{\sigma}^2(B_H)$$

Fig. 5 shows the estimated noise power $\hat{\sigma}^2(B_H)$ for JPEG Lena image with quality factor 75 in our example (PSNR=38db). The lowest position in the plot indicates the estimated horizontal block size $\hat{B}_H = B_H = 8$.

3   Experimental Results

UCID (an uncompressed color image database) [13] is a benchmark dataset for image retrieval. In our experiments, we use the UCID (version 2) for test. The database includes 1338 nature images with the size of 352×512 or 512×352. We first convert the color images into gray-scale images, and then simulate the JPEG compression scheme to create the test images with different block size and quality. The process are as follows.

– Resize the 8×8 basic table in JPEG as below by Bi-linear interpolation to obtain basic quantization matrices with different size. This is a reasonable method, because the resizing operation can preserve the order of quantization steps, *i.e.*, larger steps for high-frequency coefficients and smaller steps for DC/low-frequency coefficients. In the experiment, the sizes we employed are $4 \times 4, 8 \times 8, 16 \times 16, 32 \times 32$ and 64×64.

$$\begin{bmatrix} 16 & 11 & 10 & 16 & 24 & 40 & 51 & 61 \\ 12 & 12 & 14 & 19 & 26 & 58 & 60 & 55 \\ 14 & 13 & 16 & 24 & 40 & 57 & 69 & 56 \\ 14 & 17 & 22 & 29 & 51 & 87 & 80 & 62 \\ 18 & 22 & 37 & 56 & 68 & 109 & 103 & 77 \\ 24 & 35 & 55 & 64 & 81 & 104 & 113 & 92 \\ 49 & 64 & 78 & 87 & 103 & 121 & 120 & 101 \\ 72 & 92 & 95 & 98 & 112 & 100 & 103 & 99 \end{bmatrix}$$

– Scale those basic quantization matrices using a scaling factor just like the quality factor employed in JPEG compression to create the images with different block artifacts quality. In the experiment, the scaling factors are from 50 to 90 with a step 10, and the average PSNR on 1338 images are around $32db$ to $40db$ as shown in Table 1.

The experimental results evaluated on these test images are shown in Fig.6. Obviously, our proposed method can get very good detection results in most cases. On average, our method can achieve 89.16% accuracy, while the gradient-based method [10] just has 50.01%.

It is also observed that the larger the block sizes, the lower detection accuracy we obtain. The reason is that for a given test image, when the block size increases,

Table 1. Average values of PSNR (db) for the test images with different size

Size \ QF	50	60	70	80	90
4	32.6	33.5	34.8	36.8	40.6
8	33.1	34.0	35.2	37.0	40.7
16	33.1	34.0	35.2	37.0	40.5
32	33.0	33.8	35.0	36.7	40.3
64	32.8	33.6	34.7	36.4	40.0

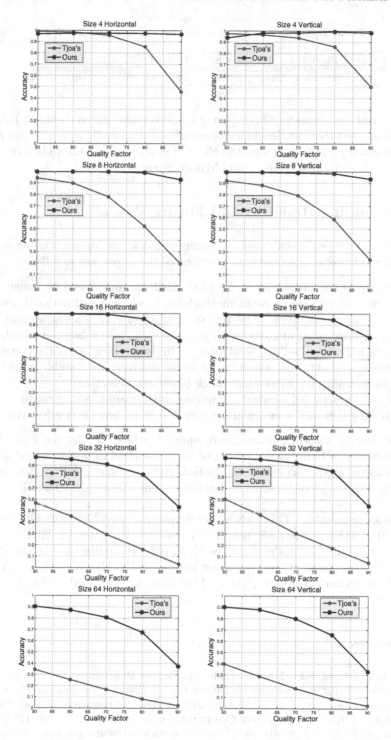

Fig. 6. Compared with gradient-based method [10]

Table 2. Number of blocks in an image with size 512×352

Size of Blocks	4×4	8×8	16×16	32×32	64×64
No. of Blocks	11264	2816	704	176	40

the number of the blocks within an image would become less as shown in Table 2, which means that there are less block boundaries can be used for estimation. Also, when the quality factor increases, the effect of block artifacts would become less. Therefore, the detection accuracy would decrease in both cases.

4   Concluding Remarks and Future Works

In this paper, we propose a novel approach for block size forensics based on the morphological operations. The contributions of the paper are as follows.

- By studying the properties of the block artifacts introduced by image compression, we propose a novel method for block size estimation based on morphological operation. The proposed method can achieve around 40% accuracy improvement compared with existing gradient-based method.
- Propose a 2×2 cross-differential filter to eliminate the effect of the content in an image, and obtain a binary image which can indicate possible boundaries within an image for subsequent block boundary detection.
- Design a structuring element combining with erosion operation to remove the bevel edges in the image and locate the block boundary effectively.

In the future, we will perform more analytic study on why the 2×2 filter work. Moreover, a blind measurement of block artifacts in images based on the morphological operations would be studied. Also, the potential applications of our method in image restoration and enhancement will be investigated.

Acknowledgements. This work is supported by NSF of Guangdong (06023191), NSFC (90604008, 60633030) and 973 Program (2006CB303104). The authors would like to thank Steven K. Tjoa for providing us with the source code in [10] for experimental comparison and thank Dr. Min Wu, Avinash Varna and Ashwin Swaminathan at Univ. of Maryland, College Park, USA, for their discussion.

References

1. Ng, T.T., Chang, S.F., Lin, C.Y., Sun, Q.: Passive-Blind image forensic. In: Multimedia Security Technologies for Digital Rights. Elsevier, Amsterdam (2006)
2. Luo, W.Q., Qu, Z.H., Pan, F., Huang, J.W.: A survey of passive technology for digital image forensics. Frontiers of Computer Science in China 1(2), 166–179 (2007)
3. Lukáš, J., Fridrich, J., Goljan, M.: Digital camera identification from sensor pattern noise. IEEE Trans. on Information Forensics and Security 1(2), 205–214 (2006)

4. Popescu, A., Farid, H.: Exposing digital forgeries in color filter array interpolated images. IEEE Trans. on Signal Processing 53(10), 3948–3959 (2005)
5. Kharrazi, M., Sencar, H., Memon, N.: Blind source camera identification. In: IEEE International Conference on Image Processing, vol. 1, pp. 709–712 (2004)
6. Swaminathan, A., Wu, M., Liu, K.J.R.: Non-intrusive forensic analysis of visual sensors using output images. In: Proc. of IEEE International Conference on Acoustics, Speech and Signal Processing, Toulouse, France, vol. 5, pp. 401–404 (2006)
7. Fan, Z., de Queiroz, R.: Identification of bitmap compression history: Jpeg detection and quantizer estimation. IEEE Trans. on Image Processing 12(2), 230–235 (2003)
8. Luo, W.Q., Qu, Z.H., Huang, J.W., Qiu, G.P.: A novel method for detecting cropped and recompressed image block. In: Proc. of IEEE International Conference on Acoustics, Speech and Signal Processing, vol. 2, pp. II-217–II-220 (2007)
9. Luo, W.Q., Wu, M., Huang, J.W.: Mpeg recompression detection based on block artifacts. In: Proc. of SPIE Electronic Imaging, Security, Steganography, and Watermarking of Multimedia Contents, vol. 6819, 68190X (2008)
10. Tjoa, S., Lin, W., Zhao, H., Liu, K.: Block size forensic analysis in digital images. In: IEEE International Conference on Acoustics, Speech and Signal Processing, vol. 1, pp. I-633–I-639 (2007)
11. Wang, Z., Bovik, A., Evan, B.: Blind measurement of blocking artifacts in images. In: IEEE International Conference on Image Processing 2000, vol. 3, pp. 981–984 (2000)
12. Wise, J., Caprio, J., Parks, T.: Maximum likelihood pitch estimation. IEEE Trans. on ASSP 24(5), 418–423 (1976)
13. Schaefer, G., Stich, M.: Ucid: an uncompressed color image database. In: Proc. SPIE Electronic Imaging, Storage and Retrieval Methods and Applications for Multimedia, vol. 5307, pp. 472–480 (2003)

Non-malleable Schemes Resisting Adaptive Adversaries

Johann Barbier[1,2] and Emmanuel Mayer[2]

[1] ESIEA, Laboratoire de Cryptologie et Virologie Opérationnelles
38, rue des docteurs Calmette et Guérin
53 000 Laval, France
[2] Centre d'Électronique de l'ARmement, Département de Cryptologie,
La Roche Marguerite, BP 57 419,
35 174 Bruz Cedex, France,
{johann.barbier,emmanuel.mayer}@dga.defense.gouv.fr

Abstract. In this paper, we focus on security models in steganography. We first recall classical security models both in cryptography and in steganography. We rely on these models to propose the definitions of the malleability-based models of security for private key steganographic schemes. We also prove that the indistinguishability-based security under the hypothesis of a chosen hidden text adaptive attack (IND-CHA2) implies malleability-based security under the same hypothesis (NM-CHA2). This connection gives us some keys to explain why many practical steganography schemes do not resist adaptive adversaries. Finally, we propose a generic construction to overcome the intrinsic vulnerability of many of them induced by the malleability property.

Keywords: malleability, adaptive adversary, security models.

Introduction

The increasing number of steganographic algorithms but also the complexity of the attacks they have to resist, strengthen the necessity to have a common methodology to evaluate the security of such algorithms. This point raises the question of the definition of the security of a steganography scheme. Three different levels of security may be considered. The first one is the security against an adversary who wants to detect the presence of hidden information inside a cover medium. This type of security is a compromise between detection and capacity, keeping in mind that "the more you hide, the more hidden information is detectable". The second one is the security against an adversary who aims at extracting hidden information that he has already detected. Trivially, extracting the hidden information with no *a priori* knowledge is equivalent to an exhaustive search applied to the secret key which parametrizes the pseudo random number generator (PRNG) dedicated to the location of the hidden bits. Finally, the last security is the classical cryptographic confidentiality security against an adversary who has access to the ciphertext which has been embedded and whose the

H.J. Kim, S. Katzenbeisser, and A.T.S. Ho (Eds.): IWDW 2008, LNCS 5450, pp. 240–253, 2009.

goal is to retrieve the plaintext. In this paper, we only deal with the detection security as the extraction security is obvious (as explained previously) and since the confidentiality security is widely investigated in cryptography.

C. Cachin [1,2,3] followed by R. Chandramouli [4,5,6] dealt with the security and the capacity of steganography algorithms through this scope. C. Cachin was the first one to define the concept of security for a steganography algorithm. Let $\mathcal{C}$ be the set of all the cover media, of distribution $P_\mathcal{C}$ and $\mathcal{S}$ the set of stego media of distribution $P_\mathcal{S}$, then the security of a steganography scheme is given by the *mutual entropy* $D_0(P_\mathcal{C}\|P_\mathcal{S})$ between $P_\mathcal{C}$ and $P_\mathcal{S}$. A steganography scheme is then said to be *ε-secure* against a passive adversary, if and only if

$$D_0(P_\mathcal{C}\|P_\mathcal{S}) = \sum_{c \in \mathcal{C}} P_\mathcal{C}(c) \log \frac{P_\mathcal{C}(c)}{P_\mathcal{S}(c)} \leq \varepsilon. \tag{1}$$

Moreover, if $\varepsilon = 0$ then the scheme is said to be *perfectly secure*. This pseudo-distance is also called *Kullbak-Liebler distance*. In other words, the security of a steganography scheme depends on the incapacity of the adversary to distinguish between two probability density functions, $P_\mathcal{C}$ and $P_\mathcal{S}$. At that point, we can notice that detection security may be define with any distribution distance, as suggested in [7].

To evaluate such a security, one usually defines the adversary that the steganographic scheme should resist. Each adversary is mapped to a *security model*, *i.e.* his capabilities and the rules he has to follow in order to attack the steganographic scheme. Numerous security models for steganography have been proposed so far. Some of them are dedicated to private key schemes [2,8,9,10] and other ones fit to public key schemes [11,3,12]. Moreover, different types of adversaries have also been considered like passives ones [2,9,12,11] or active ones [3,13]. N. Hopper's PhD. thesis [7] is perhaps one of the most complete work dealing with the security of steganography schemes. Different models and notations are rigorously defined and compared with the security of cryptography schemes. The traditional approach in security proofs is to take the designer's point of view and prove with a strong adversary, the security of a designed scheme, in a given model by using reductions to hard problems. The designer has to keep in mind that "stronger is the considered adversary, higher is the security of his scheme". The main problem with these security models is that powerful adversaries are far from the real-life adversaries who are usually considered as passive and very weak, and even sometimes do not follow the Kerckhoff's principles [14]. That issue leads to prove steganographic schemes to be perfectly secure in models with strong adversaries but these schemes are not practical since they are to much constrained.

Recently, in the same direction as A.D. Ker [15], J. Barbier [16,17] tried to reduce the gap between real-life adversaries and classical security models. He formally defined security models that correspond to specific and universal steganalyses with respect to their common sense. At the opposite of the security models above-mentioned, the adversary is very weak and so such models should not be taken into account to evaluate the security when designing a new steganographic scheme. In that context, one should take the attacker's point of view and keep in mind that "weaker is the adversary who detects a scheme, lower is the security of that scheme". The proposed models are then perfectly adapted to benchmark steganalysis all together using a common methodology. In this paper, we take the designer's point of view. First, we adapt cryptographic malleability-based models of security and define the malleability-based models of security for steganography. Then, we present a generic construction to adapt an intrinsically malleable steganographic scheme into a non-malleable one. This construction let us now consider a Chosen Hidden text Adaptive Attack (CHA2). Indeed, malleable steganographic schemes are insecure against CHA2, thus such a construction may be applied to improve their security.

The paper is organized as follows. First, using classical cryptographic models of security, we introduce in section 1 the security models in steganography and explain why an adaptive adversary is never taken into account in these models. Then, in section 2, we define malleability-based models for steganography and link them to the adaptive adversary in indistinguishability models. We also explain why practical steganographic schemes are malleable, *i.e.* are intrinsically not secure in malleability-based models and so in the indistinguishability-based models with adaptive adversary. Finally, in section 3, we propose a generic construction to overcome the intrinsic malleability of practical steganographic schemes.

1   Classical Security Models

1.1   Cryptographic Indistinguishability

We present now the classical security models. Such models are based on well-known security models in cryptography which have been introduced first by S. Goldwasser et S. Micali [18] for public key cryptography schemes, then adapted by M. Bellare *et al.* [19] for private key schemes. These models catch the difficulty that an adversary has to distinguish between pairs of ciphertexts based on the message they encrypt. They are called *indistinguishability-based* models. Moreover, we pointed out in the introduction that steganalysis may be reduced to a statistical discrimination problem, so naturally classical models of security in steganography are based on them [7,10,16]. The models we present in this paragraph aims at proving the security of a cryptographic scheme through the scope of the indistinguisability. The goal of the considered adversaries is to distinguish ciphertexts knowing the plaintexts. A formal definition of cryptographic scheme may be found in [18].

The adversary A is couple of polynomial probabilistic algorithms (A_1, A_2). Each algorithm A_i has access to an oracle $\mathcal{O}_i$. To evaluate the security, we simulate an experiment between a challenger and the adversary. At the beginning of the experiment, the challenger randomly chooses a secret key K. During the first step, A_1 can make as many queries to $\mathcal{O}_1$ he wants. At the end of this step, it stores its internal state in a variable s and generates two different messages m_0, m_1 of the same size. In the second step, the challenger randomly chooses a bit b and gives the adversary back the message m_b encrypted using the secret key K either C embedded with m and K if $b = 1$ or C otherwise. During this step, A_2 is allowed to make as many queries to $\mathcal{O}_2$ he wants excepted the decryption query with the challenge. Finally, using m_0, m_1 and s he must guess which one of the messages has been encrypted. Finally, he bets on b and returns b'.

The oracles $\mathcal{O}_1$ and $\mathcal{O}_2$ are encryption and/or decryption oracles. Encryption oracles are parametrized by K, take a plaintext as inputs and return either the ciphertext or nothing depending on the power of the adversary. Decryption oracles are parametrized by K, take a ciphertext as input and return either the plaintext or *failure*, or nothing depending on the power of the adversary. Three types of adversary are considered depending on their power:

- *IND-CPA (Chosen Plaintext Attack) adversary.* $\mathcal{O}_1$ is an encryption oracle and $\mathcal{O}_2$ always return nothing. The adversary can make $\mathcal{O}_1$ encrypt with K as many plaintexts he wants.
- *IND-CCA1 (Chosen Chiphertext Attack) adversary.* $\mathcal{O}_1$ is encryption and decryption oracle and $\mathcal{O}_2$ is an encryption oracle. During the first step, the adversary is allowed to decrypt as many ciphertexts he wants.
- *IND-CCA2 (Chosen Ciphertext Adaptive Attack) adversary.* $\mathcal{O}_1$ and $\mathcal{O}_2$ are both encryption and decryption oracles. The adversary is allowed to decrypt as many ciphertexts he wants, but he can also adapt his queries to the challenge.

Obviously, the more information the adversary has access to, the more powerful he is and the less consequences his attack has on the security of attacked scheme. Let $ATK \in \{CPA, CCA1, CCA2\}$, then the indistinguishability experiment described above with an adversary A playing ATK against a cryptographic scheme Γ and a security parameter 1^k, is noted $\mathbf{Exp}_\Gamma^{\mathbf{IND\text{-}ATK}}(\mathbf{A}, \mathbf{k})$.

1.2 Classical Models of Security in Steganography

The cryptographic indistinguishability-based models can be straight adapted for steganographic purpose. In this context, one wants to catch the ability of the adversary to distinguish between stego media and cover media. In this more constraining context, the cover media are generated by the challenger, otherwise a simple comparison between a cover medium and the challenge leads the adversary to answer correctly at each time. First, we need to recall the definition of a steganographic scheme.

Definition 1. [10] *A private key steganography scheme Σ is defined by a set C of cover media and by three polynomial algorithms :*

- *A probabilistic algorithm $\mathcal{K}$, which generates the keys. Its input is a security parameter k and its ouput is a private key K.*

- *A probabilistic embedding algorithm Emb. Its inputs are the key K, a plaintext $m \in \mathcal{M} \subset \{0,1\}^*$, a cover medium $C \in \mathcal{C}$ and its output is a stego medium $C' \in \mathcal{S}$, the set of stego media.*
- *A deterministic extraction algorithm Ext. Its inputs are the private key K, a medium C' and outputs the plaintext m if C' has been embedded using K or $\perp$, an error message.*

Remark 1. As in cryptography, this definition can be easily adapted to public key schemes. In the scope of this paper, we only deal with private key schemes. Nevertheless, all that is presented here remains true for public key schemes. In a more general context, such adaptations are not always so straight forward and should be done with lots of precautions.

The adversary A is couple of polynomial probabilistic algorithms (A_1, A_2). Each algorithm A_i has access to an oracle $\mathcal{O}_i$. At the beginning of the experiment, the challenger randomly chooses a secret key K with the generation keys algorithm $\mathcal{K}$. During the first step, A_1 can make as many queries to $\mathcal{O}_1$ he wants. At the end of this step, it stores its internal state in a variable s and generates a message m. In the second step, the challenger randomly chooses a bit b and a cover medium $C \in \mathcal{C}$. Then, it gives the adversary back either C embedded with m and K if $b = 1$ or C otherwise. During this step, A_2 is allowed to make as many queries to $\mathcal{O}_2$ he wants excepted the extraction query with the challenge. Finally, using s and m he must guess if the challenge is a stego medium or not and returns b'. $\mathcal{O}_1$ is an insertion and/or extraction oracle and $\mathcal{O}_2$ is only an insertion oracle. Insertion oracles are parametrized by K, take a cover medium and a message as inputs and return a stego medium or nothing. Extraction oracles are parametrized by K, take a stego medium and return *failure* or the embedded message, or nothing. Three types of adversary are considered depending on their power.

- *IND-PA (Passive Attack) adversary.* $\mathcal{O}_1$ and $\mathcal{O}_2$ always return nothing. This is a passive adversary.
- *IND-CMA (Chosen Message Attack) adversary.* $\mathcal{O}_1$ and $\mathcal{O}_2$ are insertion oracles. During the first step, the adversary is allowed to choose the embedded message.
- *IND-CHA1 (Chosen Hidden text Attack) adversary.* $\mathcal{O}_1$ is an extraction and insertion oracle and $\mathcal{O}_2$ is an insertion one. During the first step, the adversary is allowed to have hidden text extracted with a chosen stego media.
- *IND-CHA2 (Chosen Hidden text Adaptive Attack) adversary.* $\mathcal{O}_1$ and $\mathcal{O}_2$ are extraction and insertion oracles. The adversary is allowed to have hidden text extracted with a chosen stego media but he can also adapt his queries to the challenge.

Let $ATK \in \{PA, CMA, CHA1, CHA2\}$, then the indistinguishability game previously described, with an adversary A playing ATK against Σ and a security parameter 1^k, is denoted $\mathbf{Exp}_{\Sigma}^{\text{IND-ATK}}(\mathbf{A}, \mathbf{k})$ and is summarized in the figure 1.

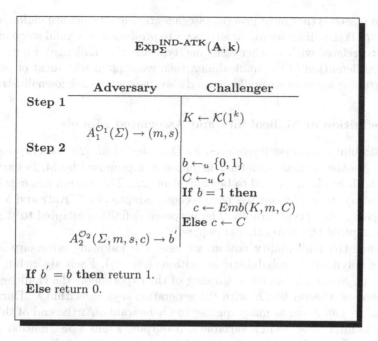

Fig. 1. Game for an IND-ATK adversary against Σ

Definition 2. [7] *We traditionally define the efficiency of the adversary by his advantage* $Adv_{\Sigma}^{IND\text{-}ATK}(A, k)$ *such that*

$$Adv_{\Sigma}^{IND\text{-}ATK}(A, k) = 2\left|\mathcal{P}r\left(Exp_{\Sigma}^{IND\text{-}ATK}(A, k) = 1\right) - \frac{1}{2}\right|,$$

where $ATK \in \{PA, CMA, CHA1, CHA2\}$.

The advantage measures the gain that the adversary obtains compared with tossing up for it. As the security is evaluated from the designer's point of view, the insecurity $InSec_{\Sigma}^{IND\text{-}ATK}(k)$ is preferred to measure the weakness of a steganography scheme. One definition of the insecurity can be found in [7].

$$InSec_{\Sigma}^{IND\text{-}ATK}(k) = \max_{A \in \mathcal{A}}\{Adv_{\Sigma}^{IND\text{-}ATK}(A, k)\}, \tag{2}$$

where $\mathcal{A}$ is the set of probabilistic polynomial attackers. Practically, the adaptive adversary (IND-CHA2) is never considered since the intrinsic malleability of many practical steganographic schemes leads them to be insecure against such an adversary.

2 Malleability-Based Models

The steganographic models are obtained by adapting and also weakening the cryptographic ones. Indeed, the IND-CCA2 model has no equivalent in steganography.

The main reason is that most practical steganographic schemes are *malleable*. The *malleability* is the difficulty for an adversary to produce a new valid stego medium which is correlated with another valid one (typically the challenge). First, we give the formal definition of the malleability, then we explain why most of practical steganographic schemes are malleable and also give some well-known illustrations.

2.1 Definition of Malleability and Associated Models

The malleability was first introduced by D. Dolev *et al.* [20] for cryptographic purpose. Another slightly different definition was proposed by M. Bellare *et al.* [21] but both has been proved to be equivalent [22]. This notion was detailed for public key cryptography but has been recently adapted by J. Katz and M. Yung [23] for private key cryptography. The proposed definition adapted to steganography is inspired by these original papers.

To define the malleability notion, we need the following adversary A. A is couple of polynomial probabilistic algorithms (A_1, A_2). Each algorithm A_i has access to an oracle $\mathcal{O}_i$. At the beginning of the experiment, the challenger randomly chooses a secret key K with the generation keys algorithm $\mathcal{K}$. During the first step, A_1 can make as many queries to $\mathcal{O}_1$ he wants. At the end of this step, it stores its internal state in a variable s. It outputs s and a distribution M over messages in the legal message space. In the second step, the challenger randomly chooses two messages m and $\tilde{m}$ according to M and a cover medium $C \in \mathcal{C}$. Then, it gives the adversary back C embedded with m and K. During this step, A_2 is allowed to make as many queries to $\mathcal{O}_2$ he wants excepted the extraction query with the challenge. Finally, using s and M it outputs a relation R and a vector of stego media, $\vec{c}$. Let $\vec{m}$ correspond to the extraction of the stego media $\vec{c}$ (*i.e.*, if $m[i]$ represents the ith component of vector $\vec{m}$, then $m[i] = Ext(K, c[i])$ for $1 \leq i \leq |\vec{m}|$). We say that a steganographic scheme is non-malleable if for every probabilistic polynomial adversary A, the probability that $R(m, \vec{m})$ is true is at most negligbly different from the probability that $R(\tilde{m}, \vec{m})$ is true. As above, different attacks are modeled by giving A_1 and A_2 access to embedding and/or extraction oracles. Let $ATK \in \{PA, CMA, CHA1, CHA2\}$, then the non-malleability game (NM) previously described, with an adversary A playing ATK against Σ and a security parameter 1^k, is denoted $\mathbf{Exp}_{\Sigma}^{\mathbf{NM\text{-}ATK}}(\mathbf{A}, \mathbf{k})$. This game is compared with $\mathbf{Rand}_{\Sigma}^{\mathbf{NM\text{-}ATK}}(\mathbf{A}, \mathbf{k})$ which is exactly the same excepted that R is evaluated with $\tilde{m}$ (*i.e.* $R(\tilde{m}, \vec{m})$ instead of $R(m, \vec{m})$). Both are summarized in figure 2.

Definition 3. *We define the efficiency of the adversary by his advantage* $Adv_{\Sigma}^{NM\text{-}ATK}(A, k)$ *such that*

$$Adv_{\Sigma}^{NM\text{-}ATK}(A, k) = |\mathcal{P}r\left(Exp_{\Sigma}^{NM\text{-}ATK}(A, k) = 1\right) - \mathcal{P}r\left(Rand_{\Sigma}^{NM\text{-}ATK}(A, k) = 1\right)|,$$

where $ATK \in \{PA, CMA, CHA1, CHA2\}$. We say that Σ is secure against an adversary A playing a NM-ATK experiment if his advantage his negligible and is secure in the NM-ATK model if it is secure against all probabilistic polynomial adversaries playing a NM-ATK experiment.

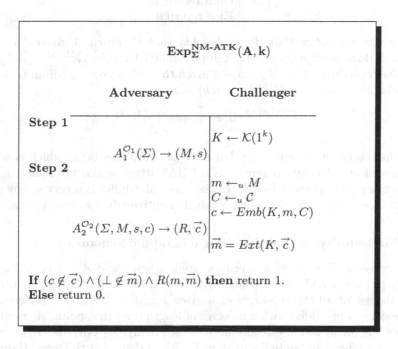

$$\mathbf{Exp}_\Sigma^{\mathbf{NM\text{-}ATK}}(\mathbf{A}, \mathbf{k})$$

Adversary	Challenger

Step 1

$$K \leftarrow \mathcal{K}(1^k)$$

$$A_1^{\mathcal{O}_1}(\Sigma) \rightarrow (M, s)$$

Step 2

$$m \leftarrow_u M$$
$$C \leftarrow_u \mathcal{C}$$
$$c \leftarrow Emb(K, m, C)$$

$$A_2^{\mathcal{O}_2}(\Sigma, M, s, c) \rightarrow (R, \vec{c})$$

$$\vec{m} = Ext(K, \vec{c})$$

If $(c \notin \vec{c}) \wedge (\bot \notin \vec{m}) \wedge R(m, \vec{m})$ **then** return 1.
Else return 0.

Fig. 2. Game for a NM-ATK adversary against Σ

Remark 2. Opposed to cryptographic models, the steganographic ones do not require that all the message in M have the same length since they are generated by the challenger and hidden in a cover medium.

For the remain of the paper we only consider adaptive adversaries (IND-CHA2 and NM-CHA2). To clearly understand why the intrinsic malleability of practical steganographic schemes are not secure against adaptive adversaries, let us first prove the following theorem.

Theorem 1

$$IND\text{-}CHA2 \Longrightarrow NM\text{-}CHA2,$$

i.e. a steganographic scheme is secure against an IND-CHA2 adversary implies that it is secure against a NM-CHA2 adversary and the reduction is efficient.

Proof. Let be A and attacker playing $Exp_\Sigma^{\mathbf{NM\text{-}ATK}}(A, k)$, we define an adversary $B = (B_1, B_2)$ as follows:

$$
\begin{array}{l|l}
B_1^{\mathcal{O}_1}(\Sigma) & B_1^{\mathcal{O}_1}(\Sigma) \\
\hline
(M,s) \leftarrow A_1^{\mathcal{O}_1}(\Sigma) & (R, \vec{c}) \leftarrow A_2^{\mathcal{O}_2}(\Sigma) \\
m \leftarrow_u M & \vec{m} = Ext(K, \vec{c}) \\
\textbf{return } (m, (M\|s)) & \textbf{If } (c \notin \vec{c}) \wedge (\perp \notin \vec{m}) \wedge R(m, \vec{m}) \\
& \textbf{then return 1} \\
& \textbf{Else return 0}
\end{array}
$$

To conclude, we notice that the probability that B returns 1 given that c has been embedded with m is exactly the probability that $Exp_\Sigma^{\text{NM-CHA2}}(A, k) = 1$ while the probability that B returns 1 given that c is a cover medium is exactly the probability that $Rand_\Sigma^{\text{NM-CHA2}}(A, k) = 1$ and so

$$
Adv_\Sigma^{\text{IND-CHA2}}(B, k) = Adv_\Sigma^{\text{NM-CHA2}}(A, k) \tag{3}
$$

∎

From that theorem, we conclude that a steganographic scheme which is not secure against an adversary playing a NM-CHA2 attack is also not secure against an adversary playing an IND-CHA2. The non-malleability is a very strong property for most steganography schemes when adaptive adversaries are considered.

2.2 Malleability of Practical Steganographic Schemes

Let us consider Σ a practical steganographic scheme and $A = (A_1, A_2)$ an adversary playing a NM-CHA2 attack. At the end of the first stage, A_1 outputs M as the set of all the messages of a given length l. Then the challenge is a cover medium embedded with a message of length l. At that point, A_2 randomly changes one bit b_i inside c to obtain $c' \neq c$. Finally, A_2 outputs (R, c') where $R(m, m')$ is true if and only if $d_H(m, m') < 2$ and $d_H(.,.)$ is the usual Hamming distance. Three cases has to be investigated:

- b_i has not been used by the embedding algorithm, and so $Ext(K, c') = m$ that implies $R(m, m)$ true.
- b_i has been used by the embedding algorithm, and is a critical bit, $i.e.$ $Ext(K, c') = \perp$ that implies the game outputs 0.
- b_i has been used by the embedding algorithm, but is not a critical bit, $Ext(K, c') = m'$ that implies $d_H(m', m) = 1$ and so $R(m, m')$ true.

The critical bits embedded crucial information needed for the extraction algorithm such as the length of the embedded message or an init vector to initialize to PRNG. Let us denote p the proportion of the critical bits. Practically, p is close to zero since most of practical steganographic schemes need only few bytes of critical bits. In that case, it is not hard to compute

$$
\mathcal{P}r\left(Exp_\Sigma^{\text{NM-CHA2}}(A, k) = 1\right) = 1 - p, \tag{4}
$$

$$
\mathcal{P}r\left(Rand_\Sigma^{\text{NM-CHA2}}(A, k) = 1\right) = l/2^l, \tag{5}
$$

and so

$$Adv_{\Sigma}^{\text{NM-CHA2}}(A, k) = |1 - p - l/2^l| \approx 1 . \tag{6}$$

Then, we can conclude that Σ is not secure in the NM-CHA2 model and thus is also not secure in the IND-CHA2 model, from theorem 1.

To be totally convinced, let us now consider an adversary $A = (A_1, A_2)$ playing an IND-CHA2 attack against Σ. During the second stage, the adversary randomly changes one bit b_i inside c to obtain c'. Since $c' \neq c$, he is allowed to submit it to the extraction oracle $\mathcal{O}_2$. He processes this step q times. We have to investigate four cases:

- c is not a stego medium then the oracle always outputs $\bot$.
- c is a stego medium and b_i has not been used by the Emb algorithm, then the oracle outputs m.
- c is a stego medium and b_i has been used by Emb, but is not critical for Ext, then the oracle outputs m' such that $d_H(m, m') = 1$.
- c is a stego medium and b_i is a critical bit, then the oracle outputs $\bot$.

Finally, the adversary returns $b' = 0$ if $\mathcal{O}_2$ outputs q error messages and $b' = 1$ otherwise. In that context, he fails only when the challenge is a stego medium and $\mathcal{O}_2$ outputs q error messages, *i.e.* he flipped q times a critical bit. The probability of such an event is

$$\mathcal{P}r(A \text{ fails }) = p^q , \tag{7}$$

and the advantage of A is then

$$Adv_{\Sigma}^{\text{IND-CHA}}(q) = 1 - p^q . \tag{8}$$

The access to the extraction oracle during the second step give the adversary a great amount of information as he is able to forge valid stego media for the secret key K which are different from the challenge. Most of practical steganographic schemes are faced to the same vulnerability.

Remark 3. One key of the success of the previous attack is that critical bits are not numerous. For instance, Outguess [24] and f5 [25] only require 32 critical bits and JPHide and JPSeek [26] need 64 of them.

3   Generic Construction for Non-malleable Schemes

From the paragraph 2, we conclude that we have to conceive steganographic schemes which are not intrinsically malleable if we want them to be robust against an adaptive adversary. To achieve this we propose a generic probabilistic construction to design steganographic schemes that output the embedded message if the stego medium is the right one and a different one otherwise.

3.1   Embedding Construction

Let us denote by Σ a practical steganography scheme, C a cover medium, K the secret key, m the message to be embedded, l the length of m, L the number of critical bits, PRNG the pseudo random number generator and H a hash function. We make the reasonnable hypothesis that m is a randomized chiphered text (to avoid trivial attacks). First, we divide C into S_u, the set of *usable bits* for embedding, and $(C \setminus S_u)$. S_u is *a priori* known and depends on the Emb algorithm. For instance for LSB embedding, S_u is the set of all LSB in the cover medium. Then, we define IV_1 by

$$IV_1 = H(l||K||C \setminus S_u). \qquad (9)$$

We are now able to initialize the PRNG with K and IV_1 and randomly select S_c a subset of S_u of size l_c, to embed the critical bits and randomly select S_m a subset of $(S_u \setminus S_c)$ of size l_m to embed the message. We now define

$$S_r = C \setminus (S_c \cup S_m), \qquad (10)$$

the subset which contains the bits which are *a priori* unused by Emb. At that point, we have defined the subset in which the critical bits (S_c) and message bits (S_m) will hold. These subsets are strongly dependent on the *a priori* unusable parts of the cover. Now, we make the embedding dependent on all bits which are neither in S_c nor in S_m. For this, we compute

$$IV_2 = H(K||S_r) \qquad (11)$$

and re-initialize the PRNG with K and IV_2 and randomly embedded the m within S_m. Now, we randomly change one bit of S_c to obtain S'_c and thus C' until

$$H(K||C') \bmod (2^L - 1) = l. \qquad (12)$$

is satisfied. Such a search is equivalent to the cupon's collector test on the L least significant bits of the output of the hash function. To have a probability of success close to 1, one needs an average of $(L2^L)$ trials.[1] The last step of the construction is probabilistic and thus may fails. The sizes of l_c and l_m are parameters of the construction and should verify

$$|l_c| \geq L \log_2 L \text{ and } |l_m| \geq l. \qquad (13)$$

Depending on the embedding algorithm, the size l_m may be either pre-defined by the construction or dynamically computed as a function of l. In order to get sure that the extraction algorithm will always output a message, we dynamically adjust the value of L if needed such that

$$|S_u| - L \log_2 L \geq 2^L. \qquad (14)$$

If equation 13 is no more satisfied, then the embedding procedure fails. Equation 14 guaranties that there are enough usable bits in S_m to embed the longest message of length $(2^L - 1)$.

---

[1] If L is too large, you may choose another unit to encode the message length and padd the message until an integer value of units.

3.2   Extraction Construction

The extraction process is symmetric. First, we obtain the length of the embedded message by computing

$$l = H(K||S).$$ (15)

As S_u is *a priori* known, one can compute

$$IV_1 = H(l||K||C \setminus S_u).$$ (16)

Then, the PRNG is initialized with K and IV_1, and S_c of size l_c is randomly selected as a subset of S_u. In the same way, S_m of size l_m is randomly selected as a subset of $(S_u \setminus S_c)$ to define

$$S_r = C \setminus (S_c \cup S_m).$$ (17)

With S_r, we calculate

$$IV_2 = H(K||S_r)$$ (18)

and re-initialize the PRNG with K and IV_2. Finally, the message is extracted from S_m by the extraction algorithm Ext using the PRNG.

Now, whatever the bit b_i we randomly change in C, the extraction oracle outputs a message of length l with probability 2^{-l}, if we consider that the values given by L least significant bits of the hash function are uniformly distributed. This construction removes the intrinsic malleability property of practical steganographic schemes.

This generic construction improves drastically the security of steganographic schemes by making them robust against adaptive adversaries. That means that if a steganographic scheme is ε-secure against a non-adaptive adversary then we are able to design at a low computational cost, a steganographic scheme that is at least ε-secure against a more powerful adversary, that is an adaptive adversary. Despite his access to an extraction oracle, an adaptive adversary does not obtain much more advantage attacking the constructed scheme than the non-adaptive adversary attacking the original scheme.

Practically, the non-malleability that we reach using this construction makes exhaustive searches on the steganographic key harder. Indeed, for each tested steganographic key the algorithm outputs a message that seems to be valid, then no particular shortcuts appear using this strategy.

4   Conclusion

The increasing number of papers dealing with security models in cryptography or in steganography but also in many fields where security is crucial and should be evaluated, magnifies the very need for rigorously defined models of security. When designing a new security scheme one wants to be trustful enough to release it for real-life applications. Security models offer a general framework to evaluate how much confident in a scheme we could be. From the adversary's point of view, many attacks may defeat the security of a given scheme. In that case, one wants to compare the efficiencies of all the proposed attacks. In the same way, security models with real-life adversaries offer a common methodology to benchmark attacks. In this paper, we tried to reduce the gap between these points of view by considering the security of practical steganographic schemes in strong models of security. We identified some weaknesses of many of them and proposed a generic construction to overcome these vulnerabilities. Such a type of construction appears to be mandatory if we require steganographic schemes that resist adaptive adversaries.

First, we recall cryptographic models of security and but also the classical ones in steganography. Then, we discussed about the malleability property and the associated models. We adapted cryptographic malleability-based models into steganographic ones in order to take into account adaptive adversaries. We also give a theorem that links the IND-CHA2 and the NM-CHA2 models. This theorem also gives us the clues to understand why most of practical steganographic schemes are not secure against IND-CHA2 adversaries. Actually, their intrinsic malleability makes them insecure in the NM-CHA2 model and thus in the IND-CHA2 model. To fill this lack and to improve the security of a given scheme, we proposed a construction that notably increase the resistance to adaptive adversaries.

Steganography security models, and more particularly private key models, have not been investigated as much as in cryptography. Some of the classical models have been defined but the hierarchy has not been yet entirely considered. Since cryptography is very close to steganography through many points of view, it may give us the keys to elaborate and deeply understand steganography models of security. This is one part of our future work.

Acknowledgments

We would like to thank S. Alt for many interesting discussions about public key cryptographic security models. Her useful comments have greatly improved our understanding of such a complex field.

References

1. Cachin, C.: An information-theoretic model for steganography. In: Aucsmith, D. (ed.) IH 1998. LNCS, vol. 1525, pp. 306–318. Springer, Heidelberg (1998)
2. Cachin, C.: An information-theoretic model for steganography. Information and Computation 192(1), 41–56 (2004)
3. Cachin, C.: Digital steganography. In: van Tilborg, H. (ed.) Encyclopedia of Cryptography and Security. Springer, Heidelberg (2005)
4. Chandramouli, R.: Mathematical theory for steganalysis. In: Proc. SPIE Security and Watermarking of Multimedia Contents IV (2002)
5. Chandramouli, R., Kharrazi, M., Memon, N.: Image steganography and steganalysis: Concepts and practice. In: Kalker, T., Cox, I., Ro, Y.M. (eds.) IWDW 2003. LNCS, vol. 2939, pp. 35–49. Springer, Heidelberg (2004)
6. Chandramouli, R., Memon, N.: Steganography capacity: A steganalysis perspective. In: Proc. SPIE, Security and Watermarking of Multimedia Contents V, Santa Clara, CA, USA, vol. 5020, pp. 173–177 (2003)

7. Hopper, N.: Toward a Theory of Steganography. PhD thesis, School of Computer Science. Carnegie Mellon University, Pittsburgh, PA, USA (2004)
8. Dedić, N., Itkis, G., Reyzin, L., Russel, S.: Upper and lower bounds on black-box steganography. In: Kilian, J. (ed.) TCC 2005. LNCS, vol. 3378, pp. 227–244. Springer, Heidelberg (2005)
9. Hopper, N., Langford, J., von Ahn, L.: Provably secure steganography. In: Yung, M. (ed.) CRYPTO 2002. LNCS, vol. 2442, pp. 77–92. Springer, Heidelberg (2002)
10. Katzenbeisser, S., Petitcolas, F.: Defining security in steganographic systems. In: Proc. SPIE Security and Watermarking of Multimedia contents IV, vol. 4675, pp. 50–56 (2002)
11. von Ahn, L., Hopper, N.J.: Public-key steganography. In: Cachin, C., Camenisch, J.L. (eds.) EUROCRYPT 2004. LNCS, vol. 3027, pp. 323–341. Springer, Heidelberg (2004)
12. Levan, T., Kurosawa, K.: Efficient public key steganography secure against adaptative chosen stegotext attacks. In: Proc. Information Hiding, 8th International Workshop, Old Town Alexandria, Virginia, USA (2006)
13. Hopper, N.: On steganographic chosen covertext security. In: Caires, L., Italiano, G.F., Monteiro, L., Palamidessi, C., Yung, M. (eds.) ICALP 2005. LNCS, vol. 3580, pp. 311–323. Springer, Heidelberg (2005)
14. Kerckhoffs, A.: La cryptographie militaire. Journal des Sciences Militaires (1883)
15. Ker, A.: The ultimate steganalysis benchmark? In: MM&Sec 2007: Proceedings of the 9th workshop on Multimedia & security, Dallas, Texas, USA, pp. 141–148. ACM, New York (2007)
16. Barbier, J., Alt, S.: Practical insecurity for effective steganalysis. In: Solanki, K., Sullivan, K., Madhow, U. (eds.) IH 2008. LNCS, vol. 5284, pp. 195–208. Springer, Heidelberg (2008)
17. Barbier, J.: Analyse de canaux de communication dans un contexte non-coopératif. Application aux codes correcteurs d'erreurs et à la stéganalyse. PhD thesis, École Polytechnique, Palaiseau, France (2007)
18. Goldwasser, S., Micali, S.: Probabilistic encryption. Journal of Computer and System Science 28, 270–299 (1984)
19. Bellare, M., Desai, A., Jokipii, E., Rogaway, P.: A concrete security treatment of symmetric encryption: Analysis of the DES modes of operation. In: Proc. 38th Symposium on Foundations of Computer Science FOC. IEEE, Los Alamitos (1997)
20. Dolev, D., Dwork, C., Naor, M.: Nonmalleable cryptography. SIAM Journal of Computing 30(2), 391–437 (2000)
21. Bellare, M., Desai, A., Pointcheval, D., Rogaway, P.: Relations among notions of security for public-key encryption schemes. In: Krawczyk, H. (ed.) CRYPTO 1998. LNCS, vol. 1462, pp. 26–45. Springer, Heidelberg (1998)
22. Bellare, M., Sahai, A.: Non-malleable encryption: equivalence between two notions and an indistinguishability-based characterization. In: Wiener, M. (ed.) CRYPTO 1999. LNCS, vol. 1666, pp. 519–536. Springer, Heidelberg (1999)
23. Katz, J., Yung, M.: Characterization of security notions for probabilistic private-key encryption. Journal of Cryptology 19(1), 67–96 (2006)
24. Provos, N.: Universal steganography (1998), http://www.outguess.org/
25. Westfeld, A.: F5-a steganographic algorithm. In: Moskowitz, I.S. (ed.) IH 2001. LNCS, vol. 2137, pp. 289–302. Springer, Heidelberg (2001)
26. Latham, A.: Steganography: JPHIDE and JPSEEK (1999), http://linux01.gwdg.de/~alatham/stego.html

An Algorithm for Modeling Print and Scan Operations Used for Watermarking

S. Hamid Amiri and Mansour Jamzad

Department of Computer Engineering, Sharif University of Technology, Tehran, Iran
s_amiri@ce.sharif.edu, jamzad@sharif.edu

Abstract. Watermarking is a suitable approach for digital image authentication. Robustness regarding attacks that aim to remove the watermark is one of the most important challenges in watermarking, in general. Several different attacks are reported that aim to make it difficult or impossible for the real owner of the digital watermarked image to extract the watermark. Some of such common attacks are noise addition, compression, scaling, rotation, clipping, cropping, etc. In this paper we address the issue of print and scan attack by introducing a method to model the scanner and printer. Then we will simulate the print and scan attack on the digital images to evaluate its robustness. In addition, we introduce how to identify the system and how to analyze the noise imposed on the digital image when it is printed and the printed version is scanned. In this approach we obtained high flexibility in analyzing the behavior of different printers and scanners. By examining the amount of degradation applied on the original watermarked image obtained after the process of scanning its printed version, we can embed the watermark in such a way that the robustness of watermark is maintained by print and scan attack. To evaluate the performance of the proposed method we used some bench marks that are available for this purpose. Our experimental results showed a high performance for our method in modeling the print and scan operations.

Keywords: Watermark, Print and scan, Image complexity, System identification, Neural network, Texture.

1 Introduction

In recent years, by development of internet and the need for secure digital communication, the need for watermarking to protect the issues of ownership and preventing forging of digital documents has become an important issue in recent years. In general in watermarking a pattern is embedded in the host digital data (image, sound, video, etc.) in such a way that the quality of digital data is not disturbed and the real owner of the data can retrieve the watermark pattern using a secret key. However, the watermark should be embedded in the host in such a way that the attacks (intentional and unintentional) could not destroy it. Print and scan attack is defined as follows: In order to remove the watermark or make it difficult to extract, the original watermarked image is printed, the printed version is scanned, and the scanned version is saved in computer. Many watermarked images are not robust with respect to this

H.J. Kim, S. Katzenbeisser, and A.T.S. Ho (Eds.): IWDW 2008, LNCS 5450, pp. 254–265, 2009.
© Springer-Verlag Berlin Heidelberg 2009

attack. The aim of this paper is to model the processes of noise addition and non-liner transforms imposed on the printed and scanned image. Then in another approach one can suggest a watermarking algorithm that will be robust with respect to this attack.

In this regard we need to analyze the changes imposed on the original watermarked image after the process of print and scan. One difficulty in this approach is that we cannot model the printer and scanner separately. That is, unless we do not create a digital version of the printed image (i.e. by scanning it), there is no way that we could compare the printed version with a digital version of it. Therefore, we have to provide one model for the processes of print and scan.

There have been several suggestions for modeling the print and scan. In [3] a model for electro photographic process in printer is introduced, where different steps of the process are described by mathematic equations. This model is then used in a method called Direct Binary Search that produces iterative halftone patterns. Although this model uses complicated mathematical equations, but it could only model a small part of printing operation. In [4] a model for print and scan is proposed by dividing the operations of print and scan into two categories: distortions related to pixel values and those related to geometric transforms, a model for print and scan is proposed. In [6] the operation of print and scan are divided into three groups such as geometric transforms (i.e. rotation, scaling and clipping), non-linear effects (i.e. Gamma correction, gamma tweaking and half toning) and noise addition. Then the effects of each group on Fourier transform coefficients have been studied and suitable coefficients to embed watermark are extracted. In [9] print and scan distortions are divided into two categories: systematic and operational. To identify systematic distortions, first the internal structure of printer and scanner are studied and then, the distortion corresponding to each subsystem is described. Operational distortions such as geometric transforms are among the distortions that are imposed by user, and have a random behavior depending on the user.

In the above mentioned papers, a system that has a behavior similar to print and scan operation has not been suggested, but they only have investigated the distortions imposed by the print and scan operations.

In this paper we try to suggest a system that can model the distortions imposed on the input image to the print and scan operations.

For modeling the print and scan operations, we assume that rotation, scaling and clipping are among the geometric transforms that happen at the end of operation. Therefore, in our modeling we focus on other operations that are imposed on input image during the print and scan operations. These operations can be divided into two groups: How noise is added to the image and the effect of non-linear transforms in output image.

To determine the noise in print and scan operations, we assume that the additive noise consists of an independent and a dependent part from the input. Since the output of print operation is not in digital format, it is not possible to determine noise in printer independent from its input. As a result we shall focus of determining the noise independent from the input in the scanner. Of course in determining the noise related to input we consider both the printer and scanner as one black box and determine the noise parameter for this black box. At the end, a non-linear system is added to the model such that we can apply the effect of non-linear transforms to our model.

The rest of paper is organized as follows, in section 2; we propose a method to determine noise that is independent from input in scanners. In section 3, we focus on specifying the noise dependent to input and how to mathematically determine the parameters of noise. Applying of non-linear effects in model will be discussed in section 4; in addition, we use non-linear systems that consider neighborhood relation for each pixel in input image. In section 5, we enumerate some criterions to evaluate our model. The results are presented in section 6. We conclude our work in section 7.

2   Specifying the Noise Independent from Input

To specify the noise that is independent from input image (I) in scanner, we scan it n times and generate images $I_1, I_2, ..., I_n$. Image $\tilde{I}$ which is average of scanned images is calculated using equation (1).

$$\tilde{I} = \frac{I_1 + I_2 + \cdots + I_n}{n} \tag{1}$$

Image $\tilde{I}$ shows the behavior of scanner for image I in average state.

In order to model independent noise from input in scanner, firstly we should calculate the mean and standard deviation of noise using the extracted samples. Standard deviation is calculated using equation (2).

$$sd(I) = \frac{std(I_1 - \tilde{I}) + std(I_2 - \tilde{I}) + \cdots + std(I_n - \tilde{I})}{n} \tag{2}$$

In equation (2), $sd(I)$ represents standard deviation of noise and std shows the standard deviation operator. On the other hand we can calculate the mean of noise with an equation similar to equation (2). In this case, according to the linear property of expectation it can be proved that the mean of noise is equal to zero.

By repeating the above operations on k images, we will have k standard deviations. Since these k standard deviations are different from each other, we can assume them as samples from a random variable. We applied the Chi-Squared probability test on the samples of this random variable to find the probability distribution of it. It was shown that they had lognormal distribution with parameters m=0,739 and v=0.269.

To find the noise which is independent from the input in the scanner, we calculated the difference of scanned images and the mean image, and then these difference values are considered as noise samples. Again, to determine the noise distribution, we used the Chi-Squared probability test. In this case, the probability distribution function was shown to be a logistic function with PDF as defined in equation (3).

In (3), μ is noise mean, and it is set to zero. Parameter s, is calculated from $s = \frac{\sqrt{3}}{\pi}\sigma$, in which σ is the noise standard deviation.

$$f(x; \mu, s) = \frac{exp\left(-\frac{x-\mu}{s}\right)}{s\left[1 + exp\left(-\frac{x-\mu}{s}\right)\right]^2} \tag{3}$$

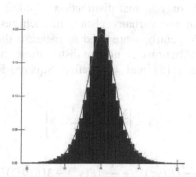

Fig. 1. Histogram of the noise extracted for an image and the logistic distribution fit on it

Fig. 1 shows the histogram of noise (i.e. random variable) extracted for an image and the logistic distribution fit on the random variables. The parameters of this fit were 0 and 1.2836.

3 Specifying the Noise Dependent to Input

We investigate the effect of print and scan operations on different gray levels to determine noise that is dependent to input. To do this, we use images with constant gray levels as input for operations. Considering the fact that gamma correction happens in scanners, gray level of pixels in scanned image changes with different gamma parameters. Thus, we printed pages having constant gray levels of 0 to 255 (256 pages) and for each page that went under the scanning process, we considered the value for gamma as 1., 1.3, 1.5, 1.8, 2., 2.3, 2.5, 2.8 and 3.

If I and I_{ps} be input image and output image for print and scan operations, respectively. After operations η_{ps} is added to input image as additive noise, we have equation (4) for noise that is dependent to the input image.

$$I + \eta_{ps} = I_{ps} \Rightarrow \eta_{ps} = I_{ps} - I, \eta_{ps} = \eta_I + \eta_D \Rightarrow \eta_D = I_{ps} - I - \eta_I \qquad (4)$$

In equation (4), η_I and η_D represent the noise that are independent from input and dependent to input image, respectively.

To specify the PDF of noise and its parameters we used the histogram of noise. Fig. 2 shows four histograms of noise for two gray levels and two gamma parameters.

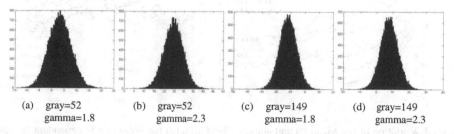

(a)	gray=52	(b)	gray=52	(c)	gray=149	(d)	gray=149
	gamma=1.8		gamma=2.3		gamma=1.8		gamma=2.3

Fig. 2. Histogram of noise that is dependent to input for two gray levels and two gamma parameters

Histograms in Fig. 2, suggest that distribution of noise is very similar to normal distribution but its mean and variance change in each case. The normality tests of Squared-Kurtosis and Negentropy are used to measure the distance between noise samples and normal distributions. In normal distribution, the values of these tests are equal to zero. Equations (5) and (6) define Squared-Kurtosis and Negentropy relations, respectively.

$$k(y) = \left[\frac{\sum (y_i - m)^4}{(N-1)\sigma^4} - 3 \right]^2 \tag{5}$$

$$J(y) = \frac{1}{12} (E(y^3))^2 + \frac{1}{48} [E(y^4) - 3(E(y^2))^2]^2 \tag{6}$$

In relations (5) and (6), y represents noise samples and m and σ show the mean and standard deviation of samples. We must be careful that for Negentropy test; the transform in equation (7) must be applied first.

$$y' = \frac{y - m}{\sigma} \tag{7}$$

In Table 1, values of these tests are shown. These values suggest that the normal distribution provides good accuracy to approximate the noise distribution.

Table 1. Results of normality tests for some noise that is dependent to input

Gray level	gamma	Squared Kurtosis	Negentropy
52	1.5	0.0025159	0.00093891
52	2.3	0.00095156	0.00012932
52	2.8	0.0027282	8.962e-005
149	1.5	0.0065468	0.00018963
149	2.3	0.021906	0.00045832
149	2.8	0.015126	0.00034508
208	1.5	0.011932	0.00057012
208	2.3	0.0021108	0.00033714
208	2.8	0.002354	0.00032484

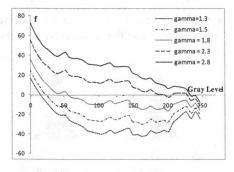 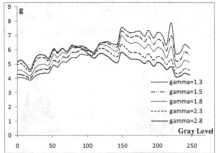

Fig. 3. Variation of the Mean (f) and standard deviation (g) for noise that is dependent to input by different gray levels and gamma parameters

As mentioned in above, parameters of noise (the mean and standard deviation) will change by different gray levels and gamma parameters. We suggest that the noise parameters can be estimated by equation (8).

$$m_\eta = f(l,\gamma) \ , \ \sigma_\eta = g(l,\gamma) \tag{8}$$

Where m_η and σ_η are the mean and standard deviation of noise, l and γ are constant gray level of input image and the gamma parameter, respectively. In Fig. 3 the extracted function for f and g are shown for some values of gamma.

As it is seen in Fig.3, the f and g functions have similar behavior with different values for gamma. Therefore, to estimate these functions, we use polynomials of order n and m as shown in equations (9) and (10).

$$f(l,\gamma) = \sum_{i=1}^{n} a_i\,(\gamma)\,l^i \tag{9}$$

$$g(l,\gamma) = \sum_{i=1}^{m} b_i\,(\gamma)\,l^i \tag{10}$$

Where, l and γ show the gray level and gamma parameter, respectively.

4   The Influence of Neighbors in Scanned Value of a Pixel

To determine the scanned value (gray level) of a pixel, in addition to the parameters discusses in previous sections, the gray level of neighboring pixels do influence on the scanned value. Print and scan are non-linear operations because of the effect of gamma tweaking, dot gain and gamma correction.

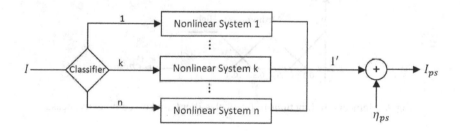

Fig. 4. General schema of the proposed model

We classify images based on their histogram and complexity and apply a non-linear system for each class to improve the accuracy of proposed model. Fig. 4 shows a block diagram of our proposed model.

To identify a system related to one class, we select some images in that class and extract the necessary samples from them for system identification. In the following, we explain classification algorithm based on histogram and image complexity and then we describe in detail the system identification method for one class of images. We use the same method of system identification in other classes.

4.1 Classification Algorithm for Images Based on Histogram and Image Complexity

For classifying images using image complexity, we use equation (11) that is based on image Quad-Tree representation [8].

$$cm(I) = \frac{n_B - 1}{n_p} \tag{11}$$

In equation (11), cm is complexity measure and n_B and n_p represent the number of blocks (leaves) in Quad-Tree and the number of pixels in image I, respectively. Higher value for cm represents images with higher complexity. This measure is calculated for one database of 1390 images and its histogram is shown in Fig. 5. cm is between 0 to 0.5 for most images. Using this measure, images are classified into low, medium and high complexity.

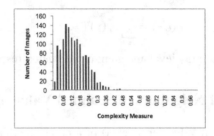

Fig. 5. Histogram of complexity measure on one image database. It is between 0 to 0.5 in most cases.

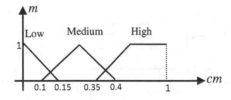

Fig. 6. Membership functions of fuzzy classifier based on complexity measure

We used a fuzzy classifier with three membership functions as shown in Fig. 6 to determine the complexity classes for images.

In this approach, the cm complexity measure is calculated for each given image, and then using membership functions of Fig. 6, the class to which the image belongs, is determined.

In order to classify the images according to their histogram, a few basis histograms are introduced, then using the distance measure in equation (12), the distance of image histogram from each base histogram is determined.

$$D_{rd}(H,H') = \frac{\sqrt{\sum_{m=0}^{255}(H_m - H'_m)^2}}{0.5\left(\sqrt{\sum_{m=0}^{255} H_m^{\,2}} + \sqrt{\sum_{m=0}^{255} H'_m{}^{2}}\right)} \tag{12}$$

Where, H_m is the mth interval in histogram H.

The base histogram that has the shortest distance to the image histogram represents the class to which the image should belong.

Fig.7, shows a few base histograms used for classification. To improve the model precision, we can increase the number of these histograms.

Therefore, we can classify images in k classes using the histogram approach and into *three* classes by complexity measure.

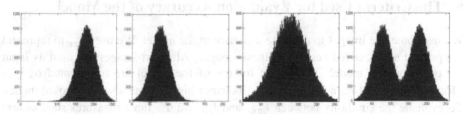

Fig. 7. Four basis histograms used to classify images based on histogram

4.2 System Identification for a Class of Images

At first a few images that lie in the class of desired system are selected and then they are printed and scanned. Using the original images and their print and scan versions the necessary samples for determining the system is obtained. Let I, I' and I_{ps} be the input, output and the scanned image, respectively. Then the relation shown in equation (13) holds.

$$I' + \eta_{ps} = I_{ps} \Rightarrow I' = I_{ps} - \eta_{ps} \tag{13}$$

After selecting the input image I for extracting samples, pixel $I'(x,y)$ is considered as output and the 3×3 neighbors of $I(x,y)$ are taken as input. For neighboring pixels that do not lie within I, the value of 255 is assigned.

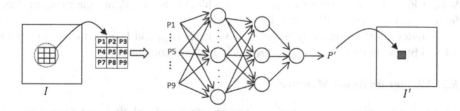

Fig. 8. Behavior of the neural network in each system to estimate non-linear operations on the input image I

After selecting the samples, the next step is to determine the system. One of the ways to determine the system is to use the adaptive filters. In these filters, the system function is defined using the steepest descent algorithm. Since the system in a non-linear one, so we shall use a non linear adaptive filter. In addition, the neural network is considered to be a non-linear adaptive filter; therefore we can use the neural network to identify the system.

We used a feed-forward neural network with 9 neurons in input, one neuron in output and two hidden layers with 15 and 9 neurons, respectively. This neural network is trained using extracted samples. Fig.8 shows an example of this neural network.

5 The Criteria Used for Evaluation Accuracy of the Model

We use some test images to evaluate accuracy of the model. We use I_{mps} to represent the printed and scanned version of these images. Also test images are used as input for our proposed model. The output images of the model are represented by I_{eps}. More similarity between I_{mps} and I_{eps} indicates higher accuracy in proposed model. To evaluate the similarity between I_{mps} and I_{eps} we use three measures such as texture based, histogram and the amount of noise added to images. In the following these measure are explained in detail.

5.1 Texture Based Measure

I_{mps} and I_{eps} are divided into 8x8 blocks and the co-occurrence matrix is calculated for each block. Then the *Energy, Entropy, Contrast* and *Homogeneity* features are calculated for the co-occurrence matrices. In addition we used two Tamura features such as *Coarseness* and *Contrast* [1]. Therefore, we construct a 6 component features vector for each block. To determine the similarity between two corresponding blocks located in the same location in two images, we use equation (14).

$$Corr = \frac{\sum_{i=1}^{6}(V_1(i) - \overline{V_1})(V_2(i) - \overline{V_2})}{\sqrt{(\sum_{i=1}^{6}(V_1(i) - \overline{V_1})^2)(\sum_{i=1}^{6}(V_2(i) - \overline{V_2})^2)}} \tag{14}$$

Where V_1 and V_2 are feature vectors for two blocks; $\overline{V_1}$ and $\overline{V_2}$ are the mean of these feature vectors. *Corr* will lie in [-1,1].

To determine the similarity between two images I_{mps} and I_{eps} the mean of *Corr* for all 8x8 blocks in these two images are calculated and compared.

5.2 Histogram Based Measure

At first the histogram of I_{mps} and I_{eps} images are calculated, then using equation (12) the relative distance between two histograms are obtained. A relative distance closer to zero indicates more similarity between two images.

5.3 The Measure Based on the Amount of Additive Noise

This measure is based on calculating the PSNR according to equation (15).

$$PSNR = 10\log_{10}\left(\frac{Max(I)^2}{mse}\right) , \quad mse = \frac{1}{mn}\sum_{i=0}^{m-1}\sum_{j=0}^{n-1}\left[I(i,j) - I'(i,j)\right]^2 \tag{15}$$

Where, I and I' are the original and the noisy images.

6 Experimental Results

We used the images shown in Fig.9 as test images to determine the preciseness of the proposed model. Such that, none of them are used in training phase of neural networks. Each of these images were used as input to the proposed model, and were printed and scanned. The printer used was a laser printer that prints by halftone method. The scanner was of flat type that uses CCD sensors.

Fig. 9. Test images used to measure the accuracy of model. Images from left to right are: Woman, Houses, Man and Butterfly.

Fig. 10. Output of the proposed model

Fig. 11. Output of the print and scan operations

In Figs 10 and 11, the output of the proposed model and the output of print and scan operations for images in Fig. 9 are shown. To compare two sets of images we used the measures suggested in section 5. Table 2 shows the result of the comparison. The 2nd column in Table 2 shows the texture similarity measure between two sets images. As seen, these two sets of images are very similar according to their texture, because in all cases, the value of equation (14) was higher than 0.95. The 3rd column of table 2 shows the relative distance between histograms of two sets of images according to equation (12).

To get a visual perception of the histogram comparison, we provide Fig. 12 that shows the histograms of original, the print and scanned version and that of model output for four test images.

In Table 2, the columns 4 and 5 show the PSNR for two sets of images. Nearly for all images, the amount of noise added to model output image, is more than the noise in image resulted from print and scan operations. Since we plan to use the proposed model to suggest a watermarking algorithm robust with respect to print and scan attack, the higher noise not only will not cause difficulty, but is an advantage too.

Table 2. Results of comparisons between output of the model and print and scan operations

Image	Texture Correlation	Relative Histogram Distance	PS Image PSNR	Output of Model PSNR
Woman	0.9577	0.1797	17.416	16.1138
Houses	0.9815	0.1794	19.683	18.9682
Man	0.9768	0.1630	19.0568	17.983
Butterfly	0.982	0.1151	18.4884	18.5071

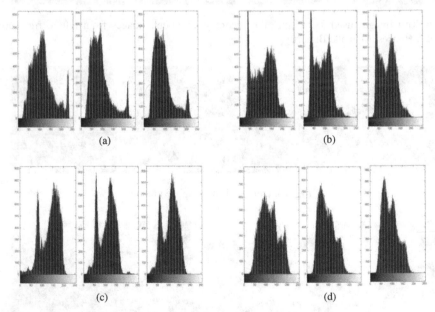

(a)   (b)

(c)   (d)

Fig. 12. Histograms of images in Figs. 9, 10 and 11. In each group left histogram is for main image, middle histogram is for output of print and scan operations and right histogram is for output of model. Histograms are for (a) Woman (b) Houses (c) Man and (d) Butterfly.

7 Conclusion

In this paper we suggested an algorithm to model the print and scan operations. Our algorithm tries to predict the modifications imposed on an input image to the print and scan operations. Then it will generate an output image that is most similar to the version that would have been produced as the result of print and scan operations. Our experimental results should that the model is highly accurate to simulate the print and scan operations.

Therefore, we can use the proposed model to analyze the behavior of print and scan operations. This analysis can be used to develop a watermarking algorithm that will be robust with respect to print and scan attack.

One of the advantages of the proposed method is that it can be used to model any printer and scanner, independent of its type. In future works, we plan to consider the role of paper type, printer tonner and manual modifications in printed paper (i.e. manual folding), etc. In addition for image classification, we can use measures other than histogram and complexity that could improve the accuracy of our model.

References

1. Howarth, P., Ruger, S.: Evaluation of Texture Features for Content-Based Image Retrieval. In: Enser, P.G.B., Kompatsiaris, Y., O'Connor, N.E., Smeaton, A., Smeulders, A.W.M. (eds.) CIVR 2004. LNCS, vol. 3115, pp. 326–334. Springer, Heidelberg (2004)
2. Hyvarinen, A., Karhunen, J., Oja, E.: Independent Component Analysis. John Wiley & Sons, New York (2001)
3. Kacker, D., Camis, T., Allebach, J.P.: Electrophotographic process embedded in direct binary search. IEEE Transactions on Image Processing 11, 243–257 (2002)
4. Lin, C., Chang, S.: Distortion Modeling and Invariant Extraction for Digital Image Print-and-Scan Process. In: International Symposium on Multimedia Information Processing, Taipei (1999)
5. Manjunath, B.S., Ma, W.Y.: Texture features for browsing and retrieval of image data. IEEE Transaction on Pattern Analysis and Machine Intelligence 18, 837–842 (1996)
6. Solanki, K., Madhow, U., Manjunath, B.S., Chandrasekaran, S.: Modeling the print-scan process for resilient data hiding. In: Proc. SPIE Security, Steganography, and Watermarking of Multimedia Contents VII, pp. 418–429 (2005)
7. Suykens, J., Vandewalle, J., de Moor, B.: Artificial Neural Networks for Modeling and Control of Non-linear Systems. Kluwer Academic Publishers, Netherlands (1996)
8. Yaghmaee, F., Jamzad, M.: Computing watermark capacity in images according to their quad tree. In: 5th IEEE International Symposium on Signal Processing and Information Technology, pp. 823–826. IEEE Press, Los Alamitos (2005)
9. Yu, L., Niu, X., Sun, S.: Print-and-scan model and the watermarking countermeasure. In: Image and Vision Computing, pp. 807–814. Elsevier, Amsterdam (2005)
10. Chen, P., Zhao, Y., Pan, J.: Image Watermarking Robust to Print and Generation Copy. In: First International Conference on Innovative Computing, Information and Control, pp. 496–500. IEEE Press, Los Alamitos (2006)
11. Song, Y.J., Liu, R.Z., Tan, T.N.: Digital Watermarking for Forgery Detection in Printed Materials. In: Shum, H.-Y., Liao, M., Chang, S.-F. (eds.) PCM 2001. LNCS, vol. 2195, pp. 403–410. Springer, Heidelberg (2001)

Space Time Block Coding for Spread Spectrum Watermarking Systems

Cagatay Karabat

Tubitak Uekae (Turkish National Research Institute Of Electronics And Cryptography),
Kocaeli, Turkey
cagatay@uekae.tubitak.gov.tr

Abstract. In this paper, we propose a novel spread spectrum image watermarking system. It employs Alamouti's space-time block coding scheme which is commonly used to improve the performance of the wireless communications. We model the digital watermarking system as a wireless communication channel and take the advantages of space-time block coding scheme. We employ $2x2$ real orthogonal design for embedding watermark sequence into the host image in discrete wavelet transform domain. The goal is to enhance the robustness of the spread spectrum watermarking system against various degradations and malicious attacks. Simulation results show that the proposed promising system considerably increases the correlation and decreases the bit error rate between the embedded and the recovered watermark sequences.

1 Introduction

The digital watermarking systems have increased their popularity nowadays with the increasing the security threads to the digital multimedia products. The watermarking process can be modeled as a communication task, in which the watermark information is transmitted over the watermark channels within the host signal. Thus, we can employ basic communications systems primitives in these systems. In the literature, several error correction codes such repetition coding, turbo coding and low density parity check coding schemes are employed in various watermarking systems in order to decrease the bit error rate (BER) and increase the correlation coefficient of the recovered watermark sequence [1]-[3]. Besides, there are some watermarking systems that employ space time codes in spatial domain to recover the embedded watermark sequence properly at the receiver [4]-[5].

In this paper, we propose to employ space-time block codes (STBC) to the spread spectrum image watermarking system in discrete wavelet transform (DWT) domain. Space time block coding schemes performs signal processing both in the spatial and temporal domain on signals by employing multiple antennas at the transmitter and the receiver to improve performance of wireless networks. We apply these wireless communications primitives to the spread spectrum watermarking system in DWT domain. We employ $2x2$ real orthogonal space-time block code (STBC) design, which is called Alamouti's coding scheme, before embedding the watermark into the host image. We demonstrate the improvements of Alamouti's scheme in comparison to the no coding case against various channel distortions and attacks. The rest of the paper is

H.J. Kim, S. Katzenbeisser, and A.T.S. Ho (Eds.): IWDW 2008, LNCS 5450, pp. 266–277, 2009.
© Springer-Verlag Berlin Heidelberg 2009

organized as follow: we briefly introduce space time block codes in Section 2. Then, we describe the proposed spread spectrum watermarking system in Section 3. In Section 4, we give the simulation results and finally we give the concluding remarks in Section 5.

2   Space Time Block Codes (STBC)

The space-time coding is employed in multiple-antenna wireless communication systems to improve the reliability of data transmission. It relies on redundancy and diversities of data copies. Among many space time coding schemes, Alamouti's scheme [6] is the first space time block code to provide full transmit diversity for systems with two transmit antennas, and is widely used due to its simple yet effective design.

In space-time coding schemes, the signal processing at the transmitter is done not only in the temporal dimension, as what is normally done in many single-antenna communication systems, but also in the spatial dimension, as shown in Fig. 1. By introducing redundancy to both dimensions, both the data rate and the performance are improved by many orders of magnitude with no extra cost of spectrum. This is also the main reason why space-time coding attracts much attention.

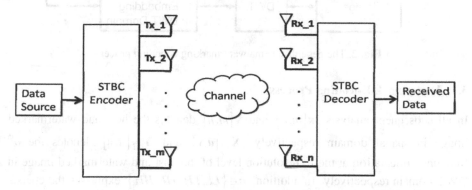

Fig. 1. General block diagram of space time block coding schemes and the channel

After modulation, information sequences are fed into the space-time encoder. The encoder maps the input data onto modulation symbols, and then sends them to a serial-to-parallel converter producing a sequence of parallel symbols. The parallel outputs are simultaneously transmitted by different antennas. For wireless mobile communications, each link from a transmit antenna to a receive antenna can be modeled by a fading channel. At the receiver, the signal at each of the receive antennas is a noisy superposition of the transmitted signals degraded by channel fading. The decoder at the receiver uses a maximum likelihood algorithm to estimate the transmitted information sequence and that the receiver has ideal channel state information on the multiple input multiple output (MIMO) channel.

3 The Proposed Watermarking System

In this section, we describe the proposed spread spectrum watermarking system in DWT domain. The main contribution of the proposed watermarking system is to employ Alamouti's STBC scheme in order to enhance the robustness of the system against channel distortions and attacks. Thus, we can reduce the bit error rate and increase the correlation coefficient of the reovered watermark at the receiver.

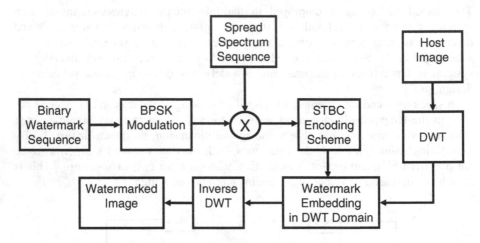

Fig. 2. The general schema watermarking encoding process

3.1 Watermark Encoding Process

In all subsequent analysis, $\mathbf{x}[m,n]$ and $\mathbf{y}[m,n]$ denotes the host and watermarked image in spatial domain respectively. $\mathbf{X}_{o,l}[u,v]$ and $\mathbf{Y}_{o,l}[u,v]$ denotes the oth frequency orientation at the lth resolution level of the host and watermarked image in DWT domain respectively. In addition, $o \in \{LL, LH, HL, HH\}$ expresses the coarse, horizontal, vertical and diagonal edges bands respectively, $l \in \{1, 2, ..., L\}$ is the resolution level and $[u,v]$ is the particular spatial location index at the resolution level l.

In the watermark encoding process illustrated in Fig. 2, the host image of size NxN is transformed into 1 level DWT domain. Then, we randomly generate the binary watermark sequence $\vec{w} = \left[w(1), w(2), ..., w(T)\right]$ of length T. We modulate this watermark sequence by using binary phase shift keying (BPSK) modulation. Next, we generate a spread-spectrum (SS) sequence $\vec{p}_c$, whose elements are from the set $\{\pm 1\}$, by using the security key K. We multiply the modulated watermark sequence with the SS sequence. Thus, we get the message sequence $\vec{x} = \left[x(1), x(2), ..., x(T)\right]$ of length T. Finally, we employ STBC encoding scheme to this message sequence which is proposed in [7]. We use $2x2$ real orthogonal design, which is known as Alamouti's coding, in order to encode the message sequence. Here, we employ 2x2 real orthogonal designs with $n = 2$ transmit antennas and $m = 2$ receive antennas. Therefore, we achieve a diversity order of $n \times m = 4$.

$$\mathbf{G_2} = \begin{bmatrix} x(1) & x(2) \\ -x(2) & x(1) \end{bmatrix} \tag{1}$$

where $\mathbf{G_2}$ is the codeword generator matrix. Thus, we generate a codeword matrix $\mathbf{C}$ as follows:

$$\mathbf{C} = \begin{bmatrix} x(1) & x(2) \\ -x(2) & x(1) \\ x(3) & x(4) \\ -x(4) & x(3) \\ . & . \\ . & . \\ x(T-1) & x(T) \\ -x(T) & x(T-1) \end{bmatrix} \tag{2}$$

Next, we embed codeword matrix elements into the coarse band coefficients in the DWT domain by using the additive-multiplicative method as follows:

$$\mathbf{Y}_{LL,1}[u,v] = \mathbf{X}_{LL,1}[u,v](1 + \gamma \mathbf{C}[u,v]) \tag{3}$$

where $\mathbf{Y}_{LL,1}[u,v]$ denotes the watermarked coarse band coefficients at the first level DWT domain, $\mathbf{X}_{LL,1}[u,v]$ denotes the host image coefficients at the first level DWT domain, $\mathbf{C}[u,v]$ denotes the codeword matrix whose elements are $\pm \bar{x}(i)$ where $i=1,2,..,T$ and γ denotes insertion strength. Finally, we obtain watermarked image by performing inverse DWT.

3.2 Watermark Decoding Process

In the watermark decoding process, we, first, compute one level DWT of both watermarked and host image as illustrated Fig. 3. Then, we determine embedded codeword bits by using the sign function as follows:

$$\mathbf{C}'[u,v] = \mathrm{sgn}\left(abs\left(\mathbf{Y}[u,v]\right) - abs\left(\mathbf{X}[u,v]\right)\right) \tag{4}$$

where $sgn(.)$ denotes sign function, $\mathbf{Y}_{LL,1}[u,v]$ denotes the watermarked coarse band coefficients at the first level DWT domain, $\mathbf{X}_{LL,1}[u,v]$ denotes the host image coefficients at the first level DWT domain, $\mathbf{C}'[u,v]$ denotes the estimated codeword matrix.

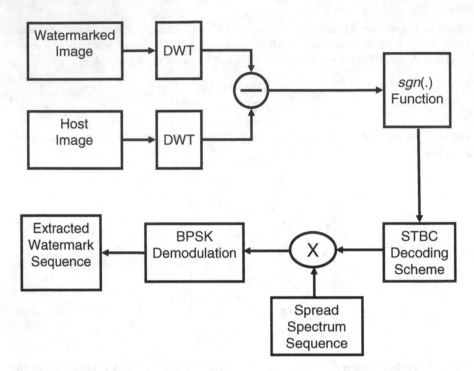

Fig. 3. The general schema of the watermark decoding process. In this schema; "-" denotes element-wise the subtraction operator and "x" denotes element-wise multiplication operator.

At the next step, we use STBC decoding scheme which is proposed in [7] to determine the message sequence $\vec{x}$. The receiver only needs to combine the received signals linearly as follows:

$$R(i) = \sum_{u=1}^{n} \sum_{v=1}^{m} \mathbf{G}'_2[u,v] \delta_k(i) \tag{5}$$

where $\mathbf{G}'_2[u,v]$ is the 2x2 matrix denoting estimated codeword sequence at time t and $t+1$ by two receive antennas, n denotes the number of transmit antennas, m denotes the number of receive antennas, $\delta_k(i)$ be the sign of $\vec{x}(i)$ in the kth row of $\mathbf{C}$. Finally, we recovered the embedded message sequence $\vec{x}'(i)$ such that:

$$\vec{x}'(i) = \arg\min_{s \in A} |R(i) - s|^2 \tag{6}$$

where $A = \{-1, 1\}$ and $\vec{x}'(i)$ denotes the recovered message sequence. Then, we regenerate SS sequence $\vec{p}_c$ by using the security key K. We multiply SS sequence $\vec{p}_c$ with the recovered message sequence $\vec{x}'$. Next, we use BPSK demodulation in order to get the embedded watermark sequence $\vec{w}$.

$$\vec{w}'(i) = \vec{x}'(i) * \vec{p}_c(i) \tag{7}$$

where $i=1,2,...,T$, $\vec{w}'$ denotes the recovered watermark sequence and $*$ denotes the element-wise multiplication of two vectors.

4 Simulation Results and Discussions

In all simulations, we use the test images of size $512x512$ and the length of watermark sequence is 256 bit. We embed the watermark sequence into the coarse band coefficients at the first level of DWT.

We assess the robustness of the proposed watermarking system against various channel distortions and attacks, i.e filtering, AWGN, JPEG compression and contrast enhancement. We calculate "Peak Signal to Noise Ratio" (PSNR) in order to objectively evaluate the perceptual quality of the watermarked image. The PSNR is a rough measure of the degradations in the host image introduced by the embedded watermark sequences as well as by the other factors. Its formulation is as follows:

$$PSNR = 10\log\frac{\mathbf{x}_{peak}^2}{\frac{1}{MN}\sum_{m=1}^{M}\sum_{n=1}^{N}(\mathbf{x}[m,n]-\mathbf{y}[m,n])^2} \tag{8}$$

where $\mathbf{x}[m,n]$ and $\mathbf{y}[m,n]$ denotes host and watermarked image respectively, M and N denotes the size of image along both dimensions.

In experiments shown in Fig. 4-Fig. 7, the insertion strength is set to 0.4 to maximize watermark power while preserving perceptual quality in the embedding process. Thus, the corresponding PSNR value is 39.60 dB which is above the minimum acceptable level of 38 dB [8] as shown in Table 1. As we increase the insertion strength, the power of embedded watermark increases, however, we may lose the perceptual quality of the watermarked image.

Table 1. The PNSR values of the watermarked image with respect to various insertion strength values

Insertion Strength	PSNR Values (dB)
0,1	51.6428
0,2	45.6222
0,3	42.1004
0,4	39.6016
0,5	37.6634
0,6	36.0798
0,7	34.7408
0,8	33.5810
0,9	32.5579
1	31.6428

We apply mean and median filtering to the watermarked image with various filter sizes to decrease the detection capability of the system in simulation shown in Fig. 4 and Fig. 5 respectively. We can state that the watermarking system is more robust against these attacks at the same filter size when we employ STBC. The *2x2* real orthogonal STBC design significantly increases the correlation coefficient between the embedded and the recovered watermark sequences. Furthermore, we compare the performance of the detectors against AWGN under various SNR values as shown in Fig. 6.The watermarked image is degraded by adding AWGN at various Signal-to-Noise Ratios (SNR) by using the below formula.

$$SNR = 10\log_{10}\left(\frac{\frac{1}{MN}\sum_{m=1}^{M}\sum_{n=1}^{N}(\mathbf{y}[m,n])^2 - \left(\frac{1}{MN}\sum_{m=1}^{M}\sum_{n=1}^{N}(\mathbf{y}[m,n])\right)^2}{\frac{1}{MN}\sum_{m=1}^{M}\sum_{n=1}^{N}(\mathbf{x}[m,n])^2 - \left(\frac{1}{MN}\sum_{m=1}^{M}\sum_{n=1}^{N}(\mathbf{x}[m,n])\right)^2}\right) \qquad (9)$$

where $\mathbf{y}[m,n]$ denotes the watermarked image, $\mathbf{x}[m,n]$ denotes the host image, M and N denotes the dimensions of the images along both directions.

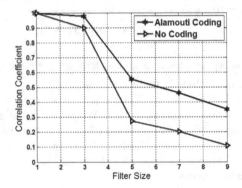

Fig. 4. Mean filter attack with various filter sizes

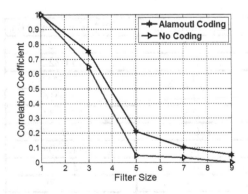

Fig. 5. Median filter attack with various filter size

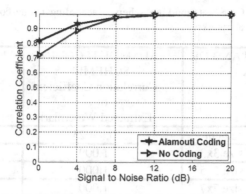

Fig. 6. Additive White Gaussian Noise attack with various SNR values

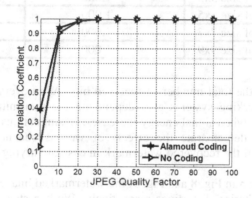

Fig. 7. JPEG compression attack with various quality factors

Then, we evaluate the performance of the detectors against JPEG compression with varying quality factors as shown in Fig. 7. The proposed watermarking system with *2x2* real orthogonal STBC designs outperforms especially at low JPEG quality factors.

We evaluate the robustness of the proposed watermarking system against noise removal filters and the contrast enhancement. Since the power of the watermark is almost equal to the noise, the adversary may employ noise removal filter with various filter sizes. The goal of the adversary is to remove and/or change the watermark embedded into the original signal. For instance; when the watermark is used to trace the illegal copies; the adversary may want to remove or change the watermark embedded into the original work. Hence, he can illegally copy and distribute the original work and nobody can trace and accuse him. The proposed watermarking system is more robust than the conventional spread spectrum watermark system against aforementioned attacks as shown in Table 2.

The BER is used as another performance metric to compare the performance of the proposed system. BER is the probability that an information bit is decoded erroneously during the watermark extraction process. It is the ratio of the number of bits

Table 2. Comparison of the watermarking schemes according to the correlation coefficient values of the extracted watermarks

DEGRADATION AND/OR ATTACKS	ALAMOUTI CODING SCHEME Correlation Coefficient of Extracted Watermark	NO CODING Correlation Coefficient of Extracted Watermark
Contrast Enhancement	0.93025	0.90125
Wiener filter of size 3x3	1.000	1.000
Wiener filter of size 5x5	1.000	1.000
Wiener filter of size 7x7	1.000	1.000
Wiener filter of size 9x9	1.000	1.000
Gaussian filter of size 3x3	1.000	1.000
Gaussian filter of size 5x5	1.000	1.000
Gaussian filter of size 7x7	1.000	1.000
Gaussian filter of size 9x9	1.000	1.000

received in error to the total number of received bits. As in every communications system, the watermarking systems also depend on the transmitted signal energy since BER decreases when signal energy increases. In simulations that are shown in Fig.8 -Fig. 12, we demonstrate the effects of various channel distortions and attacks on the BER of the recovered watermark in case of varying insertion strength values.

In simulation shown in Fig. 8 and Fig. 9, the watermarked image is exposed to *3x3* mean and median filtering distortions respectively. We can state that the proposed scheme is more robust than the no coding case at the same insertion coefficient in these simulations.

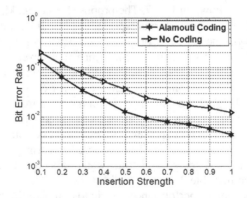

Fig. 8. *3x3* mean filter attack with various insertion strengths

Furthermore, we investigate effects the JPEG compression attack with quality factor 10 on the BER of the recovered watermark in case of varying insertion strength coefficients in Fig. 10. The proposed scheme decreases the BER of the recovered watermark sequence at all insertion strength values.

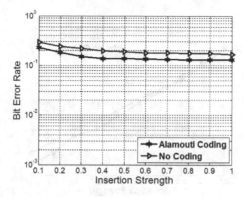

Fig. 9. *3x3* median filter attack with various insertion strengths

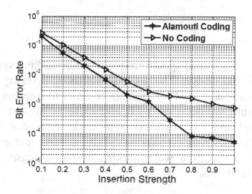

Fig. 10. JPEG compression attack at quality Factor 10 with various insertion strengths

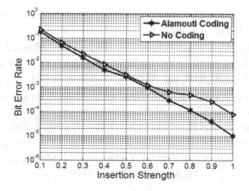

Fig. 11. AWGN attack at 4 dB with various insertion strengths

We also assses the proposed watermarking system under the AWGN channel at 4dB and 6dB SNR values in simulations shown in Fig. 11 and Fig. 12 respectively. We can state that the proposed system increses the detection capability of the system at the insertion strength value.

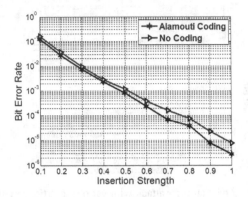

Fig. 12. AWGN attack at 6 dB with various insertion strengths

5 Conclusions

In this paper, we propose a spread spectrum image watermarking system that employs 2x2 real orthogonal STBC design. The goal is to achieve diversity when embedding watermark into the host image in DWT domain in order to improve the robustness of the system. As a result, we decrease BER and increase the correlation coefficient between the embedded and the extracted watermark sequences. We can conclude from the simulation results that employing STBC scheme enhances the detection capability of the spread spectrum watermarking system in case of aforementioned channel distortions and attacks.

References

1. Kundur, D., Hatzinakos, D.: Diversity and Attack Characterization for Improved Robust Watermarking. IEEE Trans. Img. Proc. 49, 2383–2396 (2001)
2. Rey, C., Amis, K., Dugelay, J.-L., Pyndiah, R., Picart, A.: Enhanced robustness in image watermarking using block turbo codes. In: Proc. SPIE-Security and Watermarking of Multimedia Contents V, January 2003, vol. 5020, pp. 330–336 (2003)
3. Bastug, A., Sankur, B.: Improving the Payload of Watermarking Channels via LDPC Coding. IEEE Signal Proc. Letters 11(2), 90–92 (2004)
4. Ashourian, M.: Using Space-Time Coding for Watermarking Color Images. In: Image Analysis and Recognition, Third International Conference 2006, Portugal (2006)
5. Karybali, I.G., Berberidis, K.: Improved Embedding of Multiplicative Watermarks via Space-Time Block Coding. In: EUROSIPCO 2005 (2005)

6. Alamouti, S.M.: A simple transmit diversity technique for wireless communications. Selected Areas in Communications, IEEE Journal 16(8), 1451–1458 (1998)
7. Tarokh, V., Jafarkhani, H., Calderbank, A.R.: Space-Time Block Codes from Orthogonal Designs. IEEE Transactions on Information Theory 45(5), 1456–1467 (1999)
8. Petitcolas, F.A.P., Anderson, R.J.: Evaluation of copyright marking systems. In: Proc. IEEE Multimedia Systems 1999, Florence, Italy, June 7-11, vol. 1, pp. 574–579 (1999)

Formal Analysis of Two
Buyer-Seller Watermarking Protocols

David M. Williams[1,*], Helen Treharne[1], Anthony T.S. Ho[1], and Adrian Waller[2]

[1] University of Surrey, Guildford, GU2 7XH, UK
[2] Thales Research and Technology (UK) Ltd., Reading, RG2 0SB, UK
d.m.williams@surrey.ac.uk

Abstract. In this paper we demonstrate how the formal model constructed in our previous work [1], can be modified in order to analyse additional Buyer-Seller Watermarking Protocols, identifying which specific sections of the CSP scripts remain identical and which require modification. First, we model the protocol proposed by Memon and Wong [2], an examplar of the Offline Watermarking Authority (OFWA) Model, defined in the framework by Poh and Martin [3]. Second, we model the Shao protocol [4] as an example of a protocol fitting the Online Watermarking Authority (ONWA) Model. Our analysis of the protocols reaffirms the unbinding attack described by Lei *et al.* [5] on the Memon and Wong protocol and we identify a new unbinding attack on the protocol proposed by Shao.

1 Introduction

A major benefit of digital media is the ease with which multimedia content can be duplicated and disseminated on a large scale. However, copyright owners are faced with the task of limiting these activities so that they may make a financial gain from licencing such content. Digital Watermarking has been proposed as a suitable deterrent to illegal copying/distribution.

A digital watermark is an imperceptible mark embedded into cover material designed to degrade linearly with the degradation of the content itself (e.g. compression). Embedding a digital watermark into licenced multimedia content enables copyright owners to trace piracy to the original perpetrator. However the buyer must be assured that it is only possible for the copyright owner to gather adequate evidence if and only if an illegal act has taken place.

Qiao and Nahrstedt [6] identified that watermarking schemes alone are inadequate in fulfilling the above requirement. Consequently, asymmetric fingerprinting protocols have been developed to be used in conjunction with digital watermarking schemes. The framework constructed by Poh and Martin [3] specifies the three common requirements that asymmetric fingerprinting protocols aim to satisfy and provides a classification of such protocols into four distinct

---

* The author's work is sponsored by an EPSRC Thales CASE Award.

H.J. Kim, S. Katzenbeisser, and A.T.S. Ho (Eds.): IWDW 2008, LNCS 5450, pp. 278–292, 2009.

protocol models. We will focus on just two of the models, the Online Watermarking Authority (ONWA) and Offline Watermarking Authority (OFWA) Models that together make up the set of protocols more commonly known as the Buyer-Seller Watermarking Protocols.

Our previous work [1] introduced the notion of rigorous analysis of buyer-seller watermarking protocols using a formal analysis technique used previously to check communication protocols [7]. In [1] we accurately represented a buyer-seller watermarking protocol as proposed by Ibrahim *et al.* [8] by constructing a model using the process algebra Communicating Sequential Processes (CSP) [9]. By describing our model in this manner and utilising the tool support associated with CSP we were able to conduct a thorough analysis of all the possible behaviour in the protocol. In this paper we demonstrate how our analysis, of the protocol proposed by Ibrahim *et al.* [8], is easily adapted for further analyses of various buyer-seller watermarking protocols.

We extend Poh and Martin's framework to include those buyer-seller watermarking protocols in which the buyer generates the watermark and the watermarking authority is used to verify that the watermark is well formed, such as [4] and [8]. We do not compare the ONWA and OFWA models, the framework in [3] supplies such a comparison. Instead we analyse a protocol conforming to each model simply to demonstrate that our formal analysis technique can be used for the analysis of each.

Firstly, we model the protocol proposed by Memon and Wong [2] as an example of a protocol that fits the OFWA Model. We analyse our model against a single requirement, which is a necessary test of the protocol's security. Our analysis is sufficient in reaffirming the unbinding attack described by Lei *et al.* [5]. Secondly, we model the ONWA protocol proposed by Shao [4]. We again analyse the protocol against a single requirement, which is sufficient in identifying a new attack found on the protocol.

In Section 2 we summarise the framework constructed by Poh and Martin [3]. Next we define the notation used throughout the paper in Section 3. We then analyse two buyer seller watermarking protocols in Sections 4 and 5. In in Section 6 we discuss how our scripts may be modified to analyse other protocols fitting the ONWA and OFWA models. Finally, we conclude with Section 7 giving a summary of the contributions of this paper and a discussion of further work on this subject.

2   Asymmetric Fingerprinting Protocol Framework

A detailed framework for the design and analysis of asymmetric fingerprinting protocols was proposed in [3]. According to the framework asymmetric fingerprinting protocols have the two primary requirements of: (i) Traceability and proof of illegal distribution and (ii) Framing resistance, with a third desirable requirement (iii) Anonymity and unlinkability. Asymmetric protocols fall under four models namely: (i) ZK Model, (ii) ONWA Model, (iii) OFWA Model, and (iv) TKTP Model. Buyer-Seller Watermarking Protocols fall exclusively into either of the ONWA or OFWA models. It is the ONWA and OFWA models on which this paper focuses.

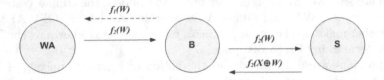

Fig. 1. Offline Watermarking Authority (OFWA) Model

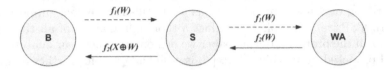

Fig. 2. Online Watermarking Authority (ONWA) Model

2.1 OFWA Model

Figure 1 shows the typical message flow of the OFWA Model. The watermarking authority WA generates the value $f_2(W)$, where f_2 is a homomorphic encryption algorithm with respect to the $\oplus$ operator, such as Paillier encryption [10], and sends this value to the buyer. The watermarking authority is regarded as offline as this message exchange between the buyer and watermarking authority can be performed independently of the watermarking insertion.

The buyer may now initiate the transaction between the buyer and the seller by forwarding the value $f_2(W)$ to the seller. It is important that f_2 is some homomorphic encryption algorithm so that the embedding may be done in the encrypted domain. The result of this embedding $f_2(X \oplus W)$ is sent to the buyer, where $X \oplus W$ denotes watermark embedding of the form: the cover material X is embedded with some mark W using some key conforming to a suitable watermark embedding scheme [11]. The marked content $X \oplus W$ is retrieved by the buyer decrypting the message using some inverse function f_2^{-1}.

The dashed arrows in Figure 1 indicate the additional message flow required when the buyer chooses the watermark, omitted in the framework by Poh and Martin. The buyer generates the value $f_1(W)$, where f_1 denotes a function of which the seller is unable to calculate the inverse f_1^{-1}. The value $f_1(W)$ is sent, by the buyer, to the watermarking authority. The watermarking authority verifies the value W by applying the inverse function f^{-1} to the received message. The message exchange then proceeds in the same manner as when the watermarking authority chooses the watermark.

2.2 ONWA Model

In the Online Watermarking Authority Model (ONWA) the buyer is only required to contact the seller but the watermarking authority is required to participate online. Figure 2 shows the typical message flow of the ONWA Model.

The watermarking authority WA generates the value $f_2(W)$, where f_2 is a homomorphic encryption algorithm and sends this value to the seller. Embedding is conducted in the encrypted domain, using a homomorphic encryption algorithm, and the result $f_2(X \oplus W)$ is sent to the buyer. The buyer retrieves the marked content $X \oplus W$ by decrypting the message using some inverse function f_2^{-1}.

The dashed arrows in Figure 2 indicate the additional message flow required when the buyer chooses the watermark. The buyer generates the value $f_1(W)$ and sends it to the seller, where f_1 denotes a function of which the seller is unable to calculate the inverse f_1^{-1}. The seller forwards $f_1(W)$ on to the watermarking authority. The watermarking authority verifies the value W by applying the inverse function f^{-1} to the received message and then proceeds in the same manner as if the watermarking authority had chosen the watermark.

3 Notation

In this section we define all abbreviations used throughout the paper for clarity of presentation.

3.1 Protocol Participants

B   : The Buyer, wishing to purchase the digital content.
S   : The Seller, whom owns/provides/distributes the digital content.
WA   : The Watermarking Authority, a trusted third party responsible for generating or verifying generated watermarks ready for embedding.
CA   : The Certification Authority, a trusted third party responsible for generating and distributing public-private key pairs along with public-key certificates. In practice a single agent may act as both the Watermarking Authority and the Certification Authority.
ARB : The Arbitrator, required in an arbitration protocol to make judgement upon a case of copyright infringement.

We use the convention of using uppercase to represent a specific value where lower case represents a variable of arbitrary value. X' denotes the cover material uniquely marked with V enabling the seller to identify exactly which entry to look up in their database once a copy has been found rather than conduct an intractable exhaustive search.

3.2 Cryptographic/Watermarking Primitives

(pk_A, sk_A) : A public-private key pair, associated with an adopted public key infrastructure (PKI), in which the public key is published and the secret key kept private to agent A.
$(\overline{pk}_A, \overline{sk}_A)$: A public-private key pair used for anonymous certification.
(pk_A^*, sk_A^*) : A *one time* public-private key pair chosen prior to each protocol run to provide anonymity/unlinkability.
$E_{pk_A}(m)$: A message m encrypted using the public key

$D_{sk_A}(m)$ belonging to agent A.
         : A cipher m may be decrypted using the relevant secret key
          belonging to agent A i.e. $D_{sk_A}(E_{pk_A}(m)) = m$.

$Cert_{CA}(pk_A)$: A X.509 compliant digital certificate [12] constructed by the
          certification authority CA such that the agent A is able to
          convince others of the validity of A's public key.

$Sign_{sk_A}(m)$: The digital signature associated with the adopted PKI.

$h(m)$: A cryptographic hash of message m such that anyone in
          possession of the message m is able to verify the hash value but
          the inverse argument does not hold. Anyone in possession of the
          hash value only is unable to construct the original message m.

$X \oplus W$: Watermark embedding of the form: the cover material X is
          embedded with some mark W using some key conforming to a
          suitable watermark embedding scheme [11].

3.3   CSP

CSP is a process algebra for describing models of interacting systems. A system model is described as a *process* (or collection of processes). CSP processes are defined in terms of the *events* that they can and cannot do. Processes interact by synchronising on events, and the occurrence of events is atomic. The set of events of a process P is denoted by αP.

Events may be compound in structure, consisting of a *channel name* and some (or none) *data values*. Thus, events have the form $c.v_1...v_n$, where c is the channel name associated with the event, and the v_i are data values. The *type* of the channel c is the set of values associated with c to produce events.

For example, the channel *comm* has type *agents* × *agents* × *messages*, where *agents* is a set of agents, involved in some message exchange that may send and receive messages over *comm*, and *messages* is the set of all possible messages that the senders may wish to transmit. The events associated with *comm* will be of the form *comm.a.b.m*, where $a \in$ *agents*, $b \in$ *agents*, and $m \in$ *messages*. The syntax of CSP provides several operators for modelling processes.

$$P ::= a \rightarrow P | c?x!v \rightarrow P | P_1 \;\square\; P_2 | \square_i P_i | S(p)$$

where a is a *synchronisation event*, c is a *communication channel* accepting inputs and sending output values, x is a data variable, v is a data value, and $S(p)$ is a process expression.

The process $a \rightarrow P$ is initially prepared to engage in an a event, after which it behaves as P. The process $c?x!v \rightarrow P$ is prepared to accept any value for x along channel c, provides v as output, and then behave as P (whose behaviour can be dependent on x). The external choice process $P_1 \;\square\; P_2$ is initially prepared to behave either as P_1 or as P_2, and the choice is resolved on occurrence of the first event. This can be generalised for an indexed set of processes: $\square_i P_i$ chooses a process from an i-indexed set of processes P.

Processes can be combined together using the parallel '∥' composition operator. When processes run in parallel they must synchronise on common events (otherwise the events can occur independently). E.g., in the parallel process:

$$a \rightarrow b \rightarrow Stop \parallel b \rightarrow c \rightarrow Stop$$

both a and c can occur independently of the other process, but the occurrence of b requires both processes to synchronise. We use an indexed parallel operator which enables us to combine the behaviour of similar processes together. During the analysis phase, we also make use of the hiding operator '\' so that we can focus on paricular events in a trace.

CSP has a theory of refinement that enables us to compare the behaviour of processes. If a process P is refined by a process Q, then all of the possible behaviours of Q must also be possible behaviours of P. In this paper we will make use of trace refinement checks: $P \sqsubseteq_T Q$.

The modelling and analysis of protocols in CSP, is supported by model checking tools, such as FDR [13]. FDR can automatically check whether a specification of a property (P) is satisfied by a proposed model (Q). If the result of a check is negative a counter example is given which provides information on the behaviour of the model which leads to the violation of the property.

4   Formal Analysis of an OFWA Protocol

4.1   A Buyer-Seller Watermarking Protocol

The Buyer-Seller Watermarking Protocol, proposed in [2], is an OFWA protocol aiming to satisfy the two primary asymmetric fingerprinting protocol requirements, namely traceability and proof of illegal distribution and framing resistance. The protocol does not attempt to provide anonymity. We use the three stage work flow illustrated in Figure 3 to perform our analysis.

We use the model checker FDR [13], which takes two processes, the protocol model and the requirement model, and verifies that the first refines the other. If the refinement check fails then the protocol does not satisfy its requirements and appropriate examples of attacks are automatically generated by FDR.

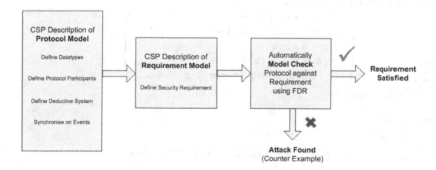

Fig. 3. Three Stage Work Flow

4.2 Protocol Model

Each protocol participant must have its own CSP process defined representing each event the agent is able to perform. Due to the restrictions on space we will not include the CSP descriptions of all the protocol participants. Instead we discuss the *BUYER* process as an appropriate example, as the buyer participates in every step of the protocol and thus every possible event is described. The processes representing the other two agents, the seller and watermarking authority, can be constructed similarly.

The buyer's watermark generation process $BUYER\_GEN(b, signed, known)$, denotes the buyer, b, performing watermarking generation in terms of three events. Upon each successful run of this process the buyer collects an encrypted watermark signed by the chosen watermarking authority $Sign_{sk_{wa}}(E_{pk_b}(wm))$ which is added to the buyer's set of *signed* encrypted watermarks.

The buyer's watermark insertion process $BUYER\_INS(b, signed, known)$, denotes the buyer, b, performing watermarking insertion in terms of four events. Upon each successful run of this process the buyer collects a piece of cover material embedded with their chosen watermark $(x' \oplus wm)$ which is added to the buyer's set of *known* watermarked material.

$$BUYER\,(b, signed, known) = \left(\begin{array}{l} \text{if } (signed = \emptyset) \text{ then} \\ \quad BUYER\_GEN(b, signed, known) \\ \text{else} \\ \quad BUYER\_GEN(b, signed, known) \\ \quad \Box \; BUYER\_INS(b, signed, known) \\ \quad \underset{s \in sellers}{\Box} \quad share.b.s?k \in known \to BUYER(b, known) \end{array} \right)$$

$$BUYER\_GEN(b, signed, known) = \\ \underset{\substack{wm \in watermarks \\ wa \in watermark\_authorities}}{\Box} \left(\begin{array}{l} comm.b.wa.Cert_{wa}(pk_b) \to \\ comm.wa.b.Sign_{sk_{wa}}(E_{pk_b}(wm)) \to \\ BUYER(b, signed \cup Sign_{sk_{wa}}(E_{pk_b}(wm)), known) \end{array} \right)$$

$$BUYER\_INS(b, signed, known) = \\ \underset{\substack{s \in sellers \\ x \in covermaterial \\ sgn \in signed \\ v \in seller\_watermarks \\ wa \in watermark\_authorities}}{\Box} \left(\begin{array}{l} comm.b.s.arg(x) \to \\ comm.b.s.Cert_{wa}(pk_b) \to \\ comm.b.s.sgn \to \\ comm.s.b.E_{pk_B}(x' \oplus wm) \to \\ BUYER(b, signed, known \cup \{x' \oplus wm\}) \end{array} \right)$$

where $x' = x \oplus v$

Fig. 4. Protocol Model: *BUYER* Process

$$REQ\_MODEL(b, s, wa, x, wm) =$$
$$\left(\begin{array}{l} SPEC_1(b, s, wa, x, wm) \\ \square \quad SPEC_2(b, s, wa, x, wm) \\ \square \quad share.b.s.(x' \oplus wm) \rightarrow ARB(b, s, wa, x, wm) \end{array} \right)$$

$$SPEC_1(b, s, wa, x, wm) =$$
$$sellerknows.evidence_1(b, s, wa, x, wm) \rightarrow SPEC_1(b, s, wa, x, wm)$$
$$\square \quad share.b.s.(x' \oplus wm) \rightarrow ARB(b, s, wa, x, wm)$$

$$SPEC_2(b, s, wa, x, wm) =$$
$$sellerknows.evidence_2(b, s, wa, x, wm) \rightarrow SPEC_2(b, s, wa, x, wm)$$
$$\square \quad share.b.s.(x' \oplus wm) \rightarrow ARB(b, s, wa, x, wm)$$

$$ARB(b, s, wa, x, wm) =$$
$$sellerknows.evidence_1(b, s, wa, x, wm) \rightarrow ARB(b, s, wa, x, wm)$$
$$\square \quad sellerknows.evidence_2(b, s, wa, x, wm) \rightarrow ARB(b, s, wa, x, wm)$$
$$\square \quad share.b.s.(x' \oplus wm) \rightarrow ARB(b, s, wa, x, wm)$$

$$REQ\_MODEL = \quad \underset{\substack{b \in buyers \\ s \in sellers \\ wa \in watermark\_authorities \\ c \in covermaterial \\ wm \in watermarks}}{\|} \quad REQ\_MODEL(b, s, wa, x, wm)$$

Fig. 5. Requirement Model: REQ_MODEL Process

$$REQ\_MODEL \sqsubseteq PROTO\_MODEL \setminus \{| \ comm \ |\}$$

Fig. 6. Refinement Check

We define the parameterised process $BUYER(b, signed, known)$, which gives the buyer, b, the choice of participating in watermark generation, watermark insertion, or illegally releasing pirated material on the *share* channel. If the buyer is not in possession of any *known* watermarked material then he must first participate in the buyer generation process $BUYER\_GEN(\dots)$. The *share* event models the real world event of a dishonest buyer releasing a pirated copy onto some file sharing network which the seller is monitoring. The full description of a buyer is given in Figure 4.

Minimal change is made to the deductive system constructed in [1] for our analysis of the Memon and Wong protocol. Our deductive system is a simplification of Roscoe's lazy spy model [14], used to analyse the Needham Schroeder (Lowe) Public Key protocol [15]. Our deductive system only allows the intelligent seller to act passively. That is, we construct an intelligent seller process $INTELLIGENT\_SELLER$ that listens in on all messages sent over the *comm* and *share* channels to build up knowledge and makes deductions using this knowledge. The intelligent seller is defined using the following three events:-

- *learn* is the event that enables the intelligent seller to build up knowledge by listening in on messages sent over the *comm* and *share* channels,

- *infer* enables the intelligent seller to deduce further knowledge by making inferences using existing knowledge,
- *sellerknows* is the event we use to observe when the intelligent seller has collected sufficient knowledge to constitute evidence.

The seller's initial knowledge, denoted by the *initialknowledge* set, consists of all agents, each agents public key, all cover material, the seller's secret key and the seller's watermarking key.

$$initialknowledge = agents \cup \{pk_a \mid a \leftarrow agents\} \cup$$
$$covermaterial \cup \{sk_{Sam}\} \cup \{wk_{Sam}\}$$

Our specification is written in the form, certain evidence of illegal file sharing may be gathered by the seller if and only if the file has been illegally shared. The evidence set defines what knowledge, learnt and deduced within the deductive system, constitutes evidence. This enables us to observe, during analysis, when an intelligent seller has gathered evidence, having defined a *sellerknows* channel upon which gathered evidence can be released.

$$Evidence = \{evidence_1(b, s, wa, x, wm), evidence_2(b, s, wa, x, wm)$$
$$\mid b \leftarrow buyers, s \leftarrow sellers, wa \leftarrow watermark\_authorities,$$
$$x \leftarrow covermaterial, wm \leftarrow watermarks\}$$

$$evidence_1(b, s, wa, x, wm) = (x' \oplus wm)$$
$$evidence_2(b, s, wa, x, wm) = Sign_{sk_{wa}}(E_{pk_b}(wm))$$

We construct our overall protocol model, by synchronising the protocol participant processes and the *INTELLIGENT_SELLER* process, which can then be verified against the protocol requirements. Figure 7, illustrates the synchronisation forming the overall protocol model *PROTO_MODEL*.

4.3   Requirement Model

We specify a single requirement which is a necessary test of the protocol's security. For the framing resistance property to be satisfied the seller should be in possession of the two pieces of evidence specified in the *evidence* set if and only if that piece of digital material has been illegally copied and shared. We describe this requirement, the second stage of our three stage work flow, by constructing the CSP process *REQ_MODEL* as illustrated in Figure 5.

The parameterised process *REQ_MODEL*(b, s, wa, x, wm) gives the choice of three possible behaviours. The processes $SPEC_1$ and $SPEC_2$ state that, if at first one piece of evidence is gathered, the alternative piece of evidence must not be gathered until the file has been illegally distributed, although the first piece of evidence may be collected over and over again. Otherwise, if the file is illegally distributed once along the *share* channel at any time, evidence gathering along with further illegal sharing of the same file may then happen arbitrarily as described by the process *ARB*. We use indexed parallel to generalise this requirement for all possible transactions.

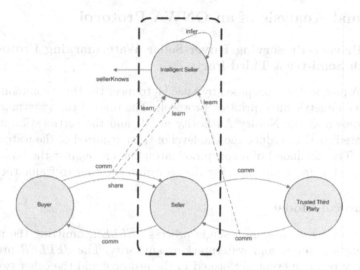

Fig. 7. Protocol Model: Synchronisation on Events

$\alpha.1$ $Bob \rightarrow Tom$: $Cert_{Tom}(pk_{Bob})$

$\alpha.2$ $Bob \rightarrow Tom$: $E_{pk_{Tom}}(Sign_{sk_{Bob}}(W_1))$

$\alpha.3$ $Tom \rightarrow Bob$: $Sign_{sk_{Tom}}(E_{pk_{Bob}}(W_1))$

$\alpha.4$ $Bob \rightarrow Sam$: $arg(X_1)$

$\alpha.5$ $Bob \rightarrow Sam$: $Cert_{Tom}(pk_{Bob})$

$\alpha.6$ $Bob \rightarrow Sam$: $Sign_{sk_{Tom}}(E_{pk_{Bob}}(W_1))$

$\alpha.7$ $Sam \rightarrow Bob$: $E_{pk_{Bob}}(X_1' \oplus W_1)$

$share$ $Bob \rightarrow Sam$: $(X_1' \oplus W_1)$

$sellerknows \rightarrow$: $Sign_{sk_{Tom}}(E_{pk_{Bob}}(W_1))$

$sellerknows \rightarrow$: $(X_2' \oplus W_1)$

Fig. 8. Counter Example of Unbinding Attack

4.4   Analysis

Lei *et al.* identified an unbinding problem inherent in the Memon and Wong protocol. The flaw becomes apparent only once the seller has intercepted at least a single item of pirated material. In this scenario, the seller is able to extract the watermark from the cover material and embed it within a second piece of cover material so as to fabricate evidence of further piracy.

Using the model checker FDR we perform a refinement check of our protocol model against our specification process. This will search through every possible state of the interpreted state machine to check whether it is possible to reach some undesirable state. When we model check our protocol model of Memon and Wong we find that the requirement is not satisfied and a counter example generated, illustrated in Figure 8, matching the unbinding problem described by Lei *et al.*

5   Formal Analysis of an ONWA Protocol

5.1   A Privacy-Preserving Buyer-Seller Watermarking Protocol with Semi-trust Third Party

An ONWA protocol was proposed by Shao [4] to meet the three common requirements of asymmetric fingerprinting protocols. The roles of the watermarking authority (known as the Notary Authority in [4]) and the certification authority are separated so as to reduce the the level of trust required of the watermarking authority. The likelihood of a conspiracy attack made against the buyer is said to be reduced as more parties must now conspire together to frame the buyer.

5.2   Protocol Model

As in Section 4.2 we discuss a single process $SELLER$, omitting the processes representing the buyer and watermarking authority. The $SELLER$ process involves every possible event performed in the protocol and the other two can be constructed similarly, consisting of only those events which they may perform.

The process $SELLER(s)$, illustrated in Figure 10, enables the seller, s, to participate in the watermark generation/insertion process, $SELLER\_GEN\_INS(s)$, or intercept illegally released pirated material on the $share$ channel.

$$
\begin{aligned}
&\alpha.1 && Bob \rightarrow Sam : arg(X_1) \\
&\alpha.2 && Bob \rightarrow Sam : Sign_{sk^*_{Bob}}(arg(X_1)) \\
&\alpha.3 && Bob \rightarrow Sam : E^{X_1}_{Tom} \\
&\alpha.4 && Sam \rightarrow Tom : E^{X_1}_{Tom} \\
&\alpha.5 && Tom \rightarrow Sam : ew \\
&\alpha.6 && Tom \rightarrow Sam : pk^*_{Bob} \\
&\alpha.7 && Tom \rightarrow Sam : S^{X_1}_{Tom} \\
&\alpha.8 && Tom \rightarrow Sam : E_{ARB} \\
&\alpha.9 && Sam \rightarrow Bob : E_{pk_{Bob}}(X'_1 \oplus W_1) \\
&\alpha.share && Bob \rightarrow Sam : (X'_1 \oplus W_1) \\
&\beta.1 && Bob \rightarrow Sam : arg(X_2) \\
&\beta.2 && Bob \rightarrow Sam : Sign_{sk^*_{Bob}}(arg(X_2)) \\
&\beta.3 && Bob \rightarrow Sam : E^{X_2}_{Tom} \\
&\beta.4 && Sam \rightarrow Tom : E^{X_2}_{Tom} \\
&\beta.5 && Tom \rightarrow Sam : ew \\
&\beta.6 && Tom \rightarrow Sam : pk^*_{Bob} \\
&\beta.7 && Tom \rightarrow Sam : S^{X_2}_{Tom} \\
&\beta.8 && Tom \rightarrow Sam : E_{ARB} \\
&sellerknows \rightarrow && : Sign_{sk_{Tom}}(E_{pk_{Bob}}(W_1)) \\
&sellerknows \rightarrow && : (X'_2 \oplus W_1)
\end{aligned}
$$

$$
\begin{aligned}
where\ E^{X_i}_a &= E_{pk_a}(ew, pk^*_{Bob}, pf_{Bob}, E_{ARB}, h(arg(X_i)), Cert_{CA}(\overline{pk}_{Bob}), S^{X_i}_{Bob}) \\
ew &= E_{pk_{Bob}}(W_1) \\
S^{X_i}_a &= Sign_{\overline{sk}_a}(ew, pk^*_{Bob}, h(arg(X_i)))
\end{aligned}
$$

Fig. 9. Counter Example of Unbinding Attack

$SELLER(s) =$

$$\square_{\substack{b \in buyers \\ x \in covermaterial \\ wm \in watermarks}} \left(\begin{array}{l} SELLER\_GEN\_INS(s, b, x, wm) \\ \square \\ share.b.s.(x' \oplus wm) \rightarrow SELLER(s) \end{array} \right)$$

$SELLER\_GEN\_INS(s, b, x, wm) =$

$$\square_{\substack{wa \in watermark\_authorities \\ arb \in arbitrators}} \left(\begin{array}{l} comm.b.s.arg(x) \rightarrow \\ comm.b.s.Sign_{sk_b^*}(arg(x)) \rightarrow \\ comm.b.s.E_{wa} \rightarrow \\ comm.s.wa.E_{wa} \rightarrow \\ comm.wa.s.ew \rightarrow \\ comm.wa.s.pk_b^* \rightarrow \\ comm.wa.s.S_{wa} \rightarrow \\ comm.wa.s.E_{arb} \rightarrow \\ comm.s.b.E_{pk_b^*}(x' \oplus wm) \rightarrow \\ SELLER(s) \end{array} \right)$$

$where \ x' \ = x \oplus v$

$E_{wa} = E_{pk_{wa}}(ew, pk_b^*, pf_b, E_{arb}, h(arg(x)), Cert_{ca}(\overline{pk}_b), S_b)$

$E_{arb} = E_{pk_{arb}}(ew, pk_b^*, pf_b, E_{arb}, h(arg(x)), Cert_{ca}(\overline{pk}_b), S_b)$

$ew \ = E_{pk_b}(wm)$

$S_b \ = Sign_{\overline{sk}_b}(ew, pk_b^*, h(arg(x)))$

$S_{wa} = Sign_{sk_{wa}}(ew, pk_b^*, h(arg(x)))$

Fig. 10. Seller Description in CSP

Our deductive system remains unchanged apart from the *initialknowledge* and *evidence* sets given below.

$initialknowledge = agents \cup \{pk_a \mid a \leftarrow agents\} \cup$
$\qquad\qquad\qquad covermaterial \cup \{sk_{Sam}\} \cup \{wk_{Sam}\}$

$Evidence = \{evidence_1(b, s, wa, x, wm), evidence_2(b, s, wa, x, wm)$
$\qquad\qquad |b \leftarrow buyers, s \leftarrow sellers, wa \leftarrow watermark\_authorities,$
$\qquad\qquad x \leftarrow covermaterial, wm \leftarrow watermarks\}$

$evidence_1(b, s, wa, x, wm) = (x' \oplus wm)$
$evidence_2(b, s, wa, x, wm) = Sign_{sk_{wa}}(ew, pk_b^*, h(arg(x)))$

We construct our overall protocol model *PROTO_MODEL*, by synchronising the protocol participant processes and the *INTELLIGENT_SELLER* process, which can then be verified against the protocol requirements.

5.3   Requirement Model

We use the same single requirement, in Figure 5, which is a necessary test of the protocol's security. The requirement differs only in what constitutes evidence, as given in the *Evidence* set defined in the deductive system above.

Table 1. CSP Sections Consistent Between Protocols

CSP Script Section	Unmodified
Cryptographic Primitives	✓
Protocol Participants	×
Message Formats	×
Deductive Rules	✓
Deductive System	✓
- Initial Knowledge	×
- Evidence Set	×
Composition	✓
Specification	✓

5.4 Analysis

When we model check our protocol model of Shao we find that the requirement is not satisfied. The flaw again occurs once the seller has intercepted at least a single item of pirated material. This is illustrated in Figure 9 by a complete run α of the protocol, followed by an associated share event. The buyer may then choose to use the same values for each variable other than the cover material, i.e. the watermark and *one time* public/private key pair (pk_{Bob}^*, sk_{Bob}^*), on a subsequent run of the protocol, β. The seller is now able to extract the watermark from the cover material and embed it within a second piece of cover material, purchased in a second run of the protocol β, so as to falsely gain evidence of further piracy.

As our previous work indicated [1], the consequence of the attack is that if a number of files are purchased, copied and illegally distributed in this manner, then the seller is only able to prove that at least one of the files has been copied and is not able to identify specifically which one(s).

6 Discussion

The CSP scripts, available at *www.cs.surrey.ac.uk/personal/pg/D.M/*, used to model and analyse the protocols, are identical in their structure consisting of seven parts. This section identifies which parts of the scripts remain unchanged between protocol analyses, and which must be modified, as illustrated in Table 1. The reader is also referred to Roscoe's CSP script(s) of the Needham Schroeder (Lowe) protocol, released in conjunction with [14], which forms the basis of our work.

All the cryptographic primitives defined in Section 3 have been included in our CSP scripts, expressed as functions. The rules that govern such functions are described by the set of deductive rules. All cryptographic primitives and associated deductive rules remain unmodified in the analyses of protocols, although only a subset may be required. If the list of cryptographic primitives is found to be incomplete then a new primitive, and associated deductive rules, can be added to the current list.

Each protocol will have differing CSP descriptions of the protocol participants. Although these processes will need to be rewritten, each buyer-seller watermarking protocol will closely follow the typical message flows, illustrated in Figures 1 and 2, thus the protocol participants will be similar in form to those in our analyses of [2] and [4], respectively. The manner in which the synchronisations between the protocol participants and the deductive system are composed is identical between protocols.

The set of possible message formats are unique to each protocol but require little extra work to that of writing the protocol participants themselves as they match identically the form of the messages used by the protocol participants.

Our reduced deductive system aims only to be sufficient in identifying the unbinding attacks discussed in our analyses. The deductive system remains unchanged for the analyses of protocols with the exception of the *initialknowledge* and *Evidence* sets. It must, however, be generalised to provide a sufficient verification of all protocol requirements.

Our single requirement is similar for every protocol analysis differing only in what constitutes evidence, as defined in the *Evidence* set. However, the set of requirements must be expanded in order to provide a sufficient verification of all protocol requirements, although each protocol may only aim to satisfy a subset of these common requirements.

7 Conclusion and Further Work

We have demonstrated how the model constructed in [1] may be adapted for further analyses of various buyer-seller watermarking protocols. Having extended the Poh and Martin's framework [3], we analysed two Buyer-Seller Watermarking Protocols, using the process algebra CSP, to identify flaws in the OFWA and ONWA protocols, proposed by Memon and Wong [2] and Shao [4] respectively. Our analyses of the protocols reaffirmed the unbinding attack described by Lei *et al.* on the Memon protocol and identified a new attack found on the protocol by Shao. We then highlighted which sections of the script require modification and which sections can remain unchanged, for the analyses of additional protocols.

Currently our deductive system restricts the intelligent seller to passive behaviour, listening in on insecure communications. This system should be made more robust by allowing other protocol participants, as well as an external intruder (man-in-the-middle), to also act maliciously and by allowing more aggressive behaviour by the intruder in line with Rosoce's lazy spy model allowing all behaviour defined in the Dolev Yao model. The set of formally specified requirements must also be extended to provide both a necessary and sufficient test of each protocol's security.

Further research is required to understand how formal analysis techniques may be used to verify protocols conforming to the remaining asymmetric fingerprinting models. In particular, it would be appropriate to study how our model may be modified in order to analyse TKTP protocols which use a trusted computing platform, rather than a watermarking authority, as a centre of trust.

Acknowledgement. The authors would like to thank Steve Schneider for his technical discussions.

References

1. Williams, D.M., Treharne, H., Ho, A.T.S., Culnane, C.: Using a formal analysis technique to identify an unbinding attack on a buyer-seller watermarking protocol
2. Memon, N., Wong, P.W.: A buyer seller watermarking protocol. IEEE Transactions on Image Processing 10(4), 643–649 (2001)
3. Poh, G.S., Martin, K.M.: A framework for design and analysis of asymmetric finger-printing protocols. In: Third International Symposium on Information Assurance and Security, pp. 457–461 (2007)
4. Shao, M.H.: A privacy-preserving buyer-seller watermarking protocol with semi-trust third party
5. Lei, C.L., Yu, P.L., Tsai, P.L., Chan, M.H.: An efficient and anonymous buyer-seller watermarking protocol. IEEE Transactions on Image Processing 13(12), 1618–1626 (2004)
6. Qiao, L., Nahrstedt, K.: Watermarking schemes and protocols for protecting right-ful ownership and customer's rights. Journal of Visual Communication and Image Representation 9(3), 194–210 (1998)
7. Roscoe, A.W., Ryan, P., Schneider, S., Goldsmith, M., Lowe, G.: The Modelling and Analysis of Security Protocols. Addison-Wesley, Reading (2001)
8. Ibrahim, I.M., Nour El-Din, S.H., Hegazy, A.F.A.: An effective and secure buyer seller watermarking protocol. In: Third International Symposium on Information Assurance and Security, pp. 21–28 (2007)
9. Hoare, C.A.R.: Communicating sequential processes. Prentice-Hall International, Englewood Cliffs (1985)
10. Paillier, P.: Public-key cryptosystems based on composite degree residuosity classes. In: Stern, J. (ed.) EUROCRYPT 1999. LNCS, vol. 1592, pp. 223–238. Springer, Heidelberg (1999)
11. Cox, I.J., Miller, L., Bloom, J.A.: Digital Watermarking. Morgan Kaufmann, San Francisco (2002)
12. Housley, R., Ford, W., Polk, W., Solo, D.: Internet x.509 public key infrastructure certificate and crl profile (1999)
13. Formal Systems. FDR 2.82. Formal Systems Ltd. (2005)
14. Roscoe, A.W.: The Theory and Practice of Concurrency. Prentice Hall, Englewood Cliffs (1998)
15. Lowe, G.: Breaking and fixing the needham-schroeder public-key protocol using fdr. In: Proceedings of the Second International Workshop on Tools and Algorithms for Construction and Analysis of Systems, pp. 147–166 (1996)

Detection of Hidden Information in Webpage Based on Higher-Order Statistics

Huajun Huang[1], Junshan Tan[1], Xingming Sun[2], and Lingxi Liu[1]

[1] Computer Science College, Central South University of Forestry & Technology,
Changsha, Hunan, 410004, China
[2] School of Computer & Communication, Hunan University, Changsha,
Hunan, 410082, China
{hhj0906,junshan_tan,sunnudt,liulx2005}@163.com

Abstract. Secret message can be embedded into letters in tags of a webpage in ways that are imperceptible to human eyes viewed with a browser. These messages, however, alter the inherent characteristic of the offset of a tag. This paper presents a new higher-order statistical detection algorithm for detecting of secret messages embedded in a webpage. The offset is used to build the higher-order statistical models to detect whether secret messages hide in tags. 30 homepages are randomly downloaded from different websites to test, and the results show the reliability and accuracy of the statistical characteristic. The probability of missing secret message decrease as the secret message increase, and it is zero, as 50% letters of tags are used to carry secret message.

Keywords: steganography; steganalysis; webpage; higher-order statistics; offset.

1 Introduction

Covert communication can use a text file or a webpage as a cover object [1-4]. The existing steganographic methods embedded secret message in a webpage by altering the order of attributes [1], embedding invisible characters [2], or switching the upper-case-lowercase states of letters in tags [3] [4].

Steganalysis mainly discusses the frangibility of the steganography algorithm, detects the stego-objects, and extracts the secret messages embedded in the objects. Up to now, there are some papers discussing about steganalysis of texts [5-9]. Edward proposed a universal steganalysis method based on language models and support vector machines to differentiate sentences modified by a lexical steganography algorithm from unmodified sentences [5]. Kot implemented a technique for the steganalysis of electronic text documents based on the similarity between the same characters or symbols [6]. Sun designed and implemented a system for text steganalysis based on the noise of the stego-text [7]. In [8] and [9], Sui proposed two steganalysis methods to detect hidden information in text file. In [10], [11] and [12], Huang proposed three steganalysis algorithms to detect hidden information in webpage.

Covert communication can use a webpage as a cover object and embed secret message in a webpage by switching the uppercase-lowercase states of letters in tags. Though the message is invisible to human eyes, it alters the inherent characteristic of the offset of a tag. In [13], Fridrich gave a RS steganalysis method to the gray image.

H.J. Kim, S. Katzenbeisser, and A.T.S. Ho (Eds.): IWDW 2008, LNCS 5450, pp. 293–302, 2009.

Use this idea, a new higher-order statistical detection algorithm is presented for detecting of secret message embedded in a webpage. The offset is used to build the higher-order statistical models to detect whether secret message hide in tags. 30 homepages are randomly downloaded from different websites to test, and the results show the reliability and accuracy of the statistical characteristic. The probability of missing secret message decrease as the secret message increase, and it is zero, as 50% letters of tags are used to carry secret message.

This paper is organized as follows. In Section 2, we give a formal description of switching the uppercase-lowercase method. Section 3 explains the basic theory and detection algorithm on detecting of hidden information in tags of a webpage based on higher-order statistics. Experiments and analysis are shown in Section 4. Finally, this paper is concluded in Section 5.

2   Formal Description of Switching the Uppercase-Lowercase Method

In this paper, we pay close attention to the embedding method that switching the uppercase-lowercase states of letters in tags. The formal description of the embedding method is given as following. Let a set $C = \{26 \text{ uppercase letters in the English alphabet}\}$, a set $A_C = \{x | A(x), x \in C\}$ be the ASCII codes of the uppercase state letters, and $c = \{26 \text{ lowercase letters in the English alphabet}\}$, a set $A_c = \{x | A(x), x \in c\}$ is the set of the ASCII codes of the lowercase state letters, where the function $A(\square)$ obtains the ASCII code of an English alphabet. Define a function $f_1(x) = x - 32$, where $x \in c$ and $f_1(x) \in C$, which switches the lowercase state of a letter in tags to uppercase. Define another function $f_0(x) = x$, where $x \in c$ and $f_0(x) \in c$, which does not switches the written state a letter in tags. The embedding progress is given as follows. Let a sequence $m = \{0,1\}^n$ denotes the n watermark bits, and $m_i \in m$ denotes i-th watermark bit in m, where $0 \le i \le n-1$. If the watermark bit $m_i = 0$, the written state of the corresponding letter in the tag is switched according to the function $f_0(x)$. If the watermark bit $m_i = 1$, the written state of the corresponding letter in the tag is switched according to the function $f_1(x)$. The embedding process will not end until $i = n-1$.

3   Detection Algorithm

3.1   The Offset of a Tag

Definition 1. The offset of a tag is the summation of the absolute value for the distance between the two adjacent letters in tags.

Let $T(x_1, x_2, ..., x_n)$ be one of tags in a webpage, and x_i denotes *i-th* letter in the English alphabet, where $0 \leq i \leq n-1$ and n is the number of letters in the tag. By intensive study, we found that: in general, all letters of the tags in regular webpage belong to one same set, which $x \in c$ or $x \in C$, where $1 \leq i \leq n$. But after embedding secret message, they will not belong to the same set. Hence, let us assume the currently modality as following. For example, there are k ($0 < k < \lceil n/2 \rceil$) pairs of adjacent letters $x_j x_{j+1}$ in the tags, which $x_j \in c$, and $x_{j+1} \in C$, where $1 \leq j \leq n$. The rest $n-k$ letters entirely fall into the set c or C. So, the tag-offset has changed during embedding secret message.

Definition 2. The offset of a tag $T(x_1, x_2, ..., x_n)$ is calculated with the discrimination function F:

$$F(T(x_1, x_2, ..., x_n)) = \sum_{i=1}^{n-1} |A(x_{i+1}) - A(x_i)|$$

$$\text{where, } x_i \in c \cup C. \quad (1)$$

The maximum distance between the two letters in the set A_c or A_C is 25. For example, the distance of "z" and "a" in ASCII code is 25, and so is "A" and "Z". But for the letters from different set, the minimum distance is 6. Such as the difference between "Z" and "a" is 6. In order to enlarge the minimum distance to at least 25, we define a set $A_c' = \{x | ASCII(x) + 20, x \in c\}$ as the offsetting set of the set A_c. Temporality, the minimum value is 26. When calculating the offset of the tag, we use set A_c' instead of A_c.

Theorem 1. $\exists T(x_1, x_2, ..., x_n)$, $\forall x_i \in c$ or $x_i \in C$, where $1 \leq i \leq n$, the offset of a tag has the minimum value.

Deduction 3. $\exists T(x_1, x_2, ..., x_n)$, for each adjacent letter $x_j x_{j+1}$, which $x_j \in c$, $x_{j+1} \in C$ or $x_j \in C$, $x_{j+1} \in c$, where $1 \leq i \leq n$, the offset of a tag has the maximum value.

3.2 High-Order Statistical Model

We define the functions f_{+1} and f_{+0} in the embedding method that switching the uppercase-lowercase states of letters in a tag. For completeness, we define a function $f_{-1}(x) = x + 32$, where $x \in A_C$ and $f_{-1}(x) \in A_c$, which changes the uppercase state of a letter in a tag to lowercase. Define another function $f_{-0}(x) = x$, where

$x \in A_C$ and $f_{-0}(x) \in A_C$, which changes the uppercase state of a letter in a tag to uppercase.

Definition 4. A Function f_{+1}, f_{+0}, f_{-1} and f_{-0} is called as a Alteration Function f .

We may wish to apply different alteration to different letter in the tag $T(x_1, x_2, ..., x_n)$. The assignment of altering to letters can be captured with a mask M , which is a n-tuple with values +1, +0,-1 and -0. The altered tag $T'(x_1, x_2, ..., x_n)$ is defined as $\left(f_{M(1)}(x_1), f_{M(2)}(x_2), ... f_{M(n)}(x_n), \right)$. The offset of a tag $T'(x_1, x_2, ..., x_n)$ is calculated with the discrimination function as following.

$$F\left(T'(x_1, x_2, ..., x_n)\right) = \sum_{i=1}^{n-1} \left| A\left(f_{M(I+1)}(x_{i+1}) \right) - A\left(f_{M(i)}(x_i) \right) \right|$$

$$x_i \in c \cup C, M_{(i)} \in \{+1, +0, -1, -0\}, \text{ where } 1 \le i \le n. \tag{2}$$

If $F\left(T(x_1, x_2, ..., x_n)\right) > F\left(T'(x_1, x_2, ..., x_n)\right)$, we call $T(x_1, x_2, ..., x_n)$ is a regular tag; if $F\left(T(x_1, x_2, ..., x_n)\right) < F\left(T'(x_1, x_2, ..., x_n)\right)$, we call $T(x_1, x_2, ..., x_n)$ is a singular tag.

We apply the alteration function f to $T(x_1, x_2, ..., x_n)$, and get three tags: R, S, U :

$$R = \left\{ T(x_1, x_2, ..., x_n) \middle| F\left(T(x_1, x_2, ..., x_n)\right) > F\left(T'(x_1, x_2, ..., x_n)\right) \right\}$$

$$S = \left\{ T(x_1, x_2, ..., x_n) \middle| F\left(T(x_1, x_2, ..., x_n)\right) < F\left(T'(x_1, x_2, ..., x_n)\right) \right\}$$

$$U = \left\{ T(x_1, x_2, ..., x_n) \middle| F\left(T(x_1, x_2, ..., x_n)\right) = F\left(T'(x_1, x_2, ..., x_n)\right) \right\}$$

Apply the non-negative alteration operation to each letter in the extracted tags, which $M_{(i)} \in \{+1, +0\}$, where $1 \le i \le n$. Calculate the offset with the **Eq.2**, and then get the tags R, S , and U . Let us denote the number of regular tags for mask M as R_M (in percents of all tags). Similarly, S_M will be denoted the relative number of singular tags. We have $R_M + S_M \le 1$. Apply the non-positive alteration operation $M_{(i)}$ which $M_{(i)} \in \{-1, -0\}$, where $1 \le i \le n$, the same as the non-negative alteration operation. We denote the number of regular tags for mask M as R_{-M} (in percents of all tags). S_{-M} will be denoted the relative number of singular tags. We have $R_{-M} + S_{-M} \leq 1$. Non-positive (non-negative) transformable functions are in charge of the transformation of letters in tags in Fig. 1. As we can be seen from the chart, the letters in the set c do not change with f_{+0}, but with f_{+1} transformed to the set C; the letters in the set C do not change with f_{-0}, but with f_{-1} transformed to the set c.

Fig. 1. Non-positive (non-negative) transformable function control the transformation of letters of tags

If a tag in a webpage did not hidden secret message, however, apply the non-positive or non-negative Alteration functions to letters in a tags will increase the offset in statistic. $\forall T\left(x_{1}, x_{2}, ..., x_{n}\right)$, if $\forall x_{i} \in c(1 \leq i \leq n)$, we have $R_{M} \neq 0$, $R_{M} >> S_{M}$, and $S_{M} \approx 0, R_{-M} \approx S_{-M} \approx 0$; if $\forall x_{i} \in C(1 \leq i \leq n)$, we have $R_{-M} >> S_{-M}$, and $S_{-M} \approx 0$, $R_{M} \approx S_{M} \approx 0$.

If a tag in a webpage embedded secret message, however, apply the non-positive or non-negative Alteration functions to the letters in the tags will decrease the offset in statistic. We have $S_{M} > R_{M}, S_{-M} > R_{-M}$, and $S_{M} > 0, S_{-M} > 0$. If all of letters in a tag are used to carry message, then $S_{M} > R_{M}, S_{-M} > R_{-M}$, $S_{M} > 0$, $S_{-M} > 0$, and $S_{M} \approx S_{-M}$, $R_{M} \approx R_{-M}$.

3.3 Detection Algorithm

Reference [3] and [4] propose the webpage steganographic algorithms base on switching the uppercase-lowercase states of letters in tags. Reference [3] presents an approach to hiding secret information by switching the initial's uppercase-lowercase states of the tags. Reference [4] embedded secret messages through switching all letters' uppercase-lowercase states of the tags. The detection algorithm performed better in the embedding algorithm proposed in [4] than [3]. In order to make it performed well when use the embedding method in [3], we give an improved detection algorithm in the following part.

Form the high-order statistical characteristics, we could judge the webpage whether embedded secret message by the relationship of R_{M}, S_{M}, R_{-M} and S_{-M}. For convenience, we just give the pseudo code of the detection algorithm as following.

Algorithm 1. Detection Algorithm

Input: Webpage file H;
Output: "T" (have secret message), "F" (no secret message)
S1. Extract the tags in H, put down in an Array;
S2. Calculate the tag-offset with **Eq.1**, write down the tag-offset's value;
S3. Alter all the letters in the tags with f_{+1}, f_{+0}, apply **Eq.2**, then write down the value;
S4. Compare the two values; get tags R, S, U. Calculate R_M and S_M;
S5. Alter all the letters in the tags with f_{-1}, f_{-0}, apply **Eq.2**, then write down the value;
S6. Compare the two values; get tags R, S, U. Calculate R_{-M} and S_{-M};
S7. From the high-order statistical characteristics and the relationship of R_M, S_M, R_{-M} and S_{-M}, output "T" or "F".

4 Experimental Results and Analysis

4.1 Experimental Results

30 homepages are randomly downloaded from different websites to test. The results of R_M and R_{-M} is plotted in Fig.2, when did not hidden message in homepages ($S_M = S_{-M} \approx 0$, so we have not marked them in the chart). As can be seen from the chart, for the webpage 4,5,6,8, $R_M \neq 0$, $R_M \gg S_M$ and $S_M \approx 0$,

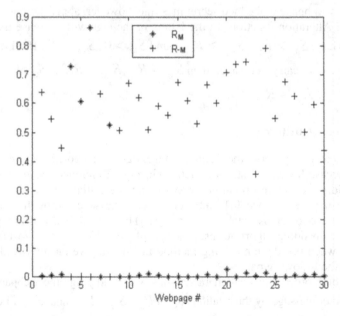

Fig. 2. R_M, R_{-M}, where none of a letter are used to hide secret information

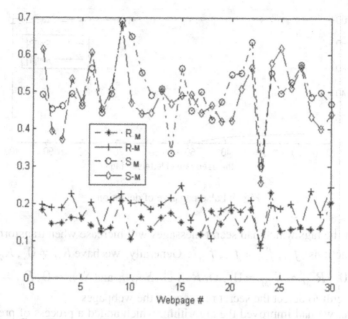

Fig. 3. R_M, S_M, R_{-M} and S_{-M}, where 100% letters are used to carry secret information

$R_{-M} \approx S_{-M} \approx 0$; as for the rest, $R_{-M} \neq 0$, $R_{-M} >> S_{-M}$, and $S_{-M} \approx 0$, $R_M \approx S_M \approx 0$. So the relationship of R_M, S_M, R_{-M} and S_{-M} meet the high-order statistical characteristic when the webpages have no secret message. The rate of false positive is 0%. R_M, S_M, R_{-M} and S_{-M}, where 100% letters are used to hide secret information was shown in Fig.3. In Fig.3, we could see that $S_M > R_M$, $S_{-M} > R_{-M}$; $S_M \approx S_{-M}$ and fluctuates at 50%; $R_M \approx R_{-M}$ and fluctuates at 15%.

We have randomly downloaded 175 homepages from different websites to test the false negative of detection. Applying the embedding methods, which proposed in [3] and [4], to the webpages which the percent of letters in the tags of webpages used to hidden message is from 10% to 100% with the increment of 10%. The stars in Fig.4 denote the false negative of detection which used the algorithm presented in [3]; the diamonds denote the false negative of detection in [4]. The results shown in Fig.4 tell us, for [3] the detected false negative is above 95, but for [4] the highest false negative is 8.5% when 10% letters in tags was used. The probability of missing secret messages decrease as the secret message increase, and it is zero, as 50% letters of tags are used to carry secret message.

4.2 Analysis

The detection algorithm can not accurately detect the webpages which hidden information used the embedding method in reference [3]. Deduction 1 said that the value of tag-offset is max when the two adjacent letters in tags is from different tags. The

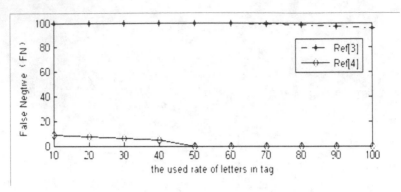

Fig. 4. False negative of detection

tag-offset of the tags which had secret messages will increase when transform the tags with the functions f_{+1}, f_{+0} or f_{-1}, f_{-0}. Generally, we have $R_M \neq 0$, $R_M \square S_M$, and $S_M \approx 0$, $R_{-M} \approx S_{-M} \approx 0$. Or $R_{-M} \square S_{-M}$, and $S_{-M} \approx 0, R_M \approx S_M \approx 0$. So, it is difficult to detect the secret message in the webpages.

Therefore, we had improved the algorithm, which added a process of pretreatment before the detection algorithm. The pretreatment uses a slide window technology showed in Fig.5. Firstly, we capture the tags' initial in order and form a vector $\vec{G} = (x_1, x_2, ..., x_n)$ for a webpage. We choose tags of 3 consecutive letters in a row (in static, the average number of the letters in tags is 2.3), then we get those groups $G_1(x_1, x_2, x_3), G_2(x_2, x_3, x_4), ..., G_{n-1}(x_{n-2}, x_{n-1}, ..., x_n)$. Then, all letters in the group has been embedded message and the detection algorithm is accurate again.

Fig. 5. Slide Window

5   Conclusion

With the popular and easily-applied steganographic tools appearing on the Internet, it is possible that terrorists use tools to communicate with their accomplices, and transmit blueprints of the next terrorist attack. How to control and destroy the activity is an urgent problem to the government, security department, and army. By further study and analysis to the steganographic algorithm that through switching the uppercase-lowercase states of letters in tags to hide messages, we found these messages, however, alter the inherent characteristic of the offset of a tag. In this paper, we propose an accurate algorithm to detect the secret messages embedded in letters of tag in a webpage. And the false negative of detection is quite low. An online suspicious webpage can be actively downloaded with a web crawler and automatically detected. The activity is in charge, which terrorists use steganographic tools to transmit the secret information hidden in the webpage in order to destroy the national security and social stability.

In the next work, we should find the general artifact and design more accurate and sensitive detection algorithm.

Acknowledgments. This paper is supported by Key Program of National Natural Science Foundation of China (No. 60736016), National Natural Science Foundation of China (No.60573045), National Basic Research Program of China (No.2006CB303000), Scientific Research Fund of Hunan Provincial Education Department (No.08B091), Key Program of Youth Fund of CSUFT (No.2008010A), and Talent Introduction Project of CSUFT (No. 104-0055).

References

1. Sun, X.M., Huang, H.J., Wang, B.W.: An Algorithm of Webpage Information Hiding Based on Equal Tag. Journal of Computer Research and Development 44(5), 756–760 (2007)
2. Zhao, Q.J., Lu, H.T.: PCA-Based Web page Watermarking. Journal of the Pattern Recognition 40, 1334–1341 (2007)
3. Shen, Y.: A Scheme of Information Hiding Based on HTML Document. Journal of Wuhan University 50, 217–220 (2004)
4. Sui, X.G., Luo, H.: A New Steganography Method Based on Hypertext. In: Proc. of Asia-Pacific Radio Science Conference, Taiwan, pp. 181–184. IEEE Computer Society Press, Los Alamitos (2004)
5. Taskiran, C.M., Topkara, U., Topkara, M.: Attacks on Lexical Natural Language Steganography Systems. In: Proceedings of the SPIE, pp. 97–105 (2006)
6. Cheng, J., Alex, C.K., Liu, J.L.: Steganalysis of Data Hiding in Binary Text Images. In: Proceedings of 2005 IEEE International Symposium on Circuits and Systems, Kobe, pp. 4405–4408. IEEE Computer Society Press, Los Alamitos (2005)
7. Luo, G., Sun, X.M., Liu, Y.L.: Research on steganalysis of stegotext based on noise detecting. Journal of Hunan University 32(6), 181–184 (2005)
8. Sui, X.G., Luo, H.: A Steganalysis Method Based on the Distribution of Space Characters. In: Proc. of the 2006 International of Communications, Circuits and Systems, Guilin, pp. 54–56. IEEE Computer Society Press, Los Alamitos (2006)
9. Liu, X.G., Luo, H., Zhu, Z.L.: A Steganalysis Method Based on the Distribution of First Letters of Words. In: Proc. of the 2006 International on Intelligent Information Hiding and Multimedia Signal, California, pp. 369–372. IEEE Computer Society Press, Los Alamitos (2006)
10. Huang, H.J., Sun, X.M., Sun, G.: Detection of Hidden Information in Tags of Webpage Based on Tag-Mismatch. In: Proc. of the 2006 International on Intelligent Information Hiding and Multimedia Signal, Haikou, pp. 257–260. IEEE Computer Society Press, Los Alamitos (2007)
11. Huang, H.J., Sun, X.M., Li, Z.S.: Detection of Hidden Information in Webpage. In: Proc. of the 4th International Conference on Fuzzy Systems and Knowledge Discovery, Kaoxiong, pp. 317–320. IEEE Computer Society Press, Los Alamitos (2007)

12. Huang, J.W., Sun, X.M., Huang, H.J.: Detection of Hiding Information in Webpages Based on Randomness. In: Proc. of The Third International Symposium on Information Assurance and Security, Manchester, pp. 447–452. IEEE Computer Society Press, Los Alamitos (2007)
13. Fridrich, J., Goljan, M., Du, R.: Reliable detection of LSB steganography in grayscale and color images. In: Dittmann, J., Nahrstedt, K., et al. (eds.) Proc. of the ACM Workshop on Multimedia and Security, Ottawa, pp. 27–30. ACM Press, New York (2001)

Secret Sharing Based Video Watermark Algorithm for Multiuser

Shangqin Xiao[*], Hefei Ling[**], Fuhao Zou[***], and Zhengding Lu[†]

Department of Computer Science, Huazhong University of Science and Technology, 430074,
Wuhan, China
lhefei@hotmail.com

Abstract. A novel video watermark algorithm based on Shamir secret sharing scheme for multiusers is proposed. It divides the field of the host video signal into multisegments, and then splits the original watermark into corresponding number shares with the Shamir's secret sharing algorithm. Each share of the watermark is embedded into the corresponding segment of the host video signal. In this scheme, only the user that has sufficient authorization can reveal the genuine watermark directly; other users could obtain shadow watermarks merely. Experimental results show the proposed watermarking scheme can offer good video quality and is robust against various degree attacks.

Keywords: Video watermarking, Secret Sharing, Robustness, Shadow Watermark.

1 Introduction

With the rapid development of the network and multimedia technique, more and more medium including images, audios and videos are transformed into digital form and widely distributed over the Internet. So the protection of intellectual property rights has been the key problem nowadays. Digital watermarking has been proposed to identify the ownership and distribution path of the digital material.

A watermark must be embedded into the data in such a way that it is imperceptible for users. Moreover, the watermark should be inaudible or statistical invisible to prevent unauthorized extraction and removal. The watermark should also have similar compression characteristics as the original signal and be robust to most of manipulations or signal processing operations on the host data. Video watermarking schemes rely on the imperfections of the human visual system (HVS) [1]. The existing watermarking techniques are classified into either spatial or transform domain techniques.

---

[*] Shangqin Xiao (1980-), male, doctoral student, research direction: watermarking, DRM.
[**] Hefei Ling (1976-), male, doctor, research direction: watermarking, DRM, cryptography, corresponds author.
[***] Fuhao Zou (1974-), male, doctor, research direction: watermarking, DRM.
[†] Zhengding Lu (1944-), male, professor, research direction: Information security, distributed system

H.J. Kim, S. Katzenbeisser, and A.T.S. Ho (Eds.): IWDW 2008, LNCS 5450, pp. 303–312, 2009.

The familiar watermark algorithms of transform domain include Discrete Cosine Transform (DCT) [2, 3], Discrete Fourier Transform (DFT) [4], Discrete Wavelet Transform (DWT) [5, 6]. Considering the applications of watermarking systems, some of them provide better imperceptibility and others focus on other benchmarks like robustness or embedding capacity.

In this paper we focus on the security of the hidden secret information and propose a secret sharing video watermarking (SSVW) scheme for sharing the watermark with multi-users. Instead of embedding the original watermark into the video like general watermarking systems, our system divides the original watermark and host video into multi-shares, embeds each share of the watermark into the corresponding part of the host video signal in temporal domain. The user having the sufficient authorization can reveal the genuine watermark directly. Otherwise, a shadow watermark can be obtained merely. Experimental results show the proposed watermarking scheme can offer good video quality and robustness under various degree attacks.

2 Pretreatment of Watermarking

In this paper, the watermark signal is a valuable binary image. Let $W = \{w(i,j) \mid w(i,j) \in \{0,1\} \wedge 1 \le i \le L_1 \wedge 1 \le j \le L_2\}$ is the binary image. At first, we convert the binary watermark image W into one dimensional sequence $W_1 = \{w_1(l) = w(i,j)\}$, $1 \le i \le L_1$, $1 \le j \le L_2$, $l = i + (j-1) \cdot L_1$. The watermark sequence W_1 is permuted into pseudo-random sequence W_2 with a private seed for eliminating correlation of adjacent element in sequence W_1 and enhancing the robustness of the watermark. For the purpose of improving the anti-attack capabilities of watermark, modulation is applied to the watermark sequence W_2. That is, the watermark W_2 is modulated into W_3 using m-sequence which is generated using encryption key k as a seed. The modulated principle is formulated as:

$$w_3(n) = w_2([n/m]k) \oplus m(n) \tag{1}$$

Where $W_3 = \{w_3(n)\}$, spreading factor is m, $w_2, w_3 \in \{0,1\}$, $0 \le k \le L_1 L_2$ and $0 \le n \le m L_1 L_2$. M-sequence has been chosen for modulation because of its good auto-correlation characteristics and secrecy. In our watermark scheme, the spreading factor selection should be under tradeoff between capacity and robustness so that it facilitates to obtain stronger robustness while having higher channel capacity.

So in the preprocess phase, watermarking is divided into multi-shares. The secret sharing scheme is based on Shamir's (t,n) threshold scheme [7], in which a secret is divided into n shares so that the shared secret can be recovered only from any t or more shares. The Lagrange interpolating polynomial $f(x)$ is adopted to partition the watermark W_3 into muliti-shares in a communication. It is said that the any t users could reveal the genuine watermark directly if $f(x)$ is a polynomial of degree $t-1$. Thus, the number of users smaller than t in the communication can't reveal the watermark. Suppose the watermark W_3 is divided into n shares and the shadow watermark (i.e., watermark share) set is A. The overall share set of the watermark is U, where the dimensions of A and U are n and t respectively. To divide the watermark W_3 into pieces S_i, we select a random m degree polynomial $a(x)$ in which $a_0,...,a_{m-1} \in Z_p$, s is constant and S_i is calculated as follows:

$$S_i = a(i) = a_0 + a_1 i + ... + a_{m-1} i^{m-1} \tag{2}$$

Each user $u_i (i = 1,2,...,n)$ has a piece of watermark $P_i = a(u_i) \bmod p$. The combination of random t shadow watermarks could reveal the genuine watermark. When the user which has k pairs of two-dimensional plane (i, s_i) wants to recover the watermark, he can compute the original watermark by the Lagrange interpolating polynomial formula as follows:

$$S = \sum_{i \in U} b_i s_i \ , b_i = \prod_{j \in U \cap j \neq i} \frac{j}{(j-i)} \tag{3}$$

3 Watermarking Based on Secret Sharing

3.1 Watermarking Embedding

In the proposed SSVW system, the watermark is composed of a number of shadow watermarks. The watermarking framework proposed in this paper is shown in figure 1:

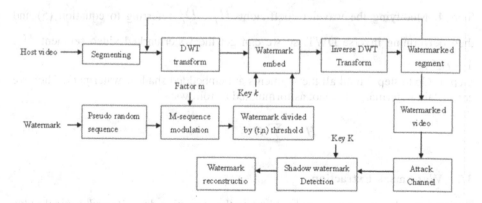

Fig. 1. Video watermarking model

The main steps performed in the proposed SSVW system are summarized below:

Step 1. The host video H is cut into n video segments $\{H_i \mid H = \sum_{i=1}^{n} H_i \wedge 1 \leq i \leq n\}$,

and segments are independent from each other. Therefore, the shadow watermark could be embedded into segment respectively.

Step 2. DWT transform is the method to decompose signal to the temporal and frequency domain. For unbalanced video signal, DWT transform is a good analysis tool to each video segment. According to the multi-resolution characteristic of DWT, select the proper wavelet family to decompose H_i up to the Lth level. The wavelet coefficients are $H_i^L, D_i^j, 0 \leq j \leq L-1$ where H_i^L is the Lth level approximate component, D_i^j are the detail parts of 1th to L^{th} level. In order to make the watermarked signal invisible, we embed the watermark into low frequency component having the highest energy of video signal by taking advantage of frequency mask effect of HVS.

Step 3. Embed the shadow watermark by adjusting the embedding strength. The proposed scheme embeds shadow watermark bit in DWT coefficient by quantization. The quantization function is described as follow:

$$Q(X_j(i)) = \left\lfloor \frac{X_j(i)}{h} \right\rfloor \tag{4}$$

Where h is quantization step, l is the length of shadow watermark, and n is the number of shadow watermark. The watermarking process can be defined as:

$$X_j'(i) = \begin{cases} Q(X_j(i)) \times h + \dfrac{h}{2}, & ifQ(X_j(i)) \bmod(2) = w_j(i) \\ Q(X_j(i)) \times h - \dfrac{h}{2}, & ifQ(X_j(i)) \bmod(2) \neq w_j(i) \end{cases}, \ (1 \leq i \leq l, 1 \leq j \leq n) \tag{5}$$

Step 4. Modifying the wavelet coefficients H_i^L, D_i^j referring to equation (5) and then performing inverse DWT, so we can get the watermarked video segment H_i', $1 \leq i \leq n$.

Step 5. Go to step 2 until all the segments are embedded shadow watermark. Then we can obtain watermarked video, as formulated as follows:

$$H' = \sum_{i=1}^{n} H_i' \tag{6}$$

3.2   Watermark Extraction

The watermark scheme proposed is blind so the extraction doesn't need original video signal. The watermark extraction process is as follows:

Step 1. Suspected video signal is divided into segments, and then performed DWT transform towards each segment. We can obtain watermarking code from the equation (7), where $\lfloor \ \rfloor$ is the floor integer function.

$$w'_j(i) = \left\lfloor \frac{X_j(i)}{h} \right\rfloor \mod 2, (1 \le i \le l, 1 \le j \le n) \tag{7}$$

Step 2. Finally, we can obtain the watermark through (t, n) threshold scheme. The watermark is recognized as the combination of shadow watermarks as follows:

$$w'(i) = ((w')^k)^{S_i} \mod p \tag{8}$$

Where p is a prime number. We should validate the shadow watermarks against cheating. Here, suppose $M = \sum_{i=1}^{t} a_i p^{i-1}$ is validation code of the shadow watermark and we choose a Hash function $h(w_i')$. All the shadow watermarks satisfy $h(w_i) < p$ and there is $M' = \sum_T h(w_i')p^{i-1}$. We can validate all the shadow water-

marks by the formulas $[\frac{M - M'}{p^{i-1}}]$. So we can get shadow watermark validate coeffi-

cient. If the result is not valid, we can choose another shadow from all the extraction watermarks, until there are n valid shadow watermarks w_i' detected and $i = 1, 2, \ldots, n$.

Step 3. Then using the predetermined (t, n) threshold scheme, which is calculated using the polynomial coefficient, the original binary watermarks are recovered. The recovery of watermark image is an inverse process of the watermark signal encoding.

$$\prod_{i=1}^{n} (w_i')^{b_i} = \prod_{i=1}^{n} ((w_i')^k)^{s_i \cdot b_i} = ((w_i')^k)^{\sum_{i=1}^{n} s_i \cdot b_i} = w', b_i = \prod_{j=1 \cap j \ne i}^{k} \frac{a_j}{a_i - a_j} \tag{9}$$

Step 4. In the scheme, the extracted watermark is visually recognizable pattern. The viewer can compare the results with the reference watermark subjectively. Also a quantitative measurement defined as NC (Normalized Correlation) [8] is used to compare embedded and extracted watermarks.

$$\rho = \sum_{i=0}^{L_1} \sum_{j=0}^{L_2} w(i,j) w'(i,j) \bigg/ \sqrt{\sum_{i=0}^{L_1} \sum_{j=0}^{L_2} w(i,j)^2} \cdot \sqrt{\sum_{i=0}^{L_1} \sum_{j=0}^{L_2} w'(i,j)^2} \tag{10}$$

Where referenced watermark is w and extracted watermark is w'. Since there are some distortions of detected watermark in tolerated degree, i.e., $\rho > 0.4$, the watermark w' is still regarded recognizable.

4 Experimental Results

In order to evaluate the visual quality and robustness against the above mentioned attacks of the proposed watermark, we have chosen threshold $(t,n) = (2,4)$. To show the simulation results and to compare with other related systems objectively, we compare the SSVW scheme with another DWT based watermarking scheme [9], which embeds an identical watermark in DWT domain by quantization.

The standard video sequence Flowergarden (mpeg2, 704×576, 6Mbit/s), Cutthroat (mpeg2, 720×470, 8Mbit/s) and the standard image Hust (gray, 43×70) were used for simulation. The performance of proposed watermarking scheme include embedding capacity or embedding rate, embedding distortion and robustness against desynchronization attacks and various frame colluded attacks.

4.1 Experiment of Perceptual Quality

The quantization step associated with watermarking strength is determined by perceptual quality. The relationship between quantization step and PSNR is shown in table 1, where the average PSNR decreases with the increasing of quantization step. In the case of $Q < 15$, the perceptual quality of video is good enough.

Table 1. Average PSNR of video

quantization step Q	2	6	10	14	18	22
PSNR of Flowergarden	45.263	40.841	38.345	34.964	29.526	27.714
PSNR of Cutthroat	47.647	44.114	41.138	37.596	32.364	30.846

As shown in figure 2, the result with $Q = 10$ reveals that the watermark does not affect the perceptual quality of original video signals from subjective perspective. This is as expected because the secret sharing model is incorporated into the SSVW scheme and the embedding process simply changes the quantization coefficient. It is well known that human visual system is less sensitive to slight quantization change. The imperceptibility of the embedded watermark is ensured.

We use PSNR (peak signal-to-noise ratio) [8] to evaluate watermarked video quality:

$$PSNR = MN \max_{m,n} I_{m,n}^2 / \sum_{m,n} (I_{m,n} - \bar{I}_{m,n})^2 \qquad (11)$$

where $M \times N$ is frame size, $I_{m,n}$ and $\bar{I}_{m,n}$ are the pixel values at point (m,n) of the original and watermarked video frame respectively. The results are illustrated in figure 3. On the whole, the PSNR curve of SSVW scheme is a little above that of the watermarking scheme proposed by Campisi. The average PSNR values of both watermarked videos are higher than 36 dB. Perceptually, the original video and the watermarked video are visually indistinguishable. It implies that the SSVW scheme is able to achieve perceptual transparency.

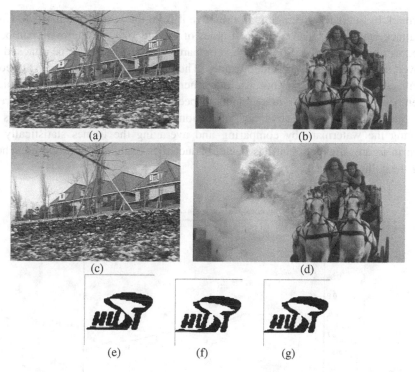

Fig. 2. Host video and watermark (a) Original video Flowergarden (6Mbit/s) (b) Original video Cutthroat (8Mbit/s) (c) Watermarked video Flowergarden (d) Watermarked video Cutthroat (e) Original watermark (f) Watermark extracted from Flowergarden (g) Watermark extracted from Cutthroat

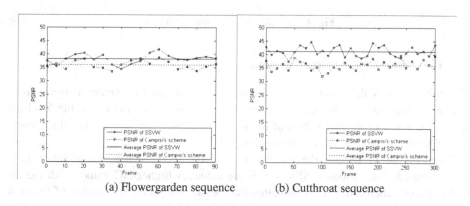

(a) Flowergarden sequence   (b) Cutthroat sequence

Fig. 3. The PSNR curve of watermarked video

4.2 Robustness with Collusion Attack

Frame averaging and collusion attack [10] is a common attack to the video watermark. When attackers collect a number of watermarked frames, they can estimate the watermark by statistical averaging and remove it from the watermarked video. Collusion attack would be done on similar frames of the video. The watermark can be regarded as noise in a video. So it will be compared with the frames to be attacked and removed the watermark in the video frames. The proposed scheme can resist to collusion attack quite well. This is because our scheme partitions a watermark into pieces and embeds them into different segments respectively, making the watermarks resistant to attacks by taking the advantage of motionless regions in successive frames and removing the watermark by comparing and averaging the frames statistically. As shown in figure 4, experiment has been conducted to evaluate the proposed scheme under the collusion attack.

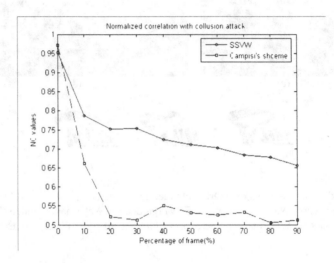

Fig. 4. NC values with collusion attack

4.3   Robustness against Other Attacks

The robustness of the proposed scheme is also investigated to common signal processing, such as addition noise, lossy compression, frame lose, and so on. Fig 5 (a)-(d) shows the result of the NC values of the watermark under the attacks including frame lose, addition noise and lossy compression attack.

From the above results, the effectiveness of the secret sharing based schemes is demonstrated. The proposed SSVW scheme achieves higher NC values with the attacks strength increasing. This indicates that before the perceptual quality of video is distorted obviously, the robustness of SSVW is much better than the other one. The overall robustness of the scheme is also improved by secret sharing. Due to the use of secret sharing to the watermark, the robustness of scheme is not only dependent on watermark embedding, but also dependent on the encryption algorithm. That means the watermark can't be removed by modulating part of efficient.

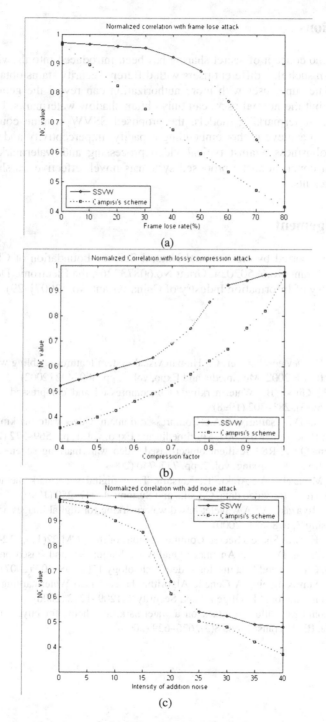

Fig. 5. Robustness of watermark (a) NC values with frame lose attack (b) NC values with lossy compression attack (c) NC values with add noise attack

5 Conclusions

In this paper, the concept of secret sharing has been introduced into the video water-marking system such that different users with different security status obtain different information. The super user with more authorization can reveal the genuine water-mark directly, but the normal users can only obtain shadow watermarks. Unlike most of the existing watermarking models, the proposed SSVW scheme could be used constructively to achieve higher embedding capacity, imperceptivity and robustness. Finally, the robustness against typical video processing and watermark attacks is highlighted. In conclusion, the proposed system is novel, effective, as shown in the experimental results.

Acknowledgement

This work is supported by the National Natural Science Foundation of China (Grant No.60803112, Grant No.60502024, Grant No.60873226); the Electronic Development Fund of Ministry of Information Industry of China (Grant No. [2007]329)

References

1. Delaigle, J.F., Devleeschouwer, C.: Human visual system features enabling watermarking. In: Proc of ICME 2002, Multimedia and Expo, vol. 2, pp. 489–492 (2002)
2. Hartung, F., Girod, B.: Watermarking of uncompressed and compressed video. Signal Processing 66(3), 283–302 (1998)
3. Simitopoulos, D., Tsaftaris, S.A.: Compressed-domain video watermarking of MPEG streams. In: Proc. of ICME 2002. Multimedia and Expo, vol. 1, pp. 569–572 (2002)
4. He, D., Sun, Q.: A RST resilient object-based video watermarking scheme. In: Proc of ICIP 2004. Image Processing, vol. 2, pp. 737–740 (2004)
5. Inoue, H., Miyazaki, A., Araki, T., Katsura, T.: A digital watermark method using the wavelet transform for video data. Circuits and Systems 4, 247–250 (1999)
6. Ejima, M., Miyazaki, A.: A wavelet-based watermarking for digital images and video. Image Processing 3, 678–681 (2000)
7. Shamir, A.: How to Share a Secret. Communications of the ACM 22(11), 612–613 (1979)
8. Wang, S., Zheng, D., et al.: An Image Quality Evaluation Method Based on Digital Watermarking. Circuits and Systems for Video Technology 17(1), 98–105 (2007)
9. Gaobo, Y.: Sun Xingming A Genetic Algorithm based Video Watermarking in the DWT Domain. Computational Intelligence and Security 2, 1209–1212 (2006)
10. Wu, Y.: Nonlinear collusion attack on a watermarking scheme for buyer authentication. Multimedia, IEEE Transactions 8(3), 626–629 (2006)

GSM Based Security Analysis for Add-SS Watermarking

Dong Zhang[1,2], Jiangqun Ni[1], Dah-Jye Lee[2], and Jiwu Huang[1]

[1] School of Information Science and Technology, Sun Yat-sen University
Guangzhou, China, 510275
[2] Department of Electrical & Computer Engineering Brigham Young University, Provo,
UT, U.S.A. 84602
zhangd@mail.sysu.edu.cn

Abstract. Attack to security of watermarking is to estimate secret keys used in watermarking communications. The security level of watermarking algorithms can be evaluated by Cramer-Rao Bound (CRB) for the estimation of secret keys. As a widely used method, Add-Spread-Spectrum (Add-SS) watermarking has drawn a lot of attention. Previous work on watermarking security is mainly based on the assumption that the host is Gaussian distributed and ignores the impact of the non-Gaussian characteristics of natural images. With the incorporation of Gaussian Scale Mixture (GSM) model for host signals, this paper presents a theoretical analysis on the security of the Add-SS watermarking algorithms. By giving the CRB and Modified Cramer-Rao Bound (MCRB) for the estimation of secret carriers under Known Message Attack (KMA) and Watermarked Only Attack (WOA), this paper also reveals the factors that may influence the security of Add-SS based watermarking algorithms. Results presented in this paper are very helpful for designing the new generation of secure and robust watermarking system.

1 Introduction

Besides robustness, invisibility, and capacity, security has become another basic constraint for digital watermarking in recent years [1]. Kerckhoff's principle [2][3] from cryptography can be employed to describe the model of watermarking security, i.e., each unit of a watermarking system (encoding/embedding, decoding/detection ...) should be declared public except its secret keys. Therefore, secret keys are the key to maintaining secured watermarking communication. For Added Spread-Spectrum (Add-SS) based watermarking algorithms, secret keys are the random numbers used to generate spread carriers. The random sequences used to dither the quantization lattice work as the secret keys in the Quantization Index Modulation (QIM) based watermarking algorithms.

The difference between security and robustness deserves to be clarified. Attack to robustness targets to increase the Bit Error Ratio (BER) of the watermarking channel while attack to security aims to gain knowledge about the secret keys [3]. Unlike the concept of robustness which deals with blind attacks, security is more critical for watermarking because it deals with intentional attacks of which the information about watermarking scheme is known to the attacker, therefore risking complete break.

H.J. Kim, S. Katzenbeisser, and A.T.S. Ho (Eds.): IWDW 2008, LNCS 5450, pp. 313–326, 2009.

Based on the knowledge available to the attacker, three different attacks can be defined: watermarked only attack (WOA), known message attack (KMA) and known original attack (KOA).

Cayre *et al*'s work in [1] is recognized as one of pioneering works in watermarking security, where they gave for the first time the definition of watermarking security and discriminated it from that of robustness. They also proposed quantifying the security of Add-SS watermarking algorithm with Fisher Information Matrix (FIM). A different approach to watermarking security was proposed by Comesaña *et al* [3]. The authors adopted Shannon's measure to the framework of watermarking security and evaluated the security level of Add-SS based watermarking with mutual information between watermarked signals and secret carriers [3]. Pérez-Freire *et al* investigated the feasible region of secret dither sequence and obtained a secure distribution of secret keys for the QIM based watermarking [5]. Ni *et al* took advantage of natural scene statistics and presented a theoretical analysis of the Add-SS based watermarking algorithm with Shannon's mutual information [6]. Practical attacking algorithms were also given in [1] and [6].

Previous work on watermarking security was mainly based on the assumption that the host is Gaussian distributed and ignored the impact of the non-Gaussian characteristics of natural images. Recently, Gaussian scale mixture (GSM) has been proposed and proved to be capable of accurately modeling natural image in wavelet domain [7]. Although the marginal distribution of natural scenes in the wavelet domain may be sharply peaked and heavy-tailed, according to the GSM model, the coefficient in a particular position is Gaussian distributed, conditioned on its scale. This paper investigates the security of Add-SS based watermarking by taking advantage of statistics of wavelet coefficients of natural image with the help of the GSM model. The CRB and Modified Cramer-Rao Bound (MCRB) are obtained and used as the measurement of watermarking security. The results will be helpful for designing the new generation of robust and secure watermarking algorithms.

The notation and conventions used in the paper are as follows: vectors and scalars are represented in capital letters and lower letters, respectively. Random variables are expressed in italic font.

The remainder of the paper is organized as follows. The GSM model and its performance are presented in Section 2. The general approaches for watermarking security are described in Section 3. The security analysis of Add-SS watermarking based on the GSM model is given in Section 4. Simulation results and analysis are included in Section 5. Finally, concluding remarks are provided in Section 6.

2 The Gaussian Scale Mixture Model of Natural Images

As mentioned in Introduction, the wavelet coefficients in one sub-band are modeled as a GSM RF (Random Field), i.e. $X = \{X_i, i \in I\}$, where I denotes the spatial index of the RF. X is formed as the product of a Gaussian random field and a random scaling variable [7], i.e.

$$X = s \cdot \mathrm{U} = \left\{ s_i \cdot U_i, i \in \mathrm{I} \right\} \tag{1}$$

Here, '=' means the equality in distribution. s_i is independent of s_j for $i \neq j$. $U \sim N(0, Q)$ is a Gaussian random field which has zero mean and covariance Q. s is a scale random variable with positive value. Conditioned on s, the probability of X is

$$P_{X|s}(X|s) = \frac{1}{(2\pi)^{\frac{N}{2}} |s^2 Q|^{\frac{1}{2}}} \exp\left(-\frac{X^T Q^{-1} X}{2s^2}\right) \quad (2)$$

where s is independent of U and can be estimated from wavelet coefficients with a Maximum Likelihood method [7].

$$\hat{s} = \arg \max_s \{\log p(X|s)\} = \sqrt{\frac{X^T Q^{-1} X}{N}} \quad (3)$$

A simplified relationship can be obtained for a scalar model of GSM. For $\forall i \neq j$, X_i is independent of X_j, and the probability of X_i can be expressed as $p_{X_i|s_i}(x_i|s_i) \sim N(0, s_i^2 \sigma_U^2)$ conditioned on s_i. The distribution of X_i is specified by its variance when s_i is given. Without the loss of generality, σ_U^2 can be assumed to be unity.

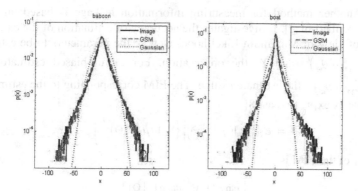

Fig. 1. Empirical marginal histogram (black lines) of wavelet coefficient in the LH1 sub-band of 'baboon' and 'boat', synthesized GSM RF samples from the same sub-band (red dashed lines) and a Gaussian with the same deviation (blue dotted lines)

Figure 1 shows the empirical histograms (black lines) of the LH1 sub-band from two natural images baboon and boat and those of the synthesized GSM RF samples from the same sub-band (red dashed lines). For comparison, a Gaussian distribution with the same standard deviation as the image sub-band is also displayed (blue dotted lines). Note that the synthesized samples capture the non-Gaussian characteristics exemplified by the high peak and heavy tails in marginal distributions.

3 Measures to Evaluate Watermarking Security

In Add-SS based watermarking algorithms, spread carriers are generated by secret keys and modulated by to-be-transferred messages before being embedded into the hosts. For pirating, knowledge of the carriers is sufficient to attack the security of watermarking, e.g. erasing/replacing the watermarking. Hence, estimation of the carriers equals to an attack to secret keys. Add-SS based watermarking algorithms are not absolutely secured, and knowledge about carriers could be leaked from communications [1]. Attacking the security is to estimate the information leakage according to multi-time observations. Security analysis is to investigate the factors which influence the information leakage of carriers.

Two methods are used to measure information leakage. One is based on Shannon's information theory. Denote Z the carriers and $h(Z)$ the entropy of carriers which measures the uncertainty of the carriers. $h(Z|Y)$ denotes the entropy of carriers if the watermarked image, Y, is known. Here, $h(\bullet)$ stands for differential entropy. The mutual information between the carriers and watermarked image can be expressed as

$$I(Z,Y) = h(Z) - h(Z|Y) \tag{4}$$

Equation (4) represents the decrease of uncertainty of carriers due to watermarking communications. In the view of an attacker, $I(Z,Y)$ is the information leakage of the carriers. Another method for measuring information leakage is based on FIM and CRB as described in [1]. It investigates the accuracy of estimation of the carriers from the point of view of Maximum Likelihood. Denote the parameters to be estimated as $\Theta = (\theta_1, \theta_2, \cdots \theta_K)^T$ and Y the observation. For an un-biased estimator, denote $\hat{\Theta} = (\hat{\theta}_1, \hat{\theta}_2, \cdots \hat{\theta}_K)^T$ the estimated value. The FIM corresponding to the estimation is a $K \times K$ matrix expressed as [8]

$$J = E\left(\left\{ \nabla_\Theta \ln p\left(Y|\Theta\right) \right\} \left\{ \nabla_\Theta \ln p\left(Y|\Theta\right) \right\}^T \right) \tag{5}$$

Each entry of the FIM is

$$J_{ij} = E\left[\frac{\partial \ln p\left(Y|\Theta\right)}{\partial \theta_i} \frac{\partial \ln p\left(Y|\Theta\right)}{\partial \theta_j} \right] \tag{6}$$

If the FIM is non-singular, the lower bound of the mean square error of estimated value is the Cramer-Rao Bound which is defined by

$$Var\left[\hat{\Theta} - \Theta\right] \geq CRB(\Theta) = tr\left(J^{-1}\right) \tag{7}$$

where '$tr(\bullet)$' means the trace operation [9]. FIM measures the information leakage and CRB evaluates the accuracy of estimation. The more information leakage of carriers, the more accurate estimation of carriers an attacker could acquire. At the same time, the security level of the watermarking algorithm will be worse.

However, FIM is not always non-singular in practice, especially in the presence of nuisance parameters. The traditional FIM is defined in equation (5). When nuisance parameters, u, exist, the likelihood is $p(Y|\Theta) = \int_{-\infty}^{\infty} p(Y|u,\Theta) p(u) du$. As this integral cannot be carried out analytically in general, it is difficult to obtain the traditional CRB [12]. A substitution is done using MCRB to measure the lower bound of the un-biased estimation [12]. MCRB is defined as

$$
MCRB(\Theta) = tr\left\{\left[E_{Y,u}\left[\left(\frac{\partial \ln p(Y|\Theta,u)}{\partial \Theta}\right)\left(\frac{\partial \ln p(Y|\Theta,u)}{\partial \Theta}\right)^{T}\right]\right]^{-1}\right\}
\tag{8}
$$

Here, MCRB uses the likelihood conditioned on Θ and u, and calculates the expectation of the observations and nuisance parameters. It has been proven that MCRB can be easily calculated in the presence of nuisance parameters [12] [13]. Although it is 'looser' than the traditional CRB, the MCRB will approach the traditional CRB when the nuisance parameters are known.

This paper uses CRB and MCRB to measure the security level of secret carriers under KMA and WOA respectively. The lower the value of CRB or MCRB is, the smaller error the estimation of carriers would have, and the lower security level the watermarking would have and vice versa.

4 Security Analysis on Add-SS Based Watermarking

4.1 Add-SS Based Watermarking Model

For Add-SS based watermarking algorithms, embedding is the addition of the water-mark signal which is the modulation of N_c carriers. The model of embedding can be expressed as

$$
Y^j = X^j + \frac{\gamma}{\sqrt{N_c}} \sum_{i=1}^{N_c} z_i a_i^j
\tag{9}
$$

where $Y^j = (y_1^j, y_2^j, \cdots, y_{N_v}^j)^T$ is the watermarked signal in the jth observation. Denote $X^j = (x_1^j, x_2^j, \cdots, x_{N_v}^j)^T$ as the host of the jth observation and z the matrix composed by all the carrier vectors. The ith vector carrier is denoted by z_i. a_i^j is employed to denote the ith bit of the embedding message in the jth observation. Here, γ is used to control the embedding strength. N_c, N_o and N_v denote the number of carriers, the number of observation and the dimension of the host in each observation, respectively. This paper assumes the same dimension is shared by the hosts and each carrier and defines the Document to Watermark Ratio as $DWR = 10\log_{10}\left(\frac{\sigma_x^2}{\gamma^2 \sigma_a^2}\right)$, where σ_x^2 is the average power of the hosts and $\gamma^2 \sigma_a^2$ is the power of the embedded watermark.

The following analysis assumes the attacker owns N_o independent observations and the hosts are characterized by the GSM model. In the interest of convenient analysis, the paper assumes BPSK is used and random messages are embedded so that a_i^j has value 1 or -1 and is independent of each other. Because attackers would acquire the host images, the result for KOA is similar to the one in [1] when the GSM model is considered. So this paper only presents the security analysis of SS based watermarking under KMA and WOA.

4.2 Known Message Attack

4.2.1 Single -Carrier

For the convenience of description, analysis of one estimated carrier is presented first. Under such condition, $N_c = 1$, and 1 bit message is embedded in each observation. According to the previous assumption, each bit of every observation is independent. For the kth dimension of the jth observation, we have $y_k^j = x_k^j + a_i^j z_{1k}$ where z_{1i} is the ith component of the carrier and a_i^j the embedded message in the jth observation. Using scalar GSM to describe the distribution of the host, $x_k^j \sim N\left(0, s_k^{j2}\sigma_U^2\right)$ in which s_k^j is the scale factor of the kth dimension of the jth host. Since the embedded messages are known, the likelihood and the log-likelihood with sample set $Y^{N_s} = \left(Y^1, Y^2, \cdots Y^{N_s}\right)$ can be expressed (10) and (11), respectively.

$$f\left(Y^{N_s}\middle|Z_1\right) = f\left(Y^1, \cdots, Y^{N_s}\middle|Z_1\right) = \prod_{j=1}^{N_s}\prod_{k=1}^{N_c}\frac{1}{\sqrt{2\pi s_k^{j2}\sigma_U^2}}\exp\left(-\frac{\left(y_k^j - \gamma a_i^j z_{1k}\right)^2}{2s_k^{j2}\sigma_U^2}\right) \tag{10}$$

$$\log f\left(Y^{N_s}\middle|Z_1\right) = \sum_{j=1}^{N_s}\sum_{k=1}^{N_c}\left[\log\frac{1}{\sqrt{2\pi s_k^{j2}\sigma_U^2}} - \frac{\left(y_k^j - \gamma a_i^j z_{1k}\right)^2}{2s_k^{j2}\sigma_U^2}\right] \tag{11}$$

The derivative of the log-likelihood of the parameter to be estimated is

$$\frac{\partial}{\partial z_{1i}}\log f\left(Y^{N_s}\middle|Z_1\right) = \gamma\sum_{j=1}^{N_s}\frac{a_i^j x_i^j}{s_i^{j2}\sigma_U^2} \tag{12}$$

Recall the description in Section 3, in which each entry of the FIM can be calculated according to (13) and (14). For diagonal entries, we have

$$FIM_{ii}\left(Z_1\right) = \int f\left(Y^{N_s}\middle|Z_1\right)\left(\frac{\partial}{\partial z_{1i}}\log f\left(Y^{N_s}\middle|Z_1\right)\right)^2 dY^1\cdots dY^{N_s} = \gamma^2\sum_{j=1}^{N_s}\frac{1}{s_i^{j2}\sigma_U^2} \tag{13}$$

For non-diagonal entries,

$$FIM_{ik}\left(Z_1\right) = \int f\left(Y^{N_s}\middle|Z_1\right)\left(\frac{\partial}{\partial z_{1i}}\log f\left(Y^{N_s}\middle|Z_1\right)\right)\left(\frac{\partial}{\partial z_{1k}}\log f\left(Y^{N_s}\middle|Z_1\right)\right)dY^1\cdots dY^{N_s}$$

$$= 0 \qquad \text{for all } i \neq k \tag{14}$$

So, we can conclude that

$$FIM\left(Z_{1}\right)=\frac{\gamma^{2}}{\sigma_{U}^{2}}\begin{bmatrix}\sum_{j=1}^{N_{s}}\frac{1}{s_{1}^{j2}} & 0 & \cdots & 0 \\ 0 & \sum_{j=1}^{N_{s}}\frac{1}{s_{2}^{j2}} & \cdots & 0 \\ \cdots & \cdots & \cdots & 0 \\ 0 & 0 & \cdots & \sum_{j=1}^{N_{s}}\frac{1}{s_{N_{e}}^{j2}}\end{bmatrix} \tag{15}$$

According to Cramer-Rao inequality, we can obtain the CRB of the estimation of one carrier under KMA as

$$CRB\left(Z_{1}\right)=tr\left(FIM\left(Z_{1}\right)^{-1}\right)=\frac{1}{\gamma^{2}}\sum_{i=1}^{N_{e}}\frac{\sigma_{U}^{2}}{\sum_{j=1}^{N_{s}}1/s_{i}^{j2}} \tag{16}$$

CRB measures the accuracy level of an un-biased estimator. From equation (16), it can be seen that the minimum square error of un-biased estimation of the carrier is influenced by the number of observations and the length of carrier. The longer the carrier is, the more difficult it is to achieve accurate estimation of the carrier; the more observations an attacker has, the more information could be acquired and a more accurate estimation can be made. However, due to the non-Gaussian characteristics of natural images as described with GSM, the relationship between CRB and the length of carriers and number of observations are non-linear.

As mentioned in [1], Cayre $et\ al$ described image hosts with the Gaussian model and presented the bound of estimation of one carrier under KMA as $tr\left(FIM\left(Z_{1}\right)^{-1}\right)_{c}=\frac{N_{v}\sigma_{x}^{2}}{\gamma^{2}N_{o}}$, where σ_{x}^{2} is the variance of the image wavelet coefficients. For the sake of fair comparison, let $\frac{1}{N_{o}N_{v}}\sum_{i=1}^{N_{e}}\sum_{j=1}^{N_{s}}s_{i}^{j2}\sigma_{U}^{2}=\sigma_{x}^{2}$ which means the equality of the variance in GSM and Gaussian models for wavelet coefficients. According to the AM-HM inequality [14]

$$\frac{1}{N}\sum_{j=1}^{N}\frac{1}{m_{j}}\geq\frac{N}{m_{1}+m_{2}+\cdots+m_{N}} \tag{17}$$

where m_{j} has positive value, we have

$$tr\left(FIM\left(Z_{1}\right)^{-1}\right)=\frac{1}{\gamma^{2}}\sum_{i=1}^{N_{e}}\frac{\sigma_{U}^{2}}{\sum_{j=1}^{N_{s}}1/s_{i}^{j2}}\leq\frac{1}{\gamma^{2}N_{o}^{2}}\sum_{i=1}^{N_{e}}\sum_{j=1}^{N_{s}}s_{i}^{j2}\sigma_{U}^{2}=tr\left(FIM\left(Z_{1}\right)^{-1}\right)_{c} \tag{18}$$

The equality is achieved when $s_{i}^{1}=s_{i}^{2}=\cdots=s_{i}^{N_{e}}$. Equation (18) shows that, with Gaussian model, the result in [1] has "over-evaluated" the security level of Add-SS watermarking system. As the GSM model can accurately describe the distribution of wavelet coefficients for natural images, a "tighter" security bound and a more accurate evaluation of security can be obtained.

4.2.2 Multi-carriers
Similarly, the result in equation (18) can be extended to the case when more than one carrier is employed. As the embedded message corresponding to each carrier is known, the log-likelihood with observation $Y^{N_{e}}=\left(Y^{1},Y^{2},\cdots Y^{N_{e}}\right)$ and FIM are

$$\log f\left(Y^{N_s}\middle| Z\right) = \sum_{j=1}^{N_s}\sum_{k=1}^{N_c}\left[\log\frac{1}{\sqrt{2\pi s_k^{j2}\sigma_U^2}} - \frac{1}{2s_k^{j2}\sigma_U^2}\left(y_k^j - \frac{\gamma}{\sqrt{N_c}}\sum_{m=1}^{N_c}a_m^j z_{mk}\right)^2\right] \tag{19}$$

$$FIM = E\left\{\left[\partial\log f\left(Y^{N_s}\middle| Z\right)\middle/\partial\left(z_1^T,\cdots,z_{N_c}^T\right)^T\right]\left[\partial\log f\left(Y^{N_s}\middle| Z\right)\middle/\partial\left(z_1^T,\cdots,z_{N_c}^T\right)^T\right]^T\right\} \tag{20}$$

Each entry of the FIM can be calculated by equation (21).

$$E(\frac{\partial\log f\left(Y^{N_s}\middle| Z\right)}{\partial z_{mn}})(\frac{\partial\log f\left(Y^{N_s}\middle| Z\right)}{\partial z_{pq}}) = \frac{\gamma^2}{\sigma_U^2 N_c}\sum_{j=1}^{N_s}\frac{a_m^j a_p^j}{s_j^{j2}}\delta_{n,q} = \frac{\gamma^2}{\sigma_U^2 N_c}FIM_{(m,n)(p,q)} \tag{21}$$

where z_{mn} is the nth dimension of the mth carrier. The location of $FIM_{(m,n)(p,q)}$ is the $(m-1)\times N_v + n^{th}$ row and the $(p-1)\times N_v + q^{th}$ column. So the FIM is

$$FIM = \frac{\gamma^2}{\sigma_U^2 N_c}\begin{bmatrix} FIM_{(1,1)(1,1)} & FIM_{(1,1)(1,2)} & \cdots & FIM_{(1,1)(N_c,N_v)} \\ FIM_{(1,2)(1,1)} & FIM_{(1,2)(1,2)} & \cdots & FIM_{(1,2)(N_c,N_v)} \\ \vdots & \vdots & & \vdots \\ FIM_{(N_c,N_v)(1,1)} & FIM_{(N_c,N_v)(1,2)} & \cdots & FIM_{(N_c,N_v)(N_c,N_v)} \end{bmatrix} \tag{22}$$

The FIM is composed of $N_c \times N_c$ partitioned matrices. Each of them is a $N_v \times N_v$ diagonal matrix. Similar to the FIM obtained in [1] under KMA, equation (22) is sensitive to the embedded messages. The value of the embedded messages will influence the existence of the inversion of FIM. Following the method used in [1], this paper also investigates the asymptotic behavior of FIM and corresponding CRB by assuming the attackers own infinite observations.

For the constant scale variables, i.e., $s_i^j = s$ $(1 \le i \le N_v, 1 \le j \le N_o)$ in every observation, the diagonal entries of equation (22) will be $\frac{1}{s^2}N_o$ $(i = 1,\cdots,N_v)$ while $\sum_{j=1}^{N_o}\frac{a_m^j a_n^j}{s^2} \to 0$ conditioned on $N_o \to \infty$. Consequently, equation (22) can be simplified as a diagonal matrix. It's easy to calculate its inverse and then find the corresponding CRB by compute the trace of the inverted matrix. This is the result given by [1] and denoted as $CRB(Z)_c$.

$$CRB(Z)_c = tr\left(FIM_c^{-1}\right) = \frac{N_c^2 N_v s^2 \sigma_U^2}{N_o \gamma^2} \tag{23}$$

For natural images, s_i^j of each host is a random variable, and each diagonal entry can be expressed as $\sum_{j=1}^{N_o}1/s_i^{j2}$ $(i = 1,\cdots,N_v)$. When infinite observations are available, the off-diagonal entries will be far smaller than the diagonal ones due to the random value of $a_m^j a_n^j$ (1 or -1). According to [10], for GSM model of natural images, the distribution of scaling variable s^2 takes the form $p(s^2) \propto 1/s^2$. As s^2 is definitely positive, it is assumed to be identically distributed without the loss of generality. According to the assumption of independence, we have

$$\sum_{j=1}^{N_c} \frac{a_m^j a_n^j}{s_i^{j2}} \xrightarrow{N_c \to \infty} N_o E\left(\frac{a_m^j a_n^j}{s_i^{j2}}\right) = 0 \tag{24}$$

So, equation (22) is simplified to a diagonal matrix which only conserves the diagonal entries. Denote the trace of this diagonal matrix's inverse as CRB_{fim}, and CRB_{FIM} the trace of the inverse of equation (22). CRB_{FIM} can be simplified as CRB_{fim} as expressed in equation (25).

$$CRB_{fim}(z) = \frac{N_c^2 \sigma_U^2}{\gamma^2} \sum_{i=1}^{N_c} \frac{1}{\sum_{j=1}^{N_c} 1/s_i^{j2}} \tag{25}$$

Experimental results also verify the equation (25). Figure 2(a) and (b) shows the comparison of CRB_{fim} and CRB_{FIM} in experiments with natural images when DWR is fixed at 10dB and 15dB. The length of carriers is fixed to 512. Hosts are randomly selected from the HL2, LH2 and HH2 sub-bands of wavelet coefficients from 8 natural images. The subscript number in the curve names denotes the number of estimated carriers (N_c). Note that the values of CRB_{fim} approaches CRB_{FIM} with the increase of the number of observations. When $\gamma^2 = 1$, equation (25), i.e., the CRB for multi-carriers will be the same as the one for single-carrier under KMA.

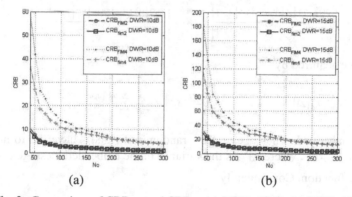

Fig. 2. Comparison of CRB_{FIM} and CRB_{fim} (a) DWR=10dB; (b) DWR=15dB

4.3 Watermarked Only Attack

For WOA, attackers only have watermarked signals. The carriers, z, and the embedded messages, a, are all unknown. Although only the carriers are the target of estimation, the embedded messages would influence the estimation of the carriers [1] [11]. Consequently, the FIM may be singular sometimes so that it's impossible to find the traditional CRB.

The work presented in [1] made constraints on the carrier for the case of a single carrier, and constructed the corresponding null-space, H, using approach proposed in [12]. The CRB of the estimated secret carrier can be obtained only if $\left(H^T \cdot FIM \cdot H\right)^{-1}$ exists. When more than one carrier were taken into account, it was necessary to assume another N_m embedded messages had been known by the attacker, and then additional $N_m \times N_c$ constraints would be required. However, these assumptions are not suggested for WOA in a strict scene. Furthermore, in most cases of practical interest, the traditional CRB cannot be carried out analytically in the presence of nuisance parameters and especially non-Gaussian noise [15]. Therefore, this paper proposes to evaluate the security of Add-SS watermarking system under WOA with MCRB.

In computing the MCRB of the estimated carriers under WOA, we define the FIM as $J_M = E_{Y,a}(\psi\psi^T)$, where

$$\psi = \partial \log p(Y^{N_c}|Z,a)\Big/ \partial \left(z_1^T, \cdots, z_{N_c}^T\right)^T \tag{26}$$

$$\left[J_M(Z)\right]_{(m,n),(p,q)} = E_{Y,a}\left[\frac{\partial \ln p\left(Y^{N_c}|Z,a\right)}{\partial z_{mn}} \frac{\partial \ln p\left(Y^{N_c}|Z,a\right)}{\partial z_{pq}}\right] \tag{27}$$

and $var\left[\hat{z}-z\right] \geq tr\left(J_M^{-1}\right) = MCRB(Z)$. The notations in $J_M(Z)$ and z are the same as the ones mentioned in section 4.2 and $\log p\left(Y^{N_c}|Z,a\right)$ is the same as in equation (18). Because

$$E_{Y,a}\left[\frac{\partial \ln p\left(Y^{N_c}|Z,a\right)}{\partial z_{mn}} \frac{\partial \ln p\left(Y^{N_c}|Z,a\right)}{\partial z_{pq}}\right] = E_a\left\{E_{Y|a}\left[\frac{\partial \ln p\left(Y^{N_c}|Z,a\right)}{\partial z_{mn}} \frac{\partial \ln p\left(Y^{N_c}|Z,a\right)}{\partial z_{pq}}\right]\right\} \tag{28}$$

$J_M(Z)$ can be acquired by

$$\left[J_M(Z)\right]_{(m,n)(p,q)} = E_a\left(\frac{\gamma^2}{\sigma_U^2 N_c}\sum_{j=1}^{N_c}\frac{a_m^j a_p^j}{s_n^{j2}}\delta_{n,q}\right) = \frac{\gamma^2}{\sigma_U^2 N_c}\sum_{j=1}^{N_c}\frac{E_a\left(a_m^j a_p^j\right)}{s_n^{j2}}\delta_{n,q} \tag{29}$$

If the embedded messages are randomized and have zero-means, we have $E_a\left(a_m^j a_p^j\right) = \sigma_a^2 \delta_{m,p}$, where σ_a^2 is the variance of the embedded message and $\delta_{m,p}$ is the Kronecker function. Consequently,

$$\left[J_M(Z)\right]_{(m,n)(p,q)} = \frac{\sigma_a^2 \gamma^2}{\sigma_U^2 N_c}\sum_{j=1}^{N_c}\frac{1}{s_n^{j2}}\delta_{m,p}\delta_{n,q} \tag{30}$$

The MCRB of estimated carriers under WOA is

$$MCRB(Z) = tr\left(J_M(Z)^{-1}\right) = \frac{\sigma_U^2 N_c^2}{\sigma_a^2 \gamma^2}\sum_{i=1}^{N_v}\frac{1}{\sum_{j=1}^{N_c} 1/s_i^{j2}} \tag{31}$$

From equation (31), it can be seen that the security level of secret carriers is linear with the variance of embedded messages.

Instead of requiring the attacker knows additional N_m embedded messages as described in [1], the approach with MCRB only requires the distribution of nuisance parameters as prior, and is closer to situation of WOA.

5 Simulation Results and Analysis

In this section, CRB and MCRB under KMA and WOA were calculated with 8 natural images (aerial, baboon, barb, boat, f16, lena, peppers and sailboat). Bi-orthogonal 9/7 wavelet was used to decompose images into 2 layers. Coefficients from HL2, LH2 and HH2 were randomly selected as hosts. The length of carriers was fixed at 512.

5.1 CRB under KMA

Figure 3(a) shows both the GSM and Gaussian based CRBs for single carrier when the DWR are set to 20dB and 15dB, respectively. With the increase of observation number, CRBs are decreased for both GSM and Gaussian model. The CRB are decreased quickly at first and slowly when more observations are known. This implies the contribution to estimation is mainly from the first few observations. Compared with the Gaussian based CRB, the GSM based one had a "tighter" security bound. This indicated that a more accurate estimation can be obtained and a more accurate evaluation of security level can be achieved with the GSM based CRB.

Figure 3(b) shows the CRBs plotted against the length of the carrier when 50 observations were available. Both the GSM and Gaussian based CRBs are drawn for comparison when the DWR are taken as 20dB and 15dB. Due to the constant host variance in the Gaussian model, its corresponding CRB is linear with respect to the length of the carrier. While in the GSM based analysis, CRB grows non-linearly with the length of carrier as a result of the random host variance as explained in Section 2. Note that the CRB growing with the length of carrier explains why using longer carriers lead to increased security.

It is observed that watermarking with higher DWR corresponds to a higher security level. Because the lower DWR indicated a higher embedded strength, the corresponding CRB has a lower value and implies a lower security level.

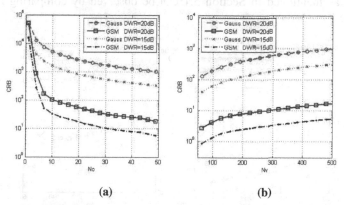

(a)   (b)

Fig. 3. The CRB for single-carrier under KMA (a) $N_v = 512$; (b) $N_o = 50$

Figures 4(a) and 4(b) show the GSM model based CRBs with 2 and 4 carriers under KMA. The DWR are fixed at 10dB and 15dB. GSM$_{fim}$ denotes the CRB calculated by equation (25). The subscript number denotes the count of estimated carriers.

For comparison purpose, the Gaussian based CRB is also shown and denoted with 'Gauss'. In all the situations of Fig 4, the GSM based analysis had a "tighter" security bound compared to the one based on the Gaussian model.

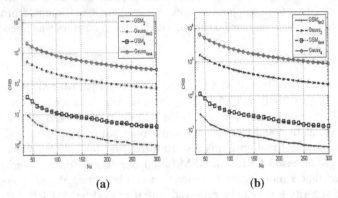

Fig. 4. The CRB for multi-carrier under KMA; the carries are set to 2 and 4 and the length of carriers is fixed at 512. (a) DWR=10dB; (b) DWR=15dB

5.2 MCRB under WOA

Figures 5(a) and 5(b) show the MCRBs of natural images under WOA when the DWR are fixed at 10dB and 15dB. $MCRB_1$, $MCRB_2$ and $MCRB_4$ denote the MCRB under the situations when one, two and four carriers are considered respectively. With an increase in carrier count, the estimation error grows and a higher security level is obtained. The estimation error is decreased if more observations are available which corresponds to a decrease of security level. The same relationship between DWR and security level as mentioned in Section 5.1 can be observed by comparing Fig. 5(a) with Fig. 5(b).

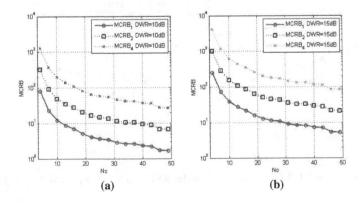

Fig. 5. The MCRB for estimated carriers under WOA (a) DWR=10dB; (b) DWR=15dB

5.3 Comparison of CRBs for Different Natural Images

Figures 6(a) and 6(b) show the comparison of CRBs for hosts from different natural images ('baboon' and 'f16') under KMA when the number of observation is fixed to 30 and DWR is 15dB. In the interest of simplicity, only one carrier is employed. The circled lines denote the CRBs based on the Gaussian model. The GSM model based CRBs are denoted by square-dashed lines ('baboon') and cross-dot-dashed lines ('f16'). Compared with 'f16', 'baboon' represented more complex details in spatial domain and its coefficients had larger variances in the wavelet sub-band. Note that 'baboon' has a larger CRB than that of 'F16' both in Figs. 6(a) and 6(b). That explains why hosts from 'baboon' provided a higher security level compared with hosts from 'F16'. Generally, the security level could be improved by using hosts with complex details for Add-SS watermarking.

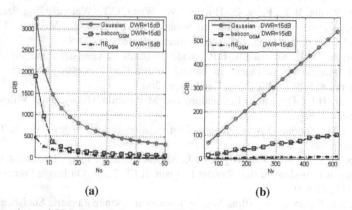

(a)   (b)

Fig. 6. Comparison of CRB from 'baboon' and 'f16' (a) length of carriers is fixed at 512; (b) 30 observations are available and fixed

6 Conclusion

This paper presents a theoretical analysis of Add-SS based watermarking security incorporating the GSM model. The security level is evaluated by the CRB and MCRB of the estimation of carriers. The results show the security level of Add-SS based watermarking is non-linear with respect to the number of observations and the length of carriers due to the non-Gaussian distribution of natural images while it is linear to the embedding strength and the variance of embedded messages. Compared with the results in [1], this paper obtains a more accurate evaluation of the security level for Add-SS based watermarking. This work will be helpful for designing the new generation of robust and secure watermarking.

Acknowledgment

The authors appreciate the supports received from NSFC (60773200, 90604008, 60633030), 973 Program (2006CB303104), NSF of Guangdong (7003722) and State Scholarship Fund (2007104754).

References

1. Cayre, F., Fontaine, C., Furon, T.: Watermarking Security: Theory and Practice. IEEE Trans. Signal Processing 53, 3976–3987 (2005)
2. Kerckhoffs, A.: La Cryptographie Militaire. Journal des Militaries 9, 5–38 (1983)
3. Comesaña, P., Péres-Freire, L., Péres-González, F.: Fundamentals of Data Hiding Security and Their Application to Spread-Spectrum Analysis. In: Barni, M., Herrera-Joancomartí, J., Katzenbeisser, S., Pérez-González, F. (eds.) IH 2005. LNCS, vol. 3727, pp. 146–160. Springer, Heidelberg (2005)
4. Cox, I.J., Miller, M.L., Bloom, J.A.: Digital Watermarking. Elsevier Science, Amsterdam (2002)
5. Pérez-Freire, L., Pérez-González, F., Furon, T., Comesaña, P.: Security of Lattice-Based Data Hiding Against the Known Message Attack. IEEE Trans. Information Forensics and Security 1(4), 421–439 (2006)
6. Ni, J.Q., Zhang, R.Y., Fang, C., Huang, J.W., Wang, C.T.: Watermarking Security Incorporating Natural Scene Statistics. In: Solanki, K., Sullivan, K., Madhow, U. (eds.) IH 2008. LNCS, vol. 5284, pp. 132–146. Springer, Heidelberg (2008)
7. Wainwright, M.J., Simoncelli, E.P.: Scale Mixtures of Gaussians and the Statistics of Natural Images. In: Advances in Neural Information Processing Systems (NIPS*99), vol. 12, pp. 855–861. MIT Press, Cambridge (2000)
8. van Trees, H.L.: Detection, Estimation, and Modulation Theory. John Wiley and Sons, Chichester (1968)
9. Cover, T.M., Thomas, J.A.: Elements of Information Theory. Wiley series in Telecommunications (1991)
10. Portilla, J., Strela, V., Wainwright, M.J., Simoncelli, E.P.: Image Denoising Using Scale Mixtures of Gaussians in the Wavelet Domain. IEEE Trans. On Image Processing 12(11), 1338–1351 (2003)
11. Amari, S.I., Cardoso, J.F.: Blind Source Separation – Semiparametric Statistical Approach. IEEE Trans. Signal Processing 45(11), 2692–2700 (1997), Special issue on neural networks
12. Stoica, P., Ng, B.C.: On the Craner-Rao Bound Under Parametric Constraints. IEEE Signal Processing Lett. 5(7), 177–179 (1998)
13. D'Andrea, N.A., Mengli, U., Reggiannini, R.: The Modified Carmer-Rao Bound and its Application to Synchronization Problems. IEEE Trans. Commun. 42, 1391–1399 (1994)
14. Chou, Y.-l.: Statistical Analysis, Holt International (1969)
15. Gini, F., Reggiannini, R., Mengali, U.: The Modified Cramer-Rao Bound in Vector Parameter Estimation. IEEE Trans. Commun. 46(1), 52–60 (1998)

Video Watermarking Based on Spatio-temporal JND Profile

Dawen Xu[1,3], Rangding Wang[2], and Jicheng Wang[1]

[1] Department of Computer Science and Technology, Tongji University, Shanghai,
201804, China
[2] CKC software lab, Ningbo University, Ningbo, 315211, China
[3] School of Electronics and Information Engineering,
Ningbo University of Technology, Ningbo, 315016, China
dawenxu@126.com

Abstract. As digital multimedia become easier to copy, exchange and modify, digital watermarking is becoming more and more important for protecting the authenticity of multimedia. The most important properties of digital water-marking techniques are robustness, imperceptibility and complexity. In this paper, a video watermarking scheme based on spatio-temporal just noticeable difference (JND) profile of human visual system (HVS) is presented. The JND profile is used to adaptively adjust the energy of the watermark during embedding. Hence, a good balance between imperceptibility and robustness is obtained. Moreover, the complete watermarking procedure in spatial domain is easy to perform. Firstly, the original video frames are divided into 3D-blocks. Spatial JND profile for motionless block and spatio-temporal JND profile for fast-motion block are estimated respectively. Then, an identical watermark is embedded in motionless block utilizing the spatial JND of HVS. Independent watermarks are embedded in fast-motion block by exploiting the spatio-temporal JND of HVS. Experimental results show the watermarked frames are indistinguishable from the original frames subjectively and the proposed video watermarking scheme is robust against the attacks of additive Gaussian noise, frame dropping, frame averaging, filtering and lossy compression.

Keywords: Spatio-temporal JND profile, human visual system (HVS), video watermarking.

1 Introduction

Nowadays, with the thriving internet technology, how to protect the copyright of the digitalized information is becoming more and more attractive. Digital watermarking technology [1]-[3] has been proposed as an effective method to protect the authenticity of the intellectual property. The principle of watermarking is to embed a digital code (watermark) within the host multimedia, and to use such a code to prove ownership, to prevent illegal copying. The watermark code is embedded by making imperceptible modification to the digital data.

Digital watermarking has focused on still images for a long time but nowadays this trend seems to vanish. More and more watermarking methods are proposed for other multimedia data and in particular for video content. To ensure an effective watermarking

H.J. Kim, S. Katzenbeisser, and A.T.S. Ho (Eds.): IWDW 2008, LNCS 5450, pp. 327–341, 2009.

scheme, two fundamental requirements need to be satisfied: invisibility and robustness. Invisibility means that the embedded watermark should be perceptually invisible; in other words its presence should not affect the video quality. Robustness represents that the watermark should be robust against intentional or unintentional operations, including lossy compression, filtering, frame dropping, frame averaging, additive noise, etc [4].

It is known that human eyes cannot sense any changes below the just noticeable distortion (JND) threshold around a pixel due to their underlying spatial/temporal sensitivity and masking properties. An appropriate JND model can significantly help to improve the performance of video watermarking schemes. The method based on the spatial JND concept has been very successful in image watermarking [5], [6]. It can be easily exported to video with a straightforward frame-per-frame adaptation. However, the obtained watermark is not optimal in terms of visibility since it does not consider the temporal sensitivity of the human eye. Until now, the spatio-temporal JND profile of HVS is not exploited in video watermarking research, according to the survey of the authors.

In this paper, the spatio-temporal JND profile is estimated to embed the maximum possible watermark energy into video signals. Since spatial properties and temporal limits of HVS are fully utilized, the watermarking scheme can achieve maximum robustness without any visible distortion in videos. Furthermore, in order to resist frame averaging, frame dropping, frame swapping and statistical analysis, two types of watermark (identical and independent watermark) are embedded in motionless and fast-motion regions of the video respectively.

The rest of the paper is organized as follow. In Section 2, an overview of the main video watermarking algorithms developed until today is given. In Section 3, the perceptual model for estimating the spatio-temporal JND profile is described. In Section 4, the watermarking embedding and extraction algorithms are introduced. In Section 5, experimental results are presented aiming at demonstrating the validity of the proposed scheme. Finally, the conclusions are drawn in Section 6.

2   Related Prior Techniques

In terms of the domain in which the watermark is embedded, video watermarking techniques can be classified as spatial-domain, transform-domain or compressed-domain. In spatial domain watermarking systems, the watermark is embedded directly in the spatial domain (pixel domain) [7]. In transform domain watermarking systems, watermark insertion is done by transforming the video signal into the frequency domain using DFT [8], DCT [9] or DWT [10]. Compressed-domain video watermarking is especially attractive. Several methods have been designed to work with specific compression standards such as MPEG-2 [11], MPEG-4 [12] and H.264 [13]. Video watermarking algorithms as those presented above do not exploit the characteristics of HVS.

Some other methods simply use implicit spatial properties of HVS, such as luminance masking, spatial masking, and contrast masking. In [14], wavelet transform is applied to the video objects and the watermark is embedded to the coefficients by weighting with a visual mask, based on spatial characteristics. In [15], the luminance sensitivity of HVS is utilized for video object watermarking. Noorkami [16] employs a human visual model adapted for a 4×4 discrete cosine transform block to increase the payload and robustness while limiting visual distortion. In this method, luminance masking and contrast masking are exploited to obtain an image-dependent quantization matrix.

In addition, some techniques exploit explicitly the frequency masking characteristics of HVS to determine the strength of the watermark. In [17], a temporal wavelet transform is applied to each video shot and the watermark, weighted according to the spatial and frequency masking characteristic of HVS, is embedded to each wavelet coefficient frame. In [18], the watermark is embedded both into INTRA and INTER MBs. A masking method is adopted to limit visual artifacts in the watermarked VOPs and to improve, at the same time, the robustness of the system.

However, none of the former methods have focused on guaranteeing temporal invisibility and achieving maximum watermark strength along the temporal direction. In [19], temporal dimension is exploited for video watermarking by means of utilizing temporal sensitivity of HVS. However, this method utilizes not the temporal masking characteristics but the temporal contrast thresholds of HVS to determine the maximum strength of watermark. Furthermore, developing more sophisticated models which include temporal masking for the encoding of video sequences remains an open research problem.

3 Spatio-temporal JND Profile

The spatial JND of each pixel is approximated by the following equation [20, 21].

$$JND_s(x, y) = f_l(x, y) + f_t(x, y) - \alpha \cdot \min\{f_l(x, y), f_t(x, y)\} \tag{1}$$

where $f_l(x, y)$ and $f_t(x, y)$ represent the visibility thresholds due to background luminance masking and texture masking; and α ($0 < \alpha < 1$) accounts for the overlapping effect in masking. We choose $\alpha = 0.3$ as suggested in [21].

$f_l(x, y)$ is derived experimentally from a subjective test and can be expressed as [22]

$$f_l(x, y) = \begin{cases} 17\left(1 - \left(\dfrac{bg(x, y)}{127}\right)^{\frac{1}{2}}\right) + 3 & if \quad bg(x, y) \leq 127 \\[3mm] \dfrac{3}{128}(bg(x, y) - 127) + 3 & if \quad bg(x, y) > 127 \end{cases} \tag{2}$$

where $bg(x, y)$ is the average background luminance, i.e.:

$$bg(x, y) = \frac{1}{32}\sum_{i=1}^{5}\sum_{j=1}^{5} p(x - 3 + i, y - 3 + j) \cdot B(i, j) \tag{3}$$

where $p(x, y)$ denotes the pixel at (x, y); $B(i, j)$ is a weighted low-pass operator as shown in Fig.1.

$f_t(x, y)$ is approximated by the following equation [23].

$$f_t(x, y) = \beta \cdot mg(x, y) \tag{4}$$

where β is the control parameter; $mg(x, y)$ denotes the maximal weighted average of luminance gradients around the pixel at (x, y) [22]. We choose $\beta = 0.117$ as the authors of [23] recommend.

$$mg(x, y) = \max_{k=1,2,3,4} \{|grad_k(x, y)|\} \tag{5}$$

$$grad_k(x, y) = \frac{1}{16}\sum_{i=1}^{5}\sum_{j=1}^{5} p(x-3+i, y-3+j) \cdot G_k(i, j) \tag{6}$$

where $G_k(i, j)$ are four directional operators as shown in Fig.2.

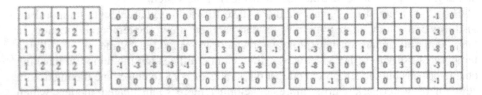

Fig. 1. Low-pass operator   **Fig. 2.** Four directional operators

Taking both spatial and temporal masking effects into consideration, the complete spatio-temporal JND is modeled as [20, 23]

$$JND_{S-T}(x, y, n) = f_d(ild(x, y, n)) \cdot JND_s(x, y, n) \tag{7}$$

where $JND_s(x, y, n)$ is the spatial JND of the n-th frame, f_d is a scale factor. The relationship between f_d and $ild(x, y, n)$ is illustrated in Fig.3. $ild(x, y, n)$ represents the average inter-frame luminance difference between the n-th and $(n-1)$-th frame.

$$ild(x, y, n) = \frac{1}{2}(p(x, y, n) - p(x, y, n-1) + bg(x, y, n) - bg(x, y, n-1)) \tag{8}$$

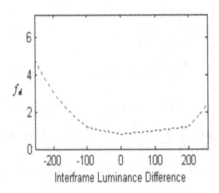

Fig. 3. Temporal masking effect

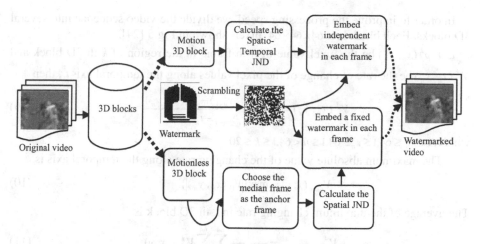

Fig. 4. Diagram of watermark embedding

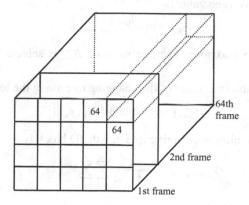

Fig. 5. 3D-blocks in video sequence

4 Proposed Method

4.1 Watermark Embedding

In order to achieve maximum robustness without any visible distortion in video signals, the spatio-temporal JND profile described in Section 3 is utilized in our watermarking scheme. In Fig.4, the watermark embedding scheme is shown.

Applying a fixed watermark to each frame in the video leads to the problem of maintaining statistical invisibility. However, applying independent watermarks to each frame also presents a problem if regions in each video frame remain little or no motion frame after frame. These motionless regions in successive video frames may be statistically compared or averaged to remove independent watermark [17]. Consequently, our scheme employs a fixed watermark for motionless regions and independent watermarks for fast motion regions. The scheme overcomes the aforementioned drawbacks associated with a fixed and independent watermarking procedure.

In order to improve the processing speed, we divide the video sequence into several 3D blocks. Each block has $64\times64\times64$ pixels shown in Fig.5 [24].

Let $p_t^k(x_n, y_n)$ be the pixel value of n-th frame in the region of k-th 3D block and $V_t^k(x_n, y_n)$ be the rate of change of the pixel values along the temporal axis t, then

$$V_t^k(x_n, y_n) = \frac{p_t^k(x_{n+1}, y_{n+1}) - p_t^k(x_n, y_n)}{t_{n+1} - t_n} \tag{9}$$

where, $1 \le x_n \le 64, 1 \le y_n \le 64, 1 \le n < 64, 1 \le k \le 20$.

The maximum absolute value of the changing rate along the temporal axis is

$$V_{max}^k(x, y) = \max_n\{|V_t^k(x_n, y_n)|\} \tag{10}$$

The average of the maximum changing rate in k-th 3D block is

$$V_{average\_max}^k = \frac{1}{64\times64}\sum_{x=1}^{64}\sum_{y=1}^{64}V_{max}^k(x, y) \tag{11}$$

Then the maximum average value is

$$B_{max} = \max_k\{V_{average\_max}^k\} \tag{12}$$

The 3D block with a maximum average value of B_{max} is selected as the fast-motion block.

The minimum absolute value of the changing rate along the temporal axis is

$$V_{min}^k(x, y) = \min_n\{|V_t^k(x_n, y_n)|\} \tag{13}$$

The average of the minimum changing rate in k-th 3D block is

$$V_{average\_min}^k = \frac{1}{64\times64}\sum_{x=1}^{64}\sum_{y=1}^{64}V_{min}^k(x, y) \tag{14}$$

The minimum average value is

$$B_{min} = \min_k\{V_{average\_min}^k\} \tag{15}$$

Accordingly, the 3D block with a minimum average value of B_{min} is selected as the motionless block.

Independent watermarks are embedded to the fast-motion block according to the following equation.

$$p_f^w(x, y, n) = \begin{cases} p_f(x, y, n) + JND_{S-T}(x, y, n) * w(x, y) & if \quad p_f(x, y, n) > JND_{S-T}(x, y, n) \\ p_f(x, y, n) & otherwise \end{cases} \tag{16}$$

where $p_f(x, y, n)$ is the pixel value of n-th frame in the fast-motion block, $p_f^w(x, y, n)$ is the watermarked pixel value, $JND_{S-T}(x, y, n)$ is the spatio-temporal JND in equation (7), $w(x, y)$ refers to the bipolar watermark. In this paper, the watermark is a binary image with size 64-by-64 shown in Fig.9. In order to improve the robustness, the watermark is scrambled by Arnold transform [25]. Then we map the binary watermark {0,1} into bipolar watermark {−1,+1}.

A fixed watermark is embedded to the motionless block described as follows:

$$p_s^w(x,y,n) = \begin{cases} p_s(x,y,n) + JND_s(x,y)*w(x,y) & if \quad p_s(x,y,n) > JND_s(x,y) \\ p_s(x,y,n) & otherwise \end{cases} \quad (17)$$

where $p_s(x,y,n)$ is the pixel value of n-th frame in the motionless block, $p_s^w(x,y,n)$ is the watermarked pixel value, $JND_s(x,y)$ is the spatial JND in equation (1). Here, we choose the median frame in the motionless block as the anchor frame. $JND_s(x,y)$ is the spatial JND of the anchor frame.

4.2 Watermark Extraction

Watermark extraction is carried out in three steps. This procedure requires both the original and the watermarked video.

1) The watermarked video should be divided into 3D blocks. The fast-motion block and the motionless block are obtained. This step is analogous to the embedding process.

2) The watermark signal can be extracted from both the fast-motion block and the motionless block. The watermark is extracted by the following formulas:

$$r(x,y) = p^*(x,y,n) - p(x,y,n) \qquad (18)$$

$$w^*(x,y) = sign(r(x,y)) \qquad (19)$$

where $p^*(x,y,n)$ is the watermarked pixel value of n-th frame in the fast-motion block or the motionless block, $p(x,y,n)$ is the original pixel value, $sign(\cdot)$ is a mathematical function that extracts the sign of a real number, $w^*(x,y)$ is the extracted bipolar watermark.

3) The extracted bipolar watermark $w^*(x,y)$ should be mapped into the binary watermark. Then inverse Arnold transform is performed to obtain the recovered watermark.

5 Experimental Results

Two kinds of tests were performed on three different video sequences for proving on one side the invisibility of the embedded watermark and on the other side the robustness against all processing which does not seriously degrade the quality of the video. The standard video sequences "Football", "Vectra" and "Foreman" were used for simulation. Each frame is of size 352×288. Only the luminance component is considered in our video watermarking process. The first 64 frames from each video are used for our tests.

5.1 Visibility Experiments

To evaluate the quality of the watermarked videos, a series of tests has been performed in which the original video and the watermarked video are displayed to the viewer. The displayed order was randomly selected as (original, watermarked) or (watermarked, original). The viewer was asked to select the video which has better quality. The embedded watermark appears perceptually undetectable and each video was selected approximately 50% of the time.

An original frame from each video is shown in Fig.6. The corresponding water-marked frames are shown in Fig. 7. In both cases, the watermarked frame appears visually identical to the original. Fig. 8 illustrates the PSNR values of 64 watermarked frames. The simulation results show that the PSNR of each watermarked frame is greater than 39dB.

| (a) "Football" (8th frame) | (b) "Vectra" (8th frame) | (c) "Foreman" (6th frame) |

Fig. 6. Original frames

| (a) "Football" (PSNR=43.79) | (b) "Vectra" (PSNR=40.34) | (c) "Foreman" (PSNR=44.70) |

Fig. 7. Watermarked frames

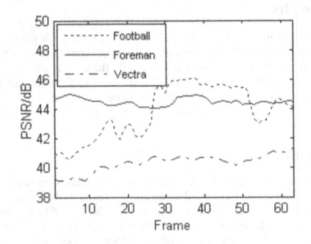

Fig. 8. PSNR of watermarked frames   **Fig. 9.** Original watermark

5.2 Robustness Experiments

Several experiments have been performed to evaluate the performance of the proposed watermarking scheme. For comparing the similarities between the original and extracted watermark signals, a normalized cross-correlation function [10] is employed as

$$NC(w, w^*) = \frac{\sum\limits_{x=0}^{N-1}\sum\limits_{y=0}^{M-1} w(x,y)w^*(x,y)}{\sqrt{\sum\limits_{x=0}^{N-1}\sum\limits_{y=0}^{M-1}[w(x,y)]^2}\sqrt{\sum\limits_{x=0}^{N-1}\sum\limits_{y=0}^{M-1}[w^*(x,y)]^2}} \tag{20}$$

where $w(x,y)$ and $w^*(x,y)$ are the original and extracted watermarks of size $N \times M$, respectively.

1) Frame Averaging and Statistical Analysis: Frame averaging is a significant attack to the video watermark. When attackers collect a number of watermarked frames, they can estimate the watermark by statistical averaging and remove it from the watermarked video [26], [27]. This scheme is more robust to frame averaging attacks since it embeds both a fixed watermark for motionless regions and independent watermarks for fast motion regions. The fixed watermark can prevent pirates from removing the watermark by statistical comparing and averaging the motionless regions in successive video frames. Independent watermarks used for fast motion regions can prevent pirates from colluding with frames from different scenes to detect and destroy the watermark.

In this experiment, we use the average of current frame and its two nearest neighbors to replace the current frame, which are formulated as:

$$p_k'(x,y) = \frac{p_{k-1}(x,y) + p_k(x,y) + p_{k+1}(x,y)}{3} \quad k = 2,3,\cdots,n-1 \tag{21}$$

We employ the fixed watermark to test the robustness and the results are shown in Table.1. It is found that the proposed scheme can resist to statistical averaging quite well.

Table 1. Results after frame averaging attack

	football	foreman	vectra
PSNR/dB	22.8956	26.5871	32.2945
NC	0.9324	0.9982	1.00
Recovered watermark			

2) Filtering: The extracted watermarks after applying median filter (3×3) are shown in Table.2. Experimental results demonstrate that the scheme is robust against median filtering ($NC > 0.9$).

Table 2. Results after median filtering attack

	football	foreman	vectra
PSNR/dB	33.5004	32.0573	32.2082
NC	0.9110	0.9577	0.9035
Recovered watermark			

3) Gaussian Noise: Gaussian noise with different strength is added to the water-marked video. Fig.10 shows the *NC* values between the original and the extracted watermarks. It is found that the video frames have visible distortion when the PSNR decreases down to 30dB, but the extracted watermark still can be distinguished (Fig.11).

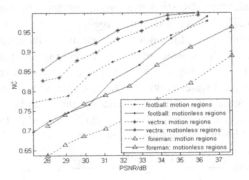

Fig. 10. NC values after noise attack

4) Frame Dropping: As a video scene consists of a large number of identical frames, frame dropping can be performed without significantly degrading the quality. How-ever, since the proposed method embeds an identical signal in each frame, it prevents the attackers from removing the watermark by frame dropping. The watermark still can be extracted even after deleting all watermarked frames, leaving only on frame. In order to remove the watermark, they should remove the whole trunk of frames, which would degrade the video quality significantly. As shown in Fig.12, the proposed method achieves high robustness against frame dropping.

5) Lossy Compression: MPEG Lossy compression is an operation that eliminates perceptually non-significant portions of the video signals. In most applications in-volving storage and transmission of digital video, a lossy compression is performed to reduce bit rates and increase efficiency. The performance of the watermarking system to survive MPEG lossy compression at very low quality is tested. Table.3 shows the results after MEPG-4 lossy compression. The tests show that the watermark is still retrievable.

(a) "Football"   8th frame (PSNR=30.2012dB)   Recovered watermark (NC=0.8421)

(b) "Vectra"   8th frame (PSNR=30.2012dB)   Recovered watermark (NC=0.9226)

(c) "Foreman"   6th frame (PSNR=31.0561dB)   Recovered watermark (NC=0.7900)

Fig. 11. Frame from videos with Gaussian noise

According to the above experimental results, the watermarking scheme is robust against a variety of common video-processing attacks such as median filtering, frame averaging, additive noise, frame dropping, and MEPG lossy compression. The extracted watermark is high similar with the original watermark.

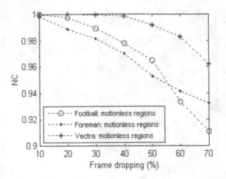

Fig. 12. Results after frame dropping attack

Table 3. Results after MEPG-4 compression attack

	football	foreman	vectra
PSNR/dB	34.9823	36.4237	36.3881
NC	0.8278	0.8550	0.9413
Recovered watermark			

For comparison purpose, two classical methods are implemented: The first method [17] exploits spatial masking, frequency masking, and temporal properties to embed an invisible watermark. The other method [4] embeds the watermark into the pixel values of a line-scanned video. The amplitude factor is varied according to spatial masking properties of HVS. The results are shown in Fig.13. The number in x-coordinate denotes *Gaussian Noise adding* (PSNR=30dB), *frame averaging*, *MPEG-4 compression*, and *frame dropping* (25%) respectively. It can be observed that the detection performance against frame averaging, MPEG-4 compression and frame dropping is improved in the proposed method.

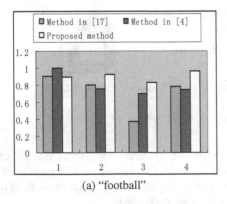

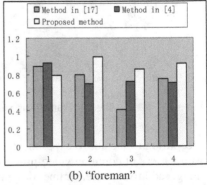

(a) "football"   (b) "foreman"

Fig. 13. Robustness (NC) comparison of several methods

5.3 Computational Complexity Analysis

The proposed watermarking scheme is performed without high computational complexity and difficulty in embedding information. Very simple operations are performed on the video, neither frequency transforms nor spatial filtering are used. The most time consuming operation is the sorting of fast-motion and slow-motion regions. The other costly operation is the computing of the spatio-temporal JND threshold. Watermark embedding/extraction are simple and easy to implement.

6 Conclusion

An important task in designing a video watermarking system is to achieve a good trade-off between invisibility and robustness. In this paper, the watermarking procedure explicitly exploits the spatio-temporal JND of HVS to guarantee that the embedded watermark is imperceptible. Watermark embedding and extraction are performed in the spatial domain directly. Furthermore, dual watermarks are embedded into fast-motion and motionless regions for robustness against the attacks of frame dropping, averaging and statistical analysis. Experimental results show that the proposed scheme can perform robustly against MPEG lossy compression, filtering, frame averaging, frame dropping and Gaussian noise attacks.

Our system shows quite good robustness to these common attacks and simultaneously obtains a good balance between imperceptibility and robustness. In the case of video watermarking, advances in understanding the psycho-visual aspects of spatio-temporal masking could make a significant impact on designing more effective watermarking schemes. However, the current video watermarking scheme is not robust to geometric attack such as rotation and scaling. Progress against this and other possible attacks will be the major direction of our future work.

Acknowledgements

This work is supported by the National Natural Science Foundation of China (No. 60672070) and the Science and Technology Project of Zhejiang Province (2007C21G2070004).

References

1. Cox, I.J., Kilian, J., Leighton, T., Shamoon, T.: Secure spread spectrum watermarking for multimedia. IEEE Transactions on Image Processing 6(12), 1673–1687 (1997)
2. Langelaar, G.C., Setyawan, I., Lagendijk, R.L.: Watermarking digital image and video data: A satate-of-the-art overview. IEEE Signal Processing Magazine 17(5), 20–46 (2000)
3. Huang, J.W., Shi, Y.Q., Shi, Y.: Embedding image watermarks in DC components. IEEE Transactions on Circuits and Systems for Video Technology 10(6), 974–979 (2000)
4. Hartung, F., Girod, B.: Watermarking of uncompressed and compressed video. Signal Processing 66(3), 283–301 (1998)

5. Li, Q., Cox, I.J.: Using perceptual models to improve fidelity and provide resistance to valumetric scaling for quantization index modulation watermarking. IEEE Transactions on Information Forensics and Security 2(2), 127–139 (2007)
6. Liu, W., Dong, L.N., Zeng, W.J.: Optimum detection for spread-spectrum watermarking that employs self-masking. IEEE Transactions on Information Forensics and Security 2(4), 645–654 (2007)
7. Huang, H.Y., Lin, Y.R., Hsu, W.H.: Robust technique for watermark embedding in a video stream based on a block-matching algorithm. Optical Engineering 47(3), 037402 (14 pages) (2008)
8. Deguillarme, F., Csurka, G., Ruanaidh, J.O., Pun, T.: Robust 3D DFT video watermarking. In: Proceedings of SPIE Security and Watermarking of Multimedia Contents, San Jose, California, vol. 3657, pp. 113–124 (1999)
9. Hsu, C.T., Wu, J.L.: DCT-based watermarking for video. IEEE Transactions on Consumer Electronic 44(1), 206–216 (1998)
10. Xu, D.W.: A blind video watermarking algorithm based on 3D wavelet transform. In: 2007 International Conference on Computational Intelligence and Security, Harbin, China, December 2007, pp. 945–949 (2007)
11. Biswas, S., Das, S.R., Petriu, E.M.: An adaptive compressed MPEG-2 video watermarking scheme. IEEE Transactions on Instrumentation and Measurement 54(5), 1853–1861 (2005)
12. Alattar, A.M., Lin, E.T., Celik, M.U.: Digital watermarking of low bit-rate advanced simple profile MPEG-4 compressed video. IEEE Transactions on Circuits and Systems for Video Technology 13(8), 787–800 (2003)
13. Zhang, J., Anthony, T.S.H., Qiu, G., Marziliano, P.: Robust video watermarking of H.264/AVC. IEEE Transactions on Circuits and Systems II: Express Briefs 54(2), 205–209 (2007)
14. Piva, A., Caldelli, R., Rosa, A.D.: A DWT-based object watermarking system for MPEG-4 video streams. In: International Conference on image processing, Vancouver, Canada, September 2000, vol. 3, pp. 5–8 (2000)
15. Bas, P., Macq, B.: A New Video-Object Watermarking Scheme Robust To Object Manipulation. In: International Conference on image processing, Thessaloniki, Greece, pp. 526–529 (2001)
16. Noorkami, M., Mersereau, R.M.: A framework for robust watermarking of h.264-encoded video with controllable detection performance. IEEE Transactions on Information Forensics and Security 2(1), 14–23 (2007)
17. Swanson, M.D., Zhu, B., Tewfik, A.H.: Multiresolution scene-based video watermarking using perceptual models. IEEE Journal on Selected Areas in Communications 16(4), 540–550 (1998)
18. Barni, M., Bartolini, F., Checcacci, N.: Watermarking of MPEG-4 video objects. IEEE Transactions on Multimedia 7(1), 23–32 (2005)
19. Koz, A., Alatan, A.A.: Oblivious spatio-temporal watermarking of digital video by exploiting the human visual system. IEEE Transactions on Circuits and Systems for Video Technology 18(3), 326–337 (2008)
20. Yang, X.K., Lin, W.S., Lu, Z.K., Ong, E.P., Yao, S.S.: Motion-compensated residue preprocessing in video coding based on just-noticeable-distortion profile. IEEE Transactions on Circuits and Systems for Video Technology 15(6), 742–752 (2005)
21. Yang, X.K., Ling, W.S., Lu, Z.K., Ong, E.P., Yao, S.S.: Just noticeable distortion model and its applications in video coding. Signal Processing: Image Communication 20(7), 662–680 (2005)

22. Chou, C.H., Li, Y.C.: A perceptually tuned subband image coder based on the measure of just-noticeable-distortion profile. IEEE Transactions on Circuits and Systems for Video Technology 5(6), 467–476 (1995)
23. Chou, C.H., Chen, C.W.: A perceptually optimized 3-D subband codec for video communication over wireless channels. IEEE Transactions on Circuits and Systems for Video Technology 6(2), 143–156 (1996)
24. Niu, X.M., Sun, S.H., Xiang, W.J.: Multiresolution watermarking for video based on gray-level digital watermark. IEEE Transactions on Consumer Electronics 46(2), 375–384 (2000)
25. Wang, C.K., Wang, J.M., Zhou, M., Chen, G.S.: ATBaM: An Arnold transform based method on watermarking relational data. In: 2008 International Conference on Multimedia and Ubiquitous Engineering, Seoul, Korea, April 2008, pp. 263–270 (2008)
26. Su, K., Kundur, D., Hatzinakos, D.: Statistical invisibility for collusion-resistant digital video watermarking. IEEE Transactions on Multimedia 7(1), 43–51 (2005)
27. Su, K., Kundur, D., Hatzinakos, D.: Spatially localized image-dependent watermarking for statistical invisibility and collusion resistance. IEEE Transactions on Multimedia 7(1), 52–66 (2005)

On the Performance of Wavelet Decomposition Steganalysis with JSteg Steganography

Ainuddin Wahid Abdul Wahab[1,2], Johann A. Briffa[1],
and Hans Georg Schaathun[1]

[1] Department of Computing, University of Surrey
[2] Faculty of Computer Science and Information Technology, University of Malaya

Abstract. In this paper, we study the wavelet decomposition based steganalysis technique due to Lyu and Farid. Specifically we focus on its performance with JSteg steganograpy. It has been claimed that the Lyu-Farid technique can defeat JSteg; we confirm this using different images for the training and test sets of the SVM classifier. We also show that the technique heavily depends on the characteristics of training and test set. This is a problem for real-world implementation since the image source cannot necessarily be determined. With a wide range of image sources, training the classifier becomes problematic. By focusing only on different camera makes we show that steganalysis performances significantly less effective for cover images from certain sources.

1   Introduction

Steganography allows a user to hide a secret message in such a way that an adversary cannot even detect the existence of the message. We are concerned with image steganography, where the secret message is represented by imperceptible modification in a cover image.

Over the last decade a wide variety of steganography techniques have appeared in the literature. In response, there have also been a wide variety of steganalysis techniques, intended to let an adversary determine whether an intercepted image contains a secret message or not. In particular, a number of techniques based on machine learning have emerged. Such techniques tend to be blind, in the sense that they do not assume any particular steganography algorithm and can usually break a variety of algorithms.

In this paper we consider a steganalysis technique due to Lyu and Farid [12]. This technique, claimed to break a variety of steganography systems including Jsteg and Outguess was published in [12].

The idea in machine learning is that the steganalysis algorithm, during a training phase, is given large sets of steganograms and natural images. These images, where the classification (steganogram or natural image) is known are used to tune the classification parameters. When an unknown image is subsequently presented to the system, it is classified according to these pre-tuned parameters.

H.J. Kim, S. Katzenbeisser, and A.T.S. Ho (Eds.): IWDW 2008, LNCS 5450, pp. 342–349, 2009.

Typically a set of professional photographs from an online database is used. However, if the algorithm is later used to analyze images from a different source, the training data may not be relevant. In particular, sensor noise from cheap cameras may have characteristics similar to an embedded message. The same could be the case for images with low quality due to poor lighting or photography skills.

In this paper we confirm this hypothesis by simulating embedding and analysis of images from different sources. We conclude that further research is needed to make the Lyu-Farid algorithm reliable for real-world applications where the image source cannot necessarily be determined. This is due to the fact that the Lyu-Farid algorithm requires detailed information about cover source.

2   The Lyu-Farid Algorithm

In this section we explain how to implement the Lyu-Farid algorithm. The algorithm uses a Support Vector Machine (SVM) classification. SVM does not operate on the images themselves. Instead a *feature vector* (i.e. a series of statistics) are extracted from the image to be used by the SVM.

The features used in the steganalysis are extracted from the wavelet domain, so we will first present the wavelet transform and then the exact features used. Subsequently we will introduce the SVM.

2.1   The Wavelet Transform

A wavelet is a waveform of limited duration with an average value of zero. One dimensional wavelet analysis decomposes a signal into basis functions which are shifted and scaled versions of a original wavelet [16]. Continuous Wavelet Transform (CWT) is usually used for time continuous signals while Discrete Wavelet Transform (DWT) is used when the signals is sampled, as in digital image processing.

Besides its usefulness in image processing, the decomposition also exhibits statistical regularities that can be exploited for certain purposes such as steganalysis. The decomposition technique used in our experiment is based on quadrature mirror filter (QMFs) [16]. The decomposition process splits the frequency space into multiple scales and orientations (vertical, horizontal, diagonal and lowpass subband). This is achieved by applying separable lowpass and highpass filter along image axes.

2.2   Wavelet Decomposition Steganalysis

Two set of statistics (basic coefficient distribution and errors in an optimal linear predictor of coefficient magnitude)are then collected. Those composed of mean, variance, skewness and kurtosis of the subband coefficient for different orientation and scales(s). The total of $12(s-1)$ for each set of statistics collected. From [13], s value is four. Based on that, 72 individual statistics are generated.

The collected statistics are then used as a feature vector to discriminate between clean (image without payload encoded into them) and stego (image with payload encoded into them) images using Support Vector Machine (SVM).

2.3   Support Vector Machine

SVM performs the classification by creating a hyperplane that separates the data into two categories in the most optimal way.

In our experiment, the SVM is employed to classify each image as either clean or stego, based on its feature vector. For our experiment, the feature vector are constructed from the wavelet decomposition process discussed previously where each image is represented with the 72 individual statistics. SVM has been shown to provide good results in [12]. Classification requires training and testing data sets. In the training set, each individual instance has a class label value and several attributes or features (feature vector). SVM will produce a model based on training set data, and then using that model to predict the class of testing set based only on their attributes.

2.4   SVM Kernel

Lyu and Farid in [12] have been using Radial Basis Function (RBF) kernel for SVM. RBF kernel function:

$$K(X_i, X_j) = \exp\left(-\gamma||X_i - X_j||^2\right), \gamma > 0$$

where γ is a kernel parameter.

According to [8], there are four basic kernel in SVM namely Linear, Polynomial, RBF and Sigmoid. The first three basic functions has been tested with the input from this experiment for clarification on why RBF used as SVM kernel in [12]. The details of the image sets or input for this experiment are discussed in section 4. Basically, the training images is the combination of images from all three devices (Table 2) while the test set is from 'Canon' and 'Sony' set of images. Both the training and test set are combination of clean and corresponding JSteg stego images (image with payload encoded using JSteg). From the result in Table 1, it can be seen that RBF really provide the best results for both set of test images. Besides that, the suggestion on how to select the best kernel in [8] also indicate that RBF is the best kernel to be used for wavelet decomposition type of data.

3   JSteg

JSteg [17] is a steganographic method for JPEG images that can be viewed as LSB steganography. It works by embedding message bits as the LSBs of the quantized DCT (Discrete Cosine Transform) coefficients. All coefficients values of '0' and '1' will be skipped during JSteg embedding process which can be perform either in sequential or random location. Quite a number of attacks have

Table 1. SVM Kernel Test

SVM Kernel	Canon			Sony		
	Linear	Polynomial	RBF	Linear	Polynomial	RBF
False Negative	2.0%	2.0%	5.0%	7.0%	19.0%	6.0%
False Positive	18.0%	19%	3.0%	1.0%	0%	0%
Detection Rate	80.0%	79.0%	92.0%	92.2%	81.0 %	94.0%

been used to defeat JSteg such as chi-square attack [1] and generalize chi-square attack [15]. Futhermore, Lyu and Farid in [12] have shown that their wavelet decomposition steganalysis technique is able to defeat JSteg.

4   Detection Experiment

To evaluate the performance of the steganalysis algorithm on JSteg, we use sample images captured using Canon digital camera, Nokia N70 mobile phone and Sony video camera. The details of devices used to capture image given in Table 2.

As in [5], cover grayscale images are used due to the fact that steganalysis are harder for grayscale images. All images were crop to the center, resulting in image sizes of 640x480 pixels and saved in JPEG image format with quality factor of 75. This can help to ensure that the image dimensions is not correlated with spatial characteristics, such as noise or local energy as what mentioned in [3].

The images keep at fixed size of 640x480 to ensure that the collected statistics will be at the same amount for each image since it has been found in [10] and [3] that detection performance is likely to suffer for smaller images.

Following the image preparation process, all the images went through the JSteg encoding process [17] to produce a set of stego images. The secret message is an image with size of 128x128. With the cover and stego images ready, the wavelet decomposition steganalysis technique conducted by using SVM [14] as classifier.

For SVM, the soft-margin parameter used with its default value of 0. The only parameter tuned to get the best result is the kernel's parameter, γ where

$$\gamma \in \{2^i\}, i \in \{0, 1, 2, 3\}$$

The total number of images used for training and testing is 400 for each set of images from different camera. Besides the three main set of images (Canon, Nokia and Sony), there is another two sets of images (Comb.200 and Comb.600) which contains a combination of images from each type of camera. These two sets, 'Comb.200' and 'Comb.600', have a total number of 400 and 1200 images for training and testing accordingly.

For reference purpose, the sixth set of images included in our experiment. It is a set of images from Greenspun's database [7] which consist of a balanced

Table 2. Device Specification

Device	Model	Resolution	Additional Info
Sony	DCR-SR42	1 MP	Digital video camera
Nokia	N70	2MP	Build in phone camera
Canon	Powershot A550	7.1MP	Canon Digital Camera

combination of indoor and outdoor images. The total number of images for training and testing for this set is 400. This database is the source of images used by Lyu in [12].

5 Discussion

Classification accuracy used in [4] and [12] to measure the performance of their proposed method while in [6] the performance were evaluated using 'detection reliability' ρ defined as $\rho = 2A - 1$, where A is the area under the Receiver Operating Characteristics (ROC) curve, also called an accuracy.

For our experiment, the results in Tables 3, 4 and 5 showing the performance of SVM using false negative and false positive rate (Table 3) together with classification precision (Table 4) and classification accuracy (Table 5). Those results confirm existing claims that te Lyu-Farid technique can be used to defeat JSteg. The detection rate for using the same source of images for training and test sets match with claims in [12]. While showing the success of Lyu-Farid technique, these results also shows that the accuracy of the technique is seriously affected by the training set used.

By using confidence interval estimation technique [2], from the results, we have computed 95.4% confidence intervals for the false negative rate for Canon (3.2%, 10.2%), Nokia (14.6%, 26.0%), and Greenspun (20.7%, 33.3%). Thus, we can confidently conclude that the steganalysis algorithm is significantly less effective for cover images from certain sources.

In [10] it has been found that JPEG image quality factor affects the steganalyzer performance where cover and stego images with high quality factors are less distinguishable than cover and stego image with lower quality. Furthermore, Böhme in [3] also found that images with noisy texture yield the least accurate stego detection. Related to those results, in our experiment, while having the same quality factor and using the same steganography technique, it has been shown that images captured using high resolution devices (Canon) are more distinguishable than cover and stego image from a low resolution device (Sony).

From [9], it has been observed that a trained steganalyzer using specific embedding technique performs well when tested on stego images from that embedding technique, but it performs quite inaccurately if it is asked to classify stego image obtained from another embedding technique. In their experiment, when steganalyzer trained solely on the Outguess (+) stego images, and asked to distinguish between cover and Outguess (+) images, it obtains an accuracy

Table 3. False Negative | False Positive (False Alarm)

Test Set	Training Set					
	Canon	Nokia	Sony	Comb.200	Comb.600	Greenspun
Canon	7.6% | 6.7%	28.0% | 2.7%	1.6% | 60.3%	2.0% | 9.0%	1.6% | 6.3%	31.0% | 2.0%
Nokia	42.3% | 3.0%	20.3% | 4.6%	15.6% | 36.6%	7.0% | 19.0%	7.3% | 11.0%	61.0% | 3.0%
Sony	64.0% | 6.7%	53.6% | 3.3%	3.3% | 4.6%	13.0% | 3.6%	8.6% | 1.3%	71.0% | 1.0%
Comb.200	40.3% | 1.0%	29.6% | 4.3%	8.0% | 30.6%	8.0% | 14.3%	9.0% | 7.6%	62.0% | 0%
Comb.600	38.0% | 0.7%	26.3% | 4.5%	7.3% | 31.3%	7.0% | 13.7%	8.6% | 3.0%	50.3% | 2.3%
Greenspun	66.0% | 9.0%	55.0% | 30.0%	6.0% | 86.0%	10.0% | 64.0%	10.0% | 61.0%	27.0% | 7.0%

Table 4. Precision

True Positive (True Positive + False Positive)

Test Set	Training Set					
	Canon	Nokia	Sony	Comb.200	Comb.600	Greenspun
Canon	93.2%	96.4%	62.0%	91.6%	94.0%	97.2%
Nokia	95.0%	94.5%	69.8%	83.0%	89.4%	92.9%
Sony	84.3%	93.4%	95.5%	96.0%	98.6%	96.7%
Comb.200	98.4%	94.2%	75.0%	86.5%	92.3%	100.0%
Comb.600	98.9%	94.2%	74.8%	87.2%	96.8%	95.6%
Greenspun	87.7%	74.0%	54.4%	67.6%	67.8%	91.3%

Table 5. Detection Rate (Accuracy)

(True Positive + True Negative) / (Total Positive + Total Negative)

Test Set	Training Set					
	Canon	Nokia	Sony	Comb.200	Comb.600	Greenspun
Canon	92.9%	84.7%	69.1%	94.5%	96.1%	83.5%
Nokia	77.4%	87.6%	73.9%	87.0%	90.9%	68.0%
Sony	64.7%	71.2%	96.1%	91.7%	95.1%	64.0%
Comb.200	79.4%	83.1%	80.7%	88.9%	91.7%	69.0%
Comb.600	80.7%	84.6%	80.7%	89.7%	94.2%	73.7%
Greenspun	78.0%	76.0%	58.0%	74.5%	75.5%	83.0%

of 98.49%. But, its accuracy for distinguishing cover images from F5 and Model Based images is 54.37% and 66.45%, respectively.

Having the same pattern, in our experiment, by covering all types of image sources while training the SVM, the technique can be seen to have a good detection rate. However, the performance decrease when the test image type is not previously in its training set. The clearest example is when SVM trained with 'Sony' images and then tested with 'Nokia' and 'Sony'. With detection rate of 96.1% for the 'Sony' test images the detection rate are lower for 'Canon' and 'Nokia' images with rates of 69.1% and 73.9% accordingly.

The Lyu-Farid technique also seems not to perform well when trained with images from higher resolution camera and then tested with lower resolution camera. For example, it can be seen when the SVM trained using images from 'Canon' and tested with images from 'Nokia' and 'Sony'. While having an accuracy of 92.9% for 'Canon' test set, the accuracy decreased to 77.4% and 64.7% respectively with 'Nokia' and 'Sony' test sets.

The number of images in training set also plays an important role to ensure that the SVM are well trained. This can be seen clearly at the differences of accuracy rate when the SVM trained using images from 'Comb.200' and 'Comb.600'. With assumption that SVM is not well trained using smaller number of images ('Comb.200' training set), the accuracy rate can be seen increased when the SVM trained with bigger number of images ('Comb.600' training set).

The problem with the above situations is the practicality for the real-world implementation of the technique. There is a huge diversity of image sources in the world today. Kharrazi in [9] has demonstrated how the computational time increases rapidly as the training set size increases. In his experiment, by having training set consists of 110 000 images, it would take more than 11 hours to design or train the non-linear SVM classifier. In our case for example, if there is new source of image found or designed, then the SVM classifier has to be retrained. If we try to cover all possible image sources, we can imagine how long it would take to retrain the SVM classifier.

Also related to image source diversity, is the attempt to train the classifier with all possible type of images in the public domain. Some researchers are trying this using a million images in the training set [9].

6   Conclusion

In our experiments we investigated the performance of Lyu-Farid steganalysis on JSteg using cover images from different sources. Our experiments show that performance claims previously made have to assume knowledge about the cover source. If the steganalyst is not able to train the SVM using cover images from a very similar source, significant error rates must be expected.

In the case of Jsteg, Lyu-Farid seems to get reasonable performance if a large and mixed training set is used (as in the Comb.600 set). A training set to enable reliable detection of any steganogram using any cover source would probably have to be enormous.

Even when we have been able to train on images from the correct source, we observe that some sources make steganalysis difficult. Images from the Nokia phone and from the Greenspun database have significantly higher false negative rates than other sources. An interesting open question is to identify optimal cover sources from the steganographer's point of view.

It may very well be possible to design steganographic algorithms whose statistical artifacts are insignificant compared to the artifacts of a particular camera. This is to some extent the case for SSIS [11] which aims to mimic sensor noise which is present (more or less) in any camera.

On the steganalytic case, quantitative results would be more interesting if real world implementation is considered. How large does the training set have to be to handle any source? Are there other important characteristics of the image which must be addressed, such as the ISO setting of the camera, lighting conditions and also indoor versus outdoor?

References

1. Westfeld, A., Pfitzmann, A.: Attacks on steganographic systems. In: IHW 1999 (1999)
2. Bhattacharyya, G.K., Johnson, R.A.: Statistical Concepts and Methods. Wiley, Chichester (1977)
3. Böhme, R.: Assessment of steganalytic methods using multiple regression models. In: Barni, M., Herrera-Joancomartí, J., Katzenbeisser, S., Pérez-González, F. (eds.) IH 2005. LNCS, vol. 3727, pp. 278–295. Springer, Heidelberg (2005)
4. Farid, H.: Detecting hidden messages using higher-order statistical models. In: International Conference on Image Processing, Rochester, NY (2002)
5. Fridrich, J., Pevný, T., Kodovský, J.: Statistically undetectable jpeg steganography: dead ends challenges, and opportunities. In: MM&Sec 2007: Proceedings of the 9th workshop on Multimedia & security, pp. 3–14. ACM, New York (2007)
6. Fridrich, J.J.: Feature-based steganalysis for jpeg images and its implications for future design of steganographic schemes. In: Fridrich, J. (ed.) IH 2004. LNCS, vol. 3200, pp. 67–81. Springer, Heidelberg (2004)
7. Greenspun, P.: Philip greenspun's home page (2008),
   http://philip.greenspun.com/
8. Hsu, C.W., Chang, C.C., Lin, C.J.: A practical guide to support vector classification. Technical report, Taipei (March 2008),
   http://www.csie.ntu.edu.tw/~cjlin/papers/guide/guide.pdf
9. Kharrazi, M., Sencar, H.T., Memon, N.: Improving steganalysis by fusion techniques: A case study with image steganography. In: Shi, Y.Q. (ed.) Transactions on Data Hiding and Multimedia Security I. LNCS, vol. 4300, pp. 123–137. Springer, Heidelberg (2006)
10. Kharrazi, M., Sencar, H.T., Memon, N.: Performance study of common image steganography and steganalysis techniques. Journal of Electronic Imaging 15(4) (2006)
11. Marvel, L.M., Boncelet, C.G., Retter, C.T.: Spread spectrum image steganography. IEEE Transactions On Image Processing 8(1), 1075–1083 (1999)
12. Lyu, S., Farid, H.: Detecting hidden messages using higher-order statistics and support vector machines. In: Petitcolas, F.A.P. (ed.) IH 2002. LNCS, vol. 2578, pp. 340–354. Springer, Heidelberg (2003)
13. Lyu, S., Farid, H.: Steganalysis using higher-order image statistics. IEEE Transactions on Information Forensics and Security 1(1), 111–119 (2006)
14. Noble, W.S., Pavlidis, P.: Gist:support vector machine,
    http://svm.nbcr.net/cgi-bin/nph-SVMsubmit.cgi
15. Provos, N., Honeyman, P.: Detecting steganographic content on the internet. CITI Technical Report, pp. 01–11 (2001)
16. Strang, G., Nguyen, T.: Wavelets and Filter Banks. Wellesley-Cambridge Press (1996)
17. Upham, D.: Jpeg-jsteg-v4, http://www.nic.funet.fi/pub/crypt/steganography

Balanced Multiwavelets Based Digital Image Watermarking*

Na Zhang[1], Hua Huang[1], Quan Zhou[2], and Chun Qi[1]

[1] School of Electronics and Information Engineering, Xi'an Jiaotong University, P.R. China
[2] National Key Lab of Space Microwave Technology, P.R. China
nzhang@stu.xjtu.edu.cn

Abstract. In this paper, an adaptive blind watermarking algorithm based on balanced multiwavelets transform is proposed. According to the properties of balanced multiwavelets and human vision system, a modified version of the well-established Lewis perceptual model is given. Therefore, the strength of embedded watermark is controlled by the local properties of the host image .The subbands of balanced multiwavelets transformation are similar to each other in the same scale, so the most similar subbands are chosen to embed the watermark by modifying the relation of the two subbands adaptively under the model, the watermark extraction can be performed without original image. Experimental results show that the watermarked images look visually identical to the original ones, and the watermark also successfully survives after image processing operations such as image cropping, scaling, filtering and JPEG compression.

Keywords: Digital watermarking, Balance multi-wavelet transform (BMW) Blind watermarking, Human visual system (HVS).

1 Introduction

With the rapid growth and widespread use of network distributions of digital media content, there is an urgent need for protecting the copyright of digital content against piracy and malicious manipulation. Under this background, watermarking systems have been proposed as a possible and efficient answer to these concerns. Watermarking is the process of embedding special mark, which is called watermark, into multimedia products such as images, voices, films, books and videos. The embedded data can later be detected or extracted from the multimedia products for the proof of ownership.

An effective watermarking system should have certain properties, such as imperceptibility, robustness, safety, and so on. Podichuk [1] uses the visual model to determine the embedding place and strength, but it is non-blind extraction. In [2], Cao JG describes a blind watermarking technique based on redundancy wavelet transformation. Pan [3] takes a binary image as watermark, embeds the same watermark into different places according to wavelet's hiberarchy. By this strategy, more information

---

* This work is supported by national high-tech research and development program ("863"program) of China under grant No, (2007AA01Z176).

H.J. Kim, S. Katzenbeisser, and A.T.S. Ho (Eds.): IWDW 2008, LNCS 5450, pp. 350–362, 2009.

of the water mark is embedded in the high frequency subband, while less in low frequency subband, but the extraction still need original image, and it only resists the JPEG compression radio no greater than 2.69. Jong [4] proposes a blind watermarking using data matrix. Watermark insertion is achieved by changing each of the corresponding coefficients in blocks of selected two sub-band areas. The results show great robustness in JPEG compression, but no other results.

In this paper, the watermark is embedded into the balanced multiwavelets field. Embedding watermark into the DWT filed is a good choice, but with the emergence of the multiwavelets, there is a new choice. Multiwavelets are the extension of scale wavelets, and the balanced multiwavelets, which offer orthogonality, symmetry and high order of approximation simultaneously, overcome the flaws caused by multiwavelets pre-filtering. Although there are plenty of papers addressing the use of wavelet transforms in watermarking, there are only a handful of papers addressing the relatively new multiwavelets transformation in watermarking applications [5], let alone balanced multiwavelets transform (BMWT) [6]. In this paper, watermark is embedded adaptively in the low and high frequency areas of the 8*8 block BMWT coefficients under a stable perceptual model, and the watermark extraction doesn't need the original image, fulfilling the blind extraction. The results show this scheme has great imperceptibility and robustness.

2 Theories and Algorithm

2.1 Multiwavelets and Balanced Multiwavelets

Multiwavelets may be considered as new conception of scalar wavelets [7]. Multiwavelets are different from scalar wavelets; they may have two or more scaling $\Phi(t)$ and wavelet functions $\Psi(t)$, while scalar wavelets have a single scaling and wavelet functions. In general, r scaling functions can be written using the vector $\Phi(t) = [\phi_1(t)\phi_2(t)\cdots\phi_r(t)]^T$, where $\Phi(t)$ is called the multiscaling function, in the same way, $\Psi(t) = [\psi_1(t)\psi_2(t)\cdots\psi_r(t)]^T$ is defined as the multiwavelet function. The two satisfies the following two-scale equation:

$$\Phi(t) = \sqrt{2} \sum_{k=-\infty}^{\infty} H_k \Phi(2t - k) \tag{1}$$

$$\Psi(t) = \sqrt{2} \sum_{k=-\infty}^{\infty} G_k \Phi(2t - k) \tag{2}$$

Where H_k and G_k are respectively the low-pass and high filter sequences which are defined as $2\,r \times r$ matrixes.

$$H_k = \begin{bmatrix} h_0(2k) & h_0(2k+1) \\ h_1(2k) & h_1(2k+1) \end{bmatrix} \tag{3}$$

$$G_k = \begin{bmatrix} g_0(2k) & g_0(2k+1) \\ g_1(2k) & g_1(2k+1) \end{bmatrix} \tag{4}$$

They should fulfill the relations: $\sum_n h_k^2(n) = 1$, $\sum_n g_k^2(n) = 1$, k=1$\Box$2.

Because the corresponding filter bands don't take the multiwavelets' approximation order, an effective signal expression should be obtained by proper pre-filter. Prefiltering may break multiwavelets' characters, while balanced multiwavelets can avoid this problem.

We define the band-Toeplitz matrix corresponding to the low-pass analysis

$$L = \begin{bmatrix} & \cdots & & & & & \\ & L_3 & L_2 & L_1 & O & O & \\ & O & L_3 & L_2 & L_1 & O & O \\ & & & & & & \cdots \end{bmatrix} \tag{5}$$

If $L^T U_1 = U_1$, the orthonormal multiwavelets system is a balanced multiwavelets system. On the assumption that BMWT has two filters channels which take as L_1、 L_2、 H_1、 H_2, therefore, BMWT has 16 subbands, shown in fig.1.

$L_1 L_1$	$L_1 L_2$	$L_1 H_1$	$L_1 H_2$	
$L_2 L_1$	$L_2 L_2$	$L_2 H_1$	$L_2 H_2$	
$H_1 L_1$	$H_1 L_2$	$H_1 H_1$	$H_1 H_2$	
$H_2 L_1$	$H_2 L_2$	$H_2 H_1$	$H_2 H_2$	

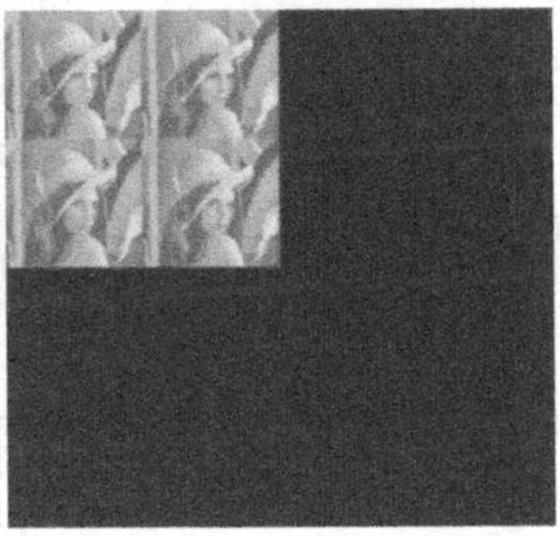

Fig. 1. Balanced multiwavelets subbands using single-level (left) and the result of lena after BMWT (right)

2.2 Perceptual Model for Balanced Multiwavelets Transforms

Because of the diversity of BMWT, there aren't uniform perception model. According to BMWT's decomposed structure, visual contrast and mask effect, this paper modifies Lewis JND model which is used in high frequency to suit for BMWT, and proposes a JND model in low frequency according to Weber's law.

Lewis [8] proposed the HVS model according to the wavelet's decomposed structure and sensitivity of the eye, which refers to three facets:

1)   The eye is less sensitive to noise in high frequency bands and in those bands having orientation of $45°$. Taking into account how the sensitivity to noise changes depends on the band that is considered, the first term is:

$$f(l,\theta) = \begin{cases} \sqrt{2}, & \theta = HH \\ 1, & \theta = 其他 \end{cases} * \begin{cases} 1, & l=0 \\ 0.32, & l=1 \\ 0.16, & l=2 \\ 0.1, & l=3 \end{cases} \qquad (6)$$

2)   The eye is less sensitive to noise in those areas of the image where brightness is high. Estimating the local brightness based on the grey level values of low pass version of the image, the second term is:

$$lu\min(l,x,y) = \frac{1}{256} I_N^{LL}(1+\frac{i}{2^{N-l}},1+\frac{j}{2^{N-l}}) \qquad (7)$$

3)   The eye is less sensitive to noise in the highly textured areas of the image. Giving a measure of the activity of texture in the neighborhood of the pixel, the third term is:

$$texture(l,x,y) = \sum_{k=1}^{N-l} 16^{-k} \sum_{\theta}^{HH,HL,LH} \sum_{i=0}^{1} \sum_{j=0}^{1} (I^{k+l,\theta}(i+\frac{x}{2^k},j+\frac{y}{2^k}))^2$$

$$+\frac{1}{16^{N-l}} var\{I_N^{LL}(1+i+\frac{x}{2^{N-l}},1+j+\frac{y}{2^{N-l}}):x=0,1;y=0,1\} \qquad (8)$$

Based on the above considerations, the JND formulation is proposed:

$$JND_n(l,\theta,x,y) = \beta * f(l,\theta) * lu\min(l.x,y)$$
$$* texture(l,x,y) \qquad (9)$$

As BMWT's decomposed characteristic is similar to scalar wavelet while there are also some differences, according to this connection, the BMWT perceptual model is proposed.

In the figure 1(a), (b), we can know that scalar wavelet decomposes different frequency in four directions, while BMWT can further decompose four approximation subbands. However, they are still in the same frequency like scalar wavelet, so we can get the first term as follows:

$$f(l,\theta)=\begin{cases}\sqrt{2},\theta=HH^1、HH^2、HH^3、HH^4\\1\,,\qquad\qquad\theta=其他\end{cases}*\begin{cases}1,\ l=0\\0.32,\ l=1\\0.16,\ l=2\\0.1,\ l=3\end{cases}\qquad(9)$$

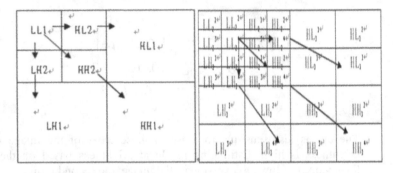

Fig. 2. The sub tree structure of wavelet (*left*) and BMW (*right*)

In figure 2, we can know that the sub tree structure of BMW and wavelet have similarities. Just as scalar wavelet, BMW's approximation band is the father node of the sub tree, and the children are in the same space location of the same direction and different scales of wavelet coefficients. Because of BMW's four approximation bands, there are four trees in BMW. According to their structure characteristic, the formulation should be modified as follows:

$$lu\min_n(l,x,y)=3+\frac{1}{256}I_N^n(1+\frac{i}{2^{N-l}},1+\frac{j}{2^{N-l}})\qquad(n\in LL_N^{1,2,3,4})\qquad(10)$$

$$texture_n(l,\theta,x,y)=\sum_{k=1}^{N-l}16^{-k}\sum_{\theta}^{HH,HL,LH}\sum_{i=0}^{1}\sum_{j=0}^{1}(I^{k+l,\theta}(i+\frac{x}{2^k},j+\frac{y}{2^k}))^2$$
$$+\frac{1}{16^{N-l}}var\{I_N^n(\{1,2\}+\frac{x}{2^{N-l}},\{1,2\}+\frac{y}{2^{N-l}})\}(n\in LL_N^{1,2,3,4})\qquad(11)$$

$$JND_n(l,\theta,x,y)=\beta*f(l,\theta)*lu\min_n(l.x,y)$$
$$*texture_n(l,x,y)(n\in LL_N^{1,2,3,4})\qquad(12)$$

Then, we can embed the watermark into the high frequency by using this JND model, we also defined a JND model which control the strength of the embedded watermark according Weber's law.

The sketch clearly shows in fig 3, human eyes are differently sensitive to different grey level. Generally eyes are most sensitive to the middle grey level, the Weber radio keeps const 0.02 in a wide middle grey level range, and non-linear descends in low and high grey level .(This paper assumes this non-linear is conic about grey level, and the contrast sensitivity is not beyond β).

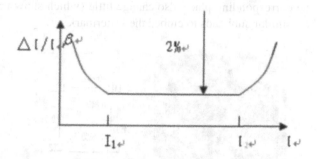

Fig. 3. Sketch of Weber's law

The BMW's low frequency represents the average luminance; it can show the grey level to some extends. We can get the term as follows:

$$JND = I(i, j) * w(i, j) \tag{13}$$

$$w(i, j) = \begin{cases} \dfrac{\beta(I(i, j) - I_1)^2}{I_1^2} + 0.02 & I(i, j) < I_1 \\ \dfrac{\beta(I(i, j) - I_2)^2}{(255 - I_2)^2} + 0.02 & I(I, J) > I_2 \\ 0.02 & \text{ot her s} \end{cases} \tag{14}$$

Fig. 4. The resulting JND profile in low frequency *(left)* and high frequency *(right)*

Fig.4 shows the result of the JND profile.

2.3 Watermarking Algorithm

After BMWT, the subbands in the same scaling are similar to each other, and the coefficients in the corresponding place also change little (which show in fig5), so we choose the highest similar subbands to embed the watermark.

Fig. 5. The approximation subband *(left)* and similarity of each other *(right)*

In this paper, the watermark is embedded in low and high frequency area of BMW. The low frequency coefficients, concentrating on the largest energy of the image, are important part of the vision and have great sensory capacity; the high frequency coefficients are the details of image, and eyes can't be conscious when embedding watermark in this area. Both of the two areas where the watermark is embedded can assure the imperceptibility and robustness.

2.3.1 Watermark Embedding Algorithm

The main steps performed in the proposed watermarking system are summarized as below:

1) A binary pseudorandom image consisting of ±1 is generated using the private embedding key k.

$$w = randn(k,l) \tag{15}$$

Where w is the watermark sequence whose length is l .

$$l = fw \times fh / 64 \tag{16}$$

2) Compute the forward-balanced multiwavelet transform (BMWT) of the 8×8 block of host image to get the subband coefficients.

3) Choose the highest similar subbands to embed the watermark. In this paper, choose each block's low frequency subband L_1L_1 、 L_2L_1 and high frequency subband L_1H_1 、 L_2H_1 to embed watermark.

4) Estimate the threshold T which is used to determinate the embedded place in low frequency area.

5) Compute the JND.
6) Perform watermark embedding using the following rules:
      If $w(i) == 1$

$$if \quad L_2L_1(i, j) - L_1L_1(i, j) > T \quad L_1L_1(i, j) = L_1L_1(i, j)$$
$$elseif \quad L_1L_1(i, j) > L_2L_1(i, j) \quad L_1L_1(i, j) = L_1L_1(i, j) + JND$$
$$elseif \quad L_1L_1(i, j) < L_2L_1(i, j) \quad L_1L_1(i, j) = L_2L_1(i, j) \qquad (17)$$
$$if \quad L_1H_1(i, j) > L_2H_1(i, j) \quad L_1H_1(i, j) = L_1H_1(i, j) + JND$$
$$elseif \quad L_1H_1(i, j) < L_2H_1(i, j) \quad L_1H_1(i, j) = L_2H_1(i, j)$$

If $w(i) == -1$

$$if \quad L_1L_1(i, j) - L_2L_1(i, j) > T \quad L_1L_1(i, j) = L_1L_1(i, j)$$
$$elseif \quad L_1L_1(i, j) > L_2L_1(i, j) \quad L_1L_1(i, j) = L_2L_1(i, j)$$
$$elseif \quad L_1L_1(i, j) < L_2L_1(i, j) \quad L_1L_1(i, j) = L_1L_1(i, j) - JND \qquad (18)$$
$$if \quad L_1H_1(i, j) > L_2H_1(i, j) \quad L_1H_1(i, j) = L_2H_1(i, j)$$
$$elseif \quad L_1H_1(i, j) > L_2H_1(i, j) \quad L_1H_1(i, j) = L_1H_1(i, j) - JND$$

7) Perform inverse-balanced multiwavelet transform (BMWT) to get the watermarked image.

2.3.2 Watermark Extraction Algorithm

1) Compute the balanced multiwavelet transform (BMWT) of the 8×8 block of watermarked image to get the subband coefficients.
2) Extract the watermark as follows:

$$if \quad L_2L_1(i, j) - L_1L_1(i, j) > T$$
$$w'(k) = 1$$
$$elseif \quad L_1L_1(i, j) >= L_2L_1(i, j) \,\&\&\, L_1L_1(i, j) - L_2L_1(i, j) > T$$
$$w'(k) = 1$$
$$elseif \quad L_2L_1(i, j) - L_1L_1(i, j) > T$$
$$w'(k) = -1 \qquad (19)$$
$$elseif \quad L_1L_1(i, j) <= L_2L_1(i, j) \,\&\&\, L_2L_1(i, j) - L_1L_1(i, j) > T$$
$$w'(k) = -1$$

$$if \quad L_1H_1(i, j) >= L_2H_1(i, j) \qquad w(k) = 1$$
$$elseif \quad L_1H_1(i, j) <= L_2H_1(i, j) \quad w(k) = -1$$

3) Detect the similarity between w and w' by the related detectors.

$$NC = \sum_{i=1}^{l} w(i)w'(i) / \sum_{i=1}^{l} w'(i)^2 \tag{20}$$

Where l is the length of the watermark; when $NC > \rho$, (ρ is an experience value), then we consider there are watermark information in the image.

The watermarked image quality is defined by PNSR; the following is the computing formulation:

$$PSNR = 10\log_{10} \frac{255^2}{MSE} \tag{21}$$

$$MSE = \frac{\sum_{i=0}^{M-1} \sum_{j=0}^{N-1} (f'(i,j) - f(i,j))^2}{MN} \tag{22}$$

M、N are the width and length of image ; $f'(i,j)$, $f(i,j)$ respectively represent the pixel value of the out image and original image in (i,j).

Because of the threshold, watermark can't be embedded in some pixels which are higher than threshold T, it will result in some watermark error when extracting. But the number is limited, which won't affect the relativity, and still make sure the right detection.

3   Experimental Results

We run experiments to valuate the imperceptibility of the proposed watermarking system using the test images in figure 6.

Fig. 6. The original images *(lake, plane, and baboon)*

The results show great imperceptibility from fig 6, 7. The PSNR are respectively 34.64, 34.73, 36.86 DB.

Fig. 7. The watermarked images *(lake, plane, and baboon)*

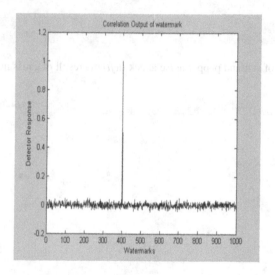

Fig. 8. The extraction result of 'baboon'

In fig 8, we generate 1000 random sequences, and the 400th is the watermark sequence.

Then we take the JEPG compression, filtering, noise, mosaic, histogram equalization and cropping attack to watermarked image of lena; the results show great robustness.

In fig 9, we take the pepper noise and gauss noise attack; the density of pepper noise is increasing from 0.01 to 0.11, and the variance of the gauss noise is from 0.005 to 0.03; the results show that it can resist those two attacks strongly.

In fig 10, we take the random noise attack and JPEG compression attack; it can resist the random noise whose density is almost 0.07, and even the quality gene is 10, the similarity is till 0.6, these indicate the great robustness in noise attack and JPEG attack.

In fig 11, we take the median filtering attack and the gauss filtering attack by using the 3*3model and 5*5 model to filter, the result shows high robustness in median filter and gauss filter attacks.

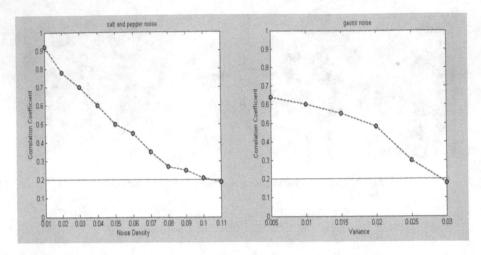

Fig. 9. The result of salt and pepper noise attack *(left)* the result of gauss noise attack *(right)*

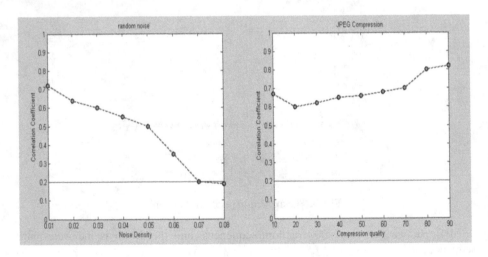

Fig. 10. The result of random noise attack *(left)* the result of JPEG compression attack *(right)*

In fig 12, we take the average filter and cropping attack; the filter uses different templates. From the left picture in fig12, we can get that it can resist such filtering, and the right picture shows high robustness in cropping.

In fig 13, the result is the histogram equalization attack; the NC is almost 0.9864. The result shows great robustness in histogram equalization attack.

Other attacks results are shown in table 1. Compared with [9] which proposes an algorithm in DWT, whose extraction needs original image, our algorithm belongs to blind detection. The algorithm shows greater result than [9], especially in geometry attacks.

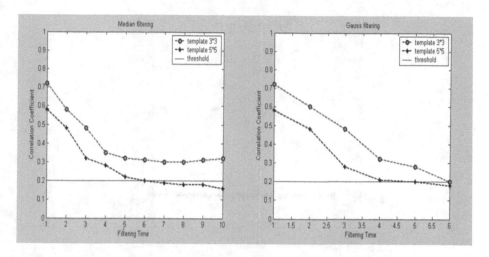

Fig. 11. The result of median filtering attack *(left)* the result of gauss filtering attack *(right)*

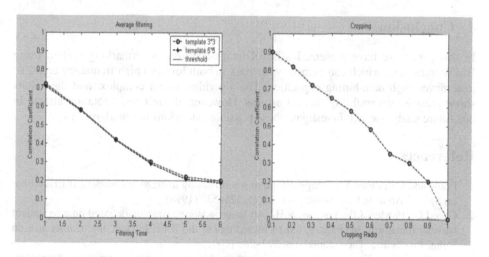

Fig. 12. The result of average filtering attack *(left)* the result of cropping attack *(right)*

Table 1. (PSNR/NC) parts of attacks

	The test image	Histogram equalization	salt and pepper noise attack(2%)	Gauss noise(σ^2=0.01)	Gamma enhanced	mosaic
This paper	Lake	23.95/0.96	20.53/0.83	19.22/0.60	16.98/0.83	23.19/0.54
	Plane	12.29/0.93	19.61/0.77	19.30/0.60	17.82/0.95	24.91/0.52
	Baboon	14.79/0.98	20.35/0.77	19.11/0.63	12.02/0.88	30.90/0.53
paper[9]	Lake	24.48/0.03	20.67/0.28	19.42/0.18	17.01/0.05	23.15/0.21
	Plane	12.331/0.13	20.90/0.24	18.75/0.31	16.79/0.08	24.31/0.28
	Baboon	14.89/0.09	20.37/0.23	18.46/0.18	22.83/0.10	31.19/0.47

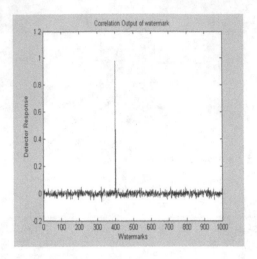

Fig. 13. The result of histogram equalization (NC=0.9864)

4 Conclusion

In this paper, we have presented a novel image-adaptive watermarking system using BMW transform, which can embed watermark in both low and high frequency areas, so it achieves high data-hiding capacities. The algorithm is not complex, and the results show great imperceptibility and robustness. However, it can't resist rotation attack. In our future study, we will investigate these possible extensions of our algorithm.

References

1. Podichuk, C.I., Zeng, W.: Image-Adaptive watermarking using visual models. IEEE Journal on Special Areas in Communications 16(4), 525–539 (1998)
2. Cao, J.G., Fowler, J.E., Younan, N.H.: An image-adaptive watermark based on a redundant wavelet transform. In: Pitas, I. (ed.) Proceedings of the IEEE International Conference on Image Processing. Thessaloniki, pp. 277–280 (2001)
3. Pan, R., Gao, Y.X.: Image watermarking method based on wavelet transform. Journal of Image and Graphics 7(7), 667–671 (2002)
4. Park, J.S., Nam, B.H.: A Blind Watermarking Using Data Matrix and Changing Coefficients in Wavelet Domain. In: Proc. of SPIE, vol. 6794, pp. 679–777 (2007)
5. Serdean, C.V., Ibrahim, M.K., Moemeniand, A., Al-Akaidi, M.M.: Wavelet and multiwavelets watermarking. IET Image Process. 1(2), 223 (2007)
6. Ghouti, L., Bouridane, A., Ibrahim, M.K., Boussakta, S.: Digital image watermarking using balanced multiwavelets. IEEE Trans. Signal Process. 54(4), 1519 (2006)
7. Lebrun, J., Vetterli, M.: Balanced Multiwavelets Theory and Design. IEEE Transactions on Signal Processing 46(4) (April 1, 1998)
8. Lewis, A.S., Knowles, G.: Image Compression Using the 2-D Wavelet Transform. IEEE Transactions on Image Processing I(2) (April 1992)
9. weiwei, W., bo, Y., guoxiang, S.: Watermarking the Lowest Approximation of Wavelet Transformation of Images. Signal Processing 17(6), December 2

A Generalised Model for Distortion Performance Analysis of Wavelet Based Watermarking

Deepayan Bhowmik and Charith Abhayaratne

Department of Electronic and Electrical Engineering, University of Sheffield
Sheffield S1 3JD, United Kingdom
{d.bhowmik,c.abhayaratne}@sheffield.ac.uk

Abstract. A model for embedding distortion performance for wavelet based watermarking is presented in this paper. Firstly wavelet based watermarking schemes are generalised into a single common framework. Then a mathematical approach has been made to find the relationship between distortion performance metrics and the watermark embedding parameters. The derived model shows that for wavelet based watermarking schemes the sum of energy of the selected wavelet coefficients to be modified is directly proportional to the distortion performance (the mean square error) measured in the pixel domain. The propositions are made using the energy conservation theorem between input signal and transform domain coefficients for orthonormal wavelet bases. Such an analysis is useful to choose the wavelet coefficients during watermark embedding procedure and to find suitable input parameters such as wavelet kernel or the choice of subband.

Keywords: Watermarking, wavelet transforms, distortion performance.

1 Introduction

With the success of transforms based image/video compression schemes, such as, JPEG, JPEG2000 and MPEG-2, the frequency domain watermarking has received a huge attention. The discrete wavelet transform (DWT) has been widely used as a multi-resolution analysis method in most frequency domain watermarking techniques. [1,2,3,4,5,6,7,8,9,10,11]. Usually the imperceptibility and robustness are considered as two major properties of any watermarking scheme. The imperceptibility is often measured by evaluating the distortion of the host image.

In this work we address the problem of modelling and analysis of embedding distortion. In the literature, such analysis have been presented by only focusing on single specific techniques [12]. Our focus in this paper is to derive a common analysis model involving an exhaustive list of wavelet based algorithms characterised by their input parameters. The main objective of the work is to derive a generalised model for distortion performance analysis of wavelet based watermarking. The generalisation of our model is based on fitting all major wavelet based watermarking schemes into a common framework, which was presented in [13].

H.J. Kim, S. Katzenbeisser, and A.T.S. Ho (Eds.): IWDW 2008, LNCS 5450, pp. 363–378, 2009.

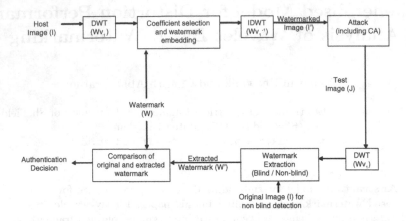

Fig. 1. Block diagram of the generalised functional modules of wavelet based watermarking schemes

In the distortion performance model first a proposition is made to show the relationship between the noise power in the transform domain and the input signal domain. Then using the above proposition a relationship is established between the distortion performance metrics and the input parameters of a given wavelet based watermarking scheme. The rest of the paper is organised with Sect. 2 presenting the generalisation of embedding schemes. Detailed mathematical analysis and the model is presented in Sect. 3 followed by experimental results in Sect. 4. Concluding remarks can be found in Sect. 5.

2   The Common Framework for Wavelet Based Watermark Embedding

There are many wavelet based watermarking schemes available in the literature. In this context a formal evaluation framework for wavelet based methods is really useful to the watermarking community. It is observed that most of the popular wavelet based watermarking schemes can be dissected in common functional blocks as shown in Fig. 1. In this paper we discuss and present the distortion performance model of wavelet based algorithms and therefore restrict our discussion to the embedding part of the watermarking schemes. In a more general form of the watermarking schemes a forward wavelet transform is applied to the target image. The wavelet coefficients are then modified according to the particular embedding procedure. The modification is done on the selected coefficients in the selected subbands. An inverse wavelet transform which is the same as the forward wavelet kernel is then applied to produce the watermarked image. The basic embedding principle for any wavelet based watermarking algorithm is the same and the modified coefficient $C'_{m,n}$ at (m, n) position, can be presented as:

$$C'_{m,n} = C_{m,n} + \Delta_{m,n} , \tag{1}$$

where $C_{m,n}$ is the coefficient to be modified and $\Delta_{m,n}$ is the modification due to watermark embedding. Based on the modification algorithms, the embedding procedures are categorised into two main types of embedding algorithms: direct coefficient modification [1,2,7,10,11] and quantisation based modification [4,6,8,3].

In the direct coefficient modification schemes, selected coefficients are directly modified based on the following generalised modification value $\Delta_{m,n}$ at (m, n) position:

$$\Delta_{m,n} = (a_1)\alpha(C_{m,n})^b W_{m,n} + (a_2)v_{m,n}W_{m,n} + (a_3)\beta C_w + (a_4)S_{m,n} , \tag{2}$$

where a_1, a_2, a_3, a_4 are the selection coefficients, $C_{m,n}$ is the coefficient to be modified, α is the watermark weighting factor, $b = 1, 2...$ is the watermark strength parameter, $W_{m,n}$ is the watermark value, $v_{m,n}$ is the weighting parameter based on pixel masking in HVS model, β is the weighting parameter in the case of fusion based scheme, C_w is the watermark wavelet coefficient and $S_{m,n}$ is any other value which is normally a function of $C_{m,n}$. In most of the algorithms watermark weighting parameters α and β are user defined to an optimal value. The watermark information $W_{m,n}$ is either generated randomly with a random seed or taken from a gray scale logo or a binary logo. As mentioned before the weighting parameter and the watermark information are always user defined, hence these are considered as constant parameters in a controlled experimental environment. Other parameters in the modification equation are a function of the wavelet coefficient $C_{m,n}$ which depends on the input image and considered as a variable here. Therefore it is observed that in all the cases the modification value is a direct function of the coefficient $C_{m,n}$ as mentioned in Table 1. This table also represents the common input parameters used in the embedding procedure and shows how different algorithms can be realised with this generalised framework. Considering a specific case, in this paper we have not chosen HVS model based watermarking scheme as our main focus is on distortion performance analysis which is different from HVS based performance metrics.

On the other hand in the case of quantisation based algorithms, the modification is based on the quantisation steps. Normally a rank order based algorithm is proposed in these type algorithms. The algorithms change the median value of a local area (typically a 3x1 coefficient window) considering the neighbouring values. The modification value $\Delta_{m,n}$ is decided based on the quantisation step δ $(-\delta \leq \Delta \leq \delta)$ within the range of the selected 3x1 window. Different functions are suggested in the literature to find the value of δ and the functions normally consist of minimum (C_{min}) and maximum (C_{max}) value of the coefficients in each selected window. A predefined weighting factor α is often used to determine the value of δ. As Δ depends on step size δ and α is user defined, the modification value Δ is typically a function of C_{min} and C_{max} in each selected 3x1 window (refer Table 1).

Table 1. Realisation of wavelet based algorithms using different combination of input parameters

Method	Selection Coeff $< a_1, a_2, a_3, a_4 >$	Subband Selection	Wavelet Kernel	Level	Reference	Δ as Function of
Direct($b = 2$)	$< 1, 0, 0, 0 >$	High	Haar	2	[1]	$f(C_{m,n})$
Direct($b = 1$)	$< 1, 0, 0, 0 >$	All	Biorthogonal	3	[2]	$f(C_{m,n})$
Direct($b = 1$)	$< 1, 0, 0, 0 >$	Low	Biorthogonal, Non-linear	3	[14]	$f(C_{m,n})$
Direct	$< 0, 0, 1, 0 >$	High	Orthogonal	4	[11]	$f(C_{m,n})$
Direct	$< 0, 0, 0, 1 >$	High	Any	2	[10]	$f(C_{m,n})$
Quantisation	-	Low	Any	2	[6]	$f(C_{min}, C_{max})$
Quantisation	-	High	Haar	1	[4]	$f(C_{min}, C_{max})$
Quantisation	-	High	Any	2	[3]	$f(C_{min}, C_{max})$

With this common generalised framework we have analysed and proposed a distortion performance model in the next section.

3 Embedding Distortion Performance Analysis

In this section a detailed discussion is carried out on the proposed model. The embedding distortion performance is usually measured by the Mean Square Error (MSE).

Definition 1. The *Mean Square Error (MSE)* or average noise power P_p in pixel domain between original image I and watermarked image I' is defined by:

$$P_p = \frac{1}{MN} \sum_{j=0}^{M-1} \sum_{i=0}^{N-1} |I(j,i) - I'(j,i)|^2 , \qquad (3)$$

where M and N are the image dimension and j and i indicate each pixel position.

In order to formulate the model we show the transformation of noise energy from frequency domain to the signal domain using Parseval's equality.

Definition 2. In the *Parseval's Equality*, the energy is conserved between an input signal and the transform domain coefficient in the case of an orthonormal filter bank wavelet base [15]. Assuming the input signal $x[n]$ with the length of $n \in Z$ and the corresponding transformed domain coefficients of $y[k]$ where $k \in Z$, according to energy conservation theorem,

$$\|x\|^2 = \|y\|^2 . \qquad (4)$$

Based on these primary definitions we build the model which consists of the following propositions and its proof.

Proposition 1. *Sum of the noise power in the transform domain is equal to sum of the noise power in the input signal for orthonormal transforms. If the input signal noise is defined by $\Delta x[n]$ and the noise in transform domain is $\Delta y[k]$ then*

$$\sum_n |\Delta x[n]|^2 = \sum_k |\Delta y[k]|^2 , \tag{5}$$

where $n \in Z$ is the length of the input signal and $k \in Z$ is the length in the transform domain, respectively.

Proof. The discrete wavelet transform (DWT) can be realised with a filter bank or lifting scheme based factoring. In both cases the wavelet decomposition and the reconstruction can be represented by a polyphase matrix [16]. The inverse DWT can be defined by a synthesis filter bank using the polyphase matrix $M'(z) = \begin{pmatrix} h'_e(z) & h'_o(z) \\ g'_e(z) & g'_o(z) \end{pmatrix}$ where $h'(z)$ represents the low pass filter coefficients and $g'(z)$ is the high pass filter coefficients and the subscripts e and o denote even and odd indexed terms, respectively. Now the transform domain coefficient y can be re-mapped into input signal x as bellow:

$$\begin{pmatrix} x_e(z) \\ x_o(z) \end{pmatrix} = \begin{pmatrix} h'_e(z) & h'_o(z) \\ g'_e(z) & g'_o(z) \end{pmatrix} \begin{pmatrix} y_e(z) \\ y_o(z) \end{pmatrix} . \tag{6}$$

Assuming Δy is the noise introduced in wavelet domain and Δx is the modified signal after the inverse transform, we can define the relationship between the noise in the wavelet coefficient and the noise in the modified signal using the following equations. From (6) we can write

$$\begin{pmatrix} x_e(z)+\Delta x_e(z) \\ x_o(z)+\Delta x_o(z) \end{pmatrix} = \begin{pmatrix} h'_e(z) & h'_o(z) \\ g'_e(z) & g'_o(z) \end{pmatrix} \begin{pmatrix} y_e(z)+\Delta y_e(z) \\ y_o(z)+\Delta y_o(z) \end{pmatrix} . \tag{7}$$

From (7) using the *Linearity* property of the Z transform of the filter coefficients and signals in the polyphase matrix we can get,

$$\begin{aligned} x_e(z) + \Delta x_e(z) &= h'_e(z)(y_e(z) + \Delta y_e(z)) \\ &\quad + h'_o(z)(y_o(z) + \Delta y_o(z)) , \\ h'_e(z)y_e(z) + h'_o(z)y_o(z) + \Delta x_e(z) &= h'_e(z)y_e(z) + h'_e(z)\Delta y_e(z) \\ &\quad + h'_o(z)y_o(z) + h'_o(z)\Delta y_o(z) , \\ \Delta x_e(z) &= h'_e(z)\Delta y_e(z) + h'_o(z)\Delta y_o(z) . \tag{8} \end{aligned}$$

Similarly $\Delta x_o(z)$ can be obtained and written as

$$\Delta x_o(z) = g'_e(z)\Delta y_e(z) + g'_o(z)\Delta y_o(z) . \tag{9}$$

Combining (8) and (9), finally we can write the polyphase matrix form of the noise in the output signal:

$$\begin{pmatrix} \Delta x_e(z) \\ \Delta x_o(z) \end{pmatrix} = \begin{pmatrix} h'_e(z) & h'_o(z) \\ g'_e(z) & g'_o(z) \end{pmatrix} \begin{pmatrix} \Delta y_e(z) \\ \Delta y_o(z) \end{pmatrix} . \tag{10}$$

Recalling the Parseval's energy conservation theorem as stated in *Definition 2.*, from (10) we can conclude that

$$\sum |\Delta x_e|^2 + \sum |\Delta x_o|^2 = \sum |\Delta y_e|^2 + \sum |\Delta y_o|^2 ,$$

$$\sum_n |\Delta x[n]|^2 = \sum_k |\Delta y[k]|^2 . \tag{11}$$

∎

Using the generalised framework, the *Proposition 1* can be applied to build the relationship between the modification energy in the coefficient domain to embed the watermark and the distortion performance metrics. In this model we made propositions for two different categories of embedding schemes, discussed in previous section.

Proposition 2. *In a wavelet based watermarking scheme, the mean square error (MSE) of the watermarked image is directly proportional to the sum of the energy of the modification values of the selected wavelet coefficients. The modification value itself is a function of the wavelet coefficients and therefore we propose two different cases based on the categorisation.*

Case A. *For the direct modification embedding method the modification is a function of the selected coefficient to be watermarked and the relationship between MSE (P_p) and the selected coefficient ($C_{m,n}$) is expressed as:*

$$P_p \propto \sum |f(C_{m,n})|^2 . \tag{12}$$

Case B. *For the quantisation based method the modification is a function of the neighbouring wavelet coefficients of the selected median coefficient to be watermarked and the relationship between MSE (P_p) and the wavelet coefficients C_{min} and C_{max} is expressed as:*

$$P_p \propto \sum |f(C_{min}, C_{max})|^2 . \tag{13}$$

Proof. In a wavelet based watermark embedding scheme the watermark information is inserted by modifying the wavelet coefficients. This watermark insertion can be considered as introducing noise in the transform domain. Hence the sum of the energy of the modification value due to watermark embedding in the wavelet domain is equal to the sum of the noise energy in the transform domain as stated in *Proposition 1.* From (1) and (5), the energy sum of the modification value $\Delta_{m,n}$ can be defined as:

$$\sum_{m,n} |\Delta_{m,n}|^2 = \sum_k |\Delta y[k]|^2 . \tag{14}$$

Similarly, the pixel domain distortion performance metrics which is represented by MSE is considered as the noise error created in the signal due to the noise

Table 2. Correlation coefficient values between sum of energy and the MSE for different wavelet kernel in various subbands

	Direct modification						Intra Subband Based					
	Haar	D-4	D-6	D-8	D-10	D-16	Haar	D-4	D-6	D-8	D-10	D-16
LL3	0.99	0.99	0.99	0.99	0.99	0.99	0.80	0.84	0.87	0.88	0.86	0.89
LH3	0.99	0.99	0.99	0.99	0.99	0.99	0.99	0.99	0.99	0.99	0.99	0.99
HL3	0.99	0.99	0.99	0.99	0.99	0.99	0.93	0.96	0.95	0.96	0.97	0.99
HH3	0.99	0.99	0.99	0.99	0.99	0.99	0.99	0.99	0.99	0.99	0.99	0.99
LH2	0.99	0.99	0.99	0.99	0.99	0.99	0.99	0.99	0.99	0.99	0.99	0.99
HL2	0.99	0.99	0.99	0.99	0.99	0.99	0.99	0.99	0.99	0.99	0.99	0.99
HH2	0.99	0.99	0.99	0.99	0.99	0.99	0.99	0.99	0.99	0.99	0.99	0.99
LH1	0.99	0.99	0.99	0.99	0.99	0.99	0.99	0.99	0.99	0.99	0.99	0.99
HL1	0.99	0.99	0.99	0.99	0.99	0.99	0.97	0.98	0.99	0.98	0.98	0.99
HH1	0.99	0.99	0.99	0.99	0.99	0.99	0.99	0.99	0.99	0.99	0.99	0.99

in wavelet domain. Therefore, the sum of the noise energy in the input signal is equal to the sum of the noise error energy P_p in the pixel domain:

$$P_p.(MN) = \sum_n |\Delta x[n]|^2 , \qquad (15)$$

where M and N are the image dimensions. Now the relationship between the distortion performance metrics MSE of the watermarked image and the coefficient modification value which is normally a function of the selected wavelet coefficients can be decided using the *Proposition 1*. Thus from (14) and (15) we can write:

$$P_p.(MN) = \sum_{m,n} |\Delta_{m,n}|^2 , \qquad (16)$$

where M and N are the image dimensions. Hence for any watermarked image, the average noise power P_p is proportional to the sum of the energy of the modification values of the selected wavelet coefficients:

$$P_p \propto \sum_{m,n} |\Delta_{m,n}|^2 . \qquad (17)$$

Now with the help of the categorisation in the generalised form of the popular wavelet based watermarking schemes as discussed in Sect. 2, a relationship is established between the error energy of the watermarked image and the selected wavelet coefficient energy of the host image. For a direct modification based algorithm, the mean square error P_p is directly proportional to the sum of the energy of the modification value Δ which is a function of wavelet coefficient value as stated below:

$$P_p \propto \sum |f(C_{m,n})|^2 . \qquad (18)$$

Similarly for the quantisation based method the mean square error depends on the neighbouring wavelet coefficient values. In this case the modification energy

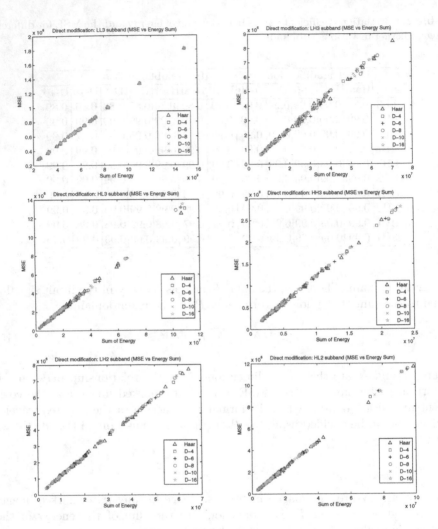

Fig. 2. Watermark embedding (Direct Modification) performance correlation plot: MSE vs. sum of energy, in different subband for individual images. Six wavelet kernels used here such as 1. Haar, 2. D-4, 3. D-6, 4. D-8, 5. D-10 and 6. D-16, respectively.

$|\Delta_{m,n}|^2$ hold an inequality due the modification range $-\delta \le \Delta_{m,n} \le \delta$:

$$|\Delta_{m,n}|^2 \le |\delta|^2 . \tag{19}$$

Therefore the upper bound of the mean square error P_p is defined by:

$$P_p \propto \sum |f(C_{min}, C_{max})|^2 . \tag{20}$$

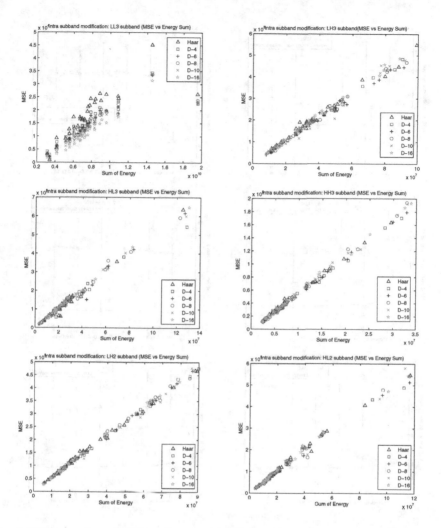

Fig. 3. Watermark embedding (Intra subband based) performance correlation plot: MSE vs. sum of energy, in different subband for individual images. Six wavelet kernels used here such as 1. Haar, 2. D-4, 3. D-6, 4. D-8, 5. D-10 and 6. D-16, respectively.

3.1 An Example of Direct Modification

Considering a specific case of the direct modification algorithm in [2] the modification value Δ is a direct function of wavelet coefficient ($\Delta_{m,n} = \alpha C_{m,n} W m, n$). Hence (18) can be modified and the MSE P_p can be expressed as:

$$P_p \propto \sum_{k=1}^{l} |C(k)|^2 , \qquad (21)$$

where $C(k)$ is the selected coefficients to be watermarked and l is the number of such selected coefficients.

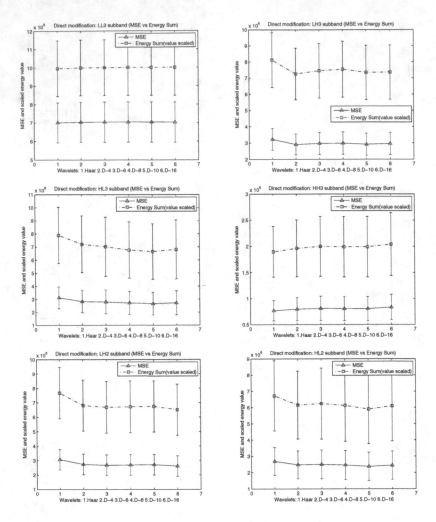

Fig. 4. Watermark embedding (Direct Modification) performance graph for different subbands. Six different wavelet kernels used here such as 1. Haar, 2. D-4, 3. D-6, 4. D-8, 5. D-10 and 6. D-16, respectively. Subbands are shown left to right and top to bottom: LL3, LH3, HL3, HH3, LH2 and HL2, respectively.

3.2   An Example of Quantisation Based Method

In an intra subband based quantisation method suggested in [6], the quantisation step δ is defined as:

$$\delta = \alpha \frac{C_{max} + C_{min}}{2} \,, \tag{22}$$

where α is the user defined weighting factor. As the modification value Δ depends on δ, with reference to (20), the relationship between the maximum limit of MSE

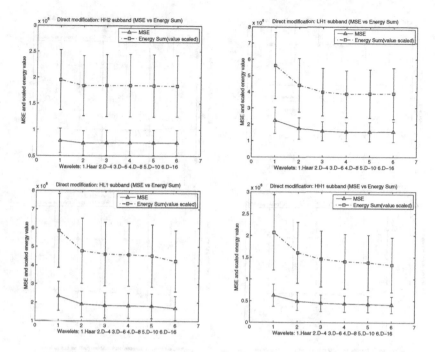

Fig. 5. Watermark embedding (Direct modification) performance graph for different subbands. Six different wavelet kernels used here such as 1. Haar, 2. D-4, 3. D-6, 4. D-8, 5. D-10 and 6. D-16, respectively. Subbands are shown left to right and top to bottom: HH2, LH1, HL1, HH1.

P_p and wavelet energy is defined by the following equation:

$$P_p \propto \sum_k (C(k)_{max} + C(k)_{min})^2 , \qquad (23)$$

where $C(k)_{max}$ and $C(k)_{min}$ are the neighbourhood coefficients of the median value and l is the number of such selected median value.

4   Experimental Simulations

The propositions made in the previous section are verified in the experimental simulations. The sum of the energy of the selected wavelet coefficients and the MSE of the watermarked image have been calculated for 30 different images with a combination of different input parameters. As the wavelet coefficients varies greatly in different subbands we have considered the performances of all subbands separately after a 3 level wavelet decomposition. Also a set of different wavelet kernels having various filter lengths are selected to perform the simulations. We simulated and studied the performance of different wavelet kernels such as Haar, Daubechies-4 (D-4), Daubechies-6 (D-6), Daubechies-8 (D-8),

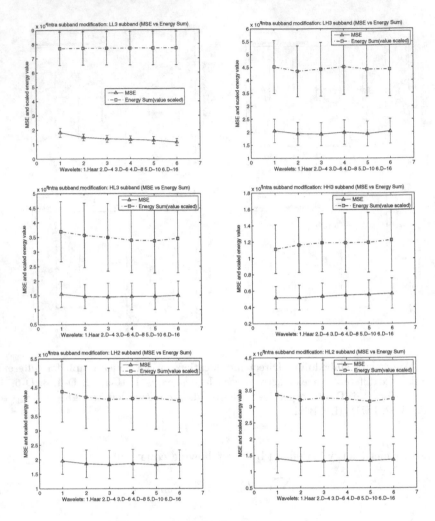

Fig. 6. Watermark embedding (Intra subband modification) performance graph for different subbands. Six different wavelet kernels used here such as 1. Haar, 2. D-4, 3. D-6, 4. D-8, 5. D-10 and 6. D-16, respectively. Subbands are shown left to right and top to bottom: LL3, LH3, HL3, HH3, LH2 and HL2, respectively.

Daubechies-10 (D-10) and Daubechies-16 (D-16) in order to verify our proposed model. Two different sets of results are obtained and displayed to verify the effects of different input parameters which are responsible for embedding distortion performance. These two sets of experimental arrangements and resulting plots are discussed separately as follows:

– In the experiment *Set 1*, the sum of energy of the selected wavelet coefficients to be modified and MSE of the watermarked image have been calculated using the same α and the same binary watermark logo for each selected

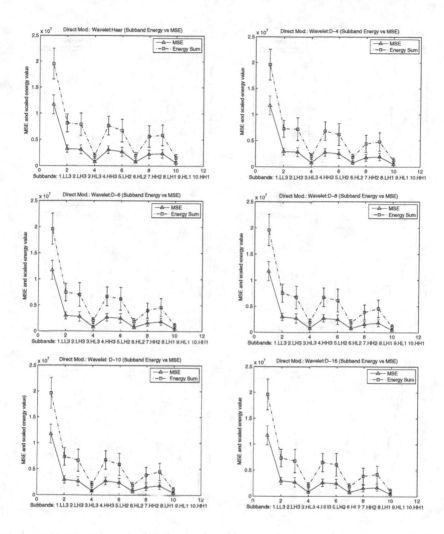

Fig. 7. Watermark embedding (Direct Modification) performance graph for various wavelets in different subband. Wavelet kernels are shown left to right and top to bottom: Haar, D-4, D-6, D-8, D-10 and D-16, respectively.

method. We have used various wavelet kernels and observed the results for each selected subbands. The correlation between MSE and the energy sum is displayed in Fig. 2 and Fig. 3 for direct modification and intra subband based embedding, respectively. The correlation coefficients are also calculated and presented in Table 2.

In another representation a set of graphs are plotted in Fig. 4 and Fig. 5 for direct modification and in Fig. 6 for intra subband based embedding. These plots present the average values of the MSE and the sum of energy for the test image set. The error bars denote the accuracy up to the 95%

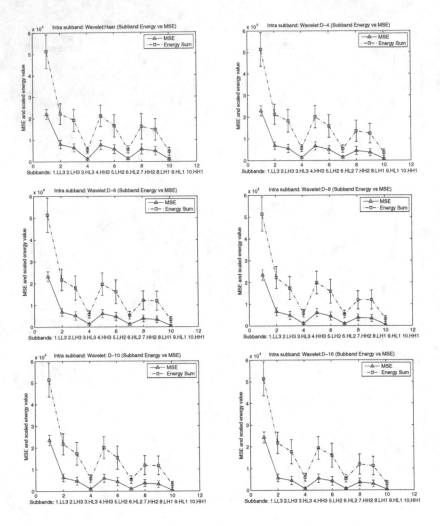

Fig. 8. Watermark embedding (Intra subband modification) performance graph for various wavelets in different subband. Wavelet kernels are shown left to right and top to bottom: Haar, D-4, D-6, D-8, D-10 and D-16, respectively.

confidence interval. For display purposes the sum of energy value was scaled, so that they can be shown on the same plot for comparing the trend.

– In the experiment *Set 2*, the performance for different subbands are plotted for each wavelet kernel in a similar fashion as mentioned in experiment *Set 1* in order to observe the trend. The direct modification results are shown in Fig. 7 and the intra subband modification methods are shown in Fig. 8. As earlier, a 95% confidence interval is considered which is denoted by the error bars.

The simulation results show a strong correlation between MSE of the watermarked image and the energy sum of the selected wavelet coefficients to be modified. It is observed that for a direct modification, the correlation coefficient value is more than 0.97 and more than 0.80 in the case of intra subband based modification, for different wavelet kernels and various selected subbands. On the other hand, a similar graph patterns are observed in Fig. 4, Fig. 5, Fig. 6, Fig. 7 and Fig. 8, which show the proportionality trend between MSE and the energy sum as proposed in the model.

These extensive simulation results strongly support the proposed model for a wide range of input images and various orthogonal wavelet kernels.

5 Conclusion

We have presented a generalised model for embedding distortion performance analysis of wavelet based watermarking schemes. With this mathematical analysis model we have achieved two different goals: generalisation of wavelet based embedding schemes and the effect of input parameters on distortion performance. We have proposed the model for orthonormal wavelet bases following the Parseval's Equality. Our model suggests that in a wavelet based watermarking scheme the MSE of the watermarked image is directly proportional to the sum of energy of the modification values of the selected wavelet coefficients. We have verified the model by evaluating the embedding distortion performance for different choices of wavelet kernels, subbands and the coefficient selections used in wavelet based watermark embedding. The experimental simulation successfully verified the proposed model.

Acknowledgments. This work is funded by BP-EPSRC Dorothy Hodgkin postgraduate award.

References

1. Xia, X., Boncelet, C.G., Arce, G.R.: Wavelet transform based watermark for digital images. Optic. Express 3(12), 497–511 (1998)
2. Kim, J.R., Moon, Y.S.: A robust wavelet-based digital watermarking using level-adaptive thresholding. In: Proc. IEEE ICIP, vol. 2, pp. 226–230 (1999)
3. Huo, F., Gao, X.: A wavelet based image watermarking scheme. In: Proc. IEEE ICIP, October 2006, pp. 2573–2576 (2006)
4. Kundur, D., Hatzinakos, D.: Digital watermarking using multiresolution wavelet decomposition. In: Proc. IEEE ICASSP, Seattle, WA, USA, May 1998, vol. 5, pp. 2969–2972 (1998)
5. Wang, H.J., Su, P.C., Kuo, C.C.J.: Wavelet-based digital image watermarking. Optics Express 3, 491–497 (1998)
6. Xie, L., Arce, G.R.: Joint wavelet compression and authentication watermarking. In: Proc. IEEE ICIP, vol. 2, pp. 427–431 (1998)
7. Barni, M., Bartolini, F., Piva, A.: Improved wavelet-based watermarking through pixel-wise masking. IEEE Trans. Image Processing 10(5), 783–791 (2001)
8. Jin, C., Peng, J.: A robust wavelet-based blind digital watermarking algorithm. Information Technology Journal 5(2), 358–363 (2006)

9. Gong, Q., Shen, H.: Toward blind logo watermarking in JPEG-compressed images. In: Proc. Int'l conf. on parallel and distributed computing, applications and technologies (PDCAT 2005), December 2005, pp. 1058–1062 (2005)
10. Feng, X.C., Yang, Y.: A New Watermarking Method Based on DWT. In: Hao, Y., Liu, J., Wang, Y.-P., Cheung, Y.-m., Yin, H., Jiao, L., Ma, J., Jiao, Y.-C. (eds.) CIS 2005. LNCS, vol. 3802, pp. 1122–1126. Springer, Heidelberg (2005)
11. Kundur, D., Hatzinakos, D.: Toward robust logo watermarking using multiresolution image fusion principles. IEEE Trans. Multimedia 6(1), 185–198 (2004)
12. Ejima, M., Miyazaki, A.: On the evaluation of performance of digital watermarking in the frequency domain. In: Proc. IEEE ICIP, October 2001, vol. 2, pp. 546–549 (2001)
13. Bhowmik, D., Abhayaratne, C.: Evaluation of watermark robustness to JPEG 2000 based content adaptation attacks. In: Proc. IET Int'l Conf. on Visual Info. Eng. (VIE 2008), July-August 2008, pp. 789–794 (2008)
14. Zhang, Z., Mo, Y.L.: Embedding strategy of image watermarking in wavelet transform domain. In: Proc. Image Compression and Encryption Technologies, vol. 4551, pp. 127–131. SPIE, San Jose (2001)
15. Vetterli, M., Kovačevic, J.: Wavelets and subband coding. Prentice-Hall Inc., Upper Saddle River (1995)
16. Daubechies, I., Sweldens, W.: Factoring wavelet transforms into lifting steps. J. Fourier Anal. Appl. 4(3), 245–267 (1998)

Multiple Watermarking with Side Information

Jun Xiao[1,2] and Ying Wang[1,2]

[1] Graduate University of the Chinese Academy of Sciences, Beijing 10049
[2] Key Lab of Network Security and Cryptology,
Fujian Normal University, Fuzhou 350007
{xiaojun,ywang}@gucas.ac.cn

Abstract. The idea of side information is introduced into multiple watermarking, and a multiple watermarking model with side information is proposed based on the Slepian-Wolf source encoding theory. In the proposed model, the correlation between the cover work and the multiple watermarks and the correlation among the watermarks themselves are considered and used as side information. The proposed model is used to improve the regular segmented multiple watermarking algorithm, and a novel watermarking algorithm is presented. Experimental results show that the improved algorithm can get a much better validity compared with the regular blind algorithm, and the efficiency of the proposed model is demonstrated. Namely, the proposed model can be used to guide the design of multiple watermarking algorithms effectively.

Keywords: multiple watermarking, side information, model, segmented watermarking, validity.

1 Introduction

As an embranchment of digital watermarking, multiple watermarking has some special applications, such as transaction tracking, and has become one of the research hotpots.

Many multiple watermarking algorithms have been proposed [1,2,3,4,5,6,7], and the existing algorithms can be divided into three classes [6,7]: re-watermarking, segmented watermarking and composite watermarking. Re-watermarking places watermarks one on top of another consecutively and the watermark signal could only be detected in the corresponding watermarked image using the former watermarked signal as the original image. At the same time, the watermark information embedded previously may be destroyed by the watermark information embedded latter, so some of the watermark information may not be extracted correctly. Segmented watermarking divides up the space available for watermarking and allocate each division to a different watermark signal. Clearly, the number of divisions limits the number of watermark signals to be embedded. Besides, when the number of the watermarks increases the size of each block decreases, and it will be hard to embed the same watermarks, still this is one of the most commonly used algorithms. Composite watermarking builds a single composite watermark from a collection of watermarks, and then

H.J. Kim, S. Katzenbeisser, and A.T.S. Ho (Eds.): IWDW 2008, LNCS 5450, pp. 379–387, 2009.

embeds the composite watermark into the cover work in a usual way. This scheme requires good signal merging methods, and its performances depend mainly on the performances of the signal merging methods, so there are few algorithms belonging to this sort.

Thus the performances of multiple watermarking algorithms need to be improved, such as the capacity, the validity and the robustness. At the same time, there are few multiple digital watermarking models, which can be used to guide the design of multiple digital watermarking algorithms.

In this paper, a side-informed multiple watermarking model is proposed based on the multiple sources encoding theory of Slepian-Wolf [8], and the model is used to improve the segmented multiple watermarking.

The rest of this paper is organized as follows. Section 2 describes the proposed multiple watermarking model. Section 3 presents the improved algorithm. Section 4 presents the experimental results. The paper is concluded in Section 5.

2   Multiple Watermarking Model with Side Information

Side-informed watermarking is developed under the elicitation of Costa's dirty paper model [9]. In side-informed watermarking, the correlation between the watermark and the cover work is used to improve the performances, and the watermark embedding process can be described as Fig.1, where m represents the watermark, C_o represents the cover work, and C_w represents the watermarked cover work.

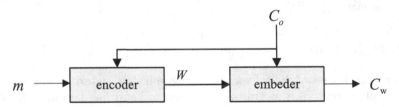

Fig. 1. Watermark embedding process of regular side-informed watermarking

But only one watermark is embedded in existing side-informed watermarking algorithms, and the idea of side information hasn't been used in multiple watermarking fully.

In fact, in Slepian-Wolf source encoding theory and Shannon's communication model with side information [8,10], which are the base of Costa's dirty paper model, there is no limit on the number of sources. Thus, a novel multiple watermarking model with side information is proposed by making full use of Slepian-Wolf source encoding theory and Shannon's communication model with side information, and the watermark embedding process is shown in Fig.2. The detailed process can be described as the following steps:

Step 1, watermarks $m[i](1 \leq i \leq N)$ are encoded using the correlation between the cover work C_o and the watermarks $m[i]$, and the encoded watermarks W_i are obtained.

Step 2, W_1 is embedded by embedder 1 using the correlation between C_o and $W_i(1 \leq i \leq N)$ and C_{W_1} is obtained.

Step 3, W_2 is embedded by embedder 2 using the correlation between C_o and $W_i(2 \leq i \leq N)$, and C_{W_2} is obtained.

Step 4, embedding the rest of the encoded watermarks $W_i(3 \leq i \leq N)$ by repeating step 3, and when all the watermarks are embedded, the watermarked cover C_W is obtained.

It's important to note that the encoders and embedders used in this model are the same with the encoders and embedders used in the regular side-informed watermarking, and any encoding and embedding method can be applied. Besidesthe watermark extracting process is similar with the regular side-informed watermarking.

According to the embedding process described above, the correlation between the cover and the watermarks and the correlation among the watermarks are used fully, and this embodies the idea of side information. Besides, many ways to use side information has already been proposed, for example, the embedding method with fixed linear correlation and the encoding method with perceptual shaping[11], and all the methods can be used in the proposed multiple watermarking model.

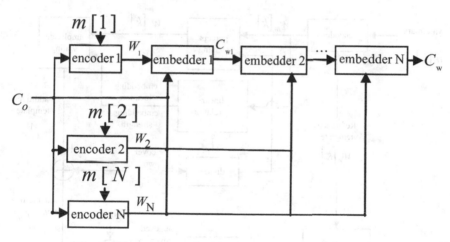

Fig. 2. Watermark embedding process of proposed model

3   Segmented Multiple Watermarking with Side Information

As described in section 1, segmented watermarking is a commonly used multiple watermarking method, and in this section, an improved segmented multiple watermarking algorithm will be proposed using the proposed model described in section 2. In the following, the regular blind segmented watermarking algorithm and the improved algorithm are introduced.

(1) regular blind segmented multiple watermarking algorithm

Suppose the cover work C_o is an image, the number of watermarks to be embedded is N, and a random sequence is used to represent a watermark. Firstly, the image C_o is divided into r divisions, and $Q = \{q_1, q_2, , ..., q_r\}$ is used to represent all the divisions, and make sure that $r \geq N$ so that all the watermarks can be embedded. Secondly, choose a division for every watermark randomly. Lastly, all the watermarks are added to the chosen divisions respectively.

When extracting the watermarks, the image to be detected is divided by applying the same method used in the embedding process, then watermarks are detected in every division respectively.

(2) segmented multiple watermarking algorithm with side information

In this algorithm, three ways of using side information are applied. The max linear correlation value between the divisions and the watermarks are used to decide the embedding location of every watermark, the Watson model [11] is used to shape the watermarks, and the watermark is embedded by fixing the linear correlation between the watermark and it's corresponding division. The watermarks are extracted by using linear correlation detector. The watermark embedding and extracting process are shown in Fig.3. In the following, the processes are described in detail.

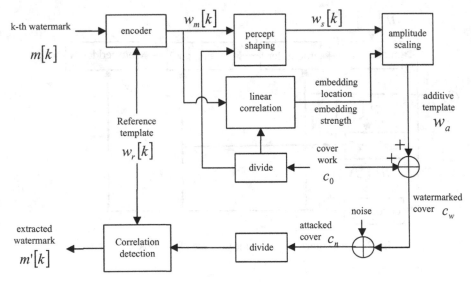

Fig. 3. Watermark embedding and extracting processes of the proposed algorithm

For convenience of description, suppose that the cover work is a gray image with size of 256×256, 128 watermarks will be embedded, and one watermark will be represented by a random sequence with size of 16×16. The watermark embedding process can be described as the following 6 steps:

Step 1, generating the watermark information templates. 128 random sequences which distribute uniformly between zero and one are generated and normalized, $w_r[k](1 \leq k \leq 128)$, is used to represent the k-th watermark sequence, and $w_r[k,i,j](1 \leq i,j \leq 16)$ represents the value at $[i,j]$ in the k-th watermark sequence. According to the value of the k-th watermark $m[k]$, the watermark sequence $w_r[k]$, is encoded according to formula (1), and the watermark information template $w_m[k]$ is obtained.

$$w_m[k] = \begin{cases} w_r[k], m[k] = 1 \\ -w_r[k], m[k] = 0 \end{cases} . \tag{1}$$

Step 2, dividing the cover image. According to the size of $w_m[k]$, the image is divided into 256 nonoverlapping blocks $C_o[t](1 \leq t \leq 256)$ with size of 16×16, and $C_o[t,x,y](1 \leq x,y \leq 16)$ is used to represent the pixel value.

Step 3, choosing the best blocks to embed the watermarks by using the correlation between the blocks and the watermark information templates as side information. For the k-th watermark information template $w_m[k]$, its linear correlations with all the 256 blocks $C_o[t](1 \leq t \leq 256)$ are computed. If $m[k] = 1$, the block which has the maximal linear correlation with $w_m[k]$ is chosen, the difference between the maximal correlation value and secondary correlation value is used as the scaling factor $\beta[k]$, and the secondary correlation value is used as the detection threshold $\tau_{lc}[k]$. If $m[k] = 0$, the block which has the minimal linear correlation with $w_m[k]$ is chosen, the difference between the minimal correlation value and the second smallest correlation value is used as the scaling factor $\beta[k]$, and the second smallest correlation value is used as the detection threshold $\tau_{lc}[k]$. The scaling factor $\beta[k]$ will be used to decide the embedding strength. For convenience of description, we suppose that the k-th block $C_o[k]$ is chosen as the best block for $w_m[k]$.

Step 4, encoding the watermark information templates with perceptual shaping [11]. 8×8 block Discrete cosine transform (DCT) is applied on the watermark information template $w_m[k]$, and $W_m[k]$ is used to represent the transformed information in the DCT domain. Watson model is used to obtain the total perceptual space $s[k]$ of $C_o[k]$. Perceptual shaping is applied on $W_m[k]$ to obtain the shapened information $W_s[k]$ according to formula (2), and then inverse DCT is applied on $W_s[k]$ to get the shapened template $w_s[k]$ in spatial domain.

$$W_s[k] = W_m[k]s[k] . \tag{2}$$

Step 5, deciding the embedding strength by using fixed linear correlation as the side information. Suppose the embedding strength of the k-th watermark is $\alpha[k]$ then the watermarked block is

$$C_w[k] = C_0[k] + \alpha[k]w_s[k] . \tag{3}$$

and the linear correlation between $C_w[k]$ and the watermark information template $w_m[k]$ is computed according to formula (4)

$$z_{lc}(w_m[k], W_w[k]) = (C_0[k] \cdot w_m[k] + w_a[k] \cdot w_m[k])/L . \tag{4}$$

where L is the pixel number of the blocks, and

$$w_a[k] = \alpha[k] w_s[k] .$$
(5)

By replacing $z_{lc}(w_m[k], C_w[k])$ with $\tau_{lc}[k] + \beta[k]$ in formula (4), the embedding strength $\alpha[k]$ is obtained as shown in formula (6).

$$\alpha[k] = \frac{L(\tau_{lc}[k] + \beta[k]) - C_0[k] \cdot w_m[k]}{w_s[k] \cdot w_m[k]} .$$
(6)

Step 6, embedding the watermark according to formula (3) with the strength obtained from formula (6) in step 5.

　　When extracting the watermarks, the location of the blocks chosen to embed the watermarks is used as the key, and the watermark extracting process is simple compared with the embedding process. Firstly, the image $C_n[t]$ to be detected is divided into blocks $C_n[t](1 \leq t \leq 256)$ applying the same method used in the embedding process. Secondly, the blocks $C_n[k]$ which may have watermarks are chosen according to the key. Then the watermarks are extracted according to formula (7) by comparing the threshold $\tau_{lc}[k]$ with $z_{lc}(w_r[k], C_n[k])$, where $z_{lc}(w_r[k], C_n[k])$ represents the linear correlation value between $C_n[k]$ and the watermark sequence $w_r[k]$.

$$m'[k] = \begin{cases} 1 & z_{lc}(w_r[k], C_n[k]) > \tau_{lc}[k] \\ \text{no watermark} & |z_{lc}(w_r[k], C_n[k])| \leq \tau_{lc}[k] \\ 0 & z_{lc}(w_r[k], C_n[k]) < -\tau_{lc}[k] \end{cases} .$$
(7)

4　Experimental Results

In multiple watermarking, how to avoid the interference of one watermark with another and obtain a good validity is one of the main problems, which is very difficult to resolve. In this section, the validity of the proposed algorithm is studied compared with the regular blind segmented algorithm.

　　In the regular blind segmented algorithm, the image is divided into 256 blocks with size of 16×16, and 128 blocks are chosen randomly as the locations to hide the 128 watermarks, and then all the watermarks are embedded with the same strength, while in the proposed algorithm, the locations and strengths are chosen according to different side information as described in section 3. In all the experiments, peak signal to noise ratio (PSNR) is used to measure the embedding distortions, and the PSNR is kept to be 40 dB. The error ratio of watermarks which can't be extracted correctly is used to measure the validity of the algorithms, and it is computed according to formula (8).

$$error ratio = \frac{the number of mistakenly extracted watermarks}{128} .$$
(8)

Considering the difference of different images and different random sequences, both the situations of random images and random sequences are tested under the same embedding distortion.

In the situation of random sequences, 1000 groups of random sequences are generated, and one group contains 128 random sequences which represent the 128 watermarks. The 1000 groups of random sequences are embedded into the gray Lena image separately using the blind algorithm and the proposed algorithm, that is to say, both of the two algorithms get 1000 watermarked Lena images. In the situation of random images, one group of sequences is generated, and it contains 128 random sequences which represent the 128 watermarks, but 1000 different gray image are chosen as the cover images. The 128 watermarks are embedded into the 1000 images separately using the blind algorithm and the proposed algorithm, that is to say, both of the two algorithms get 1000 watermarked images. Then watermarks are extracted from the watermarked images without attacks, and the error ratios are computed separately according to formula (8).

Fig.4 and Fig.5 show the distribution of the error ratios in the two situations respectively. According to Fig.4 and Fig.5, most error ratios of the proposed algorithm distributes around zero, while most error ratios of the regular blind algorithm distributes between 0.5 and 0.7. This indicates that the proposed algorithm can get a much better validity compared with the regular blind algorithm, and the effect of different images and random sequences on the ratios is bigger for the blind algorithm compared with the proposed algorithm.

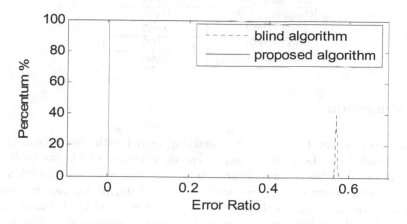

Fig. 4. Validity comparison under different watermark sequences

Besides, the robustness of the proposed algorithm is also tested. Since the validity of the blind algorithm is not very good when there is no attack, only the robustness of the proposed algorithm is given when attacks exist. Table 1 shows the average error ratios of all the watermarks. Table 1 indicates that the proposed algorithm is robust to most of the common image attacks, such as scaling, low pass filtering, med-filtering, cropping, JPEG compression and Gas noises.

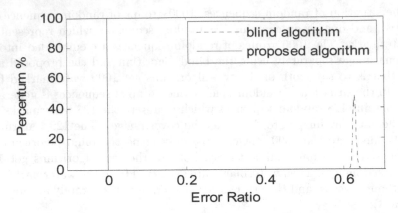

Fig. 5. Validity comparison under different cover images

Table 1. Robustness testing results of the proposed algorithm

attacks	average error ratio
Scaling (2)	0
Scaling (1/2)	0.023
Low pass filter (33)	0.257
Med filter (33)	0.047
Cropping (1/9)	0.148
JPEG (50%)	0.035
JPEG (70%)	0.092
Gas noise (0.0008)	0.098

5   Conclusion

In this paper, a novel multiple watermarking model with side information is proposed based on the multiple source encoding theory of Slepian-Wolf. The proposed model makes full use of the idea of side-informed communication, and the correlation between the cover work and the watermarks and the correlation among the watermarks themselves are considered and used. Based on the proposed model, an improved segmented multiple watermarking algorithm is presented to demonstrate the proposed model. In the improved algorithm, three ways of using side information are applied.

Experimental results of validity show that the improved algorithm can get a much better validity compared with the regular blind algorithm, and the effect of different images and random sequences on the ratios is bigger for the blind algorithm compared with the proposed algorithm. Besides, the robustness testing results show that the improved algorithm can obtain very good robustness under common attacks.

To get a better performance, more side information can be used, for example, the side information from the attacks, and this also can be included in the proposed multiple watermarking model. Thus, the proposed model can be used to guide the design of multiple watermarking algorithms effectively.

Acknowledgments. This work is supported by National Natural Science Foundation of China (No.60772155), Beijing Natural Science Foundation (No.4082029) and funds of Key Lab of Fujian Province University Network Security and Cryptology (No. 07A006).

References

1. Raval, M.S., Priti, P.R.: Discrete Wavelet Transform based Multiple Watermarking Scheme. In: 2003 Conference on Convergent Technologies for Asia-Pacific Region, vol. 3, pp. 935–938 (2003)
2. Fuhao, Z., Zhengding, L., Hefei, L.: A multiple watermarking algorithm based on CDMA technique. In: Proceedings of the 12th ACM International Conference on Multimedia, New York, pp. 424–427 (2004)
3. Khamlichi, Y.I., Machkour, M., Afdel, K., Moudden, A.: Multiple watermark for tamper detection in mammography image. WSEAS Transactions on Computers 5(6), 1222–1226 (2006)
4. Boato, G., Natale, D., Francesco, G.B., Fontanari, C.: Digital Image Tracing by Sequential Multiple Watermarking. IEEE Transactions on Multimedia 9(4), 677–686 (2007)
5. Lu, C.S., Hsu, C.Y.: Near-optimal Watermark Estimation and Its Countermeasure: Antidisclosure Watermark for Multiple Watermark Embedding. IEEE Transactions on Circuits and Systems for Video Technology 17(4), 454–467 (2007)
6. Mintzer, F., Braudaway, G.W.: If One Watermark Is Good, Are More Better? In: IEEE International Conference on Acoustics, Speech and Signal Processing, Phoenix, AZ, pp. 2067–2069 (1999)
7. Sheppard, N.P., Safavi-Naini, R., Ogunbona, P.: On Multiple Watermarking. In: Workshop on Multimedia and Security at ACM Multimedia, Ottawa, Ont, pp. 3–6 (2001)
8. Dan, J.: Information Theory and Coding, 2nd edn. University of Science and Technology of China Press, He Fei (2004)
9. Costa, M.: Writing on Dirty Paper. IEEE Transactions on Information Theory 29(3), 439–441 (1983)
10. Shannon, C.E.: Channels with Side Information at the Transmitter. IBM Journal, 289–293 (1958)
11. Cox, I.J., Miller, M.L.: Digital Watermarking. Morgan Kaufmann Publishers, San Francisco (2001)

Content Sharing Based on Personal Information in Virtually Secured Space

Hosik Sohn[1], Yong Man Ro[1], and Kostantinos N. Plataniotis[2]

[1] Department of Electrical Engineering,
Korea Advanced Institute of Science and Technology, South Korea
{sohnhosik}@kaist.ac.kr, {ymro}@ee.kaist.ac.kr
[2] Department of Electrical and Computer Engineering,
University of Toronto, Canada
{Kostas}@comm.utoronto.ca

Abstract. User generated contents (UGC) are shared in an open space like social media where users can upload and consume contents freely. Since the access of contents is not restricted, the contents could be delivered to unwanted users or misused sometimes. In this paper, we propose a method for sharing UGCs securely based on the personal information of users. With the proposed method, virtual secure space is created for contents delivery. The virtual secure space allows UGC creator to deliver contents to users who have similar personal information and they can consume the contents without any leakage of personal information. In order to verify the usefulness of the proposed method, the experiment was performed where the content was encrypted with personal information of creator, and users with similar personal information have decrypted and consumed the contents. The results showed that UGCs were securely shared among users who have similar personal information.

Keywords: UGC, Contents sharing in social media, virtual secure space, fuzzy vault.

1 Introduction

The number of user generated content (UGC) in social media has been rapidly increasing during the last several years [1], [2], [3]. Ordinary people have become producers of contents as well as consumers so called 'prosumers' , being capable of publishing their own contents on the social media such as Flickr, MySpace, FaceBook and Youtube [4], [5], [6], [7]. These social media are open space where every user is allowed to upload and consume contents freely. This open space offers better accessibility and unlimited sharing of contents among users. However, in the open space, it cannot be avoid that the uploaded UGC could be delivered to unwanted users and misused.

In order to secure contents being delivered to unwanted users or to unauthorized users, the demand of content encryption arises. But it is contrary to the principles of open space of UGC which is meant to freely share contents. In the open space such as Flickr, MySpace and Youtube, users do not know who have consumed their

H.J. Kim, S. Katzenbeisser, and A.T.S. Ho (Eds.): IWDW 2008, LNCS 5450, pp. 388–400, 2009.
© Springer-Verlag Berlin Heidelberg 2009

contents. Therefore, if users are provided with option such that they can confine or select consumers or a group of consumers for their UGC, the confidence of uploading contents on the open space would be increased. To be specific, the open space allows every user to upload and consume freely, but at the same time the mentioned option would allow creators to deliver contents to the specific consumers or a group of consumers.

The advantage of such option is that creators would not concern about the misusage of their contents by being assured that their contents are delivered to users with similar preference. Another advantage is that when uploading and downloading of contents are done based on the preference of users, i.e. the mentioned option is used, the contents can be automatically classified according to the contents preference and personal information of users, thereby assisting data and server management. However, the personal information of users such as name, gender, date of birth, religion, occupation, address and hobby should be private reflecting that users would like to hide their personal information on the Web. As a matter of fact the revealing personal information on the Web can cause various security related attacks [8].

In this paper we propose a novel scheme which allows users to keep an opportunity to generate and consume UGC securely without revealing personal information. By the proposed method, virtual secure space, which is generated with the creator's personal information, is available to only users whose personal information is well matched. In order to create the virtual secure space, we used the vault set that is an encoding result of fuzzy vault scheme [9]. By applying the fuzzy vault scheme, the personal information of users during the process of the proposed method can be kept securely.

Fuzzy vault scheme had been introduced to provide a security using biometric data such as finger print and face features [10], [11], [12]. The concept of fuzzy vault is that the vault set is generated through binding both biometric data and security key, and the generated vault set provides the secret key only if it is decoded with the identical biometric data. This paper proposes a content sharing method in a secure space by using fuzzy vault scheme.

The rest of the paper is organized as follows: Section 2 provides an overall scenario of virtual-secure content system based on user information. Section 3 describes the process of binarizing the personal information and encoding/ decoding of the vault set. The experiments and results are presented in section 4. Section 5 discusses the security issue. Finally, the paper is concluded in section 6.

2 Virtual-Secure Content System Based on User Information

In this section, the virtual secure content system is explained with user scenario. There are two choices user can make. One is to upload contents without encryption for unlimited consumption to any users. The other is our proposed method which encrypts data with secret key. In this case, content creators allow users to decrypt and consume the contents only if they have similar personal information including contents preference and profile.

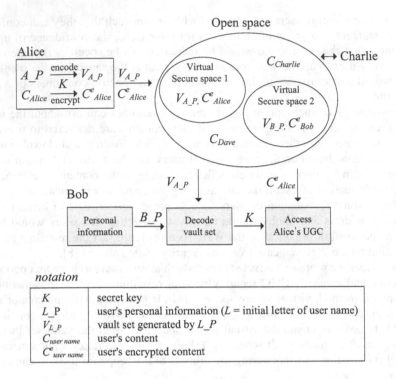

Fig. 1. The consumption process of UGC in the open space

In Fig. 1, the open space refers to the web space such as Youtube. Since the access of the contents in the open space is not restricted, users can consume contents anytime anywhere. In this space, encrypted contents (C^e_{Alice} and C^e_{Bob}) and unencrypted contents ($C_{Charlie}$ and C_{Dave}) can be uploaded according to the choices content creators made. For the encrypted content, corresponding vault set (V_{A_P} and V_{B_P}) is uploaded together. In real applications, instead of encryption, content can be visibly watermarked or restricted by providing preview only. The access to the encrypted contents are given to users according to their personal information (user profile and contents preference), and users with similar personal information can be defined as a virtual group, i.e. virtual space. This space, generated by the vault set, is only opened to users with similar personal information. Thus, the virtual space can be secured by grouping the contents according to personal information. In the view of content categorization, each virtual secure space is a category generated by the combination of personal information. Therefore, we can generate more various and flexible categories than using the current categorization method. The following will show possible scenarios that could occur.

Alice is a user (or UGC creator) who wants to provide her content to the group of people who have similar personal information. To do so, she uploads UGC (C^e_{Alice}) that is encrypted by her secret key (K) and the vault set (V_{A_P}) generated by her personal information. One possible example of her personal information would be that, she is in 'twenties', 'not married' and her favorite genre of movie is documentary movie. Her profile and content preference are the features of her personal information. Uploaded Alice's vault set creates a virtual secure space and allows accessibility to only users who have similar personal information.

Bob is a user who wants to consume Alice's content. In order to consume the content, Bob downloads Alice's vault set (V_{A_P}) and decrypts it with his personal information. For example, assume that Bob's personal information is, 'twenties', 'not married', favorite content is documentary movie and comedy show. Since his personal information is mostly overlapped with Alice's, he can decode the vault set and acquire the secret key which allows the decryption of the content. Hence, Bob can freely consume the contents inside the virtual secure space generated by Alice where this space is opened to only users who have similar personal information with Alice, like Bob. From the perspective of Alice, this can be seen as a method for limiting the accessibility to the contents consumers by creating the virtual space.

Lastly, Charlie and Dave are users who want to provide their contents to the public. The contents of Charlie and Dave ($C_{Charlie}$ and C_{Dave}) are uploaded without any encryption. This method is currently well used in most of UGC sites, where consumption of the contents is unlimited but cannot protect the contents being delivered to unwanted users.

For the scenario mentioned above, the benefit of applying fuzzy vault scheme is that, content sharing is possible even though personal information between users is not perfectly matched. And, unlike other general web sites, the proposed system does not require a centralized authority. The only function of authority in open space is to distribute personal information table (will be mentioned in section 3.1) and store contents along with vault set. If the personal information is encrypted without utilizing fuzzy vault scheme, the centralized authority is required to handle key management and distribution. Also, security of centralized authority should be guaranteed. In practice, key management is the most difficult aspects of cryptography.

Moreover, in order to compare user information among other users, encrypted personal information should be transformed to the same domain or decrypted to original information. For the case of transforming to the same domain, the authority is required that has knowledge of rules that transform encrypted personal information into the identical domain for comparison. If the encrypted personal information is decrypted to original information, the personal information is revealed at other user's side or in the open space. If the centralized authority has perfect security, the utilization of the fuzzy vault scheme is not necessary. However, in the practical standpoint, the perfect security cannot be guaranteed. And vulnerable authority has a risk of leaking all personal information at once. In the next section, the method for generating the virtual secure space is presented, where secure delivery and consumption of contents are guaranteed.

3   Realization of Virtual Space Using Personal Information

In this section, the detailed explanation of encoding and decoding procedure is presented. Vault set is generated by binding a secret key, which is used for protecting the content, and the personal information of UGC creator. In order to consume the contents, consumer should decode the vault set by his/ her personal information. If the certain number of personal information matches personal information of consumers, then the proper secret key is acquired, and finally consumers can decrypt the content. In summary, the virtual secure space, generated by the vault set, is only opened to users with similar personal information and secured by grouping the contents according to personal information. Therefore, the virtual secure space has a thread of connection with group security. We do not consider betrayal of a group member in this paper. Before explaining encoding and decoding procedure, we define the personal information item of user (binarized personal information), and describe how to generate it.

3.1 Personal Information

For the personal information used in this paper, we have referred categories from [15] for describing favorite film. In order to use personal information as an input of fuzzy vault scheme, we define items of personal information such as age, gender and information regarding favorite content. Table 1 represents the defined personal information items.

Table 1. Personal informaion table (personal information item P_i, i = 1 to 310)

Sub-category	Personal information items (PII)
Age (S_1)	$P_1 \sim P_6$
Gender (S_2)	$P_7 \sim P_8$
Marriage status (S_3)	$P_9 \sim P_{12}$
Hobbies (S_4)	$P_{13} \sim P_{35}$
Occupation (S_5)	$P_{36} \sim P_{59}$
Language (S_6)	$P_{60} \sim P_{198}$
Released year of the favorite film (S_7)	$P_{199} \sim P_{211}$
Language used in the favorite film (S_8)	$P_{212} \sim P_{223}$
Genre of the favorite film (S_9)	$P_{224} \sim P_{255}$
Source of the favorite film (S_{10})	$P_{256} \sim P_{287}$
Subjective matter of the favorite film (S_{11})	$P_{288} \sim P_{298}$
Distributor of the favorite film (S_{12})	$P_{299} \sim P_{310}$

The personal information consists of sub-categories (S_i, i=1~12) and each category consist of personal information items (PII, P_i i=1~310) that represent user information, including contents preference. User information-related PII consists of personal information, such as age, gender, occupation and hobby. Hobby [13] and occupation [14] sections were appended to define the user information. Content preference-related PII consist of information regarding favorite content, especially films. It contains items that can describe the user's favorite films such as those released year, genre, and source etc. PIIs of each one sub-category regarding user information and contents preference are presented as an example in Table 2.

Table 2. Example of personal information items

Sub-category	P_i	Personal information item	Sub-category	P_i	Personal information item
Hobby	P_{13}	Amateur science related	Genre	P_{224}	Action
	P_{14}	Animal-related		P_{225}	Adventure
	P_{15}	Arts and crafts		P_{226}	Animation
	P_{16}	Collecting		P_{227}	Avant-garde
	P_{17}	Computer-related		P_{228}	Biographical
	P_{18}	Cooking		P_{229}	Blaxploitation
	P_{19}	DIY (Do It Yourself)		P_{230}	Children
	P_{20}	Electronics		P_{231}	Comedy
	P_{21}	Film-making		P_{232}	Crime
	P_{22}	Games		P_{233}	Disaster
	P_{23}	Historical reenactment		P_{234}	Documentary
	P_{24}	Interactive fiction		P_{235}	Drama
	P_{25}	Internet-based hobbies		P_{236}	Epic
	P_{26}	Literature		P_{237}	Exploitation
	P_{27}	Model building		P_{238}	Fantasy
	P_{28}	Music		P_{239}	Film noir
	P_{29}	Observation		P_{240}	LGBT
	P_{30}	Outdoor/ nature activities		P_{241}	Horror
	P_{31}	Performing arts		P_{242}	Independent short
	P_{32}	Photography		P_{243}	Musical
	P_{33}	Physical activities		P_{244}	Pirate
	P_{34}	Toys of some sophistication		P_{245}	Romance
	P_{35}	Transportation		P_{246}	Romantic comedy
				P_{247}	Romantic drama
				P_{248}	Science fiction
				P_{249}	Screwball comedy
				P_{250}	Sports
				P_{251}	Stop-motion
				P_{252}	Thriller
				P_{253}	Conspiracy thriller
				P_{254}	War
				P_{255}	Western

To use the mentioned personal information items as an input of fuzzy vault scheme, 16 bit-value should be assigned to each item. Namely, in order for sub-items in Table 1 to be used as a input of fuzzy vault scheme, we assigned 16 bit-pseudo-random number to each personal information item. We defined this value as PII value. N number of PII values corresponding to his personal information at the encoder and decoder side is used as an input. Let the set of PII values at the encoder side be the $T=\{t_1, t_2, ...,t_N\}$, and at the decoder side be the $Q=\{q_1, q_2,...,q_N\}$.

3.2 Fuzzy Vault Encoder with Personal Information Item

Fig. 2 represents the fuzzy vault encoder using PII value. As seen in the figure, fuzzy vault encoder generates a vault set with the list of PII value and secret key as an input. The set of user's PII value is $T=\{t_1, t_2, ...,t_N\}$ which is identical to user's personal information item list, T in Fig.2. 128-bit Advanced Encryption Standard (AES) key is utilized as a secret key to protect the contents [16]. Using Cyclic Redundancy Check (CRC), 16 bit redundancy is added to the secret key. In order to generate cyclic redundancy, 16-bit primitive polynomial presented in (1) is used [17].

$$P_{CRC}(x) = x^{16} + x^{15} + x^2 + 1. \tag{1}$$

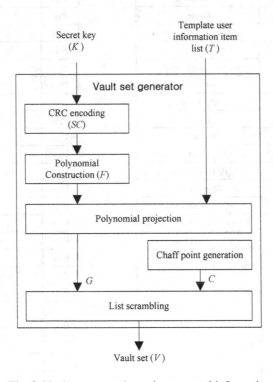

Fig. 2. Vault set generation using personal information

Through CRC encoding, 16-bit redundancy is added to 128-bit AES key so that total of 144-bit data SC is constructed. In order to construct polynomials of (2), SC is divided into non-overlapping 16-bit unit and used to generate coefficients ($C_8 \sim C_0$) of polynomial. Every operation after the construction of polynomial is under the Galois field ($GF(2^{16})$).

$$F(u) = c_8 u^8 + c_7 u^7 + \cdots + c_1 u + c_0. \tag{2}$$

A vault set, which is the encoding result, consists of genuine set G and chaff set C. The elements of genuine set is a pair of values, which is a template user information item list $T=\{t_1, t_2, ...,t_N\}$ and its projected value to polynomial $F(u)$ of equation (2). Genuine set G is expressed as equation (3).

$$G = \{(t_1, F(t_1)), (t_2, F(t_2)),...,(t_N, F(t_N))\}. \tag{3}$$

And chaff set C can be defined as equation (4).

$$C = \{(u'_1, r_1), (u'_2, r_2),...,(u'_M, r_M)\}. \tag{4}$$

Chaff set C is used to protect the genuine set securely. In the original fuzzy vault scheme, random points are used as chaff points. However, in our case, if we use the same method, adversary can easily separate chaff point and genuine points using personal information table, as in Table 1. Therefore, we define chaff set C as a set that is composed of M number of PII value $(u'_1, u'_2,...,u'_M)$ from 310 PII values in personal information table $(u'_i \neq t_j,$ for $1 \leq i \leq M,$ $1 \leq j \leq N)$, which are not used as personal information. The values satisfying equation (5) are chosen for $r_1 \sim r_M$.

$$F(u_i') \neq r_i \ \ 1 \leq i \leq M. \tag{5}$$

Then, even if the adversary (malicious attacker) knows the personal information table, the he/ she cannot distinguish between genuine set and chaff set. Finally, the vault set V is generated by scrambling the genuine set G and chaff set C.

$$V = \{(x_1, y_1), (x_2, y_2), (x_3, y_3), (x_4, y_4),...,(x_{N+M}, y_{N+M})\}. \tag{6}$$

3.3 Fuzzy Vault Decoder with Personal Information Item

The fuzzy vault decoder uses PII values an input and if more than $D+1$ number of PII values is the same as that of encoding side, the original secret key can be acquired which guarantees the successful decryption of the content. Here, D denotes the degree of polynomial in equation (2). The set of queryt user's PII value is $Q=\{q_1, q_2,...,q_N\}$, which is identical with Query user information item list Q in Fig. 3.

The element of V, (x_j, y_j) which satisfies the equation (7), is selected as a candidate point after comparing query PII values and elements of the vault set.

$$q_i = x_j, \ 1 \leq i \leq N, \ 1 \leq j \leq N+M. \tag{7}$$

If the number of candidate points is k, the next step 'Combination sets determination' generates every possible set that can select $D+1$ number of points from k number of candidate points, $C(K, D+1)$. Let each set be the $L=\{(a_0, b_0), (a_1, b_1),...,(a_D, b_D)\}$, then the polynomial is reconstructed using equation (8) for case of $C(k, D+1)$ in Lagrange interpolation block.

$$F^*(u) = \sum_{j=0}^{D} b_j f_j(u) = c_8 u^8 + c_7 u^7 + \cdots + c_1 u + c_0, \ \text{where } f_j(u) = \prod_{i=0, i \neq j}^{D} \frac{u - a_i}{a_j - a_i}. \tag{8}$$

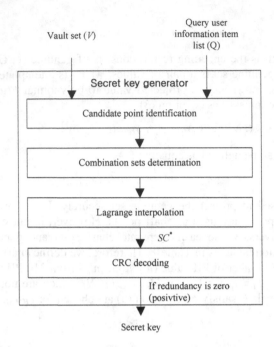

Fig. 3. Secret key generation from vault set and personal information of user

In order to construct SC^* of 144 bits data, the coefficients $C_8{\sim}C_0$ obtained by equation (8) are concatenated. Every SC^* checks the redundancy through CRC decoder. And if the redundancy is not zero, the result of CRC decoding is negative. Thus, it is impossible to acquire the proper secret key. Therefore, in the case of redundancy being zero, only 128 bits excluding LSB 16 bits are used for decrypting the content.

4   Experiment

To evaluate the effectiveness of the proposed method and demonstrate its versatility, experiments were performed to validate the consumption process of protected contents in the open space and whether virtual secure space is formed by the proposed fuzzy vault set. Experimental condition is as follows.

To simplify the experiment, we assume that each user has 12 PII values. If nine PII values are identical with each other, i.e. PII values used to encode a vault set and query user's PII values are identical, the encrypted content can be consumed. To realize this, the number of PII value is 12 in the fuzzy vault encoder as seen in Fig. 2, the degree of encoder polynomial is eight and the number of chaff point is 298. Since the degree of the polynomial is eight, 9 out of 12 of consumers PII values should be identical to that of user A in order to decode the vault set. We assumed that six users, including the content creator, are accessing the encrypted content. Detailed personal information of the six users used in the experiment is presented in Table 3.

Table 3. Private information of user A, B, C, D, E and F

Sub-category	User A	User B	User C	User D	User E	User F
S_1	P_4	P_5	P_5	P_4	P_5	P_4
S_2	P_7	P_7	P_7	P_8	P_8	P_8
S_3	P_{11}	P_{11}	P_{11}	P_{11}	P_{11}	P_{11}
S_4	P_{32}	P_{16}	P_{16}	P_{32}	P_{32}	P_{32}
S_5	P_{39}	P_{39}	P_{39}	P_{43}	P_{39}	P_{39}
S_6	P_{90}	P_{90}	P_{90}	P_{90}	P_{90}	P_{90}
S_7	P_{211}	P_{209}	P_{209}	P_{211}	P_{211}	P_{211}
S_8	P_{214}	P_{214}	P_{214}	P_{214}	P_{214}	P_{214}
S_9	P_{245}	P_{231}	P_{231}	P_{245}	P_{245}	P_{245}
S_{10}	P_{276}	P_{261}	P_{276}	P_{276}	P_{276}	P_{276}
S_{11}	P_{289}	P_{289}	P_{289}	P_{289}	P_{289}	P_{289}
S_{12}	P_{299}	P_{299}	P_{299}	P_{310}	P_{299}	P_{299}

In the experimental scenario, we assume that user A uploads his/ her encrypted content with the vault set to the open space and the rest of users consume content. As seen in Table 3, user B has seven, C has eight, user D has nine, use E has ten, and user F has eleven identical personal information with user A. We observed that whether user B, C, D, E, and F who have different personal information each other, can decrypt and consume the content of user A.

Table 4 is the decoding result of users according to the degree of the polynomial (D). We are explaining the case when D is eight. Since the number of element in vault set is the same as the number of sub-items (P_i, $1 \leq i \leq 310$) in Table 1, PII values that are different from user A's are always matched to the chaff points. Because we assumed that the number of all user's personal information is 12, the number of candidate set generated at the decoder size is always 220 ($C(12, 9)$) which is a combination of selecting 9 from 12.

In Table 4, the positive represents the number of sets when the redundancy is zero after CRC decoding at the decoder side in Fig 3. Likewise, the negative represents the number of sets when the redundancy is not zero. If the CRC decoding result is positive, the probability of a set containing the proper secret key is very high. In the case of result being negative, the set has no secret key. The decoding process is completed as soon as the secret key is found, and encrypted content is decrypted by the key.

The experimental results for each user are as follows. For user B and C, since the number of PII values identical to user A is less than nine, it is impossible to reconstruct a polynomial. Proper secret key cannot be acquired. For user D, nine PII values are identical to user A. Thus, only one set ($C(9, 9)$) out of 220 is positive. And for user E, since ten PII values are identical to user A, ten sets ($C(10, 9)$) are positive. Finally, for user F, since eleven PII values are identical to user A, fifty-five sets ($C(11, 9)$) are positive. Therefore, user D, E, and F decrypted content of user A with secret key which is a 128-bit from MSB of 144-bit SC^* in Fig 3.

Table 4. Decoding result

D	Consumer	Identical PII to user A	Candidate set	Positive set	Negative set	Authenti-cation
6	user B	7	792	1	791	Yes
	user C	8	792	8	784	Yes
	user D	9	792	36	756	Yes
	user E	10	792	120	672	Yes
	user F	11	792	330	462	Yes
7	user B	7	495	0	495	No
	user C	8	495	1	494	Yes
	user D	9	495	9	486	Yes
	user E	10	495	45	450	Yes
	user F	11	495	165	330	Yes
8	user B	7	220	0	220	No
	user C	8	220	0	220	No
	user D	9	220	1	219	Yes
	user E	10	220	10	210	Yes
	user F	11	220	55	165	Yes
9	user B	7	66	0	66	No
	user C	8	66	0	66	No
	user D	9	66	0	66	No
	user E	10	66	1	65	Yes
	user F	11	66	11	55	Yes
10	user B	7	12	0	12	No
	user C	8	12	0	12	No
	user D	9	12	0	12	No
	user E	10	12	0	12	No
	user F	11	12	1	11	Yes

This proves and validates that the virtual secure space generated by personal information of user A is opened to user D, E and F, but not to user B and C. Since user A, D, E and F have more than eight identical personal information items, the encrypted content and the vault set generated by their own personal information create a secure space where user B and C have no accessibility. In this space, user A, D, E and F can upload their contents freely and share them securely. If we apply this concept to every user in the open space, we can form various flexible groups.

5 Security Issue

In this section, we consider the possibility of personal information leakage. For the security considerations, let us assume that the adversary wants to break the virtual secure space using brute force attack simulated by our experimental environment. That is to iterate over all combinations of attributes and try to access the virtual space by randomly selecting 9 PII values from 12 sub-categories where only one PII value is selected for each sub-category. For a given personal information table, the adversary can break the virtual secure space by evaluating all combination sets $(1.94 \times 10^{13}$ sets) at maximum which takes about 1.60×10^{5} years when the degree of polynomial is 8 (Note one set evaluation time is about 0.26 sec at 3.4GHz CPU).

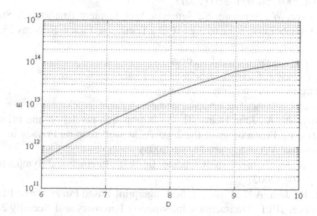

Fig. 4. Evaluation for brute force attack

The required number of evaluations to break the virtual secure space according to degree of the polynomial is presented in Fig. 4. In this figure, D denotes the degree of the polynomial and E denotes the number of evaluation to break virtual secure space.

6 Conclusion

Most of well-known and popular UGC sites allow users for unlimited sharing of contents, whereas content creator has no option for giving limitations to consumers. To solve this problem, this paper proposes UGC sharing method based on the personal information of users. The proposed method is to encrypt the content by secret key and upload it to open space with vault set that is generated by personal information of users. Since only consumers who have similar personal information to the content creator can acquire the proper secret key, content creator can limit the consumption of contents over group of consumers by his/ her intention. The first priority to take into consideration in UGC sharing based on personal information is a risk of personal information leakage. During the process of comparing personal information among users, no information is revealed to the public using fuzzy vault scheme. Moreover, the proposed system does not require a centralized authority. Since the authority does not manage users' personal information in one place, the chance of personal leakage by security attack is very low in the systemic view. With the proposed method, not only UGC creator can protect his content indirectly, but also increases the trust for consumers of the content. For further work, personal information defined in this paper can be combined with biometric data such as fingerprint and face using the proposed scheme for secured authentication system.

References

1. OECD study on the Participative Web : User Generated Content (October 3, 2007), http://www.oecd.org
2. Ames, M., Naaman, M.: Why We Tag: Motivations for Annotation in Mobile and Online Media. In: CHI 2007, pp. 971–980 (2007)
3. Loia, V., Pedrycz, W., Senatore, S.: Semantic Web Content Analysis: A Study in Proximity-Based Collaborative Clustering. IEEE Trans. on Fuzzy Systems 15, 1294–1312 (2007)
4. Youtube, http://www.youtube.com
5. Flickr, http://www.flickr.com/
6. Myspace, http://www.myspace.com/
7. Facebook, http://www.facebook.com/
8. Gross, R., Acquisti, A., John Heinz III, H.: Information revelation and privacy in online social networks. In: Proceedings of the 2005 ACM workshop on Privacy in the electronic society, Alexandria, VA, USA, pp. 71–80 (2005)
9. Juels, A., Sudan, M.: A fuzzy vault scheme. In: IEEE International Symposium, Information Theory, p. 408 (2002)
10. Nandakumar, K., Jain, A.K., Pankanti, S.: Fingerprint-Based Fuzzy Vault: Implementation and Performance. IEEE Transactions Information Forensics and Security 2(4), 744–757 (2007)
11. Wang, Y., Plataniotis, K.N.: Fuzzy Vault for Face Based Cryptographic key Generation. In: Biometrics Symposium, September 2007, pp. 1–6 (2007)
12. Lee, Y.J., Bae, K., Lee, S.J., Park, K.R., Kim, J.: Biometric key Binding: Fuzzy Vault Based on Iris Images. In: Lee, S.-W., Li, S.Z. (eds.) ICB 2007. LNCS, vol. 4642, pp. 800–808. Springer, Heidelberg (2007)
13. Wikipedia, List of hobbies, http://en.wikipedia.org/wiki/List_of_hobbies
14. U.S. Department of Labor, Bureau of Labor Statistics. Standard Occupational Classification(SOC) Major Groups, http://www.bls.gov/soc/soc_majo.htm
15. Wikipedia, List of films, http://en.wikipedia.org/wiki/Lists_of_films
16. NIST. Advanced Encryption Standard (AES) (November 2001), http://csrc.nist.gov/publications/fips/fips197/fips-197.pdf
17. Press, W.H., Teukolsky, S.A., Vetterling, W.T., Flannery, B.P.: Numerical Recipes in C, 2nd edn. Cambridge University Press, Cambridge (1992)

Evaluation and Improvement of Digital Watermarking Algorithm Based on Cryptographic Security Models

Satoshi Kaneko[1], Isao Echizen[2], and Hiroshi Yoshiura[1]

[1] Faculty of Electro-Communication, The University of Electro-Communications,
1-5-1, Chofugaoka, Chofu, 182-8585, Japan
yoshiura@hc.uec.ac.jp
[2] National Institute of Informatics,
2-1-2, Hitotsubashi, Chiyoda-ku, Tokyo, 101-8430, Japan
iechizen@nii.ac.jp

Abstract. The security of digital watermarking methods is based on the assumption that their embedding and detection algorithms are kept secret. They are thus inconvenient for public review and widespread use. A watermarking algorithm that can be open while remaining secure is thus needed. Twelve models that are analogous to cryptanalysis models are described for evaluating the security of digital watermarking. They were used to evaluate the patchwork-based watermarking algorithm and identify its weakness. Two scenarios were then developed that strengthen this algorithm. The use of these models will further the development of an open secure algorithm.

Keywords: security analysis, patchwork-based alogorithm, cryptanalysis.

1 Introduction

Digital content–such as pictures, videos, and music–is being made widely available because of its advantages over analog content. It requires less space, is easier to process, and is not degraded by aging or repeated use. A serious problem, however, is that the copyright of digital content is easily violated because the content can be readily copied and redistributed through the Internet. Digital watermarking, which helps protect the copyright of digital content by embedding copyright information into it, is a key countermeasure.

Existing digital watermarking methods [1,2,3,4,5] depend on their embedding and detection algorithms being kept secret. This means that these algorithms cannot be evaluated by third parties or widely used. A watermarking algorithm that can be open while remaining secure is thus needed to enable public review and widespread use. A security analysis of watermarking algorithms is an essential requirement for establishing an open secure algorithm. There have been several studies focusing on the security analysis of watermarking algorithms [6,7,8,9]. However, they focused on particular attack objectives and attacker

H.J. Kim, S. Katzenbeisser, and A.T.S. Ho (Eds.): IWDW 2008, LNCS 5450, pp. 401–418, 2009.

abilities and proposed no systematic model for security analysis. Moreover, the ideas presented were conceptual, and there was no proposed security analysis or improvements to specific watermarking algorithms.

In this paper, we describe 12 security analysis models for use in evaluating the strength of digital watermarking that are analogous to cryptanalysis models. We use the patchwork-based watermarking algorithm [2] as a representative spatial domain-based watermarking algorithm, clarify the effect of possible attacks by analyzing it, and describe two scenarios that strengthen it. Section 2 describes the contributions of the paper. Section 3 describes previous security analysis for digital watermarking. Section 4 describes the patchwork-based watermarking algorithm. Section 5 describes the 12 security analysis models we developed for evaluating the strength of the algorithm. Section 6 presents the results of the evaluation, and Section 7 describes the two scenarios that strengthen it. Section 8 concludes the paper with a summary of the key points and a mention of future work.

2   Contributions of This Paper

We present 12 models for analyzing the security of watermarking algorithms and that are analogous to cryptanalysis models. The models include attack objectives, attacker abilities, and evaluation criteria. The analogies are drawn by matching various aspects of digital watermarking to cryptography. More precisely, we respectively match embedded information and the watermarked image to plaintext and ciphertext and match watermark embedding and detection to encryption and decryption. We considered three kinds of attacks: identify the key, embed intended information, and embed different information from original. We established the ability of the attacker on the basis of the conditions for an embedded information attack and a watermarked image attack, which correspond to a known-plaintext attack and a chosen-ciphertext attack, respectively. We used a watermark embedder, which corresponds to an encryption oracle. We also established unconditional security and the computational complexity of the watermarking algorithm as criteria for security evaluation.

We analyzed the patchwork-based algorithm using the proposed security analysis models and identified its vulnerabilities.

- The algorithm does not provide unconditional security for the three kinds of attacks.
- The algorithm may not provide the computational complexity needed for identifying the key.
- The algorithm does not provide the computational complexity needed for identifying algorithm that corresponds to the identifying key and needed for embedding intended or different information.

We then developed two schemes for overcoming the identified vulnerabilities. Their use with the patchwork-based algorithm reduces the security of watermarking to that of cryptography. One scheme uses error detection codes, enabling it to overcome even more vulnerabilities. Since most algorithms for spatial-domain-based watermarking are extensions of the patchwork-based algorithm, the proposed models can be used to evaluate various spatial-domain-based algorithms. The models fortify the foundation on which open secure algorithms for digital watermarking can be constructed.

3   Previous Work

3.1   Security Evaluations of Watermarking Algorithms

There has been security analysis of watermarking based on the cryptographic scheme. Kalker et al.[6] matched a watermarking scheme to a communication system and discussed security against altering and removing communication data. Particularly, they indicated the existence of attacks against the watermarking scheme, which are attacks estimating the original image from the watermarked one and then estimating the image plane representing the watermarks (the watermark plane) by subtracting the original image from the watermarked one and attacks replacing the watermarked image plane with a different watermark plane (replacing attack). Moreover, they suggested requirements for countermeasures against attacks that generate the watermark plane depending on the original image and take into account the use of a watermark embedder and detector by the attacker as the abilities of the attacker. However, the details of the attacks and the effectiveness of the countermeasures were unclear because they focused on particular attack objectives and attacker abilities and presented no model for security analysis.

Zollner et al.[7] concentrated on steganography and analyzed identifiability, which means whether information is embedded or not into content. They classified the possible attacks into ones using only stego content and ones using both stego and cover content and proved that unconditional security is not ensured when subtracting cover content from stego content, which indicates one of the latter case of the attacks. They proposed a countermeasure against this vulnerability: prepare a content set by randomly and secretly selecting cover content from the content set and then embedding information into it.

Cayre et al.[8] analyzed the security of displacement-based watermarking, in which pixels are selected on the basis of a key and then replaced with watermarks. They proved that an attacker using only the watermarked image cannot identify the watermark plane if the watermarked image corresponds to the perfect cover. They also proved that, if an attacker uses both the watermarked image and watermark plane, the security of the watermarking depends on the amount of data in the watermarked image and that an attacker using both the watermarked and original images can estimate the watermark plane and use it to remove the watermarks from the image. Moreover, Cayre et al. analyzed the security of spread-spectrum-based watermarking. They classified the information the attackers can use, defined the security of the watermarking on the basis of the amount of data and calculation needed to estimate the secret parameters, and proved the existence of possible attacks by using blind-based sound source separation.

3.2 Security Models of Cryptography

Cryptanalysis assumes that a cryptographer can figure out cryptographic algorithms in detail. The analysis is done with respect to three aspects.

Attack Objective. Attack objectives include identifying the key, obtaining the algorithm, which corresponds to identifying the key, estimating plaintext from cipher text, and making an intentional change to plaintext by modifying the corresponding ciphertext.

Attacker Abilities. Attacker abilities are classified in accordance with the availability of plaintext and ciphertext. Several types of attacks have been defined by this classification, including attacks in which the attacker can use different ciphertexts, attacks in which the attacker can use different pairs of ciphertext and plaintext (known-plaintext attacks), attacks in which the attacker can select the plaintext to be encrypted (chosen-plaintext attacks), attacks in which the attacker can adaptively select plaintext on the basis of the results of encrypting previously selected plaintext (adaptive chosen-plaintext attacks), and attacks in which the attacker can select ciphertext to be decrypted and can use decrypted plaintext (chosen-ciphertext attacks). Classifications based on the conditions of availability of encryption and decryption as oracles have also been defined.

Criteria for Security Evaluation. The following criteria for evaluating cryptanalysis models have been defined.

Computational complexity: Security measured by the amount of computation needed to break a ciphertext. A cryptographic algorithm is considered insecure if the computation needed to break ciphertext encrypted by this algorithm can be done within polynomial time. The algorithm is considered secure if cannot be done within polynomial time but done within time more than polynomial time such as exponential time.

Unconditional security: Security measured by the amount of information about the plaintext that is obtainable from the data available to the attacker (ex. ciphertext). The attacker can attack with infinite computational capability. If the algorithm is considered secure in terms of unconditional security, it is also considered secure in terms of computational complexity. Unconditional security is thus at a higher level than computational complexity.

4   Assumed Watermarking Algorithm

4.1 Patchwork-Based Algorithm

In this section, we briefly describe the patchwork-based algorithm for which our proposed models for analyzing security were developed. The patchwork-based algorithm is a representative spatial-domain-based algorithm, and most spatial-domain-based watermarking schemes are extensions of it. Since the proposed models can be used for evaluating various watermarking schemes by overcoming the vulnerabilities of the patchwork-based algorithm, our analysis is specific to this algorithm. We first describe the procedures for watermark embedding and detection with the patchwork-based algorithm. For simplicity, we describe the method for one-bit watermarking. In the multi-bit case, each bit is assigned to a specific area of the image and is embedded and detected in the same way as in the one-bit case. The embedding part of the one-bit watermarking is as follows.

Watermark Embedding. Two images are used for embedding: original image $O = \{o_{i,j} \mid 1 \leq i \leq W, 1 \leq j \leq H\}$ and image $R = \{r_{i,j} = 0 \mid 1 \leq i \leq W, 1 \leq j \leq H\}$. Two sets of N different locations of pixels A and B are randomly assigned. These two sets correspond to embedded bit $b \in \{0, 1\}$; A and B are given by $A = \{(x_k, y_k) \mid 1 \leq k \leq N\}$ and $B = \{(u_k, v_k) \mid 1 \leq k \leq N\}$, which satisfy $A \cap B = \emptyset$ and $2N \leq WH$.

Watermark plane $P = \{p_{i,j} \mid 1 \leq i \leq W, 1 \leq j \leq H\}$ is then generated using the following procedure; that is, $p_{i,j}$ is determined by

$$p_{i,j} = \begin{cases} \pm\delta & \text{if } (i,j) \in A \\ \mp\delta & \text{if } (i,j) \in B \\ r_{i,j} & \text{otherwise,} \end{cases} \tag{1}$$

where $\delta(> 0)$ represents the watermark strength, and "$\pm$" and "$\mp$" respectively mean "+" and mean "$-$" when embedded bit b is 1 and "$-$" and "+" when b is 0.

Then, watermarked image $O' = \{o'_{i,j} \mid 1 \leq i \leq W, 1 \leq j \leq H\}$ is generated by adding the watermark plane to the original image: $o'_{i,j} = o_{i,j} + p_{i,j}$ for all (i,j)s.

Watermark Detection. The difference between the average value of pixels o'_{x_k,y_k}s and that of pixels o'_{u_k,v_k}s is calculated by $v = 1/N \sum_{k=1}^{N} \{o'_{x_k,y_k} - o'_{u_k,v_k}\}$. Then, b is determined by comparing v with threshold value $T(> 0)$; that is, if $v > T$, b is detected as 1; if $v < -T$, b is detected as 0; if $-T \leq v \leq T$, b is not detected.

4.2   Security of Patchwork-Based Algorithm

The use of a symmetric key in both watermark embedding and detection has been proposed as a way to improve the security of the patchwork-based algorithm [12]. This approach is similar to that used in algorithms for symmetric-key cryptography.

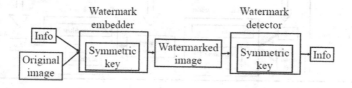

Fig. 1. Use of key in patchwork-based algorithm

More precisely, as shown in Figure 1, the locations for watermarks A and B are determined by using the symmetric key stored in the watermark embedder. In watermark detection, a watermarked pixel is identified by using the symmetric key stored in the watermark detector. Determination of the locations for the watermarks does not depend on the original image but on the symmetric key.

5   Methods of Analysis

In this section, we describe our 12 security analysis models for evaluating the security of digital watermarking and that are analogous to cryptanalysis models. Note that, given the objectives of our study, we assume that the algorithms for watermark embedding and detection are open to the public.

5.1   Correspondence of Cryptography to Digital Watermarking

Figure 2 illustrates our mapping of cryptography to digital watermarking. Plaintext and ciphertext respectively correspond to embedded information and the watermarked image, and encrypting and decrypting respectively correspond to watermark embedding and detection. The correspondence between symmetric-key cryptography and digital watermarking is shown in Table 1. Since the output of the decryptor is plaintext and that of the watermark detector is the embedded information, the correspondence of the plaintext to the embedded information is a natural interpretation of the relationship between cryptography and watermarking.

5.2   Objectives of Attacks

We assumed the following objectives based on analogy to cryptanalysis models: (I) identify key or obtain algorithm, which corresponds to identifying the

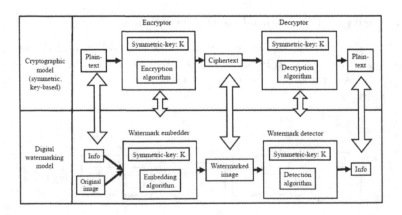

Fig. 2. Mapping cryptography to digital watermarking

Table 1. Correspondence between symmetric-key cryptography and digital watermarking

Cryptography	Plaintext	Ciphertext	Encryption	Decryption
Watermarking	Embedded information	Watermarked image	Watermark embedding	Watermark detection

key and (II) alter the watermarked image and embed different information into it. Objective (II) can be classified into two types: alter the watermarked image and embed intended information and alter the watermarked image and embed different information (not intended information). We could add a third objective: altering the watermarked image so that the embedded information cannot be detected. Although this is a critical issue for digital watermarking, we do not deal with it here as it is not markedly analogous to cryptanalysis models. Consequently, we classify attack objectives into three types.

(A) Identify key or obtain algorithm, which corresponds to identifying the key
(B) Alter the watermarked image and embed intended information into it
(C) Alter the watermarked image and embed information different from the original into it

5.3 Abilities of Attackers

We assumed attackers have three abilities.

- Attackers can obtain the watermarked image.
- Attackers can obtain the embedded information.
- Attackers can use the watermark plane.

We assume that an attacker cannot obtain the original image because digital watermarking is intended for copyright protection, so the original images are protected by the content holders. In contrast, since watermarked images are widely distributed over the Internet, an attacker can easily obtain watermarked images. If the embedded information represents the receiver's or sender's name, an attacker can recognize the content. However, if the information is encrypted, the attacker cannot recognize it. Here we assume strong conditions, meaning that the embedded information is recognizable to the attacker. The watermark plane is normally hidden by the signals of the original image. However, an attacker can identify the plane in the flat areas of the image.

We defined the conditions of availability for a watermark embedder and detector, which correspond to the encryption and decryption oracles.

Availability of watermark embedder: If an attacker can use a watermark embedder, the attacker can readily obtain both the watermarked image and the corresponding embedded information. The attacker, however, cannot generate a watermarked image where the intended information is embedded. This is because, in the cryptanalysis models, the encryption oracle does not generate ciphertext where the intended plaintext is encrypted.

Availability of watermark detector: If an attacker can use a watermark detector, the attacker can detect the watermarks in any watermarked image. The attacker can thus readily obtain both the watermarked image and the corresponding embedded information. We assumed that the attacker can only use a watermark embedder and detector and cannot obtain the keys stored in them. We assumed that the keys stored in the embedder and detector remain the same and are not changed.

5.4   Criteria for Security Evaluations

On the basis of the analogy to cryptanalysis models, we established two criteria for evaluation.

Computational complexity: Given the attacker abilities listed above, we defined computational complexity as proportional to the amount of computation needed to accomplish the objective of the attack. The watermarking algorithm is considered insecure if the computation needed is done within polynomial time. It is considered secure if it is not done within polynomial time but within time more than polynomial time such as exponential time.

Unconditional security: Again, given the attacker abilities, we defined unconditional security as proportional to the degree of objective accomplishment. The attacker can attack with infinite computational capability.

6   Security Evaluation of Patchwork-Based Algorithm

6.1   Models

On the basis of the analysis described in Section 5, we evaluated the security of the patchwork-based algorithm. We considered three attack objectives: (A) identify the key, (B) embed intended information, and (C) embed information different from the original. We also considered the four possibilities for watermark embedder and detector availability. Table 2 shows the 12 models used, which are comparable to cryptographic analysis.

Objective (A): Identify Key

Model (a): attacker can use neither embedder nor detector. An attack corresponds to a known-plaintext attack in the cryptanalysis model; it uses different pairs of embedded information and watermarked image, which respectively correspond to plaintext and ciphertext.

Model (b): attacker can use only detector. Attacker can generate any watermarked image (ciphertext) using the patchwork-based algorithm, which is open to the public, and can detect the embedded information (plaintext) from the generated image. This corresponds to an attack in which the attacker can adaptively select ciphertext on the basis of the results of decrypting the previously selected ciphertext (adaptive chosen-ciphertext attack) in the cryptanalysis model. Moreover, known-plaintext attack is possible, as in model (a).

Table 2. Models for patchwork-based algorithm evaluation

			(A) Identify key	(B) Embed intended info	(C) Embed info different from original
	Attack Objective				
	Availability Embedder	Detector			
(a)	No	No	Comparable to known-plaintext attack (uses embedded info and watermarked image)	Comparable to attack in which ciphertext corresponding to intended plaintext (embedded info and watermarked image are used) is generated	Comparable to attack in which ciphertext that does not correspond to original plaintext is generated
(b)	No	Yes	Comparable to adaptive chosen-ciphertext attack and known-plaintext attack (same as (a)-(1))	Comparable to attack in which decryption oracle is used and ciphertext corresponding to intended plaintext is generated	Comparable to attack in which decryption oracle is used and ciphertext that does not correspond to original plaintext is generated
(c)	Yes	No	Comparable to adaptive chosen-plaintext attack	Comparable to attack in which decryption oracle is used and ciphertext corresponding to intended plaintext is generated	Comparable to attack in which encryption oracle is used and ciphertext that does not correspond to original plaintext is generated
(d)	Yes	Yes	Comparable to adaptive chosen-ciphertext and adaptive chosen-plaintext attacks	Comparable to attack in which encryption and decryption oracles are used and ciphertext corresponding to intended plaintext is generated	Comparable to attack in which encryption and decryption oracles are used and ciphertext that does not correspond to original plaintext is generated

Model (c): attacker can use only embedder. Attacker can generate watermarked image (ciphertext) from any embedded information (plaintext) by using the embedder. This corresponds to an attack in which the attacker can adaptively select plaintext on the basis of the results of encrypting the previously selected plaintext (adaptive chosen-plaintext attack) in the cryptanalysis model.

Model (d): attacker can use both embedder and detector. Attacker can attack as in models (b) and (c). This corresponds to the adaptive chosen-ciphertext and adaptive chosen-plaintext attacks in the cryptanalysis model.

Objective (B): Embed Intended Information

Model (a): attacker can use neither embedder nor detector. Attacker can generate watermarked image (ciphertext) so that the intended information (plaintext) can be detected. This corresponds an attack in which the attacker generates ciphertext from the plaintext in the cryptanalysis model.

Model (b): attacker can use only detector. Attacker can use the detector in addition to attack in model (a). This corresponds to an attack in which the attacker can use decryption oracle and can generate ciphertext from the plaintext in the cryptanalysis model.

Model (c): attacker can use only embedder. Attacker can generate watermarked image (ciphertext) from any embedded information (plaintext) by using the embedder and can generate the watermarked image so that the intended information can be detected. This corresponds to an attack in which the attacker can use both the encryption and decryption oracles and can generate ciphertext from the plaintext in the cryptanalysis model.

Model (d): attacker can use both embedder and detector. Attacker can generate watermarked image (ciphertext) so that the intended information (plaintext) can be detected by using the embedder and detector. This corresponds to an attack in which the attacker can use both the encryption and decryption oracles and can generate ciphertext from the plaintext in the cryptanalysis model.

Objective (C): Embed Information Different from the Original

Model (a): attacker can use neither embedder nor detector. Attacker can analyze pairs of embedded information (plaintext) and watermarked image (ciphertext) and generate watermarked image so that information different from the original can be detected. This corresponds to an attack in which the attacker can generate ciphertext from different plaintext in the cryptanalysis model.

Model (b): attacker can use only detector. Attacker can use the detector in addition to attack in model (a). This corresponds to an attack in which the attacker can use decryption oracle and can generate ciphertext from different plaintext in the cryptanalysis model.

Model (c): attacker can use only embedder. Attacker can generate watermarked image (ciphertext) from any embedded information (plaintext) by using the embedder and can generate watermarked image so that the intended information can be detected. This corresponds to an attack in which the attacker can use the encryption oracle and can generate ciphertext from different plaintext in the cryptanalysis model.

Model (d): attacker can use both embedder and detector. Attacker can generate watermarked image (ciphertext) from any embedded information (plaintext) by using the embedder and detector. This corresponds to an attack in which the attacker can use both the encryption and decryption oracles and can generate ciphertext from different plaintext in the cryptanalysis model.

6.2   Evaluation

We evaluated the success of attacks against the patchwork-based algorithm for the 12 security analysis models described in Section 6.1. Since we measures similar to those described in previous work [6] against replacing attacks (described in Section 3.1), we neglect replacing attacks.

In all the models described, intended information can be embedded if the attacker can identify the key, and different information can be embedded if the attacker can embed intended information. Moreover, if an attacker can attack when the embedder and detector are unavailable, the attacker can carry out the same attacks when they are available. These points are the default and, for simplicity and brevity, are not described in the following individual analyses.

Objective (A): Identify Key

Model (a): attacker can use neither embedder nor detector. Since an attacker can get pairs of the watermarked image and embedded information, the attacker can attack using the following procedure.

Step 1: Create a watermark detector into which any key can be incorporated using the patchwork-based algorithm, which is open to the public.

Step 2: Detect the information in the watermarked image obtained by using the detector created in Step 1. Determine whether the key used in the detector is authentic by checking whether the detected information matches the embedded information previously obtained. If the detected information does not match the obtained information, change the key and retry the detection process. Continue the procedure until the authentic key is found.

We can evaluate the computational complexity by determining the amount of computation needed for the attack. In any model, a brute force attack is inevitably successful. Unconditional security is therefore not provided because the attack is successfully if one can attack with infinite computational capability.

Model (b): attacker can use only detector. An attacker can attack using the following procedure.

Step 1: Create a watermark embedder into which any key can be incorporated using the open algorithm and generate a watermarked image using the embedder.

Step 2: Detect the information in the watermarked image generated in Step 1 using a legitimate detector.

Step 3: Create a watermark detector in which any key can be incorporated using the open algorithm.

Step 4: Detect the information in the watermarked image generated in Step 1 using the detector created in Step 3.

Step 5: Determine whether the key used in the detector in Step 4 is authentic by checking whether the information detected in Step 2 matches that one detected in Step 4. If it does not, change the key in both the embedder in Step 1 and the detector in Step 3 and retry the detection process by using the information detected in Step 4. Continue the procedure until the authentic key is found.

Computational complexity can be evaluated whether this procedure is completed within a practical time.

Model (c): attacker can use only embedder. An attacker can attack using the following procedure.

Step 1: Generate two watermark planes on which the i-th bit of the embedded information is different by using the embedder.

Step 2: Identify the image areas where the i-th bit is embedded by comparing the two planes.

Step 3: Identify the relationship between one of the two watermark planes and the embedded information by performing this procedure over all the bits of the information.

This attack can be done within practical time proportional to the number of bits of embedded information. An attacker can thus obtain the algorithm, which corresponds to identifying the key.

Objective (B): Embedding Intended Information

Model (a): attacker can use neither embedder nor detector. An attacker can analyze the relationship between the embedded information and the watermark plane when the attacker has different pairs of them. There are two specific forms of this attack. In one, there are two watermarked images for which the i-th bit of the embedded information is different but the original images are same. In this model, an attacker can identify areas in the image where the i-th information bit is embedded by comparing the two watermarked images. In the other, there are two watermarked images in which the i-th bit of the embedded information is different, and each watermark plane can be estimated from the two images, as described in Section 5.3. In this model, an attacker can identify areas on the plane in which the i-th bit of the information is embedded by comparing the two watermark planes. This model is also effective when the original images are different.

Model (b): attacker can use only detector. The following attack is possible in addition to that in model (a).

Step 1: Generate watermark plane using the patchwork-based algorithm, which is open to the public.

Step 2: Detect the information in the generated plane by using the detector.

Step 3: Generate another watermark plane while fixing a portion of the first generated plane.

Step 4: Detect the information in the second plane. If some portion of the currently detected information matches that detected in Step 2, that portion is considered to belong to the watermark plane fixed in Step 3.

Step 5: Estimate the relationship between the embedded information and the watermark plane by repeatedly doing Steps 3 and 4.

This attack needs less computation than that of a brute force attack on the key because one can estimate the relationship between the embedded information and the watermark plane for each portion. The computational complexity therefore cannot be determined.

Objective (C): Embed Information Different from the Original

Model (a): attacker can use neither embedder nor detector. An attack that is more effective than that in model (a) of objective (B) can be made. In that model, an attacker should identify the embedded areas corresponding to each bit and do so for all bits. In this model, the attack can be done by inverting the embedded area even if the attacker can identify the embedded area corresponding to only one bit.

Model (b): attacker can use only detector. An attacker can attack using the following procedure.

Step 1: Create a watermark embedder into which any key can be incorporated using the open algorithm.

Step 2: Prepare a pair of a watermarked image and embedded information and generate a new watermarked image by embedding information different from the original into the image with twofold watermark strength by using the embedder.

Step 3: Detect the information in the generated image. The attack is successfully if the detected information differs from original. Otherwise, repeat the procedure using a different key.

6.3   Vulnerabilities

On the basis of security analysis described in Section 6.2, we summarize the vulnerabilities of the patchwork-based algorithm. First, if an attacker can attack with infinite computational capability, the attacker can identify the key by using a brute force attack on the key, as described in model (a) of objective (A). Using the identified key, the attacker can embed intended and different information. Unconditional security therefore cannot be ensured for all the models listed in Table 2.

Next, we discuss the models not discussed above, classifying them by the three attack objectives.

Objective (A): Identify Key

Model (i): computational complexity is not provided. This model corresponds to model (c) of objective (A). Using a watermark embedder, an attacker can estimate the relationship between the embedded information and the watermark plane and can obtain the algorithm, which corresponds to identifying the key.

Model (ii): computational complexity is not likely to be provided. As with the attacks in models (a) and (b) of objective (A), since there is no definition with regard to security of the patchwork-based algorithm, the amount of computation needed for an attack is uncertain. Computational complexity is therefore not likely to be provided.

Objective (B): Embed Intended Information

Model (i): computational complexity is not provided. This model corresponds to models (c) and (d) of objective (B). An attacker can identify the decipher key and can attack using the key.

Model (ii): computational complexity is not likely to be provided. The attack in model (a) of objective (B) can be made when the attacker can, by chance, get a pair of embedded information and the watermarked image that have a specific relationship. The computational complexity is not likely to be provided once one can get this pair. It is therefore not likely to be provided in this model. As with the attack in model (b) of objective (B), since there is no definition with regard to security of the patchwork-based algorithm, the amount of computation needed for the attack is uncertain. Computational complexity is therefore not likely to be provided.

Objective (C): Embed Information Different from the Original

Model (i): computational complexity is not provided. This model corresponds to models (c) and (d) of objective (C). An attacker can identify the decipher key by using the procedure described in model (c) of objective (A) and can attack using the identified key.

Model (ii): computational complexity is not likely to be provided. As with the attack in model (b) of objective (C), since there is no definition with regard to security of the patchwork-based algorithm, the amount of computational needed for the attack is uncertain. Computational complexity is therefore not likely to be provided.

7   Improvements

We propose two schemes for improving the patchwork-based algorithm.

7.1   Proposed Scheme (I)

Structure and Procedures. The vulnerabilities of the patchwork-based algorithm can be overcome by the following improvements.

(i) Algorithm is divided into a part for maintaining security and a part for embedding information into the image.

(ii) Conversion of the embedded information into a watermark plane is considered a one-to-one conversion. Moreover, this conversion is a format conversion such as that from binary digits to decimal number. It is not used in the procedure for maintaining security.

(iii) The procedure for maintaining security consists of cryptographic algorithms.

(iv) Watermarking security reduces to that of the cryptographic algorithms.

On the basis of these improvements, proposed scheme (I) can be implemented using the following simple structure. This scheme adds to the patchwork-based algorithm a process for encrypting the embedded information (See Figure 3(a)).

Watermark embedding. The information to be embedded is first encrypted using the key (K). In accordance with the patchwork-based algorithm, a watermark plane is generated on the basis of the encrypted information (C). A watermarked image is then generated by adding the watermark plane to the original image. At this time, the encrypted information and the watermark plane are in a one-to-one relationship.

Watermark detection. In accordance with the patchwork-based algorithm, the encrypted information is detected from the watermarked image. Then the encrypted information is decrypted using the same key used for watermark embedding, and the embedded information is extracted.

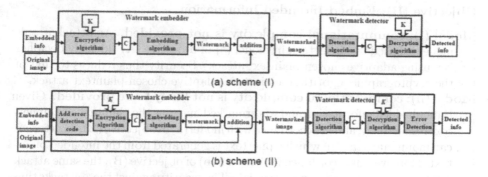

Fig. 3. Structure of proposed schemes

Security.

We evaluated the security of scheme (I) for the models described in Section 6.3.

Objective (A): Identify Key

Model (i): computational complexity is not provided. As mentioned in model (c) of objective (A) in Sections 6.1 and 6.2, the original patchwork-based algorithm is vulnerable to an adaptive chosen-plaintext attack. With scheme (I), since the original image has no effect on identifying the locations of watermarks, security against an adaptive chosen-plaintext attack when the watermarked image is considered analogous to ciphertext is similar to that against the same attack when the watermark plane is considered analogous to ciphertext. Moreover, since the watermark plane and the encrypted information are in a one-to-one relationship, security against an adaptive chosen-plaintext attack when the watermark plane is considered analogous to ciphertext is similar to that against the same attack when the encrypted information is considered analogous to ciphertext. Since security against an adaptive chosen-plaintext attack when the embedded and encrypted information are respectively considered analogous to plaintext and ciphertext is similar to that of a cryptographic algorithm, the security in this model reduces to that of the cryptographic algorithm.

Model (ii): computational complexity is not likely to be provided. An attack in model (a) of objective (A) corresponds to a known-plaintext attack in which the embedded and encrypted information are respectively considered analogous to plaintext and ciphertext. An attack in model (b) of objective (A) corresponds to an adaptive chosen-ciphertext attack in which the embedded and encrypted information are respectively considered analogous to plaintext and ciphertext. On the basis of the correspondence described in model (i) of objective (A) in this section, security in models (a) and (b) of objective (A) reduces to that of the cryptographic algorithm against known-plaintext and adaptive chosen-ciphertext attacks.

Objective (B): Embed Intended Information

Model (i): computational complexity is not provided. An attack in this
   model is based on the attack in model (c) of objective (A) (identify decipher
   key using adaptive chosen-plaintext attack). Security thus reduces to that of
   the cryptographic algorithm against an adaptive chosen-plaintext attack.

Model (ii): computational complexity is not likely to be provided. Given
   the embedded information and watermarked image as the plaintext and ciphertext respectively, and in accordance with model (a) of objective (B), an attacker
   can mount an attack in which ciphertext is generated from the intended plaintext. Moreover, in accordance with model (b) of objective (B), the same attack
   can be done using the decryption oracle. The security against these attacks thus
   reduce to that of the cryptographic algorithm against the same attack given the
   correspondence described in model (i) of objective (A) in this section.

Objective (C): Embed Information Different from the Original

Model (i): computational complexity is not provided. The security in this
   model reduces to that of the cryptographic algorithm against an adaptive
   chosen-plaintext attack as in model (i) of objective (B) in this section.

Model (ii): computational complexity is not likely to be provided. In
   the attack of model (b) of objective (C), the attacker embeds information
   different from the original into the original watermarked image with twofold
   watermark strength by using the open algorithm and an arbitrary key. The
   algorithm of a legitimate detector is likely to detect ciphertext C' different
   from the original when this attack is executed. In the cryptographic scheme,
   plaintext can be decrypted from arbitrary ciphertext. Once C' is detected,
   information different from the original can be decrypted by the detection
   algorithm. Therefore, embedding information different from the original is
   likely to be successful.

Scheme (I) overcomes the vulnerabilities of several of the models: (A) identify
decipher key, (B) embed intended information, and (C) (i) embed information
different from the original. It does not overcome the vulnerabilities of model (C)
(ii). We therefore propose scheme (II).

7.2   Proposed Scheme (II)

Structure and Procedures. Scheme (II) uses error-detection code to overcome the vulnerabilities of model (C) (ii). Figure 3(b) shows the structure of the
scheme. We assume that the algorithm for the error-detection code is open to
the public, the same as that of the watermarking algorithm.

Watermark embedding. Error-detection code is added to the embedded information, and the embedding information with error detection code is encrypted using a key (K). In accordance with the patchwork-based algorithm,
   a watermark plane is generated on the basis of the encrypted information
   (C). The procedure then proceeds in the same way as for scheme (I).

Watermark detection. As with scheme (I), the encrypted information is decrypted using the same key used for watermark embedding, and the embedded information, with the error-detection code, is extracted. Any alteration of the embedded information is identified by using the error-detection code.

Security. For an attack of the model (C) (ii) to be successful against scheme (II), the watermarked image must pass through the error-detection scheme of a legitimate detector. That is, a pair of embedded information and the corresponding error-detection code must be decrypted by a legitimate key. If the embedded information is I and the corresponding error-detection code as $E(I)$, an attack is successful if an attacker who knows nothing about the key can generate ciphertext so that $I|E(I)$ can be decrypted using the key. The security against this attack depends on the security of the cryptographic and error-detection algorithms and the length of the error-detection code

8   Conclusion

We described 12 security analysis models that are analogous to cryptanalysis models and designed for evaluating the security of digital watermarking. They take into account the attack objectives, the attacker abilities, and the security level. We used them to identify the vulnerabilities of the patchwork-based algorithm and proposed two schemes that improve its security against three types of attacks: identify decipher key, embed intended information, and embed different information. The use of these models will further the development of an open secure algorithm, Future work will focus on analyzing the security provided by these two schemes against a removal attack, one of the essential requirements for watermarking security, on analyzing the security provided by other watermarking algorithms such as the spread spectrum-based algorithm, and on establishing improved schemes.

References

1. Cox, I., et al.: Digital Watermarking. Morgan Kaufmann Pub., San Francisco (2001)
2. Bender, W., et al.: Techniques for data hiding. IBM Syst. J. 35(3&4), 313–336 (1996)
3. Cox, I., et al.: A secure, robust watermark for multimedia. In: Proc. Information Hiding First International Workshop, pp. 185–206 (1996)
4. Bas, P., et al.: Geometrically invariant watermarking using feature points. IEEE Trans. Image Processing 11(9), 1014–1028 (2002)
5. He, D., et al.: A RST resilient object-based video watermarking scheme. In: Proc. 2004 International Conference on Image Processing, pp. 737–740 (2004)
6. Kalker, T., et al.: Considerations on watermarking security. In: Proc. Fourth IEEE Workshop on Multimedia Signal Processing, pp. 201–206 (2001)
7. Zollner, J., et al.: Modeling the security of steganographic systems. In: Aucsmith, D. (ed.) IH 1998. LNCS, vol. 1525, pp. 344–354. Springer, Heidelberg (1998)

8. Cayre, F., et al.: Watermarking Security: Theory and Practice. IEEE Trans on Signal Processing 53(10), 3976–3987 (2005)
9. Cayre, F., et al.: Watermarking Attack: Security of WSS Techniques. In: Cox, I., Kalker, T., Lee, H.-K. (eds.) IWDW 2004. LNCS, vol. 3304, pp. 171–183. Springer, Heidelberg (2005)
10. Schneier, B.: Applied Cryptography. John Wiley & Sons Inc., Chichester (1995)
11. Goldreich, O.: Foundations of Cryptography. Cambridge University Press, Cambridge (2001)
12. Furukawa, J.: Secure Detection of Watermarks. IEICE Trans. E87-A, (1), 212–220 (2004)

A Color Image Watermarking Scheme in the Associated Domain of DWT and DCT Domains Based on Multi-channel Watermarking Framework

Jiang-bin Zheng and Sha Feng

School of Computer, Northwestern Polytechnic University, Xi'an 710072, China
zhengjb@nwpu.edu.cn
fengsha_2005123@sina.com

Abstract. In this paper, a new watermarking scheme based on the multi-channel image watermarking framework is proposed, which is mainly used to resist geometric attacks. The multi-channel image watermarking framework generates a watermarking template from one of image channels data, and then embeds this watermarking template into another image channel. Self-synchronization can be obtained in the procedure of detecting watermarking, because both the watermarking template and watermarking image undergo a similar geometric attack. The watermarking scheme we propose in this paper generates watermarking templates from one image channel data to which we have done DWT and DCT transformation and embeds these watermarking templates into another image channel data to which we have done DWT and DCT transformation. This scheme is robust to some geometric attacks and common signal processing such as geometric distortion, rotation, JPEG compression and adding noise ect.. Experimental results demonstrate that this watermarking scheme produced by the multi-channel watermarking framework achieves a high performance of robustness.

Keywords: Multi-channel framework; image watermarking; geometric attacks; DWT; DCT.

1 Introduction

Digital watermarking is mainly used for copyright protection. Now the proposed watermarking schemes are robust to common signal processing such as compression, adding noise and filter ect., but many schemes are fragile to geometric attacks such as rotation, scaling, and cropping ect., because minor geometric manipulation to the watermarked image could dramatically reduce the receiver ability to detect watermark. The watermarking schemes against geometric attacks contain two generations: the first generation watermarking schemes and the second generation watermarking schemes.

The first generation watermarking schemes can be categorized into five kinds: the watermarking schemes based on the image registration technique, the watermarking schemes based on the exhaustive search, the watermarking schemes based on invariant transform, the watermarking schemes based on template insertion and the watermarking schemes based on the self-reference watermarking.

H.J. Kim, S. Katzenbeisser, and A.T.S. Ho (Eds.): IWDW 2008, LNCS 5450, pp. 419–432, 2009.

The watermarking schemes based on the image registration technique [1] estimate the geometric deformation parameters by matching the original image and the attacked watermarking image.

The watermarking schemes based on the exhaustive search [2] define the scope of various distorted parameters and its resolutions, by testing all assemble of parameters to find the best geometric transform parameters. This approach can work well if a watermarking image undergoes an affine transformation with little scale parameters, otherwise it will be time consuming.

The watermarking schemes based on invariant transform embed watermarking into a geometrically invariant domain [3], normalization domain [4], or Zernike transform domain [5]. The authors in [6,7] embed watermarking into Fourier–Mellin domain which provides a translation, rotation and scaling invariant. An image that undergoes an affine transformation has a unique normalized image under a certain image size, the authors in [8,9,10] embed watermarking into a normalization image. In [11], the authors propose a watermarking scheme based on wavelet transformation. The schemes in [12] and [13], embed the watermarking in an affine-invariant domain by using generalized Radon transformation, and Zernike moment, respectively. The authors in [14] propose a multi-bit image watermarking algorithm using local Zernike moments. The watermarking schemes based on invariant transform are robust against global affine transformation, but they are often vulnerable to cropping attack or un-affine geometric transform attacks.

The schemes based on template insertion embed artificial reference template into an image to identify the geometrical transform and then recover watermarking according to the embedded reference template. The authors in [15,16] embed two orthogonal lines into DFT domain. After detecting these two line length and angle, they can solve parameters of an affine transformation that the image undergoes.

The watermarking schemes based on the self-reference watermarking are the ones which watermarking themselves are identifiable, so they do not rely on synchronous templates, but on their own autocorrelation to get synchronous. One kind of watermarking uses auto-correlation function to get synchronization [17], and the other do not need to get synchronization, such as cirque watermarking [18].

The second generation watermarking schemes which can be called the feature-based watermarking schemes are based on image content. This kind of watermarking schemes embed the watermarking in the geometrically invariant image features, the watermarking can be detected without synchronization error. Recently, image feature based watermarking schemes have drawn much attention for its high robustness to geometric attacks. The scheme shown in paper [19] chooses feature points which are stable to geometric transformation by using the Harris detector, and performs a Delaunay tessellation on these points to embed the watermarking with a classical additive scheme. Another two examples shown in papers [20] and [21] extract feature points by using Mexican hat wavelet, and embed several copies of the watermarking into the disks centered at the feature points. The authors in [22] propose an image watermarking scheme based on invariant regions of scale-space representation.

All of the above mentioned schemes supposed that a watermarked image undergoes an affine transformation attack, in fact a watermarking image often undergoes other types of geometrical attacks besides affine transformation. The above mentioned schemes are suitable for affine transformation, but these schemes will be helpless when a watermarked image undergoes other kinds of geometrical attacks such as cropping, projective transformation and other special geometrical distortions. We also find that most watermarking schemes embed a watermarking signal into a single channel of an image such as a luminance channel. We have proposed a multi-channel watermarking framework, and designed a spatial domain [23] and a DWT domain [24] watermark based on this framework. In this paper we design a association of DWT and DCT domains watermarking scheme based on this framework to demonstrate the advantages of this algorithm. By using this multi-channel framework, the embedded watermarking template and watermarking image undergo the similar geometrical attacks. So synchronization is not necessary in the procedure of detecting a watermark even if the watermarked image undergoes a special geometric distortion attacks.

The rest of this paper is organized as follows. In section 2, we introduce the multi-channel watermarking framework. In section 3, we give an association of DWT and DCT domains example based on the multi-channel watermarking framework. In section 4, we show the experimental results. And the conclusion and feature work are followed in the last section 5.

2 Multi-channel Watermarking Framework

The watermarking schemes which embed watermarking into single image channel such as luminance channel or single color channel are the most common methods against general geometric attacks. When the watermarking images are exposed to geometric attacks, the attacked watermarking signal is quite different from the original embedded watermarking signal unless we synchronize these two signals, so correlation detection is often difficult to check watermarking correctly when the watermarking image undergoes geometric attacks. In order to synchronize these two watermarking, we must estimated the geometrical transformation that a watermarking image undergoes correctly. But there are few solutions to this problem even if we suppose the watermarking image undergoes simple geometrical transformation such as affine transformation. We propose a multi-channel watermarking framework in Fig. 1, and it can synchronize these two signals automatically without resolving geometric transformation which the watermarking image undergoes.

In this section, we explain the embedding and detecting processes of multi-channel watermarking framework, and expound the advantages of multi-channel framework.

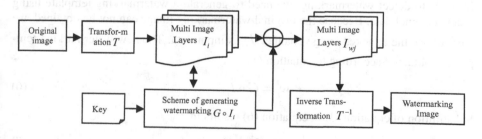

Fig. 1. Multi-channel Watermarking Framework

2.1 Embedding Watermarking Based on Multi-channel Watermarking Framework

The embedding procedure of the multi-channel watermarking framework is shown in Fig. 1. From Fig. 1, we can see the first step we should do is to decompose the original image into several channels. And then we design watermarking template generation function G, and a watermarking template w_t is obtained by using equation (1).

$$w_t = G \circ I_i \tag{1}$$

The generated watermarking template w_t should be satisfied with normal distribution and with zero mean, i.e. $w_t \sim N(0, \delta^2)$, and image channel I_i data are input parameters of function G. This generated watermarking template is embedded into one or more channels of original image by using (2)

$$I_{wj} = I_j + w_t \tag{2}$$

where I_j is another image channel data.

2.2 Detecting Watermarking Based on Multi-channel Watermarking Framework

If the image channel I_i which generate watermarking and the watermarking image channel I_{wj} undergo a geometric transformation attack D, the results are denoted as follows

$$I_i' = D \circ I_i \tag{3}$$

$$I_{wj}' = D \circ I_{wj} \tag{4}$$

Substitution of equation (2) in equation (3) yields

$$I_{wj}' = D \circ I_{wj} = D \circ (I_j + w_t) = D \circ I_j + D \circ w_t \tag{5}$$

In order to detect watermarking, we need to generate a watermarking template using image channel data. Because the original watermarking image can not be obtained, we have to use the attacked image channel I_i' as input data. The new watermarking template is obtained according to equation (1)

$$w_t' = G \circ I_i' \tag{6}$$

Substitution of equation (3) in equation (6) results in

$$w_t' = G \circ I_i' = G \circ D \circ I_i \tag{7}$$

If both the watermarking template generation scheme G and geometric transformation D are satisfied with $G \circ D = D \circ G$ and $w_t^{'} \sim N(0, \delta^{'2})$, substituting $G \circ D$ into equation (7) yields

$$w_t^{'} = G \circ D \circ I_i =^{\Theta} D \circ G \circ I_i = D \circ w_t \qquad (8)$$

By comparing equation (4) and equation (8), we prove that both watermarked image and generated watermarking template undergo the same geometric transformation attack D.

In order to detect a watermarking template signal, a correlation between the attacked watermarking image and the watermarking template is calculated

$$Cor = I_{wj}^{'} \otimes w_t^{'} \qquad (9)$$

Substitution of equation (5,8) into equation (9) yields

$$
\begin{aligned}
Cor = I_{wj}^{'} \otimes w_t^{'} &= [D \circ I_j + D \circ w_t] \otimes w_t^{'} \\
&= [D \circ I_j] \otimes w_t^{'} + [D \circ w_t] \otimes [D \circ w_t]
\end{aligned} \qquad (10)
$$

Because of $w_t^{'} \sim N(0, \delta^{'2})$, the correlation Cor can be approximated as

$$Cor \approx | w_t^{'} |^2 = \delta^{'2} \qquad (11)$$

If the image which need to be detected is with watermarking in itself, we can get a big value of the correlation Cor, otherwise the value of the correlation Cor will be very small.

In order to detect watermarking, most of the watermarking schemes choose to synchronize a watermarking signal with a watermarking image. These methods need obtain the inverse geometric transformation D^{-1} that the watermarking image undergoes. The inverse geometric transformation is quite difficult to solve when the geometric transformation does not have an analytic expression. The above mathematical process of multi-channel watermarking framework resolve this problem well. When a watermarked image undergoes a geometric transformation attack, the generated watermarking template follows the transformation and consequently performs a self-synchronization. Multi-channel watermarking framework need only satisfy with $G \circ D = D \circ G$. So it is easy to design watermarking algorithm and achieve a high performance of geometrical robustness.

3 Watermarking Framework

We have proposed a spatial domain [23] and a DWT domain [24] watermarking schemes based on the multi-channel framework. The spatial domain watermarking scheme is robust to geometric attacks, but fragile to common signal processing. The DWT domain watermarking scheme can resist geometric attacks and some signal processing, but it is vulnerable while adding noise. So we propose the watermarking scheme in the associated domain of DWT and DCT domains based on the multi-channel framework which can resolve the problems the DWT domain watermarking scheme can not overcome. The embedding and detection procedures of this watermarking scheme are shown in figure 2 and 3 respectively.

3.1 DWT and DCT Transformation

First of all, we decompose the original image into the image channels Y, C_r and C_b. Next we perform DWT transformation of three image channels Y, C_r and C_b respectively, and get the *HH* frequency coefficients of the image channels C_r and C_b which are denoted as C_{rhh} and C_{bhh} respectively. In fact, The *HH* frequency coefficients are not the

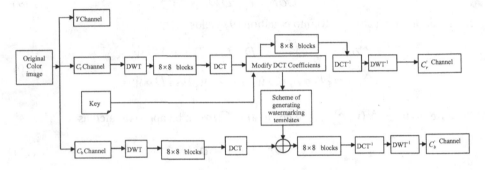

Fig. 2. Watermarking Embedding Procedure

best choice for watermarking embedding, because they might simply be removed for they contain the lowest energy of DWT decomposition, thus the watermarking can be effectively removed while keeping the perceptual quality to some extent. We can also use some other frequency coefficients such as *HL* or *LH*. And we have reason to believe that the proposed watermarking scheme in this paper will achieve better results in use of other frequency coefficients, if the experimental results which are obtained by using *HH* frequency coefficients meet our requirements.

$$C_{rhh}=HH\{DWT[C_r]\} \tag{12}$$

$$C_{bhh}=HH\{DWT[C_b]\} \tag{13}$$

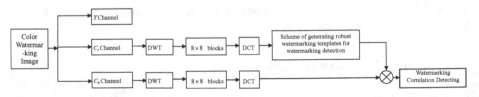

Fig. 3. Watermarking Detection Procedure

After DWT transformation, we divide C_{rhh} and C_{bhh} into 8×8 pixel blocks which do not superpose one another. We denote the number of the pixel blocks with N and the pixel blocks with r_i and b_i (i=1,2,...,N). Then we perform DCT transformation to these pixel blocks respectively.

$$B_i = DCT(b_i) \tag{14}$$

$$R_i = DCT(r_i) \tag{15}$$

3.2 Generating Watermarking Template

We can calculate the watermarking templates directly using the pixel blocks B_i or R_i ($i=1,2,...N$) of the original color image as the input data. However, this kind of algorithm can not control the properties of the watermarking templates because it is a derivation of the original image data. In order to control the properties of the watermarking templates, we generate the watermarking templates w_i ($i=1,2,...N$) satisfied with normal distribution with zero mean or a pseudo-random-noise sequence and one of these templates is shown in Fig 6-a. Authors in [25] proposed a method to construct a watermarking based on a chaotic trajectory, because of its controlled low-pass properties. Then we modify the value of the pixel blocks R_i ($i=1,2,...N$) of original color image by using equation (16) to embed the watermarking templates. This method which modify the pixel blocks R_i ($i=1,2,...N$) to embed the watermarking templates generated firstly make the scheme more robust than directly generating watermarking templates from the pixel blocks R_i ($i=1,2,...N$), though they are similar.

$$R_i(x, y) =$$

$$\{R_i(x, y) - [R_i(x, y) \bmod \beta] + \frac{\beta}{2}\} + sign[w_i(x, y)] * \{\frac{\beta}{4} + sign[w_i(x, y)] \cdot \delta\}$$

$$(i=1,2,...N) \tag{16}$$

where the β and δ are constant, and δ should be $0 \leq \delta < \frac{\beta}{4}$.

3.3 Embedding Watermarking Template

We embed the generated watermarking templates into the pixel blocks B_i ($i=1,2,...N$) of original color image by using equation (17).

$$B_i(x, y) = B_i(x, y) + \alpha \cdot w_i(x, y) \, (i=1,2,...N) \tag{17}$$

where the α is a constant which can be adjusted according to the complicated degree of the image texture.

3.4 Detecting Watermarking Scheme

Before detecting watermarking template, we perform the DWT and DCT transformation to the watermarking image by the methods used in 3.1. We denote the number of the 8×8 pixel blocks as N_0. Next we calculate the watermarking templates $w_i'(i=1,2,...N_0)$ by using equation (18).

$$w_i'(x,y) = \begin{cases} 1 & |[R_i'(x,y) \bmod \beta] - (\dfrac{3\beta}{4}+\delta)| < Thres \\ -1 & |[R_i'(x,y) \bmod \beta] - (\dfrac{\beta}{4}-\delta)| < Thres \\ 0 & \text{otherwise} \end{cases}$$

$$(i=1,2,...N_0) \tag{18}$$

where the β and δ are same as the equation (16), and *Thres* is a predetermined constant.

Define C_i^0, C_i^1 and C_i^{-1} $(i=1,2,...N_0)$. The C_i^0 is the number of pixels in watermarking template if $w_i'(x,y)=0$. The C_i^1 is the number of pixel if $w_i'(x,y)=1$. The C_i^{-1} is the number of pixels if $w_i'(x,y)=-1$.

We can determine the whether the watermarking exists in an image by calculating the correlation between the generated watermarking templates $w_i'(i=1,2,...N_0)$ and the pixel blocks $B_i(i=1,2,...N)$. We calculate the average values of the correlation E_i, F_i and $M_i(i=1,2,...N_0)$ by using equation (19,20,21).

$$E_i = \frac{\sum\limits_{i=1}^{N_0} B_i'(x,y)w_i'(x,y)}{C_i^1}, \quad \underset{\equiv}{} w_i'(x,y)=1 \tag{19}$$

$$F_i = \frac{\sum\limits_{i=1}^{N_0} B_i'(x,y)w_i'(x,y)}{C_i^{-1}}, \quad \underset{\equiv}{} w_i'(x,y)=-1 \tag{20}$$

$$M_i = |E_i - F_i|(i=1,2,...N_0) \tag{21}$$

We calculate the average value e by using equation (22) where c is the total number of the 8×8 pixel blocks with the value $M_i \neq 0 (i=1,2,...N_0)$. And if the condition equation (23) is meet, we think that the watermarking exist in image channel, otherwise there is no watermarking.

$$e = \frac{\sum\limits_{i=1}^{N} M_i}{c} \quad (i=1,2,...N_0) \tag{22}$$

$$e \geq \gamma \qquad (23)$$

where γ is a constant.

4. Experiments

In this section, we investigate the performance of the watermarking schemes shown in section 3 to geometric attacks. The original image we use is Lena image with size of 256*256. The watermarking template we design is shown in Fig. 6-a. In order to obtain the better experimental results, we have tried a lot of data to determine the parameters. Finally we carry on the algorithm with parameters $\beta = 8$, $\delta = 1$, $\alpha = 20$, $Thres = 1$, $\gamma = 1.0000$, and the experimental results meet our requirements.

We judge the image perceptual quality through two methods: visual observation and image PNSR. As showed in Fig.6-b, we find that the watermarking in the watermarking image is almost invisible. As illustrated in Table 1, we calculate the PSNR respectively for three image channels Y, C_r and C_b of the Lena image, from Table 1, we can find that the brightness channel which occupy the most energy of an image obtain the highest PSNR value. The PSNR values meet our requirements.

We test the robustness of this watermarking scheme to resist noise. We add Gaussian noise to the Lena image. The results are shown in Fig. 4. We find that the biggest noise strength the watermarking scheme can resist is about 4.0%. If the added noise strength is larger than 4.0%, the threshold $\gamma = 1.0000$ will be failed to detect the existence of the watermarking.

We also test the robustness of this watermarking scheme to rotation transformation and we rotate the image using bi-cubic interpolation. The original image and the parameters are the same as above. The results are shown in Fig. 5. We find that the proposed watermarking scheme is robust to rotation and it is succeed for all of the detection angles we choose. From Fig. 5, we find that the detection results are better at 90, 180 and 270 degrees than other degrees, because the watermarking templates we use are symmetrical square with white and black patches, and are the same as the original templates when they rotate 90, 180 and 270 degrees.

We test the watermarking image using a wide class of special geometrical transformation which are provided by Adobe Photoshop including Shear, Zigzag, Pinch, Polar, Twirl, Wave, Ripple, Spherize, Cropping and Affine transform etc. (shown in Fig. 6). From Fig. 6, we can see that the image is distorted seriously, but the common geometric distortion attacks are minor in order to retain the image information. If we successfully detect the watermarking after the image is distorted seriously, we will easily detect the watermarking from the images which undergo minor attacks. Especially, watermarking image is attacked by JPEG compression (Shown in Table 2). From the experiments results which are shown in Fig. 6 and Table 2 we find that all of detection results e0 are bigger than the predetermined threshold $\gamma = 1.0000$, and meantime, all of detection results e1 are smaller than this predetermined threshold.

On the other hand, the watermarking template is generated from image channel data, so we must test detection results when there is no watermarking in an image. When there is no watermarking in an image, we should get a right judgment also. We test

Lena image without embedding watermarking and we find that the proposed water-marking scheme is robust to the above mentioned attacks and it is succeed for all of the detection attacks we choose. The experimental results (e1 shown in Table 2, Fig. 4, Fig. 5 and Fig. 6) show that right judgments can be gain by using condition equation (23).

Table 1. The PSNR value of the Lena image

Image channels	PSNR (dB)
Y	53.0727
Cb	24.6051
Cr	39.4123

Table 2. Detect results of JPEG compression

Compression quality coefficient	Lena image		Results
	e0	e1	
7	2.4258	0.1128	succeed
8	3.9790	0.1295	succeed
9	4.3692	0.1525	succeed

The Lena image is attacked by JPEG compression provided by Adobe Photoshop. The e0 are detection results using equation (22) of the image with watermarking, and e1 are the detection results using equation (22) of the image without watermarking.

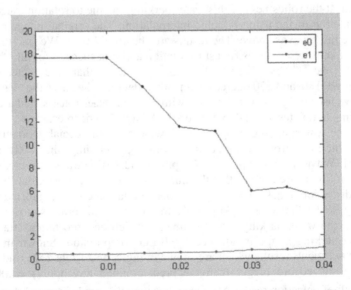

Fig. 4. The detection results while adding Gaussian noise provided by Adobe Photoshop to the Lena image. The e0 are detection results using equation (22) of the images with watermarking, and e1 are the detection results using equation (22) of the images without watermarking.

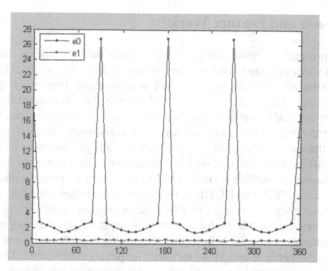

Fig. 5. The detection results when a Lena image is rotated from 0^0 to 360^0. The e0 are detection results using equation (22) of the images with watermarking, and e1 are the detection results using equation (22) of the images without watermarking.

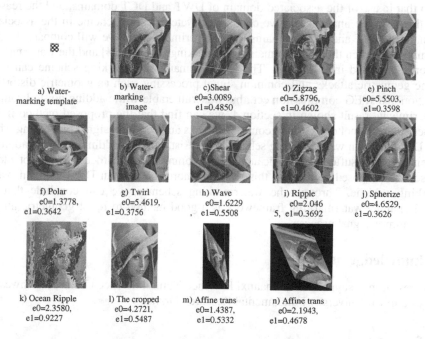

a) Watermarking template

b) Watermarking image

c) Shear
e0=3.0089,
e1=0.4850

d) Zigzag
e0=5.8796,
e1=0.4602

e) Pinch
e0=5.5503,
e1=0.3598

f) Polar
e0=1.3778,
e1=0.3642

g) Twirl
e0=5.4619,
e1=0.3756

h) Wave
e0=1.6229
e1=0.5508

i) Ripple
e0=2.046
5, e1=0.3692

j) Spherize
e0=4.6529,
e1=0.3626

k) Ocean Ripple
e0=2.3580,
e1=0.9227

l) The cropped
e0=4.2721,
e1=0.5487

m) Affine trans
e0=1.4387,
e1=0.5332

n) Affine trans
e0=2.1943,
e1=0.4678

Fig. 6. Fig. a is the generated watermarking template with size of 1*1 white and black patches, white and black pixels are the watermarking signal -1 and 1 respectively. Fig. b is the watermarking Lena image with size of 256*256. The Fig. c~k are obtained by using Adobe Photoshop 'distort' functions. The e0 are detection results using equation (22) of the images with watermarking, and e1 are detection results using equation (22) of the images without watermarking.

5 Conclusion and Feature Work

This paper presents a novel robust image watermarking scheme based on the multi-channel watermarking framework for resisting both geometric attacks and common signal processing attacks. The multi-channel watermarking framework generates the watermarking templates from one channel and embeds them into another channel. Because both the embedded watermarking and watermarking templates undergo the same attacks, watermarking can be detected without synchronizing the two watermarks or restoring the attacked watermarking image. Therefore a high robustness is obtained.

In the proposed scheme, we do DWT transformation to the decomposed channels, and get the *HH* frequency coefficients to do DCT transformation. Then we embed or detect watermarking after DWT and DCT transformation to the image channels. Although we use the *HH* frequency coefficients of DWT decomposition which occupy the lowest energy and are not the best choice for watermarking embedding, the experimental results meet our requirements. So we believe that the proposed watermarking scheme in this paper will achieve better results if we use other frequency coefficients. The fact that we obtained good experimental results in use of *HH* frequency coefficients of DWT decomposition proves the superiority of the multi-channel watermarking framework.

We have tried some other methods before the experiment in this paper , and we find the experimental results in use of only DWT domain or only DCT domain are worse than that in use of the associated domain of DWT and DCT domains, and the reasons for this is in the analysis. So we choose the watermarking scheme in the associated domain of DWT and DCT domains to do experiment. Next we will compare the performance between the DWT domain watermarking scheme [24] and the watermarking scheme proposed in this paper. The DWT domain watermarking scheme can resist some geometric attacks and common signal processing such as geometric distortion, rotation and JPEG compression ect., but it is vulnerable while adding noise. From the experimental result shown in section 4, we can find that the proposed method in this paper can resist not only some geometric attacks and common signal processing which the DWT domain watermarking scheme can resist, but also adding noise. In addition, when the image suffers geometric attacks and common signal processing except adding noise, the proposed method in this paper is more robust than DWT domain watermarking scheme. Through the watermarking schemes, we can conclude that the multi-channel watermarking framework is a good method to resist geometric attacks and common signal process.

Acknowledgement

This research is supported by Shaanxi Province Natural Science Grants, Northwestern Polytechnical University Fundamental Research Grants.

References

1. Brown, G.L.: A Survey of Image Registration Techniques. ACM Computing Surveys 24(4), 325–376 (1992)
2. Kutter, M.: Watermarking resisting to translation, rotation and scaling. In: Proc. SPIE, November 1998, vol. 3528, pp. 423–431 (1998)

3. Huang, X.Y., Luo, Y., Tan, M.S.: A Image Digital Watermarking based on DWT in Invariant Wavelet Domain. In: IEEE Fourth International Conference on Image and Graphics, October 2007, pp. 329–334 (2007)
4. Song, Q., Zhu, G.X., Luo, H.J.: Image Watermarking Based on Image Nornalization. In: Proc. of Int. Sym. on Comm. and Info. Tech., May 2005, pp. 1454–1457 (2005); Ling, H.F., Lu, Z.D., Zou, F.H., Yuan, W.G.: A Geometrically Robust Watermarking Scheme Based on Self-recognition Watermarking Pattern. In: IEEE Workshop in Multimedia Comm. and Computing, June 2006, pp. 1601–1604 (2006)
5. Singhal, N., Lee, Y.-Y., Kim, C.-S., Lee, S.-U.: Robust Image Watermarking Based on Local Zernike Moments (June 2007)
6. O'Ruanaidh, J.J.K., Pun, T.: Rotation, scale and translation invariant digital image watermarking. In: Proceedings of International Conference on Image Processing, October 1997, vol. 1, pp. 536–539 (1997)
7. Lin, C.Y., Wu, M., Bloom, J.A., Cox, I.J., Miller, M.L., Lui, Y.M.: Rotation, scale, and translation resilient watermarking of images. IEEE Trans. Image Process. 10(5), 767–782 (2001); Setyawan, I.: Geometric Distortion in Image and Video Watermarking: Robustness and Perceptual Quality. Ph D Thesis, Delft University of Technology (2004)
8. Li, Z., Kwong, S., Wei, G.: Geometric moment in image watermarking. In: Proceedings of the 2003 International Symposium on Circuits and Systems, May 2003, vol. 2, pp. II-932–II-935 (2003)
9. Dong, P., Galatsanos, N.P.: Affine transformation resistant watermarking based on image normalization. In: Proceedings of International Conference on Image Processing, June 2002, vol. 3(24-28), pp. 489–492 (2002)
10. Alghoniemy, M., Tewfik, A.H.: Geometric invariance in image watermarking. IEEE Trans. Image Process. 13(2), 145–153 (2004)
11. Senthil, V., Bhaskaran, R.: Wavelet Based digital image watermarking with robustness against on geometric attacks. In: Conference on Computational Intelligence and Multimedia Applications, December 2007, vol. 4, pp. 89–93 (2007)
12. Simitopoulos, D., Koutsonanos, D.E.: Robust image watermarking based on generalized radon transformations. IEEE Transactions on Circuits and 101 Systems for Video Technology 13(8), 732–745 (2003)
13. Kim, H.S., Lee, H.K.: Invariant image watermarking using zernike moments. IEEE Transactions on Circuits and Systems for Video Technology 13(8), 766–775 (2003)
14. Singhal, N., Lee, Y.-Y., Kim, C.-S., Lee, S.-U.: Robust image watermarking based on local Zernike moments. IEEE Trans. on Multimedia signal processing, 401–404 (October 2007)
15. Kang, X., Huang, J., Shi, Y.Q., Lin, Y.: A DWT-DFT composite watermarking scheme robust to both affine transform and JPEG compression. IEEE Trans. on Circuits and Systems for Video Technology 13(8), 776–786 (2003)
16. Pereira, S., Pun, T.: Robust template matching for affine resistant image watermarks. IEEE Trans. on Image Processing 9(6), 1123–1129 (2000)
17. Ling, H.F., Lu, Z.D., Zou, F.H., Yuan, W.G.: A Geometrically Robust Watermarking Scheme Based on Self-recognition Watermarking Pattern. In: IEEE Workshop in Multimedia Comm. and Computing, June 2006, pp. 1601–1604 (2006)
18. Setyawan, I.: Geometric Distortion in Image and Video Watermarking: Robustness and Perceptual Quality. Ph D Thesis, Delft University of Technology (2004)
19. Bas, P., Chassery, J.M., Macq, B.: Geometrically invariant watermarking using feature points. IEEE Trans. on Image Processing 11(9), 1014–1028 (2002)
20. Tang, C.W., Hang, H.M.: A feature-based robust digital image watermarking scheme. IEEE transactions on Signal Processing 51(4), 950–958 (2003)

21. Weinheimer, J.: Towards a Robust Feature-Based Watermark. University of Rochester (2004)
22. Seo, J.S., Yoo, C.D.: Image watermarking based on invariant regions of scale-space representation. IEEE Trans. on see also Acoustics, Speech, and Signal Processing 54, 1537–1549 (2006)
23. Zheng, J.B., Feng, D.D., Zhao, R.C.: A Multi-channel framework for image watermarking. In: IEEE Proceeding of the 4th Int. Conf. on Machine Learning and Cyber., vol. 4, pp. 3096–3100 (2005)
24. Jiangbin, Z., Sha, F.: A color image multi-channel DWT domain watermarking algorithm for resisting geometric attacks. In: International Conference on Machine Learning and Cybernetics (2008)
25. Cotsman, G., Pitas, I.: Chaotic watermarks for embedding in the spatial digital image domain. In: Proceedings ICIP 1998, Chicago, IL, October 1998, vol. 2, pp. 432–436 (1998)

An Efficient Buyer-Seller Watermarking Protocol Based on Chameleon Encryption

Geong Sen Poh and Keith M. Martin

Information Security Group, Royal Holloway, University of London,
Egham, Surrey, TW20 0EX, United Kingdom
{g.s.poh,keith.martin}@rhul.ac.uk

Abstract. Buyer-seller watermarking protocols are designed to deter clients from illegally distributing copies of digital content. This is achieved by allowing a distributor to insert a unique watermark into content in such a way that the distributor does not know the final watermarked copy that is given to the client. This protects both the client and distributor from attempts by one to falsely accuse the other of misuse. Buyer-seller watermarking protocols are normally based on asymmetric cryptographic primitives known as homomorphic encryption schemes. However, the computational and communication overhead of this conventional approach is high. In this paper we propose a different approach, based on the symmetric Chameleon encryption scheme. We show that this leads to significant gains in computational and operational efficiency.

1 Introduction

The buying and selling of multimedia content such as movies and songs through the Internet raises a number of important security issues, one of which is how to address illegal distribution of content by clients. One mechanism is to *deter* them from this act by providing a transparent method for the distributor to *trace* the perpetrator. Fingerprinting schemes [2,11] are one such method. These schemes allow the distributor to insert a unique *watermark* into content that identifies a client when a suspicious copy is found. However, since the distributor knows the watermark, a dishonest client can claim that illegal copies of the content are actually distributed by the distributor. Conversely, it is possible for the distributor to frame an honest client by inserting the watermark into content and distributing copies of it. Buyer-seller watermarking protocols [3,7,9,13,15] were proposed to address these issues.

The main idea behind these protocols is that the distributor must not know the final marked copy given to the client. A client will thus not be able to claim that a copy of the content found on the Internet is distributed by the distributor. Similarly, the distributor will not be able to frame the client. One well-established way to achieve this is by combining a linear watermarking scheme [4] with asymmetric cryptographic primitives known as homomorphic encryption schemes, such as Okamoto-Uchiyama [16] and Paillier [17]. Using these two building blocks, the watermark is embedded into content by the distributor in encrypted form. This

H.J. Kim, S. Katzenbeisser, and A.T.S. Ho (Eds.): IWDW 2008, LNCS 5450, pp. 433–447, 2009.
© Springer-Verlag Berlin Heidelberg 2009

approach can be made effective, but is computationally expensive and requires high bandwidth, with the size of the encrypted content being larger than the original content.

Our Contribution. In this paper we modify this approach by deploying a significantly different type of cryptographic primitive in order to establish a buyer-seller watermarking protocol that has lower computational overhead and requires less network bandwidth. The primitive we use is the symmetric-based Chameleon encryption scheme proposed by Adelsbach *et al.* [1]. An interesting characteristic of this primitive is that the watermark embedding operation happens during the decryption operation conducted by the client. This results in the distributor needing only to symmetrically encrypt the content without embedding any client watermark. This completely differs from the conventional approach [15,9,3,5,7,8,13], where the distributor embeds the client watermark in the encrypted domain, which is computationally expensive. In addition, the conventional approach normally employs a trusted watermark authority to generate client watermarks and hence needs to be online whenever client watermarks are requested, whereas the trusted third party in the protocol based on Chameleon encryption is only involved during registration. The subsequent content watermarking and distribution sessions between the distributor and the client do not require the involvement of this trusted third party.

Organisation of the Paper. In Section 2 we describe some of the basics of buyer-seller watermarking protocols. We discuss related work in Section 3 and the new approach in Section 4.

2   Basic Notions

In this section we describe basic notions that are important for the construction and analysis of our proposal.

Objects. We define a content space $\mathcal{X} \subseteq \mathbb{R}$. *Content* in a content space $(X \in \mathcal{X})$ is a vector of real numbers $X = (x_1, \ldots, x_n)$. Each vector element of content x_i, $1 \leq i \leq n$, is also a real number in the range of $[0, z]$, where z is the maximum value allowed for x_i. We further define a watermark space $\mathcal{W} \subseteq \mathbb{R}$. A *watermark* in a watermark space $(W \in \mathcal{W})$ is a vector of real numbers $W = (w_1, \ldots, w_n)$. The range of values that each watermark element can take depends on the algorithm that is used to generate them. For examples, a watermark can be *bipolar*, which means $w_i \in \{-1, 1\}$, or a watermark may be chosen from a normal (Gaussian) distribution, as in a spread spectrum watermarking scheme [4]. Finally, we define a text space $\mathcal{T} \subseteq \{0, 1\}^*$. A text *text* $\in \mathcal{T}$ is required, for example, to describe the purchase order between a distributor and a client.

Buyer-Seller Watermarking Protocols. These are protocols that provide content distribution between two parties, in which the owner (or buyer) of content can be traced in a fair and secure manner if copies of this content are found to being illegally distributed, where the precise notion of *fair and secure* is explained below.

Parties Involved. A buyer-seller watermarking protocol involves a *distributor* D, who provides (or sells) content to *client* C, while *arbiter* A settles disputes between the distributor D and client C. A special trusted third party may also be involved. In many buyer-seller watermarking protocols [15,9,13], this role is played by a *watermark certification authority* (*WCA*), who is responsible for generating and certifying client watermarks. We will only require a trusted *key distribution centre* (KDC), who generates key materials for the distributor and client. We assume that the distributor D and the client C do not trust one other. We also assume that KDC and A will not conspire with D and/or C.

Threats. The main security threats are:

- *Distributors.* A distributor D may frame a client C. This happens when D inserts a unique watermark that matches C's identity into copies of the content and distributes this widely. Later D can accuse C of illegal content distribution by extracting this watermark from these copies.
- *Clients.* There are two main threats:
  - A client C may try to remove the watermark in the marked content.
  - A client C may redistribute copies of content given by the distributor D, and later deny this fact when confronted by the distributor D.

Security Properties. These motivate the three main security properties of a buyer-seller watermarking protocol [9,13]:

- *Traceability.* The identity of a legitimate, but dishonest, client C who illegally distributes content can be traced by the distributor D.
- *Framing Resistance.* An honest client C cannot be falsely accused of illegal distribution by the distributor D.
- *Non-repudiation of Redistribution.* A dishonest client C who has redistributed illegal copies of content cannot refute this fact. This allows the distributor D to prove the illegal act of C to a third party. In this case framing resistance is a prerequisite since, without this property, C can claim that it was D who redistributed copies of content.

Some existing protocols [9,13] include the protection of client's privacy as an additional security property, but in this paper we will only focus on providing the three fundamental properties. In addition, we assume that there exists a secure channel that allows two parties to authenticate one another and exchange messages in a secure manner. In practical terms, this can be achieved by distributing content under a TLS/SSL [6] session.

Public Key Assumptions. As for the conventional approach, we assume the existence of a supporting Public Key Infrastructure which permits an initial registration phase *before* the instantiation of the protocol, where all involved parties register with a fully trusted certificate authority either through an online or manual process. The certificate authority verifies the identity of each of the registrants, such as the distributor D and the client C, and issues them certified key pairs (e.g. signing and verification keys sk_D, pk_D for a digital signature scheme [12]). Two parties who wish to communicate can now use these key pairs to authenticate one another or to generate digital signatures. We denote $S_Y \Leftarrow \mathsf{Sig}_{sk_Y}(M)$ as the signing operation conducted by party Y on message M with signing key sk_Y, resulting in signature S_Y. Similarly we denote $\{\mathtt{Valid}, \mathtt{Invalid}\} \Leftarrow \mathsf{Ver}_{pk_Y}(S_Y)$ as the verification operation conducted by any relevant party on signature S_Y with verification key pk_Y.

3   Related Work

3.1   The Conventional Buyer-Seller Watermarking Approach

Buyer-seller watermarking protocols were first proposed by Qiao and Nahrstedt [21] and later improved by Memon and Wong [15]. Ju et al. [9] then presented a protocol that also protects client's privacy. Several buyer-seller protocol variants have been constructed since then, including the protocols proposed in [3,5,7,8,13]. These protocols are all based on homomorphic encryption schemes [17] and many also require a trusted third party in the form of a WCA.

We explain the core mechanism employed in these protocols by using SSW embedding for the watermarking and Paillier homomorphic encryption.

Cox et al. SSW Embedding [4]. Assuming that we have content $X = (x_1, \ldots, x_n)$ and watermark $W = (w_1, \ldots, w_n)$, the embedding operations in Cox et al.'s scheme has the simple structure:

$$x_i^W = x_i + \rho w_i \quad 1 \le i \le n,$$

where ρ is a real number, which is determined by the robustness and fidelity requirements of the watermarking algorithm.

Paillier Homomorphic Encryption Scheme [17]. This encryption scheme represents one of the most common schemes deployed in recent proposals of buyer-seller watermarking protocols such as in [5,8]. It requires a large composite integer $m = pq$, where p and q are large primes, and $\gcd(m, \phi(m)) = 1$. Here $\gcd(,)$ denotes *greatest common divisor* and $\phi(.)$ the *Euler's totient function*. Also, let $g \in \mathbb{Z}_{m^2}^*$ be an element of order m. The public key is (m, g) and the private key is (p, q). Given content $x < m$ and a randomly chosen $r < m$, encryption is computed as:

$$E(x) = g^x r^m \bmod m^2.$$

For decryption, given $E(x) < m^2$, we have $D(E(x)) = x$.

Homomorphic Properties. The encryption scheme $E(x)$ has the additive homomorphic property, in which, given content $x_1, x_2 < m$,

$$E(x_1)E(x_2) \bmod m^2 = E(x_1 + x_2) \bmod m^2.$$

Combining SSW and Paillier. Content watermarking in the encrypted domain is performed as follows:

$$E(x_i)E(\rho w_i) \bmod m^2 = E(x_i + \rho w_i) \bmod m^2, \quad 1 \leq i \leq n.$$

In order to embed watermark $W = (w_1, \ldots, w_n)$ into content $X = (x_1, \ldots, x_n)$, every element of the watermark, w_i, and every element of the content, x_i, will need to be encrypted using the homomorphic encryption scheme. The combining of SSW and Paillier thus allows a watermark, which is generated by WCA, to be embedded by the distributor without this distributor knowing the watermark. As a result, the distributor will not be able to frame a client, and a client cannot claim that a found illegal copy is distributed by the distributor. This is the essence of the conventional approach.

Remark 1. An alternative approach is provided by asymmetric fingerprinting schemes [19,20,18], which avoid the need for a trusted WCA at the cost of higher computation and communication requirements.

3.2   Chameleon Encryption

Adelsbach *et al.* proposed the Chameleon encryption scheme in [1]. The intriguing property of this scheme is that it only involves modular addition and watermarking happens simultaneously during decryption when a client decrypts content using his watermarked key material.

Overview. The basic idea behind Chameleon encryption is as follows:

1. Encryption and decryption keys are generated. The encryption key, known as the master table MT, contains a vector of real numbers in the same space as the content X. The decryption key, known as the user table UT, is a *slightly different* version of MT, consisting of elements of MT reduced by a small real number.
2. In order to encrypt content X, elements in the master table MT are selected and added to the elements in X.
3. Decryption is computed by selecting the elements from UT and subtracting these elements from X. Since UT is slightly different from MT, the decryption will introduce a small "error" (watermark) into the decrypted content, thus making this content unique to the holder of UT.

We adapt the notation of [1] to describe the four phases of the scheme in more detail.

Setup. In this phase, three types of table are generated:

- **Master table** MT: This is a vector of real numbers denoted by $MT = (mt_1, \ldots, mt_L)$, where $L = 2^b$ and b a positive integer. Note that $MT \in \mathcal{X}$ (the content space). Each mt_α, $1 \leq \alpha \leq L$, is randomly selected from $[0, z]$, where z is the maximum value allowed for mt_α.

- **Fingerprint tables:** These are denoted $FT^{(1)}, \ldots, FT^{(N)}$, where N is the number of clients. Each $FT^{(i)} = (ft_1^{(i)}, \ldots, ft_L^{(i)})$, $1 \leq i \leq N$, is a vector of real numbers. It is required that $ft_\alpha^{(i)} = \frac{1}{s} w_\alpha^{(i)}$, $1 \leq \alpha \leq L$, where s is a small positive integer and $w_\alpha^{(i)}$ is an element in watermark $W^{(i)} = (w_1^{(i)}, \ldots, w_L^{(i)})$. The watermark $W^{(i)}$ is generated based on a watermarking scheme such as the SSW scheme described in [11]. Elements of a fingerprint table are used to watermark elements of the master table in order to generate a user table for the client.
- **User tables:** Each user table is a vector of integers $UT^{(i)} = (ut_1^{(i)}, \ldots, ut_L^{(i)})$, $1 \leq i \leq N$. It is generated as $UT^{(i)} = MT - FT^{(i)}$, where $ut_\alpha^{(i)} = mt_\alpha - ft_\alpha^{(i)} \bmod p$ for $1 \leq \alpha \leq L$, p a sufficiently large integer. Elements of a user table are used for fingerprinting and decryption of content.

Encryption. Content $X = (x_1, \ldots, x_n)$ is encrypted based on the master table MT. We assume that the value of s has been determined. In order to encrypt content, a random session key K_r is generated using a pseudo-random number generator (PRNG). This session key K_r is then used as an input to a *pseudo-random sequence generator* (PRSG), which is a special type of PRNG which outputs a pseudo-random integer sequence R with bit length $n \cdot s \cdot b$ (n is the number of elements representing the content, and b the positive integer used to determine L). Using R, $n \cdot s$ elements, $(k_1, \ldots, k_{n \cdot s})$, are randomly selected from the master table MT. Finally, the encrypted content $E = (e_1, \ldots, e_n)$ is:

$$e_\beta = x_\beta + \sum_{j=1}^{s} k_{s(\beta-1)+j} \bmod p \text{ for } 1 \leq \beta \leq n. \tag{1}$$

As an example, let $s = 4$, $R = (9, L, 8, 7, 56, \ldots, 10)$, $k_1 = mt_9$, $k_2 = mt_L$, $k_3 = mt_8$ and $k_4 = mt_7$. The encryption of the content's first element x_1 will be $e_1 = (x_1 + mt_9 + mt_L + mt_8 + mt_7) \bmod p$.

Joint Fingerprinting and Decryption. Fingerprinting and decryption of content is carried out by first selecting the watermarked elements in the user table $UT^{(i)}$ based on the identical R used for encryption. We denote elements selected from $UT^{(i)}$ as $(k_1^f, \ldots, k_{n \cdot s}^f)$. Following the previous example, let $s = 4$ and $k_1^f = ut_9^{(i)}$, $k_2^f = ut_L^{(i)}$, and so on. The watermarked content $X^{(i)} = (x_1^{(i)}, \ldots, x_n^{(i)})$ is obtained during decryption of the encrypted content $E = (e_1, \ldots, e_n)$ as follows:

$$x_\beta^{(i)} = e_\beta - \sum_{j=1}^{s} k_{s(\beta-1)+j}^f \bmod p \text{ for } 1 \leq \beta \leq n. \tag{2}$$

Continuing from the previous example, decryption of the first element is:

$$x_1^{(i)} = (e_1 - (ut_9^{(i)} + ut_L^{(i)} + ut_8^{(i)} + ut_7^{(i)})) \bmod p$$
$$= (x_1 + ft_9^{(i)} + ft_L^{(i)} + ft_8^{(i)} + ft_7^{(i)}) \bmod p.$$

Remark 2. Recall that for the generation of the fingerprint table we require that $ft_\alpha^{(i)} = \frac{1}{s}w_\alpha^{(i)}$, for $w_\alpha^{(i)}$ an element of a watermark $W^{(i)} = (w_1^{(i)}, \ldots, w_n^{(i)})$. So for the example shown above, we have $ft_9^{(i)} = \frac{1}{s}f_9^{(i)}$, and so on. This gives us $x_1 + \frac{1}{4}w_9^{(i)} + \frac{1}{4}w_L^{(i)} + \frac{1}{4}w_8^{(i)} + \frac{1}{4}w_7^{(i)}$ statistically equivalent to $x_1 + w_1^{(i)}$, in which $w_1^{(i)} = (\frac{1}{4}w_9^{(i)} + \frac{1}{4}w_L^{(i)} + \frac{1}{4}w_8^{(i)} + \frac{1}{4}w_7^{(i)})$ is a watermark element conforming to the distribution of the underlying watermarking scheme. Due to this, robustness of the watermark embedded into content is equal to that of the watermarking scheme used.

Detection. If we denote $(g_1^{(i)}, \ldots, g_{s \cdot n}^{(i)})$ as a vector where $g_\beta^{(i)} = ft_\beta^{(i)}$, (for example, $g_1^{(i)} = ft_9^{(i)}$, $g_2^{(i)} = ft_L^{(i)}$), then a watermark $\widehat{W}^{(i)} = (\widehat{w}_1^{(i)}, \ldots, \widehat{w}_n^{(i)})$ can be extracted from a found copy of content $\widehat{X}^{(i)} = (\widehat{x}_1^{(i)}, \ldots \widehat{x}_n^{(i)})$ using the original content $X = (x_1, \ldots, x_n)$ as follows:

$$\widehat{w}_\beta^{(i)} = x_\beta - \widehat{x}_\beta^{(i)} = \sum_{j=1}^{s} g_{s(\beta-1)+j}^{(i)} \text{ for } 1 \le \beta \le n. \tag{3}$$

Again, using the example given previously, this can be shown as $\widehat{w}_1^{(i)} = x_1 - \widehat{x}_1^{(i)} = x_1 - (x_1 + ft_9^{(i)} + ft_L^{(i)} + ft_8^{(i)} + ft_7^{(i)})$. This extracted watermark $\widehat{W}^{(i)}$ is then compared with the original watermark $W^{(i)}$ to measure their similarity by [4]:

$$\text{Sim}(\widehat{W}^{(i)}, W^{(i)}) = \frac{\widehat{W}^{(i)} W^{(i)}}{\sqrt{\widehat{W}^{(i)} \widehat{W}^{(i)}}}.$$

If $\text{Sim}(\widehat{W}^{(i)}, W^{(i)}) > t$, where t is a predetermined threshold, it means that the detection of the watermark is successful.

Remark 3. As mentioned in [1], the modulo operator of all the above operations only allows computation of integers but MT, FT and X are all based on real numbers. This issue can be addressed by representing a real value as an integer by scaling. As an illustrative example, 15.687 can be represented as 15687 or 1568700, depending on the requirement of the scheme. Hence we say that there is a one-to-one mapping from $[0, z] \in \mathbb{R}$ to $[0, Z] \in \mathbb{Z}$.

4 A Buyer-Seller Watermarking Protocol Based on Chameleon Encryption Scheme

In this section we show that the Chameleon encryption scheme described in the previous section can be deployed in the Buyer-Seller watermarking scenario to achieve a four-phase protocol with new and desirable properties.

Key Distribution. In this phase, the distributor D and the client C register their details with a key distribution centre KDC based on the key pairs they obtained during the initial registration phase conducted by the certificate authority. Upon receiving the registration requests from D and C, the KDC runs the **Setup** phase of the chameleon encryption scheme to produce a master table MT^{D_C}, a client fingerprint table FT^C and a client user table UT^C. Both tables are signed by KDC, resulting in two signatures $S_{MT} \Leftarrow \text{Sig}_{sk_{KDC}}(MT^{D_C})$ and $S_{UT} \Leftarrow \text{Sig}_{sk_{KDC}}(UT^C)$. The master table MT^{D_C} is then passed to D. This master table is intended for D to encrypt content specifically for C. The user table UT^C is passed to C. The client C and the distributor D can then use their respective user table and master table to conduct many content watermarking and distribution sessions without further contacting the KDC. This means key distribution is a one-time process. Figure 1 illustrates the transactions between KDC with D and C.

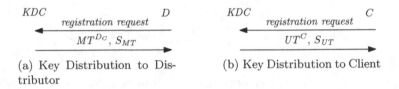

(a) Key Distribution to Distributor

(b) Key Distribution to Client

Fig. 1. Keys Distribution

Content Watermarking and Distribution. In this phase the client C requests content X from the distributor D, and D watermarks content X and sends the marked content to C. We describe the protocol steps as follows:

1. Client C sends a purchase request to the distributor D.
2. Upon receiving the request, D prepares a purchase agreement $text_p$, which contains details such as description of content, date of purchase, price and licensing terms.
3. Client C checks the purchase agreement and signs it. This is represented as $S_C \Leftarrow \text{Sig}_{sk_C}(text_p)$. The signature S_C is sent to D.
4. Distributor D verifies S_C and generates a client's watermark V^C. This watermark V^C is embedded into content X. We denote content with V^C embedded as $X^v = (x_1^v, \ldots, x_n^v)$. The reason for embedding V^C is so that D is able to trace and match C with the client's record from copies of content D has found. In this case D can generate and embed V^C using any watermarking schemes. After that, D runs the **Encryption** phase of the Chameleon encryption scheme. Recall from Section 3.2 that this means D generates K_r using a PRNG, and K_r is then used as an input to a PRSG to generate a pseudo-random sequence R. Elements of master table MT^{D_C} are chosen based on R to encrypt X^v, as shown in (1). We denote the encrypted marked content as E^v. This is then signed together with the purchase agreement $text_p$ and C's signature S_C. The signing process is represented as $S_E \Leftarrow \text{Sig}_{sk_D}(E^v, S_C, text_p)$. Signature S_E, K_r and E^v are then sent to C.

5. The client C verifies S_E and runs the **Joint Fingerprinting and Decryption** phase of the Chameleon encryption scheme. This means C uses K_r as input to the PRSG to generate R, and elements of his user table UT^C are selected based on R to decrypt E^v. At the same time of the decryption, a second watermark W^C is embedded into this content, as shown in (2). We denote the final decrypted content with two embedded watermarks as X^{vw}.

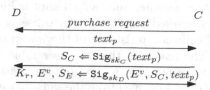

Fig. 2. Content Watermarking and distribution

Identification. When a suspicious copy of content $\widehat{X}$ is found, the distributor D tries to detect watermark V^C based on the watermarking scheme deployed. If the watermark is detected, then D proceeds to match V^C to the client's identity from their records.

Dispute Resolution. After identifying C based on the watermark V^C, the distributor D may wish to further prove to a third party that C has illegally distributed copies of content. In this case, D sends X^v, K_r, E^v, $\widehat{X}$, $text_p$, S_E and S_C to trusted arbiter A. After A verifies the signatures to confirm the agreement between C and D on the content, A extracts $\widehat{W}$ from $\widehat{X}$ based on (3). After that A gives $\widehat{W}$ to KDC and requests KDC to run $\mathtt{Sim}(\widehat{W}, W^C)$. If the watermark is detected, KDC retrieves the associated identity of C and sends the information to A for A to decide whether C is guilty or not.

4.1 Security

We analyse the security of the protocol that deploy Chameleon encryption based on the security properties presented in Section 2. This relies primarily on the security of the underlying primitives.

Traceability. Traceability is assured since D can trace its content to C through the watermark V^C, and also that KDC can detect watermark W^C to identify C. In addition, C has no knowledge of W^C and thus cannot simply remove the watermark from content. This depends on the security of the underlying watermarking scheme, which is measured by its robustness and collusion resistance, and the security of the Chameleon encryption scheme [1]. In the following we discuss the assurance of traceability based on these two primitives.

- *Watermarking Scheme.* One well-established watermarking scheme is the SSW scheme proposed in [4,11]. Both the existing buyer-seller watermarking protocols [3,7,9,13,15] and the Chameleon encryption scheme are based on this scheme. The reason for choosing SSW schemes is that according to [14], they are suitable for watermarking application to trace content as they are experimentally tested to be robust against many signal processing operations such as compression. They are also robust against multiple watermarking. SSW schemes are also resistant against many clients colluding using their contents to wipe off a watermark. This property is known as *collusion resistance* and was formalised in [11].

- *C attempts to remove W^C based on knowledge of the user table UT^C of the Chameleon encryption scheme.* Such an attempt is identical to removing a watermark W from marked content, which subsequently means defeating the robustness and collusion resistance of the watermarking scheme as previously discussed. The reason is that the user table UT^C is generated by subtracting the fingerprint table $FT^{(i)}$ from the master table MT^{Dc}, and the master table MT^{Dc} is derived from the same space as content X, that is $MT^{Dc} \in \mathcal{X}$. Similarly, the elements in the fingerprint table $FT^{(i)}$ have the same statistical properties of a watermark generated from a watermarking scheme. Although the value of the elements in $FT^{(i)}$ have been fractioned by s, as mentioned in [1], this can be compensated by generating a higher number of elements for $FT^{(i)}$, which means having a larger value for L. So, given $MT^{Dc} \in \mathcal{X}$ and $FT^{(i)} \in \mathcal{W}$, the user table UT^C is statistically identical to a marked content, and our earlier statement holds.

  We also note that Chameleon encryption has been shown to be *semantically secure*, which means that an attacker, who is not any of the parties involved in the protocol, learns nothing from the encrypted content [1].

Note that traceability may be improved if a more effective watermarking scheme or watermark coding method is devised, such as the recent improvement on watermark coding for Chameleon encryption proposed in [10]. We also note that the distributor D has the flexibility of using any watermarking scheme for embedding V^C. Here we assume both watermarks V^C and W^C to be using the same SSW scheme since it is known to be robust to multiple watermarking.

Framing Resistance. The distributor D will not be able to frame the client C if the following two cases are true:

- *Case 1.* D has *no knowledge* of the final marked copy of content given to C, unless C distributed this copy illegally.
- *Case 2.* Once an illegal copy is found and the watermark W^C is detected, this watermark is of no value to D except for proving the guilt of the client on the particular marked content where the watermark W^C is detected.

Case 1. The distributor D may try to learn the watermark W^C based on his master table MT^{Dc}. This is equivalent to learning a watermark W generated using a watermarking scheme based on only the original content X. Recall that $X \in \mathcal{X}$ and $MT^{Dc} \in \mathcal{X}$. This means that the master table MT^{Dc} has identical statistical distribution as content X. Similarly, W^C has the same distribution as a watermark W generated using a watermarking scheme. Hence guessing W^C given MT^{Dc} is akin to guessing W with only possession of the original content

X. This brings us to the identical framing resistant settings of the conventional buyer-seller watermarking protocols [3,7,9,13,15], where the seller has the original content X, but the watermark W is generated by other parties.

Case 2. This is required so that D cannot extract the watermark from the found copy and embed this watermark into other more valuable content (the *unbinding problem* [13]). This holds, since there is a binding statement that binds the marked copy X^{vw} received by the client (and all similar copies where W^C can be detected) to the purchase agreement $text_p$. This is achieved through the signature S_E generated on the marked copy, client's signature and the agreement. Thus, assuming that the Chameleon encryption scheme, the digital signature scheme and the watermarking scheme are secure, framing resistance is assured as long as the marked copy X^{vw} and the user table UT^C are kept secret by C.

Non-repudiation of Redistribution. The client C cannot repudiate that he or she illegally distributed content because:

- there exists a signature S_C binding C to $text_p$, and $text_p$ binds X^{vw},
- when the watermark $\widehat{W}$, which can be matched to C, is detected, the identity of C is revealed by KDC.

Based on this information the arbiter A demonstrates that C is guilty. Note that the purchase agreement $text_p$ plays an important role as it contains description of content that will eventually be used to confirm the illegal content found $\widehat{X}$ is indeed a copy of the original X (which is similar to X^{vw}). Thus, assuming the security of the underlying Chameleon encryption and digital signature schemes, client C cannot deny having illegally distributed copies of content if an arbiter A with information gathered from D and KDC demonstrates that this is the case based on the verified signature S_C and the detected watermark $\widehat{W}$.

Thus, assuming the security of the underlying primitives, traceability, framing resistance and non-repudiation of redistribution are all provided, the protocol based on Chameleon encryption satisfies the security requirements of a secure buyer-seller watermarking scheme.

4.2   Efficiency

Computation and Storage. We use the following criteria to evaluate the computational performance and storage requirements of the protocol:

- The computational requirement for encryption and decryption of content.
- The size of encrypted content and key storage needed for D and C.

For the computational requirement, our evaluation is based on the encryption and decryption mechanisms applied in each protocol. We will (reasonably) assume that the existing buyer-seller watermarking protocols [3,7,9,13,15] deploy the homomorphic encryption scheme of Paillier [17]. As for storage requirements, the size of the encrypted content is measured. This affects the bandwidth during transmission from D to C. We also measure the size of storage required to store the key materials (i.e. storage space required for the homomorphic encryption in the existing protocols compare to the key tables in the Chameleon encryption scheme).

In Table 1, **BSW** denotes conventional buyer-seller watermarking protocols and **CE** denotes the protocol based on Chameleon encryption. We further denote n as the number of elements of content $X = (x_1, \ldots, x_n)$, $m = pq$ where p and q are distinct large prime, b the bits length of content element x_i, while $|y|$ denotes the bits length of y. In addition, recall that from the discussion on Chameleon encryption in Section 3.2, we have L the number of entries in MT. Finally, for $x_i \in [0, z]$, the alphabet Z represents the scaled integer value of z (which is required due to the modular operation as discussed in Section 3.2).

For time complexity evaluation (**Bits Ops**), the most expensive computation for **BSW** are modular exponentiation under the Paillier homomorphic encryption scheme. These have the complexity of $\mathcal{O}(|m|^2)$. As the computation is repeated n times to encrypt each content element x_i, the computation is $\mathcal{O}(|n||m|^2)$ bit operations for **BSW**. On the other hand **CE** requires just $\mathcal{O}(|n||Z|)$ bit operations, since it involves only modular addition. This is because to encrypt or decrypt all content elements which can have the maximum size of Z, n modular additions are required. Thus the computational advantage of **CE** is substantial. Note that for simplicity, we have used the conventional square-and-multiply algorithm for measuring the complexity of modular exponentiation. It is known that there are methods to increase the efficiency of computation. One will be to pre-process some of the key material.

As for measuring **Key Size** and **Encrypted Content Size**, we borrow a simple but practical example from [1]. Given a content image with $n = 10000$ significant coefficients, where each coefficients is of length 16 bits, we have $Z = 2^{16}$. Suppose $L = 8 \cdot Z$ then $L = 8 \cdot 2^{16} = 2^{19}$, where $L = 8 \cdot Z$ achieves the statistical quality required for the master table MT^{D_C} as mentioned in [1]. Then the size of the tables MT^{D_C} and UT^C (**key size**) for D and C in **CE** is $L \cdot |Z|$ $= 8 \cdot 2^{16} \cdot 16 = 2^{23}$ bits $= 1$ MByte, which can be acceptable in terms of current storage capacity. Similarly, the encrypted content size for **CE** can be calculated as $n \cdot |Z| = 10000 \cdot 16$ bits $= 20$ KByte. This is less than the encrypted content size of **BSW**, since irrespective of the 16 bits of the coefficients, the size of an

Table 1. Efficiency Comparisons

Parameters	BSW	CE
Bits Ops		
Encryption (D)	$\mathcal{O}(|n||m|^2)$	$\mathcal{O}(|n||Z|)$
Decryption (C)	$\mathcal{O}(|n||m|^2)$	$\mathcal{O}(|n||Z|)$
Key Size		
D	$|m|$	$L \cdot |Z|$
C	$|m|$	$L \cdot |Z|$
Encrypted Content Size	$n \cdot |m|$	$n \cdot |Z|$

encrypted content efficient depends on $|m|$, where current security requirements suggest that $|m| = 1024$. So the encrypted content size will be $10000 \cdot 1024$ bits ≈ 1.3 MByte. Hence the protocol based on **CE** requires less network bandwidth when the encrypted content is being transmitted.

In summary, we observe that **CE** is computationally more efficient than the conventional approach, but come at the expense of greater key storage requirements.

Operational Efficiency. In addition to computational efficiency, the adoption of the Chameleon encryption primitive only require D and C to obtain the key materials from the trusted KDC during registration. Subsequently C can request content from D many times, which is secured based on the user table UT^C and the master table MT^{D_C}, without further involving the KDC. This is beneficial compared with protocols such as [13,15], where for each request of content by C in a session, a new watermark W^C needs to be generated by a trusted watermark authority. Thus the **CE** approach enjoys the operational advantage of being primarily a two-party protocol during the main operational phases, while the conventional approach typically requires three parties.

4.3   Alternative Approaches

We note that there are alternative ways in which this Chameleon encryption approach to buyer-seller watermarking could be deployed:

- **CE** *with always online KDC.* If the KDC is always online, and key storage is an issue, it is possible to design the protocol so that there is no need for D and C to store the master table MT^{D_C} and the user table UT^C, respectively. This is achieved by KDC holding all three tables. When C wishes to get content, D contacts KDC with authenticated information of C. Then KDC generates the pseudo-random sequence R and uses the sequence to select the elements from MT^{D_C} and UT^C. Every s elements are added together. These newly selected and added elements from MT^{D_C} (and UT^C) become the encryption key (and joint fingerprinting and decryption key) for D (and C). Following the previous example in our efficiency analysis, the key size is $n \cdot |Z| = 10000 \cdot 16$ bits $= 20$ KByte, which is much smaller than the size of the master and user tables.
- **CE** *with reduced client KDC contact.* If contacting KDC by the client C is an issue during key distribution, the protocol can be modified so that the user table UT^C is delivered to C through D. This can be done by KDC encrypting and signing the user table UT^C based on conventional cryptographic primitives, and giving UT^C to D, who then forwards the table to C. In this case, C only needs to communicate with D for the entire execution of the protocol.

The **CE** approach would thus seem to offer some interesting trade-offs between storage restrictions and operational contact with a trusted third party, allowing it to be tailored to different application environments.

4.4   Application Scenario

We close by briefly describing an example of an application scenario in which the advantages of the **CE** approach apply. Consider a content distribution environment where clients purchase content (such as digital songs and movies) from a content provider using a portable entertainment device such as an iPhone. Since the client will wish to have instant access to the content after purchase, efficient communication between the content provider and the client is crucial. In this scenario, it will clearly be disadvantageous if it is always necessary to contact a trusted third party for each content distribution session. The limited *communication* requirements of the **CE** approach presented in Section 4 would seem to be more appropriate than those of the conventional approach [13,15] for dynamic mobile environments of this type. In addition, the *computation efficiency* of the **CE** approach is beneficial to this type of scenario for faster access to content, although issues concerning constrained computational power is indeed diminishing over time due to the advancement of computer technologies. Similarly, the limitation of the **CE** approach on key storage may not be an issue over time due to the same reason.

5   Conclusion

We have discussed a new approach to the design of efficient buyer-seller watermarking protocols, based on Chameleon encryption. This approach offers some significant advantages over the conventional approach to the design of such protocols. The efficiency gains come from the fact that Chameleon encryption is a symmetric primitive that does not rely on the computationally intensive operations normally associated with public key encryption, and from the integration of the watermarking and decryption operations. We believe that the applicability of this approach merits further investigation.

References

1. Adelsbach, A., Huber, U., Sadeghi, A.-R.: Fingercasting-Joint Fingerprinting and Decryption of Broadcast Messages. In: Batten, L.M., Safavi-Naini, R. (eds.) ACISP 2006. LNCS, vol. 4058, pp. 136–147. Springer, Heidelberg (2006); Also, Technical Report detailing the ACISP 2006 paper
2. Blakley, G.R., Meadows, C., Purdy, G.B.: Fingerprinting Long Forgiving Messages. In: Williams, H.C. (ed.) CRYPTO 1985. LNCS, vol. 218, pp. 180–189. Springer, Heidelberg (1986)
3. Choi, J.-G., Sakurai, K., Park, J.-H.: Does It Need Trusted Third Party? Design of Buyer-Seller Watermarking Protocol without Trusted Third Party. In: Zhou, J., Yung, M., Han, Y. (eds.) ACNS 2003. LNCS, vol. 2846, pp. 265–279. Springer, Heidelberg (2003)
4. Cox, I.J., Kilian, J., Leighton, T., Shamoon, T.: Secure Spread Spectrum Watermarking for Multimedia. IEEE Trans. on Image Processing 6(12), 1673–1687 (1997)

5. Deng, M., Preneel, B.: On Secure and Anonymous Buyer-Seller Watermarking Protocol. In: 2008 International Conference on Internet and Web Applications and Services. IEEE Computer Society Press, Los Alamitos (2008)
6. Dierks, T., Rescorla, E.: The TLS Protocol Version 1.1. In: RFC 4346 (2006)
7. Goi, B.-M., Phan, R.C.-W., Yang, Y., Bao, F., Deng, R.H., Siddiqi, M.U.: Cryptanalysis of Two Anonymous Buyer-Seller Watermarking Protocols and an Improvement for True Anonymity. In: Jakobsson, M., Yung, M., Zhou, J. (eds.) ACNS 2004. LNCS, vol. 3089, pp. 369–382. Springer, Heidelberg (2004)
8. Ibrahim, I.M., Nour El-Din, S.H., Hegazy, A.F.A.: An Effective and Secure Buyer-Seller Watermarking Protocol. In: Third International Symposium on Information Assurance and Security (IAS 2007), pp. 21–26. IEEE Computer Society Press, Los Alamitos (2007)
9. Ju, H.S., Kim, H.J., Lee, D.H., Lim, J.I.: An Anonymous Buyer-Seller Watermarking Protocol with Anonymity Control. In: Lee, P.J., Lim, C.H. (eds.) ICISC 2002. LNCS, vol. 2587, pp. 421–432. Springer, Heidelberg (2003)
10. Katzenbeisser, S., Skorić, B., Celik, M.U., Sadeghi, A.-R.: Combining Tardos Fingerprinting Codes and Fingercasting. In: Furon, T., Cayre, F., Doërr, G., Bas, P. (eds.) IH 2007. LNCS, vol. 4567, pp. 294–310. Springer, Heidelberg (2008)
11. Kilian, J., Leighton, F.T., Matheson, L.R., Shamoon, T.G., Tarjan, R.E., Zane, F.: Resistance of Digital Watermarks to Collusive Attacks. Technical Report TR-585-98, Princeton University, Department of Computer Science (1988), ftp://ftp.cs.princeton.edu/techreports/1998/585.ps.gz
12. RSA Labs. RSA Signature Scheme with Appendix - Probabilistic Signature Scheme (2000), ftp://ftp.rsasecurity.com/pub/rsalabs/rsa_algorithm/nessie_pss.zip
13. Lei, C.-L., Yu, P.-L., Tsai, P.-L., Chan, M.-H.: An Efficient and Anonymous Buyer-Seller Watermarking Protocol. IEEE Trans. on Image Processing 13(12), 1618–1626 (2004)
14. Ray Liu, K.J., Trappe, W., Wang, Z.J., Wu, M., Zhao, H.: Multimedia Fingerprinting Forensics for Traitor Tracing. EURASIP Book Series on Signal Processing and Communication, vol. 4. Hindawi Publishing Corporation (2005)
15. Memon, N., Wong, P.W.: A Buyer-Seller Watermarking Protocol. IEEE Trans. on Image Processing 10(4), 643–649 (2001)
16. Okamoto, T., Uchiyama, S.: A New Public-Key Cryptosystem as Secure as Factoring. In: Nyberg, K. (ed.) EUROCRYPT 1998. LNCS, vol. 1403, pp. 308–318. Springer, Heidelberg (1998)
17. Paillier, P.: Public-key Cryptosystems Based on Composite Degree Residuosity Classes. In: Stern, J. (ed.) EUROCRYPT 1999. LNCS, vol. 1592, pp. 223–238. Springer, Heidelberg (1999)
18. Pfitzmann, B., Sadeghi, A.-R.: Coin-Based Anonymous Fingerprinting. In: Stern, J. (ed.) EUROCRYPT 1999. LNCS, vol. 1592, pp. 150–164. Springer, Heidelberg (1999)
19. Pfitzmann, B., Schunter, M.: Asymmetric Fingerprinting. In: Maurer, U.M. (ed.) EUROCRYPT 1996. LNCS, vol. 1070, pp. 84–95. Springer, Heidelberg (1996)
20. Pfitzmann, B., Waidner, M.: Anonymous Fingerprinting. In: Fumy, W. (ed.) EUROCRYPT 1997. LNCS, vol. 1233, pp. 88–102. Springer, Heidelberg (1997)
21. Qiao, L., Nahrstedt, K.: Watermarking schemes and protocols for protecting rightful ownerships and customer's rights. Journal of Visual Communication and Image Representation 9(3), 194–210 (1998)

First Digit Law and Its Application to Digital Forensics
(Invited Paper)

Yun Q. Shi

Department of Electrical and Computer Engineering
New Jersey Institute of Technology
Newark, NJ 07102, USA
shi@njit.edu

Abstract. Digital data forensics, which gathers evidence of data composition, origin, and history, is crucial in our digital world. Although this new research field is still in its infancy stage, it has started to attract increasing attention from the multimedia-security research community. This lecture addresses the first digit law and its applications to digital forensics. First, the Benford and generalized Benford laws, referred to as first digit law, are introduced. Then, the application of first digit law to detection of JPEG compression history for a given BMP image and detection of double JPEG compressions are presented. Finally, applying first digit law to detection of double MPEG video compressions is discussed. It is expected that the first digit law may play an active role in other task of digital forensics. The lesson learned is that statistical models play an important role in digital forensics and for a specific forensic task different models may provide different performance.

1 Introduction

In our digital age, digital media have been being massively produced, easily manipulated, and swiftly transmitted to almost anywhere in the world at anytime. While the great convenience has been appreciated, information assurance has become an urgent and critical issue faced by the digital world. The data hiding, cryptography, and combination of both have been shown not sufficient in many applications. Digital data forensics, which gathers evidence of data composition, origin, and history, is hence called for. Although this new research field is still in its infancy stage, it has started to attract increasing attention from the multimedia-security research community. The research topics cover detection of image forgery, including image splicing; classification of computer graphic images from photographic images; identification of imaging sensor, including digital cameras and digital scanners; detection of single JPEG compression from double JPEG compression, and single sampling from double sampling, to name a few. It is observed that statistical models play an important role in all of these digital forensic tasks.

In this lecture, Benford's law, well-known in statistics and applied to detect fraud in accounting data, is first presented in Section 2. In Section 3, two discoveries are reported. One is that the block discrete cosine transform (block DCT, or simply BDCT) coefficients follow the Benford law. Another is that the first digits of one-time JPEG quantized BDCT coefficients obey a different logarithm law, called

H.J. Kim, S. Katzenbeisser, and A.T.S. Ho (Eds.): IWDW 2008, LNCS 5450, pp. 448–453, 2009.

generalized Benford's law, while that of two-time JPEG quantized (with two different quantization steps) BDCT coefficients do not. Section 4 demonstrates that these two discoveries have found applications in determination if a given BMP image has been JPEG compressed or not; if it has been JPEG compressed once, how the corresponding JPEG Q-factor can be determined. Detection of double JPEG compression and MPEG detection have been addressed in Sections 5 and 6, respectively. The conclusion is drawn in Section 7.

2 Benford's Law

Benford's law, also known as the first digit law, discovered by Newcomb in 1881 [1] and rediscovered by Benford in 1938 [2], states that the probability distribution of the first digits, x ($x = 1, 2, ..., 9$), in a set of natural numbers is logarithmic. Specifically, the distribution of the first digits follows

$$p(x) = \log_{10}(1 + \frac{1}{x}), x = 1, 2, ..., 9 \tag{1}$$

where $p(x)$ stands for probability of x. A statistical explanation and analysis of this law is given by Hill [3]. Benford's law has been used to detect income evasion and fraud in accounting area [4,5].

3 JPEG Compression and First Digit Law

In practice, JPEG compression is the most popularly utilized international image compression standard nowadays. In the JPEG compression, an input image is first divided into consecutive and non-overlapped 8x8 blocks. The 2-D DCT is applied to each block independently. Afterwards, the BDCT coefficients are quantized according to JPEG quantization table. The quantized coefficients are zigzag scanned and entropy encoded. Here, only the non-zero AC components of both the BDCT coefficients and the JPEG quantized coefficients are considered.

3.1 BDCT Coefficients and Benford's Law

It has been found that the first digit distribution of the BDCT coefficients follows Benford's law closely [6]. Specifically, with 1338 uncompressed images in UCID [7], the average Chi-square divergence between the actual first digit distribution and the Benford law is less than 1%. This does not come with surprise because it is well-known that the AC BDCT coefficients can be modeled by the generalized Laplacian distribution [8] and the exponential distribution guarantees Benford's law [3].

3.2 Quantized BDCT Coefficients and Generalized Benford's Law

It is shown [6] that the first digit distribution of quantized BDCT coefficients also follow logarithm law, however, with the probability of the first digit 1 higher than that specified by Benford's law, and the probability of the first digit 9 lower than that specified by Benford's law. That is, the distribution becomes steeper after quantization. This skewness of the distribution towards the lower digits is in fact caused by the rounding process in quantization. It is demonstrated that the larger the quantization step, the skewer the distribution. One possible parameterized logarithm model is given as follows.

$$p(x) = N \log_{10}(1 + \frac{1}{s + x^q}), x = 1, 2, ..., 9 \qquad (2)$$

where N is a normalization factor which makes $p(x)$ a probability distribution, s and q are model parameters to precisely describe the distributions for different images and different JPEG compression Q-factors. This model is also referred to as generalized Benford's law. This is because when $s=0$ and $q=1$, Formula (2) becomes equal to Formula (1), meaning that the Benford law is just a special case of this distribution model.

3.3 Double JPEG Compression and Generalized Benford's Law

If a JPEG compressed image is decompressed, and the decompressed image goes through another JPEG compression with a different Q-factor from that used in the first JPEG compression, which is referred to as double JPEG compression, it is found [6] that the first digit distribution of the double JPEG compressed image will deviate from the generalized Benford law. The degree of deviation depends on the relation between the first and the second Q-factors.

These discoveries presented above have laid down the foundation for detection of JPEG compression history of BMP images, double JPEG and MPEG compression, which will be presented in Sections 4, 5 and 6, respectively.

4 Detection of JPEG Compression History for BMP Images

Based on the Benford law and the generalized Benford law introduced above, the procedures have been developed to detect if a given BMP image has been JPEG compressed or not; if the image has been compressed once, what the utilized Q-factor is in the JPEG compression. The idea is as follows [6]. First, apply JPEG compression with Q-factor equal to 100 to the given BMP image; then examine the first digit distribution of the compressed image to see if it follows the generalized Benford law or not. If yes, then the given BMP image has never been JPEG compressed before or it has been JPEG compressed with Q-factor 100. Otherwise, the given BMP image has been JPEG compressed before with a Q-factor different from 100. If it has been compressed once, the Q-factor can be determined by reducing step-by-step the Q-factor used in quantization gradually until a good fitting of the generalized Benford law has been observed. This firstly encountered Q-factor is the answer, and the accuracy depends on the step selected in testing. It is shown that this scheme outperforms the prior-art [9], which uses a different model to measure JPEG quantization artifacts.

5   Detection of Double JPEG Compression

It is well-known that double JPEG compression has been an important factor that needs be carefully considered and examined in JPEG steganalytic work. Moreover, from digital forensics point of view, it is a legitimate question to ask if a given JPEG image has been compressed once or more than once. Note that by double JPEG compression, we mean here the two quantization processes involving two different Q-factors. Two pioneering works in detecting double JPEG compressions [10,11] are based on the same observation. That is, double JPEG compression will cause some periodic pattern in the histogram of JPEG quantized block DCT coefficients. Based on first digit distribution in JPEG compression discussed in Section 3, a new method to detect double JPEG compression has been recently reported [12]. There, more advanced detection results have been reported, indicating that different statistical models may bring out different performance for the same forensic task.

Although it has been shown that the first digit distribution of AC JPEG quantized BDCT coefficients, referred to as JPEG coefficients, follows closely to, or deviate from, the generalized Benford law depending on whether single JPEG compression or double JPEG compression has been applied, it is hard to use only one threshold to measure the closeness of the actual distribution with respect to the generalized Benford law in order to classify the double compression from single compression with high detection rate. This seems caused by the fact that the generalized Benford law is image dependent, which is rather different from the Benford law, which is image independent. Therefore, the machine leaning framework is needed for this task in order to boost classification accuracy. That is, the probabilities of the nine first digits of AC quantized BDCT coefficients are used as features, a classifier such as Fisher linear discriminator (FLD) or support vector machine (SVM) is trained first, and the trained classifier is then used for testing. It has been shown that using first digit distribution of mode-based AC quantized BDCT coefficients is more efficient than using first digit distribution of all AC quantized BDCT coefficients. This is because mode-based first digit distribution is more sensitive to artifact caused by double JPEG compression. The superior performance in detecting double JPEG compression to that contained in [10,11] has been reported in [12].

6   Detection of Double MPEG Compression

By double MPEG compression it means the two MPEG compressions resulting in alteration of video bit rate. Obviously, from digital forensics point of view, i.e., gaining evidence for data history and composition, it is necessary to be able to detect double MPEG compression. For instance, the originality of a video presented as evidence in court can be denied if it is found that the video has been doubly compressed. Furthermore, in reality, video forgery such as manipulation made on frame, and/or frame omission and addition is likely followed by an additional MPEG compression because of the huge amount of data involved. Therefore, double MPEG compression may be able to reveal the possible trace of forgery.

It is well-known that video MPEG compression is more complicated than image JPEG compression. In MPEG video, there are I, P, and B three different types of frames. I-frames are intra-frame coded, while P- and B-frames are inter-frame coded. Changes made in I-frames will influence corresponding P- and B-frames, and the modification in P-frames may change relevant B-frames as well. In [13], the authors applied their work originally developed for double JPEG compression to double MPEG compression. That is, they examine if the I-frames of a video sequence have been double compressed or not. Although it is the first piece of work in double MPEG compression, this work by nature is applicable to variable bit rate (VBR) mode only. Namely, their method is not workable for constant bit rate (CBR) mode, which is known to have the quantization steps varying on the macro-block (MB) level in order to keep constant bit rate during the MPEG compression for different video content. It is noted that the CBR mode is used more popularly than the VBR mode these days. Hence, the more capable double MPEG detection schemes are called for.

In [14], the first digit distribution of AC MPEG coefficients in I-, P-, and B-frames are examined. It is found that the discoveries reported in Section 3 are generally valid for these three different types of frames with both VBR and CBR modes. The deviation of the first digit distribution of non-zero AC coefficients in the CBR mode from the generalized Benford law due to double MPEG compression is often less obvious than that in the VBR mode though. Therefore, for the same reason described for detecting double JPEG compression, the machine learning framework has to be used in order to achieve high detection rate in detecting double MPEG compression. The probabilities of nine first digits of non-zero AC MPEG coefficients together with the three fitting parameters provided by Matlab curve fitting toolbox form a 12-D feature vector. The SVM classifier with poly kernel is used in classification. Furthermore, it is proposed to use group of pictures (GOP) as a unit for the machine leaning framework to determine double MPEG compression or not. For a video sequence, a majority voting strategy based on all of GOP's is used. Satisfactory detection performance has been reported with respect to 10 widely used video sequences in both VBR and CBR modes.

7 Discussion and Conclusion

A   JPEG compression detection for a given BMP image, double JPEG compression detection and double MPEG compression belong to digital forensics, an emerging and needed research field in our digital age.

B   Statistical model plays an important role in forensic research. Different models may lead to different performances. Therefore, study and research on effective statistical model to the issue faced is crucial in forensic research.

C   First digit distribution is shown to be effective in JPEG compression detection for a given BMP image, and detection of double JPEG or MPEG compression.

D   According to B, however, more advanced statistical model should be researched.

E   In some cases, where one single threshold on some statistical quality may not perform satisfactorily due to complicated reality, the machine learning framework with multi-dimensional feature vectors, modern classifier, and training process with a sufficient training data may improve detection performance dramatically.

F   It is expected that the first digit law will find other applications in digital forensics.

Acknowledgement

The author acknowledges fruitful technical discussions with and significant contributions made by Drs. Dongdong Fu, Wen Chen, Chunhua Chen, and Wei Su; Professors Hong Zhang and Jiwu Huang; and doctoral candidate Mr. Bin Li.

References

1. Newcomb, S.: Note on the frequency of use of the different digits in natural numbers. American Journal of Mathematics 4(1/4), 39–40 (1881)
2. Benford, F.: The law of anomalous numbers. Proc. Amer. Phil. Soc. 78, 551–572 (1938)
3. Hill, T.P.: A statistical derivation of the significant-digit law. Statistical Science 10, 354–363 (1996)
4. Nigrini, M.: The detection of income evasion through an analysis of digital distributions. Ph.D Thesis, Department of Accounting, University of Cincinatti (1992)
5. Durtschi, C., Hillison, W., Pacini, C.: The effective use of Benford's law to assist in detecting fraud in accounting data. Journal of Forensic Accounting V, 17–34 (2004)
6. Fu, D., Shi, Y.Q., Su, W.: A generalized Benford's law for JPEG coefficients and its applications in image forensics. In: SPIE Electronic Imaging: Security, Steganography, and Watermarking of Multimedia Contents, San Jose, CA, USA (January 2007)
7. Schaefer, G., Stich, M.: UCID - An Uncompressed Colour Image Database. Technical Report, School of Computing and Mathematics, Nottingham Trent University, U.K (2003)
8. Reininger, R.C., Gibson, J.D.: Distributions of the two dimensional DCT coefficients for images. IEEE Trans. on Commun. COM-31, 835–839 (1983)
9. Fan, Z., Queiroz, R.L.: Identification of bitmap compression history: JPEG detection and quantizer estimation. IEEE Transaction on Image Processing 12(2) (February 2003)
10. Popescu, A.: Statistical tools for digital image forensics., Ph. D. thesis, Dartmouth College (December 2004)
11. Lukas, J., Fridrich, J.: Estimation of primary quantization matrix in double compressed JPEG images. In: Proc. of DFRWS 2003, Cleveland, OH, USA, August 5-8 (2003)
12. Li, B., Shi, Y.Q., Huang, J.: Detecting double compressed JPEG Image by using mode based first digit features. In: IEEE International Workshop on Multi-media Signal Processing, MMSP 2008, Queensland, Australia (October 2008)
13. Wang, W.H., Farid, H.: Exposing digital forgeries in video by detecting double MPEG compression. In: ACM Multimedia and Security Workshop, Geneva, Switzerland (2006)
14. Chen, W., Shi, Y.Q.: Detection of double MPEG video compression using first digits statistics. In: International Workshop on Digital Watermarking (IWDW 2008), Busan, Korea (November 2008)

Digital Camera Identification from Images – Estimating False Acceptance Probability
(Invited Paper)

Miroslav Goljan

Dept. of Electrical and Computer Engineering, SUNY Binghamton,
Binghamton, NY 13902-6000, USA
mgoljan@binghamton.edu

Abstract. Photo-response non-uniformity noise present in output signals of CCD and CMOS sensors has been used as fingerprint to uniquely identify the source digital camera that took the image. The same fingerprint can establish a link between images according to their common source. In this paper, we review the state-of-the-art identification method and discuss its practical issues. In the camera identification task, when formulated as a binary hypothesis test, a decision threshold is set on correlation between image noise and modulated fingerprint. The threshold determines the probability of two kinds of possible errors: false acceptance and missed detection. We will focus on estimation of the false acceptance probability that we wish to keep very low. A straightforward approach involves testing a large number of different camera fingerprints against one image or one camera fingerprint against many images from different sources. Such sampling of the correlation probability distribution is time consuming and expensive while extrapolation of the tails of the distribution is still not reliable. A novel approach is based on cross-correlation analysis and peak-to-correlation-energy ratio.

1 Introduction

Digital cameras became affordable commodity for almost everyone. Tens of millions of them have been produced and sold every year. Billions of images are taken and stored in digital form. Along with the scene content, they contain auxiliary data in file headers. But even if the file header is stripped off, the pixels data contain some traces of signal and image processing that can be used in forensics analysis. Image forensics aims to reveal information about the source camera, its brand and model, camera setting, amount of zoom, exposure, time and date, to detect image forgeries and manipulations, reverse-engineer cameras and more. For example, the work of Khanna *et al.* addresses the problem of classification of imaging sensor types [1], [2], Swaminathan *et al.* recognizes color filter arrays and interpolation methods [3], Popescu and Farid introduced a large number of image forensic tools [4] that can reveal forgeries. Forensic analysis of this kind is in an early stage of development but increasing interest of research community speeds up the progress. One of the most reliable methods was proposed by Lukáš *et al.* [9] and further explored by Chen *et al.* [10] and others [11] that is capable of identifying the exact digital camera the image was taken with

H.J. Kim, S. Katzenbeisser, and A.T.S. Ho (Eds.): IWDW 2008, LNCS 5450, pp. 454–468, 2009.
© Springer-Verlag Berlin Heidelberg 2009

(source identification). There are some situations, when such information is a vital piece of evidence in crime investigation. One of them is child pornography where linking photos to the suspect camera can provide a strong evidence for prosecution or steer the investigation. Applications like this require very low probability of wrong accusation. This paper addresses the problem of false camera identification and aims to improve error control while lowering the cost and demand on computations.

One of the challenging aspects is the large number of cameras that have to be uniquely distinguished in an analogy with human fingerprinting. A list of requirements on a camera identifier (*camera fingerprint*) is the following

- high dimensionality (to cover the large number of cameras)
- uniqueness (no two cameras have the same fingerprint)
- stability over time and typical range of physical conditions under which cameras operate
- robustness to common image processing, including brightness, contrast, and gamma adjustment, filtering, format conversions, resampling and JPEG compression
- universality (virtually all digital cameras have it).

On the one hand, this list may not be complete; on the other hand, some requirements may be relaxed if necessary.

In digital image watermarking, an invisible signal (watermark) is inserted in the image to carry some information. This information can be used for owner identification, for an evidence of image authenticity and integrity, for media fingerprinting, or to carry auxiliary data inseparably from the image pixels for other applications. There is a trade-of between watermark robustness and the amount of data it can carry. Image forensics, in contrast to image watermarking, cannot plant fingerprints into existing images. The only option is to explore existing signals that are produced in cameras during image acquisition and on-board signal processing. Fortunately, Photo-Response Non-Uniformity (PRNU) of imaging sensors (CCD, CMOS, and their modern derivatives) is an ideal source of such fingerprinting watermark that is already inherently present in almost all pictures imaging sensors produce. PRNU is caused by material impurities and imperfections in CCD and CMOS manufacturing. Dark content images, such as those taken at night with low light exposure, are not as much affected by PRNU while they may contain dark current [13]. Both these signals resemble noise, and, together, they do exhibit the desirable properties listed above.

Once we know that a fingerprint exists, using it for camera sensor identification (CSI) consists of two tasks. One is *fingerprint estimation*, for which we may have the functional camera or a set of images that were positively taken with that camera. The second is *fingerprint detection* or testing the hypothesis that the camera fingerprint is present in the image under investigation. We shortly reiterate the estimation and detection parts in the next section. We choose the simplest form without improvements found in recent publication of Chen *et al.* [12]. We will discuss the detector properties, the problem of setting detection threshold, and our new approach in Sections 3 and 4. Experiments are presented in Section 5. The summary concludes the paper.

2 Fingerprint Estimation and Detection

Because all images from one camera should contain the same fingerprint, a naïve but a fair representation of the camera fingerprint is a pixel-wise average of a number of images taken with the camera. However, simple averaging is not the best choice. First, such average would contain a significant portion of images' content, so it is better to work with noise residuals. Second, simple averaging does not take into account the fact that PRNU is modulated by the amount of light that falls on the imaging sensor. The maximum likelihood estimate derived by Chen *et al.* [12] for the multiplicative model of PRNU [13] is therefore a better camera fingerprint. In the next paragraph, we adopt necessary notation and concepts from the referenced paper [12].

2.1 Estimation

Let the grayscale (or one color – red, green, or blue) image be represented with an $m \times n$ matrix $\mathbf{I}[i, j]$, $i = 1, ..., m$, $j = 1, ..., n$, of integers (pixel intensities) ranging from 0 to 255. To eliminate the image content and to increase signal-to-noise ratio for the fingerprint as the signal of interest, the first step is noise extraction. Noise residual $\mathbf{W}$ is obtained by Wiener filtering the image in wavelet domain. We prefer the filter according to [14] for its performance and high speed implementation. Its computational complexity is linear in the number of pixels. (For example, applying the filter to a 4 Mpixel gray scale image on a PC with Pentium 4, 3.4 GHz takes around 2.25 sec, 8 Mpixel image around 4.5 sec.) The only parameter of the filter is variance σ^2 of the stationary noise that is being separated. Denoting the denoised image as $F(\mathbf{I})$, $\mathbf{W} = \mathbf{I} - F(\mathbf{I})$. We start with a multiplicative model of PRNU $\mathbf{K}$,

$$\mathbf{W} = a\mathbf{I}\mathbf{K} + \Xi, \tag{1}$$

where a is a constant, Ξ is a noise term representing all random noises. Throughout this paper, matrices and vectors are in bold font and the operations between them, such as multiplication or ratio of matrices, will always be understood as element-wise. This model is simpler than the one in the reference [12], where an attenuation matrix is considered instead of the constant a. With this complexity reduction in the model, the performance of CSI may slightly drop. What we will gain by that is a much simpler detection part later (no need for a correlation predictor and for more images associated with it) and the system will be more amenable to error analysis. At this point, we omit the property of dark current that behaves the same way as PRNU for fixed image intensities. We will address this issue partially later with an attenuation function.

The maximum likelihood estimation of PRNU $\hat{\mathbf{K}}$ (together with dark current) from a set of N images $\mathbf{I}_1, ..., \mathbf{I}_N$ originated from one camera is given by formula (2).

$$\hat{\mathbf{K}} = \frac{1}{S} \sum_{k=1}^{N} \mathbf{W}_k \mathbf{I}_k, \quad S = \sum_{k=1}^{N} (\mathbf{I}_k)^2, \tag{2}$$

up to a multiplicative constant. The noise residuals in the formula are calculated independently as $\mathbf{W}_k = \mathbf{I}_k - F(\mathbf{I}_k)$ for each $k = 1, ..., N$. At this point, $\hat{\mathbf{K}}$ becomes our *camera fingerprint*.

Images that are brighter than others contribute more to the sum in (2) because PRNU is more pronounced in them. If we have the camera at hand the best images we can obtain for PRNU estimation are images of high luminance, (obviously) no saturation and uniform content (flat fields). The mean square error of this estimate increases with N.

2.2 Detection

The very basic scenario for the camera identification problem is the following. All we have is an image in a format produced by a digital camera, not further processed in other ways than lossless conversions, and one camera fingerprint obtained by the estimation in Section 2.1. Let I denote the image matrix that represents pixel values. The question is whether or not the image originated from the camera. At this point, we assume that the question is equivalent to deciding whether or not the image contains the camera fingerprint. This leads to a binary hypothesis test. As we did before, we apply the host signal rejection by noise extraction from the image data I.

Let $W = I - F(I)$ be the noise residual of the image. The binary hypothesis test contains noise-only hypothesis H_0 and fingerprint presence hypothesis H_1,

$$H_0: \ W = \Xi \,,$$
$$H_1: \ W = I\hat{K} + \Xi \,. \tag{3}$$

The optimal detector under the assumption that the noise term Ξ is a sequence of i.i.d. random variables with unknown variance is the normalized correlation

$$\rho = corr\left(I\hat{K}, W\right). \tag{4}$$

The decision is obtained by comparing ρ to a decision threshold ρ_{th}.

In Neyman-Pearson hypothesis approach, the decision threshold is set so that the false acceptance probability will not exceed a certain level α. False acceptance (FA) occurs when hypothesis H_0 is true but we decide H_1, while false rejection (FR) occurs when we accept H_0 when H_1 is true. In our scenario, FA occurs when the camera fingerprint is declared to be present in the image while it is not, FR occurs when the presence of the fingerprint is missed. To satisfy the desired level α for FA probability, P_{FA}, we need to estimate the pdf f_0 of the test statistics (4) under H_0, $f_0(x) = \Pr(x|H_0)$. This typically requires evaluation of (4) for a large amount of images coming from other cameras than the fingerprint K. Relations between probability of false acceptance P_{FA} or false rejection P_{FR} and the threshold ρ_0 are given by equations (5).

$$P_{FA} = \int_{x > \rho_0} f_0(x)dx \,, \ P_{FR} = \int_{x \le \rho_0} f_1(x)dx \,. \tag{5}$$

3 System Improvements and Generalization – Previous Art

The first publication on PRNU based camera identification by Lukáš et al. in 2005 [15] spurred a lot of research and publications [16], [12], [17]. Some adjusted the method to address other devices. Its applicability to scanners was tested by Khanna et al. [18],

Sankur *et al.* studied cell phone cameras that produce highly compressed JPEG images [19], [20] is devoted to camcorders. Other work deals with various image processing generalizing the detection part of CSI for cropped and scaled images, including digitally zoomed ones [11], and for printed images involving very small rotation and nonlinear distortion [21]. PRNU often survives image processing, such as JPEG compression, noise adding, filtering, or gamma correction, unless the PSNR is too low. Surprisingly, no improvement has been done at the noise extraction stage, i.e., the SNR between the fingerprint and the noise residual **W** seems to be hard to improve. This is namely true when computational complexity is an issue. Other characteristics than PRNU have also been explored, dust specs on the sensor protective glass of SLR cameras [22], optical deficiencies of lenses [23], or CFA artifacts [24], [3].

The demand for minimizing error rates has been the motivation for some previous work. Chen *et al.* [10] introduced a correlation predictor and modified the fingerprint detection in terms of prediction error. By modeling the prediction error and consequently pdf $f_1(x) = \Pr(x|\mathrm{H}_1)$ as identically distributed Gaussian variables, this construction allows for satisfying estimation of the FR probability. Only this innovative modification takes into account the prior knowledge of image **I** and corresponds to a slightly different scenario in which one image is fixed and the camera fingerprints are what is being randomly chosen.

As noted in Section 2, to evaluate false acceptance probability as a function of the threshold on the normalized correlation (4) one may need a large amount L of "randomly chosen" images and calculate ρ for each of them and every time a new fingerprint is taken into the hypothesis test. The FA probability is then estimated by the false alarm rate, which is the function of the decision threshold x,

$$FAR(x) = \frac{1}{L} \sum_{i,\rho_i \geq x}^{L} 1. \tag{6}$$

The problem with such sampling becomes apparent when we are after a very small probability $\alpha \ll 1$. The amount of images needed is of the order of $L \approx 1/\alpha$. The smaller the number L, the less reliable the FA probability estimate is. Early attempts to model this probability with Generalized Gaussian pdf had limited success [9]. The biggest obstacle was instability of the shape of samples (6), a bias and skewness in their distribution. More light into this problem was shed in [10] in the section about preprocessing $\hat{\mathbf{K}}$. The noise term in (1) almost always contains periodic signals and structured noise responsible for small positive correlation between noise residuals of images from different cameras. After realizing the reason for this effect, the importance of separating such unwanted signals from the PRNU estimate became eminent.

Filler *et al.* proposed in his recent study [25] to characterize $\hat{\mathbf{K}}$ from (2) with a set of features and utilize them in a different forensic application – identification of camera brands and camera models. Once we admit that image noise residuals systematically contain a signal that is the same or similar in all images from more cameras the hypothesis test changes its character. We will come back to this in the next section.

One significant structured noise with periodicities is called *linear pattern* of the camera fingerprint and is defined through *zero-mean operation* on $\hat{\mathbf{K}}$. Elements of $ZM(\hat{\mathbf{K}})$ for all $i=1,.. , m$ and $j=1,.. , n$ are

$$ZM(\hat{\mathbf{K}})[i, j] = \hat{\mathbf{K}}[i, j] - \frac{1}{m}\sum_{i=1}^{m}\hat{\mathbf{K}}[i, j] - \frac{1}{n}\sum_{j=1}^{n}\hat{\mathbf{K}}[i, j] + \frac{1}{mn}\sum_{i=1, j=1}^{m,n}\hat{\mathbf{K}}[i, j], \qquad (6)$$

which makes the mean of every column and every row of the matrix equal to zero. The linear pattern is

$$LP(\hat{\mathbf{K}}) = \hat{\mathbf{K}} - ZM(\hat{\mathbf{K}}). \qquad (7)$$

It is the color interpolation (demosaicing) in cameras equipped with color filter arrays (CFA) and row-wise and column-wise operations in signal processing of the imaging sensor what is responsible for *linear pattern*. A slight rounding error due to limited bit-depth in missing colors computations is all it takes to cause this phenomenon. Typically, the energy of the linear pattern $LP(\hat{\mathbf{K}})$ is about an order smaller than the energy of $ZM(\hat{\mathbf{K}})$. However, it does influence pdf f_0 and FA probability markedly as we will also see in the next sections. Suddenly, pdf f_0 behaves nicely, Gaussian model fits very well. We can estimate the FA probability by fitting the model through a smaller number of sampled data (6) and computing the right tail probability of the Gaussian pdf known as the Q-function.

Another preprocessing step proposed earlier is Wiener filtering in the Fourier domain in order to suppress any high magnitudes in the spectrum. The exact type of images, amount of lossy compression, or type of cameras for which this filtering really improves CSI, is yet to be determined. The importance of removing the linear pattern from camera fingerprints have been emphasized in publications ([10],[12]). Despite of that, it is still being omitted in some papers ([26][Bayram]) resulting in poor performance of camera identification for some cameras and image formats or their processing.

From now on, $ZM(\hat{\mathbf{K}})$ is what we call the *camera fingerprint*.

4 Normalized Correlation Abandoned

We want to choose the FA probability for our tests, determine the threshold ρ_0 once, and evaluate the decision possibly many times. There is no problem with this plan unless the camera fingerprint estimate changes. Such change will require re-evaluation of the threshold, which can be time consuming or even infeasible. This problem is not new. In the work on CIS for cropped and scaled images [11], we handled similar problem. Having a fixed FA probability, it was not guaranteed that the threshold for resized images had to stay the same. The need for a more stable relation between FA probability and the decision threshold led to the introduction of Peak to Correlation Energy (PCE) ratio as a replacement for normalized correlation detector. We demonstrate that PCE is much more suitable detection statistic, even for the basic problem of camera identification, than the normalized correlation. This may be surprising when the correlation was derived as the optimal detector. But assumptions on the model, which our hypothesis test (3) is based on, may not be satisfied. The assumption of independence of Gaussian variables (the noise term Ξ) is one culprit. Properties of PCE are especially useful when a periodic signal common to images from various cameras (like the linear pattern) enter the image noise residuals $\mathbf{W}$ as well as $\mathbf{W}_1, ..., \mathbf{W}_k$.

It can be shown that the expected sample variance of the normalized correlation between two identically distributed independent random signals of length k is inversely proportional to k (for large k). It is also now true for correlation between noise residuals and camera fingerprint (multiplied by $\mathbf{I}$) under hypothesis H_0 since they are close to being perfectly independent. Thus, a change of the fingerprint size from k_1 pixels to k_2 pixels will cause an expected change of the threshold from ρ_{th} to $\rho_{\text{th}}\sqrt{k_1/k_2}$. We could therefore normalize ρ by the same factor and keep the threshold unchanged if it is just the camera resolution what changes. Our advocacy for PCE comes from elsewhere. We show that the introduction of a periodic signal (like the linear pattern) in both the image noise residue and the camera fingerprint increases correlation ρ, possibly triggering a FA (if ρ_{th} is not adjusted), while PCE drops in such situation (affecting the threshold for PCE very little).

First, we introduce notation and definitions on one-dimensional vectors of real numbers $\mathbb{R}^n$. Their later generalization to two-dimensional matrices is straightforward: one more index is added and a sum over one index is replaced with a sum over both indices. The following definitions apply to centered vectors. Vector $a = (a_1, a_2, ..., a_n)$ is centered if the sum of all elements is zero. If a vector is not centered, its sample mean must be subtracted from all its elements before applying these definitions. Such approach makes formulas simpler compared to general definitions that do not assume before-hand centralization.

Let a and b be two centered vectors in $\mathbb{R}^n$, and operation $\oplus$ is modulo n addition in $\mathbf{Z}_n$. The *circular cross-correlation* is defined as

$$c(k) = \frac{1}{n}\sum_{i=1}^{n} a_i b_{i \oplus k}, \quad k = 0, ..., n-1. \tag{8}$$

Normalized circular cross-correlation between a and b is

$$C(k) = \frac{\displaystyle\sum_{i=1}^{n} a_i b_{i \oplus k}}{\sqrt{\displaystyle\sum_{i=1}^{n} a_i^2 \sum_{i=1}^{n} b_i^2}}, \quad k = 0, ..., n-1. \tag{9}$$

It is the scalar product of **a** and **b** circularly shifted by offset k divided by the norms of **a** and **b**.

Peak to Correlation Energy (PCE) ratio is the squared correlation divided by sample variance of the circular cross-correlations,

$$PCE_0(\mathbf{a}, \mathbf{b}) = \frac{c^2(0)}{\dfrac{1}{n - |\mathcal{A}|}\sum_{k, k \notin \mathcal{A}} c^2(k)}. \tag{10}$$

where $\mathcal{A}$ is a small square area around zero where a peak correlation is expected for correlated vectors and $|\mathcal{A}|$ is its cardinality. Index 0 at *PCE* is meant to distinguish it from definition in reference [11] where it includes a search for signal shift (or cropping). In that paper, PCE has the maximum over all circular cross-correlations in its numerator. Here, we basically follow the definition according to Kumar and Hassebrook [27]. We point out that PCE does not change if correlation c is replaced with normalized correlation C. The denominator from (9) cancels out when substituted into (10).

From the Central Limit Theorem, the cross-correlation values for independent vectors follow the Gaussian distribution. We demonstrate in Figure 1 that cross-correlations between $I\hat{K}$ and W for $k \notin \mathcal{A}$ are also well approximated using the Gaussian distribution. Connecting PCE with P_{FA} then needs the following assumption.

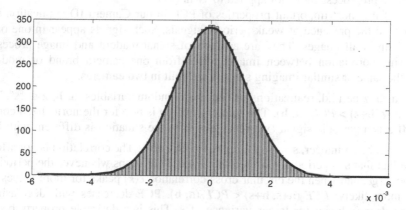

Fig. 1. Gaussian fit of cross-correlations for a 1-Mpixel image

Assumption: The mean of squared normalized correlation between image noise residual and other than correct camera fingerprints (hypothesis H_0) can be estimated as the mean squared correlation of image noise residual and the correct but shifted fingerprint modulated by the image intensities.

Although this assumption may not be perfectly satisfied, it offers a good insight into experimental evaluation of CSI. All such correlations appear in the denominator of PCE_0 definition (10). Our new detection statistic based on PCE becomes

$$\varphi = PCE_0\left(ZM\left(I\hat{K}\right), W\right). \tag{11}$$

Assuming the pdf in Figure 1 is zero mean, the new detection threshold φ_{th} has the following analytical relationship to the probability of false alarm,

$$\varphi_{th} = \left[Q^{-1}\left(P_{FA}\right)\right]^2, \tag{12}$$

where Q is the complementary cumulative density function of a normal random variable $N(0,1)$, i.e., Q^{-1} is a scaled inverse error function. Notice that this threshold does not depend on signals length (number of pixels), while ρ_{th} does. Computation complexity is not an issue. The cross-correlation (8) is implemented via Fast Fourier Transform and the normalization as it is in the normalized correlation, or in (9), is not needed. Another saving comes when removing the linear pattern from the signals. Zero-mean operation needs to be applied just on one of the two signals because the following holds,

$$PCE_0\left(ZM\left(I\hat{K}\right), ZM\left(W\right)\right) = PCE_0\left(ZM\left(I\hat{K}\right), W\right). \tag{13}$$

The proof is easy, all that has to be show is that all $c(k)$ in (10) are equal on both sides of (13). The same equation does not hold for the normalized correlation due to changing vector norms, i.e. $corr(ZM(a), ZM(b)) \neq corr(ZM(a), b)$. We thus explained why zero-mean preprocessing is not applied to W in (11).

One of the most important properties of PCE in our Camera ID application is its response to the presence of weak periodic signals. Such signals appear in one or another form in all images. They are artifacts of signal readout and image processing increasing correlation between image noises from one camera brand or model or when the same or similar imaging sensors are built in two cameras.

Let a, b, z be i.i.d. realizations of Gaussian random variables, a, b, $z \in \mathbb{R}^n$. Then $PCE_0(a+z, b+z) > PCE_0(a, b)$. The same inequality is true for the normalized correlation. If z is a periodic signal (we rename it as s) the situation is different with PCE. Let $l = m/n$ be an integer, $s = (r_1, r_2, ..., r_m)^l \in \mathbb{R}^n$, $m>1$. The correlation is not affected by the fact that s is periodic. On the other hand, PCE drops whenever the period m is not too large, and when the circular cross-correlation $c(k)$ peaks for more values of k. Then more likely, $PCE_0(a+s, b+s) < PCE_0(a, b)$. PCE decreases with decreasing m and the drop is larger for larger variance of s. This is a desirable property because signals with small periods (such as below 100) cannot be as unique as those with no periodicity and thus should not be part of camera fingerprints. If any periodicities are present, we wish they do not trigger positive identification.

In an ideal case, we may be able to estimate P_{FA} for one single hypothesis test by inverting (12),

$$P_{FA} \approx Q\left(\sqrt{\varphi}\right). \tag{14}$$

However, a presence of some unknown weak signals that may be hidden in fingerprints of different cameras (still causing false alarms) would cause an error in the estimate (14). Large experimental tests reveal that (14) is a good practical estimate if we adjust it conservatively by a correction factor. But a cleaner solution lies in further de-correlation of all camera fingerprints – our future research topic.

5 Experiments

We have run several experiments to support our theoretical results. The first one compares behavior of the normalized correlation and the PCE detection statistics when evaluated for every $i = 1, ..., N$ during estimation of camera fingerprint from i images of a "flat" scene. We ran the experiments twice, first with zero-mean (6) preprocessing and then without it and repeated them for two cameras, Canon G2, never compressed 4-Mpixel images of a twilight sky, $N=100$, and Fuji E550, JPEG compressed 6-Mpixel images of a blue sky, $N=80$.

In the plots, Figures 2-5 (left), we see that without zero-mean preprocessing the normalized correlation is slightly larger and is always steadily increasing with N. At the same time Figures 2-5 (right) show that PCE is much smaller when zero-mean preprocessing did not remove the linear pattern and it is not increasing for N larger than some N_0, $N_0 \approx 30$ for never compressed images and $N_0 \approx 1$ for JPEGs.

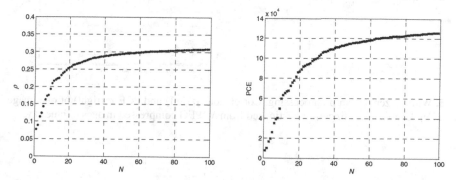

Fig. 2. Correlations ρ (left) and PCE φ (right) of one image with a Canon G2 camera fingerprint estimated from N uncompressed images of cloudless sky

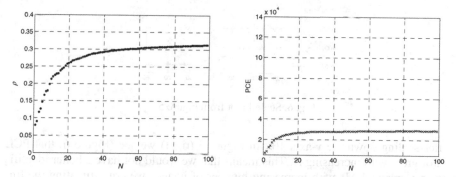

Fig. 3. Without zero-mean preprocessing. Correlations ρ (left) and PCE φ (right) of one image with a Canon G2 camera fingerprint estimated from N uncompressed images of cloudless sky.

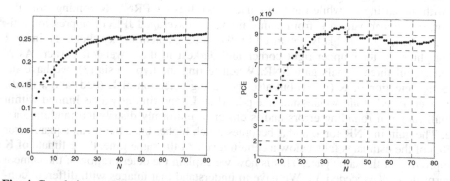

Fig. 4. Correlations ρ (left) and PCE φ (right) of one image with a Fuji E550 camera fingerprint estimated from N JPEG compressed images of blue sky

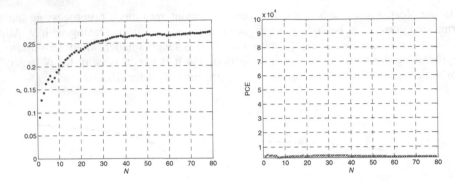

Fig. 5. Without zero-mean preprocessing. Correlations ρ (left) and PCE φ (right) of one image with a Fuji E550 camera fingerprint estimated from N JPEG compressed images of blue sky.

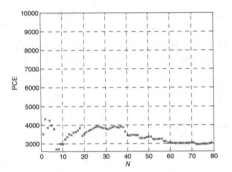

Fig. 6. Scaled plot from Figure 5

After scaling down the y-axis 10× in Figure 5 (right) we see (Figure 6) that PCE does not grow with increasing N. This means that we would not have a better identification in terms of FAR with increasing number of images we are estimating the fingerprint from at all (!). This is a very surprising observation that we can now explain.

The estimate $\hat{\mathbf{K}}$ from (2) contains at least 3 kinds of signals that do not average out with increasing N (while random noise does). It is the PRNU $\mathbf{K}$ coming from the sensor, true linear pattern $\mathbf{L}$ from demosaicing, and averaged JPEG compression artifacts $\mathbf{J}$. The last one is likely enhanced by the same gradient in sky shots; the images are brightest in their upper-right corner and darkest in the lower-left corner. As N increases, all three signals gain SNR. When testing the hypothesis (3), we have increasing detection of $\mathbf{K}$ (good) but also increasing detection of $\mathbf{L}$ (bad) and $\mathbf{J}$ (bad). Moreover, PRNU follow Gaussian-like pdf while $\mathbf{L}$ (and similarly $\mathbf{J}$) is limited within a small range of rounding errors and is closer to uniformly distributed random variable. The gain in SNR for $\mathbf{L}$ and $\mathbf{J}$ becomes much higher once the mean square error (MSE) of the estimates falls below a certain level. At the same time, the estimate of $\mathbf{K}$ is improving at the same pace. This is how we explain the deterioration of the camera fingerprint as N exceeded 38. We have to understand that images with different content, as well as with different compression factors, may contain less similar artifacts $\mathbf{J}$

and the deterioration would not show up. This is what we see in Figure 3. Zero-mean preprocessing is very effective but other filtering is necessary to better remove JPEG compression artifacts $\mathbf{J}$.

These experimental examples show three things.

a) The normalized correlation does not tell much about camera identification performance in terms of detection errors. The threshold for correlation has to be adjusted every time the camera fingerprint is changed.

b) PCE is likely behaving in relation with FA probability. The threshold for PCE can stay fixed even though the camera fingerprint changes.

c) Processing of the camera fingerprint with the zero-mean operation (6) is highly important. Further filtering when JPEG compression is present is desirable. Wiener filtering in the Fourier domain as proposed earlier by Chen *et al.* may be the right answer.

The second experiment was to verify the relation (14). We employed a large scale test with 100,050 images downloaded from the Flickr image sharing web site. Images were in their native resolution, camera fingerprints estimated from randomly chosen 50 (only) images. The signals went through RGB-to-luminance conversion before correlating. Beside zero-mean preprocessing and filtering the camera fingerprints in Fourier domain, we included two correction steps: intensity attenuation function in the term $\mathbf{I\hat{K}}$ replacing it with $att(\mathbf{I})\hat{\mathbf{K}}$ and cutting off all saturated pixels from the images. These pixels were identified automatically using a simple thresholding filter combined with a constraint on the minimum number 2 of such pixels in the closest neighborhood. The parameter for the denoising filter F was $\sigma^2 = 9$.

For each of the 667 cameras, 150 images from different cameras were randomly chosen to calculate PCE statistics φ. The resulting histogram is in Figure 7 (left). To see how it compares to the ideal case, we evaluated PCE for 100,050 pairs of random signals with normal Gaussian pdf and plotted next to it in Figure 7 (right).

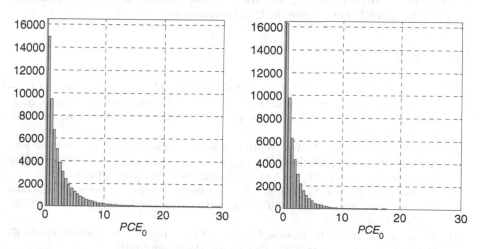

Fig. 7. Histogram of PCE for full size images under H_0 (left), for simulated Gaussian distributed correlations (right)

Comparing the two histograms, the tail is heavier in Figure 7 (left). It suggests that our decision threshold φ_{th} may have to be adjusted by adding a small correction factor. As the conclusion of this section, we argue that the threshold φ_{th} does not have to be re-evaluated for every camera fingerprint. A certain correction factor may be needed to improve the estimate (14). Other measures may include additional filtering of the noise residuals. This will be a subject of our following-up research.

6 Summary

After reviewing the method of camera sensor identification by unique photo-response non-uniformity, we propose to replace the normalized correlation detector with peak to correlation energy ratio. This way, the detection threshold will not vary with varying signal length, different cameras and their on-board image processing nearly as much as for normalized correlation detector. We estimate the probability of FA directly from the threshold set on PCE, which reduces otherwise high demand for large testing needed to set up the threshold for normalized correlation used before.

We show that the linear pattern strongly limits the performance of camera sensor identification if not removed from camera fingerprint estimates. Larger normalized correlation may not necessarily mean smaller probability of FA even if evaluated for the same camera.

Acknowledgements

The work on this paper was supported by the AFOSR grant number FA9550-06-1-0046. The U.S. Government is authorized to reproduce and distribute reprints for Governmental purposes notwithstanding any copyright notation there on. The views and conclusions contained herein are those of the authors and should not be interpreted as necessarily representing the official policies, either expressed or implied, of Air Force Research Laboratory, or the U.S. Government.

References

[1] Khanna, N., Mikkilineni, A.K., Chiu, G.T.C., Allebach, J.P., Delp, E.J.: Forensic Classification of Imaging Sensor Types. In: Proc. SPIE, Electronic Imaging, Security, Steganography, and Watermarking of Multimedia Contents IX, San Jose, CA, vol. 6505, pp. 0U–0V (2007)

[2] Gou, H., Swaminathan, A., Wu, M.: Robust Scanner Identification based on Noise Features. In: Proc. SPIE, Electronic Imaging, Security, Steganography, and Watermarking of Multimedia Contents IX, San Jose, California, vol. 6505, pp. 0S–0T (2007)

[3] Swaminathan, A., Wu, M., Liu, K.J.R.: Nonintrusive Component Forensics of Visual Sensors Using Output Images. IEEE Transactions on Information Forensics and Security 2(1), 91–106 (2007)

[4] Propescu, A.C., Farid, H.: Statistical Tools for Digital Forensic. In: Fridrich, J. (ed.) IH 2004. LNCS, vol. 3200, pp. 128–147. Springer, Heidelberg (2004)

[5] Popescu, A.C., Farid, H.: Exposing Digital Forgeries by Detecting Traces of Resampling. IEEE Transactions on Signal Processing 53(2), 758–767 (2005)

[6] Popescu, A.C., Farid, H.: Exposing Digital Forgeries in Color Filter Array Interpolated Images. IEEE Transactions on Signal Processing 53(10), 3948–3959 (2005)

[7] Farid, H.: Exposing Digital Forgeries in Scientific Images. In: Proc. ACM Multimedia & Security Workshop, Geneva, Switzerland, pp. 29–36 (2006)

[8] Popescu, A.C., Farid, H.: Exposing Digital Forgeries by Detecting Duplicated Image Regions. Technical Report, TR2004-515. Dartmouth College, Computer Science (2004)

[9] Lukáš, J., Fridrich, J., Goljan, M.: Digital Camera Identification from Sensor Pattern Noise. IEEE Transactions on Information Forensics and Security 1(2), 205–214 (2006)

[10] Chen, M., Fridrich, J., Goljan, M.: Digital Imaging Sensor Identification (Further Study). In: Proc. SPIE, Electronic Imaging, Security, Steganography, and Watermarking of Multimedia Contents IX, San Jose, California, vol. 6505, pp. 0P–0Q (2007)

[11] Goljan, M., Fridrich, J.: Camera Identification from Scaled and Cropped Images. In: Delp, E.J., et al. (eds.) Security, Forensics, Steganography, and Watermarking of Multimedia Contents X, vol. 6819, p. 68190E (2008)

[12] Chen, M., Fridrich, J., Goljan, M., Lukáš, J.: Determining Image Origin and Integrity Using Sensor Noise. IEEE Transactions on Information Security and Forensics 3(1), 74–90 (2008)

[13] Healey, G., Kondepudy, R.: Radiometric CCD Camera Calibration and Noise Estimation. IEEE Transactions on Pattern Analysis and Machine Intelligence 16(3), 267–276 (1994)

[14] Mihcak, M.K., Kozintsev, I., Ramchandran, K.: Spatially Adaptive Statistical Modeling of Wavelet Image Coefficients and its Application to Denoising. In: Proc. IEEE Int. Conf. Acoustics, Speech, and Signal Processing, Phoenix, Arizona, vol. 6, pp. 3253–3256 (1999)

[15] Lukáš, J., Fridrich, J., Goljan, M.: Determining Digital Image Origin Using Sensor Imperfections. In: Proc. SPIE, Image and Video Communications and Processing, San Jose, California, vol. 5685, pp. 249–260 (2005)

[16] Lukáš, J., Fridrich, J., Goljan, M.: Detecting Digital Image Forgeries Using Sensor Pattern Noise. In: Proc. SPIE, Electronic Imaging, Security, Steganography, and Watermarking of Multimedia Contents VIII, San Jose, California, vol. 6072, pp. 0Y1–0Y11 (2006)

[17] Goljan, M., Chen, M., Fridrich, J.: Identifying Common Source Digital Camera From Image Pairs. In: Proc. ICIP 2007, San Antonio, Texas (2007)

[18] Khanna, N., Mikkilineni, A.K., Chiu, G.T.C., Allebach, J.P., Delp, E.J.: Scanner Identification Using Sensor Pattern Noise. In: Proc. SPIE, Electronic Imaging, Security, Steganography, and Watermarking of Multimedia Contents IX, San Jose, CA, vol. 6505, pp. 1K–1 (2007)

[19] Sankur, B., Celiktutan, O., Avcibas, I.: Blind Identification of Cell Phone Cameras. In: Proc. SPIE, Electronic Imaging, Security, Steganography, and Watermarking of Multimedia Contents IX, San Jose, California, vol. 6505, pp. 1H–1I (2007)

[20] Chen, M., Fridrich, J., Goljan, M.: Source Digital Camcorder Identification Using CCD Photo Response Non-uniformity. In: Proc. SPIE, Electronic Imaging, Security, Steganography, and Watermarking of Multimedia Contents IX, San Jose, California, vol. 6505, pp. 1G–1H (2007)

[21] Goljan, M., Fridrich, J., Lukáš, J.: Camera Identification from Printed Images. In: Delp, E.J., et al. (eds.) Security, Forensics, Steganography, and Watermarking of Multimedia Contents X, vol. 6819, p. 68190I (2008)

[22] Dirik, A.E., Sencar, H.T., Husrev, T., Memon, N.: Source Camera Identification Based on Sensor Dust Characteristics. In: Proc. IEEE Workshop on Signal Processing Applications for Public Security and Forensics, Washington, DC, pp. 1–6 (2007)

[23] Choi, K.S., Lam, E.Y., Wong, K.K.Y.: Automatic source camera identification using the intrinsic lens radial distortion. Optics Express 14(24), 1551–1565 (2006)

[24] Bayram, S., Sencar, H.T., Memon, N., Avcibas, I.: Source camera identification based on CFA interpolation. In: Proc. ICIP 2005. IEEE International Conference on Image Processing, pp. 69–72 (2006)

[25] Filler, T., Fridrich, J.: Using Sensor Pattern Noise for Camera Model Identification. In: Proc. ICIP 2008, San Diego, California, pp. 12–15 (2008)

[26] Sutcu, Y., Bayram, S., Sencar, H.T., Memon, N.: Improvements on Sensor Noise Based Source Camera Identification. In: Proc. IEEE, International Conference on Multimedia and Expo, pp. 24–27 (2007)

[27] Kumar, B.V.K.V., Hassebrook, L.: Performance measures for correlation filters. Applied Optics 29(20), 2997–3006 (1990)

Design of Collusion-Resistant Fingerprinting Systems: Review and New Results
(Invited Paper)

C.-C. Jay Kuo

University of Southern California

To protect the copyright of a media file, one idea is to develop a traitor tracing system that identifies its unauthorized distribution or usage by tracing fingerprints of illegal users (*i.e.*, traitors). There exists a simple yet effective attack that can break a naive traitor tracing system easily known as the collusion attack. The design of a fingerprinting system that is robust against collusion attacks has been intensively studied in the last decade. A review of previous work in this field will be given. Most previous work deals with collusion attacks with equal weights or its variants (*e.g.*, the cascade of several collusion attacks with equal weights). Consequently, the weight of each colluder is a constant throughout the collusion process. The collusion attack on continuous media such as audio and video with *time-varying* weights is simple to implement. However, we are not aware of any effective solution to this type of attacks. To address this problem, we first show that this can be formulated as a multi-user detection problem in a wireless communication system with a time-varying channel response. Being inspired by the multi-carrier code division multiaccess (MC-CDMA) technique, we propose a new fingerprinting system that consists of the following modules: 1) codeword generation with a multi-carrier approach, 2) colluder weight estimation, 3) advanced message symbol detection. We construct hiding codes with code spreading followed by multi-carrier modulation. For colluder weight estimation, we show that the colluder weight estimation is analogous to channel response estimation, which can be solved by inserting pilot signals in the embedded fingerprint. As to advanced message symbol detection, we replace the traditional correlation-based detector with the maximal ratio combining (MRC) detector and the parallel interference cancellation (PIC) multiuser detector. The superior performance of the proposed fingerprinting system in terms of user/colluder capacity and the bit error probability (BEP) of symbol detection is demonstrated by representative examples.

H.J. Kim, S. Katzenbeisser, and A.T.S. Ho (Eds.): IWDW 2008, LNCS 5450, p. 469, 2009.
© Springer-Verlag Berlin Heidelberg 2009

Author Index